D1752380

Edited by
Evgeny Katz

Biomolecular Information Processing

Related Titles

Katz, Evgeny (Ed.)

Molecular and Supramolecular Information Processing

From Molecular Switches to Logic Systems

2012

ISBN: 978-3-527-33195-6

Katz, Evgeny (Ed.)

Information Processing Set

2 Volumes
(comprising "Biomolecular Information Processing" and "Molecular and Supramolecular Information Processing")

2012

ISBN: 978-3-527-33245-8

Samori, P., Cacialli, F. (Eds.)

Functional Supramolecular Architectures

for Organic Electronics and Nanotechnology

2011

ISBN: 978-3-527-32611-2

Feringa, B. L., Browne, W. R. (Eds.)

Molecular Switches

Second, Completely Revised and Enlarged Edition

2011

ISBN: 978-3-527-31365-5

Cosnier, S., Karyakin, A. (Eds.)

Electropolymerization

Concepts, Materials and Applications

2010

ISBN: 978-3-527-32414-9

Matta, C. F. (Ed.)

Quantum Biochemistry

2010

ISBN: 978-3-527-32322-7

Wolf, E. L.

Quantum Nanoelectronics

An Introduction to Electronic Nanotechnology and Quantum Computing

2009

ISBN: 978-3-527-40749-1

Stolze, J., Suter, D.

Quantum Computing

A Short Course from Theory to Experiment

2008

ISBN: 978-3-527-40787-3

Helms, V.

Principles of Computational Cell Biology

From Protein Complexes to Cellular Networks

2008

ISBN: 978-3-527-31555-0

Edited by Evgeny Katz

Biomolecular Information Processing

From Logic Systems to Smart Sensors and Actuators

WILEY-VCH

WILEY-VCH Verlag GmbH & Co. KGaA

The Editor

Prof. Dr. Evgeny Katz
Clarkson University
Department of Chemistry
and Biomolecular Science
8, Clarkson Avenue
Potsdam, NY 13699-5810
USA

Cover
The cover page picture was designed by Dr. Vera Bocharova (Clarkson University) and represents artistic vision of the chapter "Bioelectronic Devices Controlled by Enzyme-Based Information Processing Systems" by Evgeny Katz.

■ All books published by **Wiley-VCH** are carefully produced. Nevertheless, authors, editors, and publisher do not warrant the information contained in these books, including this book, to be free of errors. Readers are advised to keep in mind that statements, data, illustrations, procedural details or other items may inadvertently be inaccurate.

Library of Congress Card No.: applied for

British Library Cataloguing-in-Publication Data
A catalogue record for this book is available from the British Library.

Bibliographic information published by the Deutsche Nationalbibliothek
The Deutsche Nationalbibliothek lists this publication in the Deutsche Nationalbibliografie; detailed bibliographic data are available on the Internet at <http://dnb.d-nb.de>.

© 2012 Wiley-VCH Verlag & Co. KGaA, Boschstr. 12, 69469 Weinheim, Germany

All rights reserved (including those of translation into other languages). No part of this book may be reproduced in any form – by photoprinting, microfilm, or any other means – nor transmitted or translated into a machine language without written permission from the publishers. Registered names, trademarks, etc. used in this book, even when not specifically marked as such, are not to be considered unprotected by law.

Composition Laserwords Private Limited, Chennai, India

Printing and Binding Markono Print Media Pte Ltd, Singapore

Cover Design   Adam Design, Weinheim

Print ISBN: 978-3-527-33228-1
ePDF ISBN: 978-3-527-64550-3
ePub ISBN: 978-3-527-64549-7
mobi ISBN: 978-3-527-64551-0
oBook ISBN: 978-3-527-64548-0

Printed in Singapore
Printed on acid-free paper

Contents

Preface *XIII*
List of Contributors *XV*

1 **Biomolecular Computing: From Unconventional Computing to "Smart" Biosensors and Actuators – Editorial Introduction** *1*
Evgeny Katz
References *5*

2 **Peptide-Based Computation: Switches, Gates, and Simple Arithmetic** *9*
Zehavit Dadon, Manickasundaram Samiappan, Nathaniel Wagner, Nurit Ashkenasy, and Gonen Ashkenasy
2.1 Introduction *9*
2.2 Peptide-Based Replication Networks *10*
2.2.1 Template-Assisted Replication *10*
2.2.2 Theoretical Prediction of the Network Connectivity *11*
2.2.3 *De novo* Designed Synthetic Networks *12*
2.3 Logic Gates within Ternary Networks *13*
2.3.1 Uniform Design Principles of All Two-Input Gates *13*
2.3.2 OR Logic *14*
2.3.3 AND Logic *15*
2.3.4 NAND Logic *15*
2.3.5 XOR Logic *15*
2.4 Symmetry and Order Requirements for Constructing the Logic Gates *16*
2.4.1 Symmetry and Order in Peptide-Based Catalytic Networks *16*
2.4.2 How Symmetry and Order Affect the Replication of RNA Quasispecies *17*
2.5 Taking the Steps toward More Complex Arithmetic *19*
2.5.1 Arithmetic Units *19*
2.5.2 Network Motifs *20*

2.6	Experimental Logic Gates	21
2.6.1	OR Logic	21
2.6.2	NOT, NOR, and NOTIF Logic	21
2.6.3	Additional Logic Operations	23
2.7	Adaptive Networks	24
2.7.1	Chemical Triggering	24
2.7.2	Light Triggering	24
2.7.3	Light-Induced Logic Operations	25
2.8	Peptide-Based Switches and Gates for Molecular Electronics	28
2.9	Summary and Conclusion	29
	Acknowledgments	30
	References	30
3	**Biomolecular Electronics and Protein-Based Optical Computing**	**33**
	Jordan A. Greco, Nicole L. Wagner, Matthew J. Ranaghan, Sanguthevar Rajasekaran, and Robert R. Birge	
3.1	Introduction	33
3.2	Biomolecular and Semiconductor Electronics	34
3.2.1	Size and Speed	34
3.2.2	Architecture	36
3.2.3	Nanoscale Engineering	37
3.2.4	Stability	38
3.2.5	Reliability	38
3.3	Bacteriorhodopsin as a Photonic and Holographic Material for Bioelectronics	40
3.3.1	The Light-Induced Photocycle	40
3.3.2	The Branched Photocycle	42
3.4	Fourier Transform Holographic Associative Processors	42
3.5	Three-Dimensional Optical Memories	45
3.5.1	Write, Read, and Erase Operations	46
3.5.2	Efficient Algorithms for Data Processing	48
3.5.3	Multiplexing and Error Analysis	50
3.6	Genetic Engineering of Bacteriorhodopsin for Device Applications	51
3.7	Future Directions	53
	Acknowledgments	54
	References	54
4	**Bioelectronic Devices Controlled by Enzyme-Based Information Processing Systems**	**61**
	Evgeny Katz	
4.1	Introduction	61
4.2	Enzyme-Based Logic Systems Producing pH Changes as Output Signals	62

4.3	Interfacing of the Enzyme Logic Systems with Electrodes Modified with Signal-Responsive Polymers *64*	
4.4	Switchable Biofuel Cells Controlled by the Enzyme Logic Systems *68*	
4.5	Biomolecular Logic Systems Composed of Biocatalytic and Biorecognition Units and Their Integration with Biofuel Cells *70*	
4.6	Processing of Injury Biomarkers by Enzyme Logic Systems Associated with Switchable Electrodes *74*	
4.7	Summary and Outlook *77*	
	Acknowledgments *78*	
	References *78*	

5 Enzyme Logic Digital Biosensors for Biomedical Applications *81*
Evgeny Katz and Joseph Wang
5.1 Introduction *81*
5.2 Enzyme-Based Logic Systems for Identification of Injury Conditions *82*
5.3 Multiplexing of Injury Codes for the Parallel Operation of Enzyme Logic Gates *85*
5.4 Scaling Up the Complexity of the Biocomputing Systems for Biomedical Applications – Mimicking Biochemical Pathways *89*
5.5 Application of Filter Systems for Improving Digitalization of the Output Signals Generated by Enzyme Logic Systems for Injury Analysis *94*
5.6 Conclusions and Perspectives *96*
Acknowledgments *98*
Appendix *98*
References *99*

6 Information Security Applications Based on Biomolecular Systems *103*
Guinevere Strack, Heather R. Luckarift, Glenn R. Johnson, and Evgeny Katz
6.1 Introduction *103*
6.2 Molecular and Bio-molecular Keypad Locks *104*
6.3 Antibody Encryption and Steganography *108*
6.4 Bio-barcode *113*
6.5 Conclusion *114*
Acknowledgments *114*
References *114*

7 Biocomputing: Explore Its Realization and Intelligent Logic Detection *117*
Ming Zhou and Shaojun Dong
7.1 Introduction *117*
7.2 DNA Biocomputing *119*

7.3	Aptamer Biocomputing *121*
7.4	Enzyme Biocomputing *124*
7.5	Conclusions and Perspectives *128*
	References *129*

8	**Some Experiments and Models in Molecular Computing and Robotics** *133*
	Milan N. Stojanovic and Darko Stefanovic
8.1	Introduction *133*
8.2	From Gates to Programmable Automata *133*
8.3	From Random Walker to Molecular Robotics *139*
8.4	Conclusions *142*
	Acknowledgments *143*
	References *143*

9	**Biomolecular Finite Automata** *145*
	Tamar Ratner, Sivan Shoshani, Ron Piran, and Ehud Keinan
9.1	Introduction *145*
9.2	Biomolecular Finite Automata *146*
9.2.1	Theoretical Models of a Molecular Turing Machine *146*
9.2.2	The First Realization of an Autonomous DNA-Based Finite Automaton *150*
9.2.3	Three-Symbol-Three-State DNA-Based Automata *155*
9.2.4	Molecular Cryptosystem for Images by DNA Computing *157*
9.2.5	Molecular Computing Device for Medical Diagnosis and Treatment *In Vitro* *159*
9.2.6	DNA-Based Automaton with Bacterial Phenotype Output *161*
9.2.7	Molecular Computing with Plant Cell Phenotype *163*
9.3	Biomolecular Finite Transducer *167*
9.4	Applications in Developmental Biology *172*
9.5	Outlook *176*
	References *178*

10	***In Vivo* Information Processing Using RNA Interference** *181*
	Yaakov Benenson
10.1	Introduction *181*
10.1.1	Regulatory Pathways as Computations *181*
10.1.2	A Computation Versus a Computer *182*
10.1.3	Prior Work on Synthetic Biomolecular Computing Circuits *182*
10.2	RNA Interference-Based Logic *183*
10.2.1	General Considerations *183*
10.2.2	Logic Circuit Blueprint *184*
10.2.3	Experimental Confirmation of the Computational Core *188*
10.3	Building the Sensory Module *189*
10.3.1	Direct Control of siRNA by mRNA Inputs *191*

10.3.2	Complex Transcriptional Regulation Using RNAi-Based Circuits	194
10.4	Outlook	195
	References	197

11 Biomolecular Computing Systems 199

Harish Chandran, Sudhanshu Garg, Nikhil Gopalkrishnan, and John H. Reif

11.1	Introduction	199
11.1.1	Organization of the Chapter	199
11.2	DNA as a Tool for Molecular Programming	200
11.2.1	DNA Structure	200
11.2.2	Review of DNA Reactions	200
11.3	Birth of DNA Computing: Adleman's Experiment and Extensions	203
11.3.1	NP-Complete Problems	203
11.3.2	Hamiltonian Path Problem via DNA Computing	204
11.3.3	Other Models of DNA Computing	204
11.3.4	Shortcomings and Nonscalability of Schemes Using DNA Computation to Solve NP-Complete Problems	204
11.4	Computation Using DNA Tiles	205
11.4.1	TAM: an Abstract Model of Self-Assembly	205
11.4.2	Algorithmic Assembly via DNA Tiling Lattices	206
11.4.2.1	Source of Errors	206
11.4.3	Algorithmic Error Correction Schemes for Tilings	207
11.5	Experimental Advances in Purely Hybridization-Based Computation	209
11.6	Experimental Advances in Enzyme-Based DNA Computing	212
11.7	Biochemical DNA Reaction Networks	217
11.8	Conclusion: Challenges in DNA-Based Biomolecular Computation	218
11.8.1	Scalability of Biomolecular Computations	218
11.8.2	Ease of Design and Programmability of Biomolecular Computations	220
11.8.3	*In Vivo* Biomolecular Computations	220
11.8.4	Conclusions	220
	Acknowledgments	221
	References	221

12 Enumeration Approach to the Analysis of Interacting Nucleic Acid Strands 225

Satoshi Kobayashi and Takaya Kawakami

12.1	Introduction	225
12.2	Definitions and Notations for Set and Multiset	226
12.3	Chemical Equilibrium and Hybridization Reaction System	227
12.4	Symmetric Enumeration Method	230

12.4.1	Enumeration Graph	*230*
12.4.2	Path Mappings	*231*
12.4.3	Enumeration Scheme	*232*
12.4.4	An Example of Enumeration Scheme – Folding of an RNA Molecule	*233*
12.4.5	Convex Programming Problem for Computing Equilibrium	*235*
12.5	Applying SEM to Nucleic Acid Strands Interaction	*236*
12.5.1	Target Secondary Structures	*237*
12.5.2	Introducing Basic Notations	*237*
12.5.3	Definition of Enumeration Graph Structure	*239*
12.5.4	Associated Weight Functions	*241*
12.5.5	Symmetric Properties	*242*
12.5.6	Complexity Issues	*242*
12.6	Conclusions	*243*
	References	*244*

13 Restriction Enzymes in Language Generation and Plasmid Computing *245*
Tom Head

13.1	Introduction	*245*
13.2	Wet Splicing Systems	*246*
13.3	Dry Splicing Systems	*249*
13.4	Splicing Theory: Its Original Motivation and Its Extensive Unforeseen Developments	*252*
13.5	Computing with Plasmids	*253*
13.6	Fluid Memory	*254*
13.7	Examples of Aqueous Computations	*255*
13.8	Final Comments about Computing with Biomolecules	*260*
	References	*261*

14 Development of Bacteria-Based Cellular Computing Circuits for Sensing and Control in Biological Systems *265*
Michaela A. TerAvest, Zhongjian Li, and Largus T. Angenent

14.1	Introduction	*265*
14.2	Cellular Computing Circuits	*267*
14.2.1	Genetic Toolbox	*267*
14.2.1.1	Engineered Gene Regulation	*267*
14.2.1.2	Quorum Sensing	*269*
14.2.2	Implementations	*269*
14.2.2.1	Oscillators	*269*
14.2.2.2	Switches	*270*
14.2.2.3	AND Logic Gates	*270*
14.2.2.4	Edge Detector	*271*
14.2.2.5	Complex Logic Functions with Multiple Strains	*272*
14.2.3	Transition to *In Silico* Rational Design	*273*

14.2.4	Transition from Enzyme Computing to Bacteria-Based Biocomputing *274*	
14.3	Conclusion *276*	
	Acknowledgments *277*	
	References *277*	

15	**The Logic of Decision Making in Environmental Bacteria** *279*	
	Rafael Silva-Rocha, Javier Tamames, and Víctor de Lorenzo	
15.1	Introduction *279*	
15.2	Building Models for Biological Networks *281*	
15.3	Formulation and Simulation of Regulatory Networks *283*	
15.3.1	Stochastic Versus Deterministic Models *284*	
15.3.2	Graphical Models *285*	
15.4	Boolean Analysis of Regulatory Networks *285*	
15.4.1	Translating Biological Networks into Logic Circuits *286*	
15.4.2	Integration of Regulatory and Metabolic Logic in the Same Boolean Circuit *287*	
15.4.3	From Digital Networks to Workable Models *288*	
15.5	Boolean Description of m-xylene Biodegradation by *P. putida* mt-2: the TOL logicome *289*	
15.5.1	Narrative Description of the TOL Regulatory Circuit *291*	
15.5.2	Deconstruction of the Ps–Pr Regulatory Node into Three Autonomous Logic Units *292*	
15.5.3	Formalization of Regulatory Events at the Upper and Lower TOL Operons *294*	
15.5.4	3MB Is the Endogenous Signal Carrier through the Domains of the TOL Network *296*	
15.5.5	The TOL Logicome *296*	
15.6	Conclusion and Outlook *298*	
	Acknowledgments *299*	
	References *299*	

16	**Qualitative and Quantitative Aspects of a Model for Processes Inspired by the Functioning of the Living Cell** *303*	
	Andrzej Ehrenfeucht, Jetty Kleijn, Maciej Koutny, and Grzegorz Rozenberg	
16.1	Introduction *303*	
16.2	Reactions *304*	
16.3	Reaction Systems *305*	
16.4	Examples *307*	
16.5	Reaction Systems with Measurements *310*	
16.6	Generalized Reactions *312*	
16.7	A Generic Quantitative Model *315*	
16.8	Approximations of Gene Expression Systems *316*	

16.9	Simulating Approximations by Reaction Systems	*318*
16.10	Discussion *319*	
	Acknowledgments *321*	
	References *321*	

17 Computational Methods for Quantitative Submodel Comparison *323*
Andrzej Mizera, Elena Czeizler, and Ion Petre

- 17.1 Introduction *323*
- 17.2 Methods for Model Decomposition *324*
- 17.2.1 Knockdown Mutants *324*
- 17.2.2 Elementary Flux Modes *325*
- 17.2.3 Control-Based Decomposition *325*
- 17.3 Methods for Submodel Comparison *327*
- 17.3.1 Mathematically Controlled Model Comparison *327*
- 17.3.2 An Extension of the Mathematically Controlled Comparison *328*
- 17.3.3 Local Submodel Comparison *329*
- 17.3.4 A Quantitative Measure for the Goodness of Model Fit Against Experimental Data *329*
- 17.3.5 Quantitative Refinement *330*
- 17.3.6 Parameter-Independent Submodel Comparison *331*
- 17.3.7 Model Comparison for Pathway Identification *332*
- 17.4 Case Study *332*
- 17.4.1 A Biochemical Model for the Heat Shock Response *332*
- 17.4.2 Control-Based Decomposition *334*
- 17.4.3 The Knockdown Mutants *335*
- 17.4.4 Local Comparison of the Knockdown Mutants *336*
- 17.4.5 Parameter-Independent Comparison of the Mutant Behavior *337*
- 17.4.6 Pathway Identification for the Phosphorylation-Driven Control of the Heat Shock Response *341*
- 17.5 Discussion *342*
  - Acknowledgments *343*
  - References *343*

18 Conclusions and Perspectives *347*
Evgeny Katz
References *349*

Index *351*

Preface

The use of biomolecular systems for processing information, performing logic operations, computational operations, and even automata performance is a rapidly developing research area. The entire field was named with the general buzzwords, "biomolecular computing" or "biocomputing." Exciting advances in the area include the use of various biomolecular systems including proteins/enzymes, DNA, RNA, DNAzymes, antigens/antibodies, and even whole biological (usually microbial) cells operating as "hardware" for unconventional computing. The most important feature of the biocomputing systems is their operation in biochemical and even biological environment. Many different applications of these systems, in addition to unconventional computation, are feasible; while their biosensor/biomedical use is obviously one of the most important applications. Interfacing of biological systems with "smart" biosensors, signal responsive materials, and bioelectronic devices is of highest importance for future developments in the area of biomolecular computing. The various topics covered highlight key aspects and the future perspectives of biomolecular computing. The book discusses experimental work done by biochemists and biologists and theoretical approaches developed by physicists and computer scientists. The different topics addressed in this book will be of high interest to the interdisciplinary community active in the area of unconventional biocomputing. It is hoped that the collection of the different chapters will be important and beneficial for researchers and students working in various areas related to biochemical computing, including biochemistry, materials science, computer science, and so on. Furthermore, the book is aimed to attract young scientists and introduce them to the field while providing newcomers with an enormous collection of literature references. I, indeed, hope that the book will spark the imagination of scientists to further develop the topic.

Finally, the Editor (E. Katz) and Publisher (Wiley-VCH) express their thanks to all authors of the chapters, whose dedication and hard work made this book possible, hoping that the book will be interesting and beneficial for researchers and students working in various areas related to unconventional biocomputing. It should be noted that the field of biocomputing extends to the fascinating area of chemical molecular and supra-molecular synthetic systems which consideration was outside the scope of the present book. This complementary area of molecular computing (based on synthetic rather than natural biological molecules) is covered in another new book of Wiley-VCH: ***Molecular and Supramolecular Information Processing: from Molecular Switches to Logic Systems*** – E. Katz, Editor. Both the books are must for the shelves of specialists interested in various aspects of molecular and biomolecular information processing.

Potsdam, NY, USA *Evgeny Katz*
October 2011

List of Contributors

Largus T. Angenent
Cornell University
Department of Biological and
Environmental Engineering
214 Riley-Robb Hall
Ithaca, NY 14853
USA

Gonen Ashkenasy
Ben Gurion University of the
Negev
Department of Chemistry
P.O. Box 653
84105 Beer Sheva
Israel

Nurit Ashkenasy
Ben Gurion University of the
Negev
Department of Materials
Engineering

and

The Ilse Katz Institute for
Nanoscale Science and
Technology
84105 Beer Sheva
Israel

Yaakov Benenson
ETH Zurich
Department of Biosystems
Science and Engineering
Mattenstrasse 26
4058 Basel
Switzerland

Robert R. Birge
University of Connecticut
College of Liberal Arts and
Science
Department of Chemistry
55 North Eagleville Road
Unit 3060
Storrs, CT 06269-3060
USA

Harish Chandran
Duke University
Department of Computer Science
P.O. Box 90129
Durham, NC 27708-0129
USA

Elena Czeizler
University of Helsinki
Faculty of Medicine
Tukholmankatu 8B
00014 Helsinki
Finland

Zehavit Dadon
Ben Gurion University of the Negev
Department of Chemistry
P.O. Box 653
84105 Beer Sheva
Israel

Shaojun Dong
Changchun Institute of Applied Chemistry
State Key Laboratory of Electroanalytical Chemistry
Chinese Academy of Sciences
5625 Renmin Street
Changchun 130022
P.R. China

Andrzej Ehrenfeucht
Department of Computer Science
University of Colorado at Boulder
430 UCB
Boulder, CO 80309-0430
USA

Sudhanshu Garg
Duke University
Department of Computer Science
P.O. Box 90129
Durham, NC 27708-0129
USA

Nikhil Gopalkrishnan
Duke University
Department of Computer Science
P.O. Box 90129
Durham, NC 27708-0129
USA

Jordan A. Greco
University of Connecticut
College of Liberal Arts and Science
Department of Chemistry
55 North Eagleville Road
Unit 3060
Storrs, CT 06269-3060
USA

Tom Head
Binghamton University
Department of Mathematics
Binghamton, NY 13902-6000
USA

Glenn R. Johnson
Air Force Research Laboratory
Airbase Sciences Division
Tyndall Air Force Base, FL 32403
USA

Evgeny Katz
Clarkson University
Department of Chemistry and Biomolecular Science
8 Clarkson Avenue
Potsdam, NY 13699-5810
USA

Takaya Kawakami
University of Electro-Communications
Department of Computer Science
1-5-1 Chofugaoka
Chofu, 182-8585 Tokyo
Japan

Ehud Keinan
Technion – Israel Institute of Technology
Schulich Faculty of Chemistry
Technion City
32000 Haifa
Israel

and

The Scripps Research Institute
Department of Molecular Biology and the Skaggs Institute for Chemical Biology
10550 North Torrey Pines Road
La Jolla, CA 92037
USA

Jetty Kleijn
LIACS
Leiden University
2300 RA Leiden
The Netherlands

Satoshi Kobayashi
University of Electro-Communications
Department of Computer Science
1-5-1 Chofugaoka
Chofu, 182-8585 Tokyo
Japan

Maciej Koutny
School of Computing Science
Newcastle University
Claremont Tower
Claremont Road
NE1 7RU
Newcastle upon Tyne
UK

Zhongjian Li
Cornell University
Department of Biological and Environmental Engineering
214 Riley-Robb Hall
Ithaca, NY 14853
USA

Víctor de Lorenzo
Systems Biology Program
Centro Nacional de Biotecnología – CSIC
c/ Darwin, 3
Campus de Cantoblanco
28049 Cantoblanco-Madrid
Spain

Heather R. Luckarift
Air Force Research Laboratory
Airbase Sciences Division
Tyndall Air Force Base, FL 32403
USA

and

Universal Technology Corporation
1270 N. Fairfield Road
Dayton, OH 45432
USA

Andrzej Mizera
Åbo Akademi University
Department of Information Technologies
Turku Centre for Computer Science
Joukahaisenkatu 3-5A
20520 Turku
Finland

Ion Petre
Åbo Akademi University
Department of Information
Technologies
Turku Centre for Computer
Science
Joukahaisenkatu 3-5A
20520 Turku
Finland

Ron Piran
Technion – Israel Institute of
Technology
Schulich Faculty of Chemistry
Technion City
32000 Haifa
Israel

and

The Scripps Research Institute
Department of Molecular Biology
and the Skaggs Institute for
Chemical Biology
10550 North Torrey Pines Road
La Jolla, CA 92037
USA

Sanguthevar Rajasekaran
University of Connecticut
College of Liberal Arts and
Science
Department of Computer Science
and Engineering
371 Fairfield Road
Unit 2155
Storrs, CT 06269-3060
USA

Matthew J. Ranaghan
University of Connecticut
College of Liberal Arts and
Science
Department of Molecular and
Cell Biology
91 North Eagleville Road
Unit 3125
Storrs, CT 06269-3060
USA

Tamar Ratner
Technion – Israel Institute of
Technology
Schulich Faculty of Chemistry
Technion City
32000 Haifa
Israel

John H. Reif
Duke University
Department of Computer Science
P.O. Box 90129
Durham, NC 27708-0129
USA

Grzegorz Rozenberg
LIACS
Leiden University
2300 RA Leiden
The Netherlands

and

Department of Computer Science
University of Colorado at Boulder
430 UCB
Boulder, CO 80309-0430
USA

Manickasundaram Samiappan
Ben Gurion University of the Negev
Department of Chemistry
P.O. Box 653
84105 Beer Sheva
Israel

Sivan Shoshani
Technion – Israel Institute of Technology
Schulich Faculty of Chemistry
Technion City
32000 Haifa
Israel

Rafael Silva-Rocha
Systems Biology Program
Centro Nacional de Biotecnología – CSIC
c/ Darwin, 3
Campus de Cantoblanco
28049 Cantoblanco-Madrid
Spain

Darko Stefanovic
University of New Mexico
Department of Computer Science and Center for Biomedical Engineering
MSC01 1130
1 University of New Mexico
Albuquerque, NM 87131
USA

Milan N. Stojanovic
Columbia University
Departments of Medicine and Biomedical Engineering
630 W 168th St.
New York, NY 10032
USA

Guinevere Strack
Air Force Research Laboratory
Airbase Sciences Division
Tyndall Air Force Base, FL 32403
USA

and

Oak Ridge Institute for Science and Education
Belcamp, MD 21017
USA

Javier Tamames
Systems Biology Program
Centro Nacional de Biotecnología – CSIC
c/ Darwin, 3
Campus de Cantoblanco
28049 Cantoblanco-Madrid
Spain

Michaela A. TerAvest
Cornell University
Department of Biological and Environmental Engineering
214 Riley-Robb Hall
Ithaca, NY 14853
USA

Nathaniel Wagner
Ben Gurion University of the Negev
Department of Chemistry
P.O. Box 653
84105 Beer Sheva
Israel

Nicole L. Wagner
University of Connecticut
College of Liberal Arts and
Science
Department of Molecular and
Cell Biology
91 North Eagleville Road
Unit 3125
Storrs, CT 06269-3060
USA

Joseph Wang
University of California – San
Diego
Department of NanoEngineering
9500 Gilman Dr.,
La Jolla, CA 92093
USA

Ming Zhou
Changchun Institute of
Applied Chemistry
State Key Laboratory of
Electroanalytical Chemistry
Chinese Academy of Sciences
5625 Renmin Street
Changchun 130022
P.R. China

1
Biomolecular Computing: From Unconventional Computing to "Smart" Biosensors and Actuators – Editorial Introduction
Evgeny Katz

Chemical computing [1] as a subarea of unconventional computing [2] has achieved tremendous development in the past two decades, driven mostly by the idea of making revolutionary changes in computing technology. While the conventional silicon-based electronic technology comes to the physical limit of miniaturization [3], chemical systems might operate at the level of single molecules, bringing information processing systems from the present microsize to novel nanosize [4]. Even more importantly, chemical systems can perform massively parallel computational operations with involvement of as many as 10^{23} molecules, resulting in a speed of information processing presently impossible in silicon-based computers [5]. Motivated by these ideas from computer science, chemists designed sophisticated switchable molecules and supramolecular complexes to perform logic operations and mimic computing systems [6]. Complex chemical reactions with unusual kinetics (e.g., oscillating diffusional systems – Belousov–Zhabotinsky reactions) [7] were suggested as media performing computing operations [8]. Extensive research in the area of reaction–diffusion computing systems [9] resulted in the formulation of conceptually novel circuits performing information processing with the use of subexcitable chemical media [10]. Novel conceptual approaches required for the usage of new chemical "hardware" were designed, resulting in algorithms potentially capable of solving "hard-to-solve" computational problems, thus demonstrating potential advantages of the novel unconventional chemical computing systems over classic silicon-based systems. The present state of the art of the unconventional chemical computing was summarized in the recent Wiley-VCH book: *"Molecular and Supramolecular Information Processing: From Molecular Switches to Logic Systems,"* E. Katz – Editor.

It should be noted, however, that chemical systems designed for information processing usually suffer from two major problems: (i) They are very difficult to prepare – in other words – the synthetic processes required for their preparation are so complex that only a few laboratories are able to prepare and study the switchable molecules operating as the chemical computing "hardware." This problem is technical rather than conceptual, and it could be solved at the present level of technology if the molecular computing elements find real applications. (ii) The main challenge in further development of chemical information processing systems is scaling up their complexity assembling individual logic gates in logic networks [11]. Impressive results have recently been achieved in this direction [6]. Combination of chemical logic gates in small groups or networks resulted in simple computing devices performing basic arithmetic operations such as half-adder/half-subtractor or full-adder/full-subtractor [12]. Integration of several functional units in a molecular structure resulted in multisignal responses to stimuli of various chemical or physical natures, thus allowing different logic operations or even simple arithmetic functions to be performed within a single multifunctional molecule [13]. Despite the progress achieved, assembling complex systems from individual chemical components is very limited and presently achieved only for very small networks incomparable with silicon-based electronic chips. The chemical computing units performing logic operations [6] and functioning as auxiliary "devices" (e.g., memory units [14], multiplexers/demultiplexers [15]) are very difficult for integration in functional networks. In other words, each chemical unit might be a perfect computing element, but the integration of this element with other similar elements for their concerted operation is extremely difficult. The difficulty in the interconnectivity of chemical elements in computing networks mostly originates from incompatibility of the chemical input and output signals. The product of the preceding chemical reaction frequently cannot be used as a reagent for the following chemical step. Even more problematic is the use of chemical switchable systems activated by physical signals (such as light [16], magnetic [17] or electrical field [18]) since these signals operating as inputs cannot be reproduced by the chemical reactions and cannot be used for interconnecting several chemical steps in a functional network. This is already a conceptual problem that limits the practical application of chemical systems, keeping them mostly at the level of single units, being scientific "toys" rather than practical devices. It is not surprising that these kinds of molecules were not used by Nature in living systems, where interconnectivity between chemical steps is critically important for their concerted operation, being the base of life.

Many of the problems hardly addressable by synthetic chemical systems can be solved naturally by utilization of biomolecular systems [19, 20]. The emerging research field of biocomputing, based on application of biomolecular systems for processing chemical information, has achieved higher complexity of information processing while using much simpler chemical tools, because of the natural specificity and compatibility of biomolecules [21]. Different biomolecular tools, including proteins/enzymes [20, 22], DNA [19, 23], RNA [24], and whole cells [25], were used to assemble computing systems processing biochemical information. Arithmetic functions, for example, full-adder, were realized using RNA as the information processing biomolecular tool [26]. Deoxyribozymes with various catalytic abilities toward DNA assemblies were applied to extend the computing options provided by DNA-based systems [27]. RNA-based computing systems exploit the biological regulatory functions of RNA in cells, thus allowing operation of cells as "biocomputers" programed by artificially designed biomolecular ensembles [28]. Recently pioneered DNA molecules with biocatalytic properties mimicking enzyme functions, called *DNAzymes* [29], were extensively used to carry out logic operations [30]. These briefly mentioned biomolecular computing systems represent a rapidly developing research field, and they are already covered by comprehensive review articles, for example, on DNA [31], RNA [32], and DNAzyme [33] biocomputing.

The present book summarizes the diverse subareas of biomolecular computing including (i) various aspects of protein/enzyme information processing systems – Chapters 2–7 (ii) DNA/RNA-based computing systems – Chapters 8–13; (iii) application of whole biological (mostly microbial) cells for biocomputing – Chapters 14–16; as well as (iv) general computational aspects of biomolecular computing – Chapter 17. Chapter 18 offers conclusions and perspectives for the biomolecular computing research area summarized by the Editor.

The variety of the systems described in the book and their possible applications are really impressive. While some of the biochemical systems, particularly represented by DNA computing, follow the general trend of unconventional computing, pretending to bring up novel computational chemical "devices" and algorithms competing with conventional silicon-based computers [34], other systems, mostly represented by enzyme-based assemblies, are directed to "noncomputational" applications, which are more related to "smart" biosensors [35] and bioactuators [36]. Biomolecular systems can perform various automata operations [37], particularly illustrated by the tic-tac-toe game [38]. Much more complex robotic functions of biocomputing systems are also feasible [39]. However, the main expected shorter term practical benefit of biomolecular computing systems is their ability to process biochemical information received in the form of chemical inputs directly from biological systems, offering the possibility to operate in biological environments [40], for biomedical/diagnostic [35] and homeland security applications [41]. Biomolecular logic gates and their networks can recognize various biomarkers associated with diseases [42] or injuries [43] and generate a biomedical conclusion in the binary form "YES"/"NO" upon logic processing of the biomarker concentration patterns. The produced binary output can be extended to a chemical actuation resulting in drug release or bioelectronic system activation controlled by logic conclusions derived from the information processed by biomolecular systems [44]. This research direction will certainly result in tremendous contribution to future personalized medicine [45]. Biochemical systems activated by several chemical input signals processed via logic circuitry implemented in the biochemical assembly can activate/inactivate various bioelectronic devices [46], for example, electrodes [47], biofuel cells [48], and field-effect transistors [49] (Figure 1.1), thus contributing to the next level of sophistication of bioelectronics [50] (Figure 1.2). Chemical signal processing through biocatalytic or biorecognition reactions might be applicable in information security systems performing encoding and encrypting operations as well as providing hiding of information in steganography applications [51] (Figure 1.3). Biocomputing systems can also be used as a part of signal-responsive "smart" materials with functions controlled by logically processed biochemical signals. Various nanostructured materials, including switchable membranes [52]

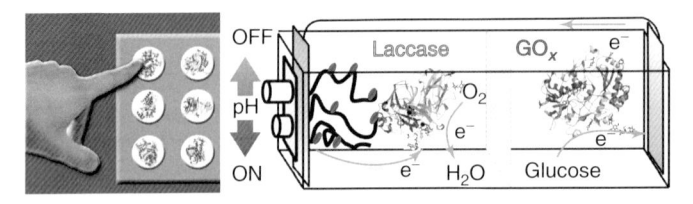

Figure 1.1 A cartoon illustrating biocomputing control over a switchable biofuel cell producing electrical power on demand upon receiving signals processed through an enzyme-based logic system (see [48] for details).

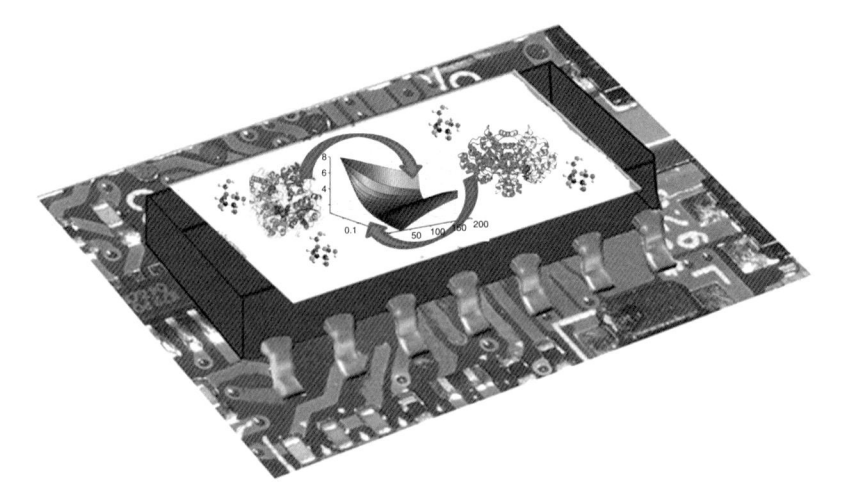

Figure 1.2 An artistic vision of the integration of biomolecular systems with bioelectronic devices (see [46] for details).

Figure 1.3 A cartoon outlining application of a biorecognition information processing system for data security, encoding, and steganography (see [51c] for details).

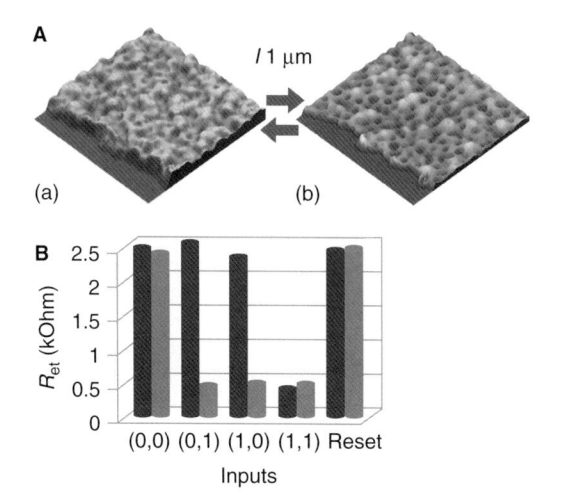

Figure 1.4 The signal-responsive membrane associated with an indium tin oxide (ITO) electrode and coupled with the enzyme-based AND (dark gray bars)/OR (light gray bars) logic gates. (A) Atomic force microscope (AFM) topographic images (10×10 μm^2) of the membrane with closed (a) and open (b) pores. (B) The electron transfer resistance, R_{et}, of the switchable interface derived from the impedance spectroscopy measurements obtained upon different combinations of the input signals. (Adapted from 52, with permission; Copyright American Chemical Society, 2009.)

(Figure 1.4), can benefit from built-in logic implemented via biocomputing gates and networks [52, 53].

The variety of systems inspired by biology and their possible applications are really unlimited, and the combination of computer science, biomolecular science, material science, and electronics will result in novel scientific and technological advances in this multidimensional research area. The present book aims at summarizing the achievements in this rapidly developing multifaceted research area providing background for further progress and helping in understanding of various aspects in this complex scientific field.

References

1. Katz, E. (Guest ed.) (2011) *Isr. J. Chem.*, **51** (1), 13–14, and review articles wherein.
2. (a) Calude, C.S., Costa, J.F., Dershowitz, N., Freire, E., and Rozenberg, G. (eds) (2009) *Unconventional Computation*, Lecture Notes in Computer Science, Vol. 5715, Springer, Berlin; (b) Adamatzky, A., De Lacy Costello, B., Bull, L., Stepney, S., and Teuscher, C. (eds) (2007) *Unconventional Computing 2007*, Luniver Press.
3. (a) Freebody, M. (2011) *Photonucs Spectra*, **45**, 45–47; (b) Rupp, K. and Selberherr, S. (2010) *Proc. IEEE*, **98**, 351–353; (c) Rupp, K. and Selberherr, S. (2011) *IEEE Trans. Semicond. Manufact.*, **24**, 1–4; (d) Powell, J.R. (2008) *Proc. IEEE*, **96**, 1247–1248; (e) Choi, C. (2004) *New Scientist*, **182**, 12–12.

4. (a) Stadler, R., Ami, S., Joachim, C., and Forshaw, M. (2004) *Nanotechnology*, **15**, S115–S121; (b) De Silva, A.P., Leydet, Y., Lincheneau, C., and McClenaghan, N.D. (2006) *J. Phys. Condens. Mater.*, **18**, S1847–S1872.
5. Adamatzky, A. (2004) *IEICE Trans. Electron.*, **E87C**, 1748–1756.
6. (a) de Silva, A.P., Uchiyama, S., Vance, T.P., and Wannalerse, B. (2007) *Coord. Chem. Rev.*, **251**, 1623–1632; (b) de Silva, A.P. and Uchiyama, S. (2007) *Nat. Nanotechnol.*, **2**, 399–410; (c) Szacilowski, K. (2008) *Chem. Rev.*, **108**, 3481–3548; (d) Credi, A. (2007) *Angew. Chem. Int. Ed.*, **46**, 5472–5475; (e) Pischel, U. (2007) *Angew. Chem. Int. Ed.*, **46**, 4026–4040; (f) Pischel, U. (2010) *Aust. J. Chem.*, **63**, 148–164; (g) Andreasson, J. and Pischel, U. (2010) *Chem. Soc. Rev.*, **39**, 174–188.
7. (a) Belousov, B.P. (1959) *Collection of Abstracts on Radiation Medicine in 1958*, Medicine Publishing, Moscow, pp. 145–147; Field, R.J. and Burger, M. (translated eds) (1985) *Oscillations and Traveling Waves in Chemical Systems*, John Wiley & Sons, Inc., New York. (in Russian); (b) Epstein, I.R. (2006) *Proc. Natl. Acad. U.S.A.*, **103**, 15727–15728.
8. (a) Lebender, D. and Schneider, F.W. (1994) *J. Phys. Chem.*, **98**, 7533–7537; (b) Tóth, A., Gáspár, V., and Showalter, K. (1994) *J. Phys. Chem.*, **98**, 522–531; (c) Tóth, A. and Showalter, K. (1995) *J. Chem. Phys.*, **103**, 2058–2066.
9. Adamatzky, A. (2011) *J. Comput. Theor. Nanosci.*, **8**, 295–303.
10. (a) Adamatzky, A., Costello, B., Bull, L., and Holley, J. (2011) *Isr. J. Chem.*, **51**, 56–66; (b) Costello, B.D. and Adamatzky, A. (2005) *Chaos Solitons Fractals*, **25**, 535–544.
11. Wagner, N. and Ashkenasy, G. (2009) *Chem. Eur. J.*, **15**, 1765–1775.
12. (a) Pischel, U. (2007) *Angew. Chem. Int. Ed.*, **46**, 4026–4040; (b) Brown, G.J., de Silva, A.P., and Pagliari, S. (2002) *Chem. Commun.*, 2461–2463.
13. (a) Liu, Y., Jiang, W., Zhang, H.-Y., and Li, C.-J. (2006) *J. Phys. Chem. B*, **110**, 14231–14235; (b) Guo, X., Zhang, D., Zhang, G., and Zhu, D. (2004) *J. Phys. Chem. B*, **108**, 11942–11945; (c) Raymo, F.M. and Giordani, S. (2001) *J. Am. Chem. Soc.*, **123**, 4651–4652.
14. (a) Chatterjee, M.N., Kay, E.R., and Leigh, D.A. (2006) *J. Am. Chem. Soc.*, **128**, 4058–4073; (b) Baron, R., Onopriyenko, A., Katz, E., Lioubashevski, O., Willner, I., Wang, S., and Tian, H. (2006) *Chem. Commun.*, 2147–2149; (c) Galindo, F., Lima, J.C., Luis, S.V., Parola, A.J., and Pina, F. (2005) *Adv. Funct. Mater.*, **15**, 541–545; (d) Bandyopadhyay, A. and Pal, A.J. (2005) *J. Phys. Chem. B*, **109**, 6084–6088; (e) Pina, F., Lima, J.C., Parola, A.J., and Afonso, C.A.M. (2004) *Angew. Chem. Int. Ed.*, **43**, 1525–1527.
15. (a) Andreasson, J., Straight, S.D., Bandyopadhyay, S., Mitchell, R.H., Moore, T.A., Moore, A.L., and Gust, D. (2007) *J. Phys. Chem. C*, **111**, 14274–14278; (b) Amelia, M., Baroncini, M., and Credi, A. (2008) *Angew. Chem. Int. Ed.*, **47**, 6240–6243; (c) Perez-Inestrosa, E., Montenegro, J.M., Collado, D., and Suau, R. (2008) *Chem. Commun.*, 1085–1087.
16. (a) Katz, E. and Shipway, A.N. (2005) in *Bioelectronics: from Theory to Applications*, Chapter 11 (eds I. Willner and E. Katz), Wiley-VCH Verlag GmbH, Weinheim, pp. 309–338; (b) Bonnet, S. and Collin, J.P. (2008) *Chem. Soc. Rev.*, **37**, 1207–1217; (c) Ashton, P.R., Ballardini, R., Balzani, V., Credi, A., Dress, K.R., Ishow, E., Kleverlaan, C.J., Kocian, O., Preece, J.A., Spencer, N., Stoddart, J.F., Venturi, M., and Wenger, S. (2000) *Chem. Eur. J.*, **6**, 3558–3574; (d) Thanopulos, I., Kral, P., Shapiro, M., and Paspalakis, E. (2009) *J. Mod. Opt.*, **56**, 1–18.
17. (a) Hsing, I.M., Xu, Y., and Zhao, W.T. (2007) *Electroanalysis*, **19**, 755–768; (b) Katz, E., Baron, R., and Willner, I. (2005) *J. Am. Chem. Soc.*, **127**, 4060–4070; (c) Katz, E., Sheeney-Haj-Ichia, L., Basnar, B., Felner, I., and Willner, I. (2004) *Langmuir*, **20**, 9714–9719.
18. (a) Zheng, L. and Xiong, L. (2006) *Colloids Surf. A*, **289**, 179–184; (b) Riskin, M., Basnar, B., Katz, E., and Willner, I. (2006) *Chem. Eur. J.*, **12**, 8549–8557; (c) Riskin, M., Basnar, B., Chegel,

V.I., Katz, E., Willner, I., Shi, F., and Zhang, X. (2006) *J. Am. Chem. Soc.*, **128**, 1253–1260.
19. (a) Xu, J. and Tan, G.J. (2007) *J. Comput. Theor. Nanosci.*, **4**, 1219–1230; (b) Soreni, M., Yogev, S., Kossoy, E., Shoham, Y., and Keinan, E. (2005) *J. Am. Chem. Soc.*, **127**, 3935–3943; (c) Stojanovic, M.N. and Stefanovic, D. (2003) *Nat. Biotechnol.*, **21**, 1069–1074.
20. Katz, E. and Privman, V. (2010) *Chem. Soc. Rev.*, **39**, 1835–1857.
21. (a) Saghatelian, A., Volcker, N.H., Guckian, K.M., Lin, V.S.Y., and Ghadiri, M.R. (2003) *J. Am. Chem. Soc.*, **125**, 346–347; (b) Ashkenasy, G. and Ghadiri, M.R. (2004) *J. Am. Chem. Soc.*, **126**, 11140–11141.
22. (a) Sivan, S. and Lotan, N. (1999) *Biotechnol. Prog.*, **15**, 964–970; (b) Sivan, S., Tuchman, S., and Lotan, N. (2003) *Biosystems*, **70**, 21–33; (c) Deonarine, A.S., Clark, S.M., and Konermann, L. (2003) *Future Generation Comput. Syst.*, **19**, 87–97; (d) Ashkenazi, G., Ripoll, D.R., Lotan, N., and Scheraga, H.A. (1997) *Biosens. Bioelectron.*, **12**, 85–95; (e) Unger, R. and Moult, J. (2006) *Proteins*, **63**, 53–64.
23. Ezziane, Z. (2006) *Nanotechnology*, **17**, R27–R39.
24. (a) Win, M.N. and Smolke, C.D. (2008) *Science*, **322**, 456–460; (b) Rinaudo, K., Bleris, L., Maddamsetti, R., Subramanian, S., Weiss, R., and Benenson, Y. (2007) *Nat. Biotechnol.*, **25**, 795–801; (c) Ogawa, A. and Maeda, M. (2009) *Chem. Commun.*, 4666–4668.
25. (a) Simpson, M.L., Sayler, G.S., Fleming, J.T., and Applegate, B. (2001) *Trends Biotechnol.*, **19**, 317–323; (b) Li, Z., Rosenbaum, M.A., Venkataraman, A., Tam, T.K., Katz, E., and Angenent, L.T. (2011) *Chem. Commun.*, **47**, 3060–3062.
26. Lederman, H., Macdonald, J., Stefanovic, D., and Stojanovic, M.N. (2006) *Biochemistry*, **45**, 1194–1199.
27. Stojanovic, M.N., Mitchell, T.E., and Stefanovic, D. (2002) *J. Am. Chem. Soc.*, **124**, 3555–3561.
28. Win, M.N. and Smolke, D.D. (2008) *Science*, **322**, 456–460.
29. Schlosser, K. and Li, Y. (2009) *Chem. Biol.*, **16**, 311–322.
30. (a) Li, T., Wang, E., and Dong, S. (2009) *J. Am. Chem. Soc.*, **131**, 15082–15083; (b) Elbaz, J., Shlyahovsky, B., Li, D., and Willner, I. (2008) *ChemBioChem*, **9**, 232–239; (c) Moshe, M., Elbaz, J., and Willner, I. (2009) *Nano Lett.*, **9**, 1196–1200.
31. Stojanovic, M.N., Stefanovic, D., LaBean, T., and Yan, H. (2005) in *Bioelectronics: From Theory to Applications*, Chapter 14 (eds I. Willner and E. Katz), Wiley-VCH Verlag GmbH, Weinheim, pp. 427–455.
32. Benenson, Y. (2009) *Curr. Opin. Biotechnol.*, **20**, 471–478.
33. Willner, I., Shlyahovsky, B., Zayats, M., and Willner, B. (2008) *Chem. Soc. Rev.*, **37**, 1153–1165.
34. Shapiro, E. and Benenson, Y. (2006) *Scientific Am.*, 45–51.
35. (a) Wang, J. and Katz, E. (2011) *Isr. J. Chem.*, **51**, 141–150; (b) Wang, J. and Katz, E. (2010) *Anal. Bioanal. Chem.*, **398**, 1591–1603.
36. Strack, G., Bocharova, V., Arugula, M.A., Pita, M., Halámek, J., and Katz, E. (2010) *J. Phys. Chem. Lett.*, **1**, 839–843.
37. (a) Adar, R., Benenson, Y., Linshiz, G., Rosner, A., Tishby, N., and Shapiro, E. (2004) *Proc. Natl. Acad. U.S.A.*, **101**, 9960–9965; (b) Pei, R.J., Matamoros, E., Liu, M.H., Stefanovic, D., and Stojanovic, M.N. (2010) *Nat. Nanotechnol.*, **5**, 773–777.
38. Macdonald, J., Li, Y., Sutovic, M., Lederman, H., Pendri, K., Lu, W.H., Andrews, B.L., Stefanovic, D., and Stojanovic, M.N. (2006) *Nano Lett.*, **6**, 2598–2603.
39. (a) Stojanovic, M.N. and Stefanovic, D. (2011) *J. Comput. Theor. Nanosci.*, **8**, 434–440; (b) Stojanovic, M.N. (2011) *Isr. J. Chem.*, **51**, 99–105.
40. Kahan, M., Gil, B., Adar, R., and Shapiro, E. (2008) *Phys. D*, **237**, 1165–1172.
41. Chuang, M.-C., Windmiller, J.R., Santhosh, P., Valdés-Ramírez, G., Katz, E., and Wang, J. (2011) *Chem. Commun.*, **47**, 3087–3089.
42. (a) May, E.E., Dolan, P.L., Crozier, P.S., Brozik, S., and Manginell, M. (2008) *IEEE Sens. J.*, **8**, 1011–1019; (b) von

Maltzahn, G., Harris, T.J., Park, J.-H., Min, D.-H., Schmidt, A.J., Sailor, M.J., and Bhatia, S.N. (2007) *J. Am. Chem. Soc.*, **129**, 6064–6065.

43. (a) Pita, M., Zhou, J., Manesh, K.M., Halámek, J., Katz, E., and Wang, J. (2009) *Sens. Actuat. B*, **139**, 631–636; (b) Manesh, K.M., Halámek, J., Pita, M., Zhou, J., Tam, T.K., Santhosh, P., Chuang, M.-C., Windmiller, J.R., Abidin, D., Katz, E., and Wang, J. (2009) *Biosens. Bioelectron.*, **24**, 3569–3574; (c) Windmiller, J.R., Strack, G., Chuan, M.-C., Halámek, J., Santhosh, P., Bocharova, V., Zhou, J., Katz, E., and Wang, J. (2010) *Sens. Actuat. B*, **150**, 285–290; (d) Bocharova, V., Halámek, J., Zhou, J., Strack, G., Wang, J., and Katz, E. (2011) *Talanta*, **85**, 800–803.

44. Privman, M., Tam, T.K., Bocharova, V., Halámek, J., Wang, J., and Katz, E. (2011) *ACS Appl. Mater. Interfaces*, **3**, 1620–1623.

45. (a) Phan, J.H., Moffitt, R.A., Stokes, T.H., Liu, J., Young, A.N., Nie, S.M., and Wang, M.D. (2009) *Trends Biotechnol.*, **27**, 350–358; (b) Fernald, G.H., Capriotti, E., Daneshjou, R., Karczewski, K.J., and Altman, R.B. (2011) *Bioinformatics*, **27**, 1741–1748.

46. Katz, E. (2011) *Isr. J. Chem.*, **51**, 132–140.

47. Privman, M., Tam, T.K., Pita, M., and Katz, E. (2009) *J. Am. Chem. Soc.*, **131**, 1314–1321.

48. Katz, E. and Pita, M. (2009) *Chem. Eur. J.*, **15**, 12554–12564.

49. Krämer, M., Pita, M., Zhou, J., Ornatska, M., Poghossian, A., Schöning, M.J., and Katz, E. (2009) *J. Phys. Chem. C*, **113**, 2573–2579.

50. Willner, I. and Katz, E. (eds) (2005) *Bioelectronics: from Theory to Applications*, Chapter 14, Wiley-VCH Verlag GmbH, Weinheim, pp. 427–455.

51. (a) Halámek, J., Tam, T.K., Chinnapareddy, S., Bocharova, V., and Katz, E. (2010) *J. Phys. Chem. Lett.*, **1**, 973–977; (b) Strack, G., Ornatska, M., Pita, M., and Katz, E. (2008) *J. Am. Chem. Soc.*, **130**, 4234–4235; (c) Kim, K.-W., Bocharova, V., Halámek, J., Oh, M.-K., and Katz, E. (2011) *Biotechnol. Bioeng.*, **105**, 1100–1107.

52. Tokarev, I., Gopishetty, V., Zhou, J., Pita, M., Motornov, M., Katz, E. and Minko, S. (2009) *ACS Appl. Mater. Interfaces*, **1**, 532–536.

53. (a) Minko, S., Katz, E., Motornov, M., Tokarev, I., and Pita, M. (2011) *J. Comput. Theor. Nanosci.*, **8**, 356–364; (b) Pita, M., Minko, S., and Katz, E. (2009) *J. Mater. Sci.: Mater. Med.*, **20**, 457–462; (c) Motornov, M., Zhou, J., Pita, M., Tokarev, I., Gopishetty, V., Katz, E., and Minko, S. (2009) *Small*, **5**, 817–820; (d) Motornov, M., Zhou, J., Pita, M., Gopishetty, V., Tokarev, I., Katz, E., and Minko, S. (2008) *Nano Lett.*, **8**, 2993–2997.

2
Peptide-Based Computation: Switches, Gates, and Simple Arithmetic

Zehavit Dadon, Manickasundaram Samiappan, Nathaniel Wagner, Nurit Ashkenasy, and Gonen Ashkenasy

2.1
Introduction

Boolean algebra deals with the "0" and "1" values, utilized as presentations of *false* and *true* statements, respectively. The Boolean logic thus provides simple and concise means to describe the output of chemical processes that depend on more than one factor. Historically, the study of chemical logic gates has started with the design of small organic molecules that perform the desired functions when triggered by simple entities such as protons, hydroxyls, or metal ions [1, 2]. Since the first demonstration by de Silva, many different chemical logic systems were developed, including metal–organic complexes, peptides, and DNA. The response of these systems to additional chemical triggers, as well as to electrochemical and light triggers, has been demonstrated quite frequently [3–7]. Interestingly, many of the recently described logic operations, and also the more complex arithmetic units, were designed based on pursuing dynamic processes instead of simple binding to one operating molecule [8–24]. Beyond the basic demonstrations, scientists were able to exploit their new gates as smart devices that control various applications such as catalysis, sensing analytes, drug delivery, and even *in vivo* transcription. In a related line of research, modules of large biochemical systems have also been described to function as Boolean entities, in studies that used the "top-down" screening for such operations [25–31]. This chapter describes primarily our own research, in which we use synthetic networks made of small proteins as tools for performing chemical computations. We do so by practicing both the bottom-up approach, using individual molecules as gates, and the top-down approach, for which the entire molecular network is used to perform the desired functionality.

Biology can serve as inspiration and can provide chemists with the design principles for engineering networks of interacting and replicating molecules. These can potentially be used as controllable tools for studying systems behavior. Toward this aim, different research groups have designed and characterized dynamic combinatorial libraries [32–38] and replication networks made of nucleic acids (DNA and RNA) [39–42], peptides [43–46], and small organic molecules [47–50].

The catalytic principles of all of these molecular families have been utilized to devise networks that consist of two, three, four, and larger numbers of molecules [51]. The observed molecular replication within such networks is a consequence not only of autocatalysis but also of cross-catalytic processes in which a template molecule is used to enhance the formation of different molecules, usually its own mutants [51].

We discuss below our recent studies directed at understanding the dynamics of catalytic networks and the ways one can take advantage of this knowledge in order to identify Boolean operations in relatively large chemical systems, as well as for *de novo* design of small replication networks that perform specific functions. Chapter 3 describes the "recipe" for constructing peptide-based replication networks. It should be noted here that unlike the case of nucleic acids for which the interaction rules are very well understood based on the Watson–Crick base pairing, for the peptide-based computation, we had to define these interaction rules at the single residue level [46]. We have then used that information for virtually simulating their kinetics. It is shown how the networks can be manipulated in order to facilitate molecular replication through all Boolean logic operations and how the catalytic pathways can be wired together to perform more complex computational modules. Beyond just simulations and simple experiments, we show that while in principle all the gates may be constructed, symmetry and order constraints limit the types of logic that may be practically achieved. Furthermore, we discuss the formation of logic gates within adaptive networks that respond to changes in the environment (pH, salt, and light) and the first steps toward realization of switching and gating molecular electronic devices using peptides.

2.2
Peptide-Based Replication Networks

2.2.1
Template-Assisted Replication

The ability to design relatively large and complex molecular replication networks offers model systems for understanding the principles of self-organization and functional characteristics of complex systems. In our *de novo* designed systems, the template-assisted ligation reaction shown in Figure 2.1a [52] serves as the functional element that wires the networks. This catalytic process employs a dimeric template, which binds noncovalently to fragments of the products and enhances their ligation. In such a process, the template does not necessarily catalyze its own formation, via an autocatalytic process, but rather the formation of another molecule (a "cross-catalytic" process), which in turn can operate as a template for reactions within the network medium. The implementation of autocatalysis and cross-catalysis in the network can lead to different catalytic cycles (shown in general form in Figure 2.1b), depending on the specific properties of each of the molecules.

2.2 Peptide-Based Replication Networks

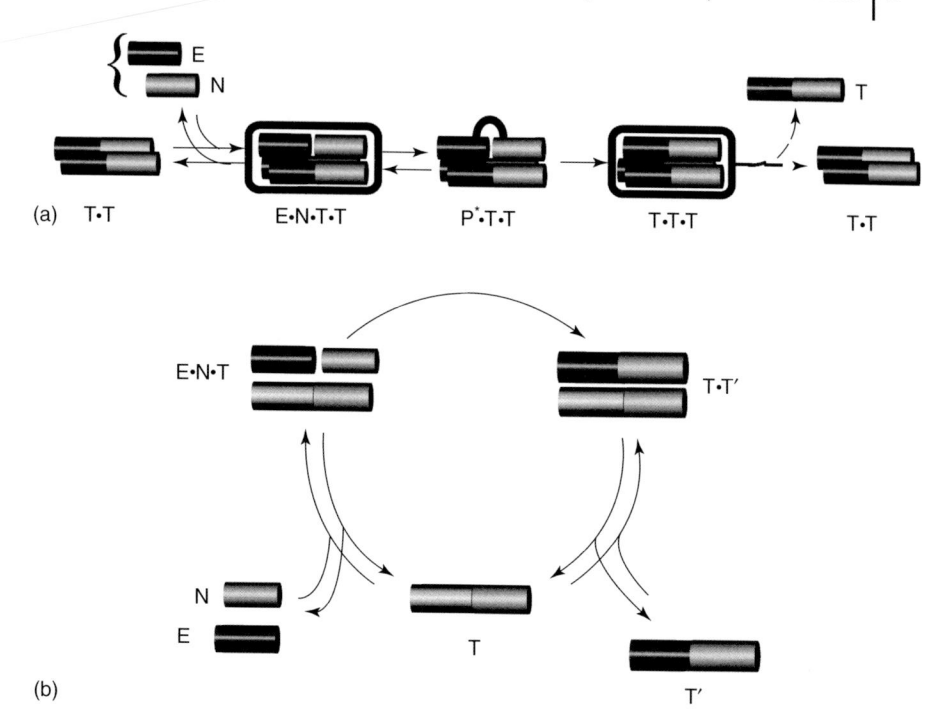

Figure 2.1 Template-assisted ligation reactions that connect two nodes in replication networks. (a) Schematic description of an autocatalytic process, in which a dimeric template catalyzes the coupling of two substrate molecules to form product–template complex, which can then dissociate into a dimeric template and a free monomer. T, full length template molecule, E and N, electrophilic and nucleophilic substrates, respectively; and P*, intermediate product obtained during replication. (b) Operational catalytic pathways within a network. When T and T' are equal, the cycle describes autocatalytic processes; otherwise, the cycle describes cross-catalytic processes.

2.2.2
Theoretical Prediction of the Network Connectivity

We have shown that the selectivity in template-directed fragment condensation reactions can be rationalized based on the relative stability differences between competing intermediates and products ensembles (Figure 2.1a; black boxes). Using such data, it might be possible to determine in advance the network pathways, by estimating the differences in the stability of all reasonable template–product ensembles. The design and study of peptide replicating systems is based on a simple protein folding motif, namely, the α-helical coiled coil, for which the factors contributing to the thermodynamic and kinetic stability of the helical ensembles are reasonably well understood. Many research articles have been published during the past years [53–56] describing the effects of mutations at the recognition interface on the coiled-coil stability and consequently on controlling peptide-based self-replication. One of the methods showed the possibility of affecting the coiled-coil stability by changing the amino acids at the *e* and *g* positions along the helix [53, 54], suggesting a thermodynamic scale (as $\Delta\Delta G$ relative to the control interaction in Ala–Ala mutants) for the coupling energies.

Ashkenasy *et al.* [46] have used the cited method and reported on the design, graph estimation, and analyses of a relatively simple synthetic molecular system. In order to study a chemical network of manageable size, a peptide array composed of eighty-one 32-residue coiled coils was derived by substituting glutamic acid, alanine, and lysine residues at four *e* and *g* heptad positions (*g*8, *e*13, *g*15, and *e*20) that line a portion of the coiled-coil recognition interface. Using the scoring algorithm, with some reasonable assumptions, the change in stability ($-\Delta\Delta G$) for possible trimeric products was calculated by summing the energetics of six pairwise cross-strand *g* to *e′* interactions and using a fixed threshold value. Analysis of all the plausible template-directed pathways in the coiled-coil matrix resulted in a nonrandom graph with 25 nodes joined by 53 directed edges, including 5 autocatalytic pathways. This graph was clustered and hierarchical, such as cellular networks, but unlike the cellular networks possessed mirror symmetry due to peptides with different residue permutations along the recognition interface that offer the same combinations of interacting residues.

2.2.3
De novo Designed Synthetic Networks

The validity of the above design principles was experimentally analyzed by testing a nine-node network, comprising a main segment of the graph, for its capacity to establish the predicted network connectivity (Figure 2.2) [46]. The efficient operational processes were found in the "network experiment" by following the kinetics of formation of nine molecules (T_{1-9}), which were produced when one nucleophilic (N) and nine electrophilic (E_{1-9}) fragments reacted simultaneously. The network connectivity of each of the templates/products was inferred from their ability to enhance their own formation and/or the formation of other mutant(s).

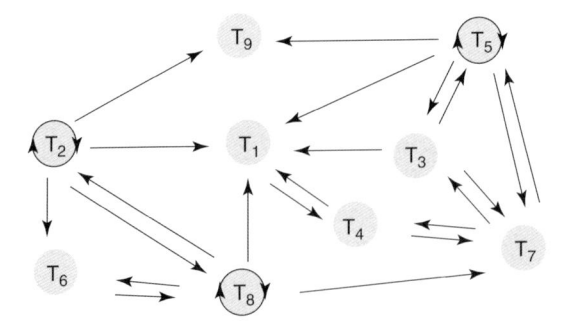

Figure 2.2 Experimental replication network made of nine replicating peptide molecules. Each node in the graph represents a distinct peptide template or product, with the edges (arrows) designating template-assisted ligation pathways pointing from the template to the product; arrows circling a template represent autocatalytic activity.

In agreement with the scoring analysis, the experimental network showed almost all the predicted pathways. This study shows that even a relatively small molecular network seems to mimic some of the graph architectural and basic dynamic features commonly associated with much larger complex systems.

2.3 Logic Gates within Ternary Networks

2.3.1 Uniform Design Principles of All Two-Input Gates

We have recently formulated a general mechanism describing simultaneous auto- and cross-catalysis within replication networks (Eqs. (2.1a) and (2.1b)) [57]. In this description, like in the mechanism shown in Figure 2.1a, the dimers T_jT_k are the only catalytically active species, while there are given propensities for association of monomers to form the dimers and trimers. In addition to the catalytic reactions, each of the reactants E_i combines directly with N to form the templates T_i in a slower background reaction (Eq. (2.1a)).

$$E_i + N + T_jT_k \underset{<a>_{ijk}}{\overset{a}{\rightleftharpoons}} E_iNT_jT_k \overset{b_{ijk}}{\underset{<f>}{\longrightarrow}} T_iT_jT_k \overset{f_{ijk}}{\underset{<f>}{\rightleftharpoons}} T_i + T_jT_k \overset{d_{jk}}{\underset{<d>}{\rightleftharpoons}} T_i + T_j + T_k \quad (2.1a)$$

$$E_i + N \overset{g_i}{\longrightarrow} T_i \quad (2.1b)$$

In Eq. (2.1a), the diffusion-limited association is described by the rate constants a, $<d>$, and $<f>$; the ligation step is described by the rate constants b_{ijk}; and the dissociation is described by the rate constants $<a>_{ijk}$, d_{jk}, and f_{ijk}. The g_i parameters in Eq. (2.1b) represent the rate constants of the background template-free ligation reactions. Such a system of reactants and templates, with autocatalysis and mutual cross-catalysis of various strengths, forms a molecular network. A productive network connection is evident by an efficient catalytic process leading to a template, while the interruption of a connection from one template to another corresponds to a very high value (greater by several orders of magnitude) for the specific dissociation constant $<a>$.

On the basis of the wealth of molecular information present in the synthetic networks, and using the above mechanism, we were able to mimic all the basic Boolean logic functions [58]. The construction of all basic two-input logic gates, such as OR, AND, IFNOT (INHIBIT), NOR (Not OR), NAND (Not AND), XOR (eXcluded OR), XNOR (Not XOR or eXcluded NOR), and NIFNOT (Not IFNOT), was performed by activating or interrupting individual network connections with respect to the formation of T_3 molecules (Figure 2.3). The design was based on the following principles: an output is taken as "1" if T_3 is formed quickly via template-assisted catalysis or as "0" if T_3 is formed slowly solely because of the

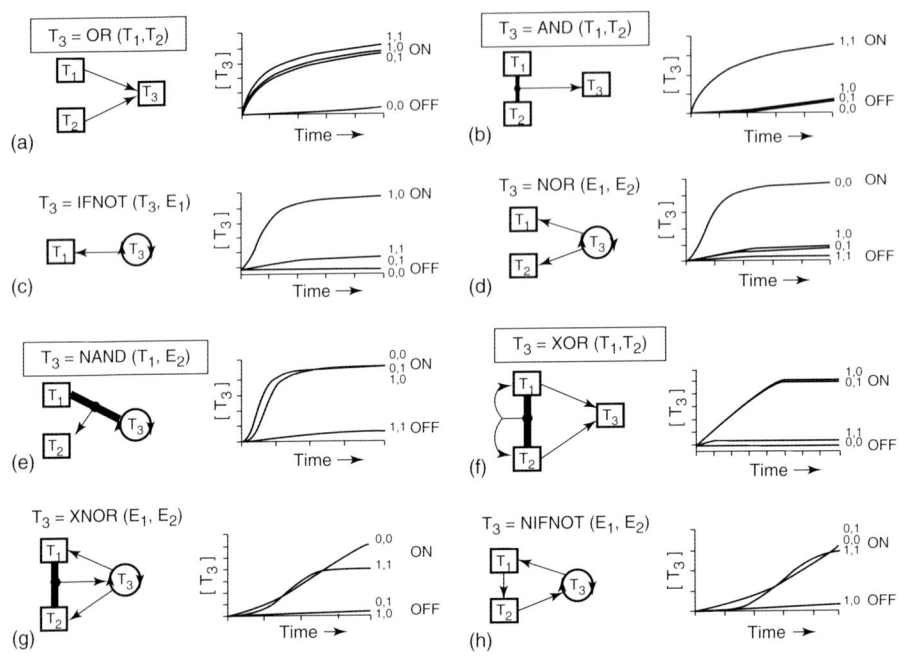

Figure 2.3 Graphical descriptions (left) and dynamic simulations (right) of all two-input logic gates, affecting the replication of T_3. The resulted logic operation for each of the studied network is given within panels a – h. An arrow from T_j to T_i depicts the cross-catalytic pathway in which the template dimer T_jT_j catalyzes the reactants E_i and N to form T_i. An arrow from the line joining T_j and T_k to T_i depicts the cross-catalytic pathway in which the template dimer T_jT_k catalyzes the reactants E_i and N to form T_i. Circular arrows depict autocatalytic pathways, and bold arrows represent stronger catalytic pathways. In the dynamic simulations, each graph shows the production of T_3 as a function of time for four possible combinations of two input variables.

noncatalyzed background reaction. For each gate, the production of T_3 for four possible combinations of two input variables has been followed.

In the following sections, the design, construction, and simulation of four different logic gates (Figure 2.3, black boxes) are described as an example for the above: (i) OR and (ii) AND – which are basic and simple logic gates required in various complex logic devices, computational modules, and network motifs; (iii) NAND (Not AND) – a central gate in logic gates theory since all possible logic operations may be constructed from its combinations; and (iv) XOR (eXcluded OR) – which participates in various logic devices and computational modules.

2.3.2
OR Logic

The OR gate (Figure 2.3a) can be performed within a network when both T_1 and T_2 are capable of catalyzing the formation of T_3, that is, each of the dimers T_1T_1 and T_2T_2 can act as catalysts for the formation of T_3. But, the dimers cannot act as autocatalysts for their own formation or in the back cross-catalytic reactions $T_3T_3 \rightarrow T_1$ and $T_3T_3 \rightarrow T_2$. Hence, in computer simulation of the OR gate, initial high concentrations of either T_1 or T_2, or both, were taken as the "1" inputs and enabled the fast production of T_3 (Figure 2.3a, right panel). When no template is provided ("0, 0" input), T_3 is produced by the slow background reaction and then via some catalysis once small amounts of T_1 and T_2 are produced, so the output is relatively negligible but not entirely zero. The "experimental" conditions in the simulations led to slight differences between the "1, 0" and "0, 1" cases, while the "1, 1" case is slightly stronger since both these pathways are active in parallel.

2.3.3
AND Logic

The AND gate (Figure 2.3b) allows for catalysis of T_3 via the T_1T_2 heterodimer, while none of the other auto- or cross-catalytic pathways are allowed. Here too, the computer simulation took initial high concentrations of either T_1 or T_2 as the "1" inputs. Unlike the OR gate case, introducing only one of the templates in the "1, 0" or "0, 1" case results in slow T_3 production. Only the "1, 1" combination results in fast production of the latter (Figure 2.3b, right panel).

2.3.4
NAND Logic

The NAND (Not AND) gate's output is "1" when neither or either of the inputs are present, and "0" when both the inputs are present (Figure 2.3e). In the NAND gate constructed here, T_3 is a good autocatalyst, while T_3 and T_1 are both required for more efficient heterodimeric cross-catalysis leading to the formation of T_2. The inputs are the initial concentrations of T_1 and E_2, an electrophile that competes with E_3 for binding N. If both T_1 and E_2 are present, the cross-catalytic pathway will be dominant; otherwise the autocatalytic pathway dominates and produces T_3. The key to successful design of this NAND gate is to accelerate the competing cross-catalytic pathway $T_1T_3 \rightarrow T_2$. For the "1, 1" case, the output is very low, while in the other three cases, it is high (Figure 2.3e, right panel).

2.3.5
XOR Logic

The XOR (Excluded OR) gate (Figure 2.3f) is not a trivial gate; its output is "1" when either one of the inputs is present, but is "0" when neither or both of the inputs are present. The XOR gate implemented here is constructed similarly to the OR gate, but contains two additional competing heterodimeric catalytic pathways (Figure 2.3f). Specifically, each of the homodimers T_1T_1 and T_2T_2 catalyzes the formation of T_3, while the heterodimer formed between T_1 and T_2 (T_1T_2) catalyzes E_1 and N, and E_2 and N, to form T_1 and T_2, respectively. Since the two latter catalytic pathways (shown in Figure 2.3f by thicker lines) are more efficient than the two former pathways, the production of T_1 and T_2 outraces the production of T_3 if both T_1 and T_2 are initially present at high concentrations. The successful XOR design is achieved by weakening the forward catalytic pathways $T_1T_1 \rightarrow T_3$ and $T_2T_2 \rightarrow T_3$ while augmenting the backward competing catalytic pathways $T_1T_2 \rightarrow T_1$ and $T_1T_2 \rightarrow T_2$. Computer simulation of the gate (Figure 2.3f, right panel) reveals the expected "1" output for the "1, 0" and "0, 1" cases. The formation of T_3 is relatively slow in this case because of the background production of T_1 and T_2. The output of the "1, 1" case is "0," since T_1 and T_2 are formed quite early via the competing cross-catalytic pathways, inhibiting T_3 production.

These examples show how basic auto- and cross-catalytic processes may be connected and combined to form logic gates and simple arithmetic units.

2.4
Symmetry and Order Requirements for Constructing the Logic Gates

2.4.1
Symmetry and Order in Peptide-Based Catalytic Networks

While we have shown earlier how all logic gates may be constructed in principle, symmetry constraints limit the types of logic that may be practically constructed for lower order catalytic systems, a conclusion with far-reaching implications regarding the centrality of higher order catalysis in molecular self-organization and Systems Chemistry [59].

In principle, the basic elements that form the functional modules within networks may be symmetric or asymmetric and can consist of first, second, or higher order catalytic reactions. The analysis of the reaction mechanisms such as in Eqs. (2.1a) and (2.1b) showed that, due to symmetry constraints, and reasonable chemical assumptions, at least second-order catalysis is necessary in order to form cooperative and asymmetric sequentially linked units [59, 60]. We have shown that unlike first-order catalysis – where only a monomer template is required to catalyze the reactants, the second-order catalysis – in which homo- or heterodimer templates enhance catalysis – is inherently useful for the construction of both cooperative and asymmetric-linked network elements.

To show this limitation, it is necessary to understand the basic symmetry of the rate constants and intermediate complexes in the catalytic equations. First-order catalysis may be described by the following set of equations:

$$E_i + N + T_j \underset{<a>_{ij}}{\overset{a}{\rightleftharpoons}} E_iNT_j \xrightarrow{b_{ij}} T_iT_j \underset{<d>}{\overset{d_{ij}}{\rightleftharpoons}} T_i + T_j \qquad (2.2a)$$

$$E_i + N \xrightarrow{g_i} T_i \qquad (2.2b)$$

where, in analogy to Eqs. (2.1a) and (2.1b), a and $<d>$ are the diffusion-limited rate constants of association, the rate constant b_{ij} describes the ligation, $<a>_{ij}$

and d_{ij} describe the dissociation of the intermediates, and g_i represents the rate constants of the background template-free ligation reactions. We note that d must be symmetric, that is, $d_{ij} = d_{ji}$ for all i, j, since T_iT_j is equivalent to T_jT_i. However, $<a>$ need not be symmetric, and in general, $<a>_{ij} \neq <a>_{ji}$, since E_iNT_j is not the same as E_jNT_i.

Second-order catalysis, described by Eqs. (2.1a) and (2.1b), follows a different set of symmetries. Here, d follows the same symmetry as before: $d_{jk} = d_{kj}$ for all j, k. f is also symmetric, $f_{ijk} = f_{ikj} = f_{jki} = f_{jik} = f_{kij} = f_{kji}$ for all i, j, k, since $T_iT_jT_k$, $T_iT_kT_j$, $T_jT_kT_i$, $T_jT_iT_k$, $T_kT_iT_j$, and $T_kT_jT_i$ are all equivalent. Similarly, $<a>$ follows the subsymmetry $<a>_{ijk} = <a>_{ikj}$ for all i, j, k. Apart from this, cross-catalysis is in general not symmetric, since $E_iNT_jT_j$ is not the same as $E_jNT_iT_i$ and $T_iT_jT_j$ is not the same as $T_jT_iT_i$.

We have further assumed that the rate constants describing these reactions obey reasonable chemical assumptions. The ligation constants, b_{ij} (or b_{ijk}) and g_i, are assumed to be roughly proportional and dependent only on reactant (i.e., independent of template) $b_{ij} \approx b_i \approx k_{bg}(i)g_i$. Furthermore, we assumed based on experimental systems that the dissociation constants $<a>$ and d, in first-order catalysis, and $<a>$ and f, in second-order catalysis, are proportional (at least by order of magnitude), where the constant of proportionality depends only on i (reactant) and not on j or k (template). We call this the "Broken Lego Property"; roughly, it is equivalent to assuming that the relative bond strengths of approximately matching lego pieces, where some are partially broken, are determined only by the broken ones [58].

The importance of this mathematical analysis is the following: the basic sequential linked units or their equivalent are required building blocks of complex molecular networks, since their asymmetry and modularity allow for their concatenation with other linked units to form more complex network elements [2, 10, 61, 62]. The cooperative cross-catalytic logic gates also play a crucial role in the complexification of networks and Systems Chemistry, since all possible logic gates may be constructed from linkable NAND or NOR gates. Since these gates are constructed from the NOT and AND or NOT and OR gates, and the one-input linkable NOT gate is constructed from the NAND, they also require higher order catalysis.

2.4.2
How Symmetry and Order Affect the Replication of RNA Quasispecies

One open question in the study of synthetic chemical networks has to do with the validity of the models and learned information for targeting challenges that are relevant to biology. In order to start probing this question, we have looked recently for simple but "more biological" models, such as models that can account for replication of very simple species – for example, RNA viruses.

The quasispecies model describes processes related to the Darwinian evolution of certain self-replicating entities within the framework of physical chemistry, the origin of life and viral evolutionary dynamics [63]. Catalytic reactions are relevant to the origin of life and early molecular evolution, since facilitating autocatalysis and cross-catalysis may lead to mutations and, consequently, to dynamic evolution [64–66]. As discussed earlier for peptide networks, interacting catalytic reactions of various reaction orders were shown to form catalytic networks of increasing complexity, and certain crucial features of network complexity were found to require at least second-order catalysis.

We have recently developed a quasispecies approach for analyzing mutation and selection in catalytic reactions of varying order [67]. We discussed how the error catastrophe that reflects the transition from localized to delocalized quasispecies population is affected by catalytic replication of different reaction orders. Specifically, the second-order mechanisms lead to a discontinuity in the mean fitness of the population at the error threshold (Figure 2.4). This is in contrast to the behavior of the first-order, autocatalytic replication mechanism considered in the standard quasispecies model. This suggested that quasispecies models with higher order of replication mechanisms produce discontinuities in the mean fitness, and hence, in the viable population fraction, at the error threshold, while lower order replication mechanisms yield a continuous mean fitness function.

We have suggested that this work is important for understanding RNA (viral) evolution methodologically and practically. The methodology was practiced by identifying specific behavioral and dynamical features in chemical or biochemical networks and thus may be common to other systems. It may be useful for studying

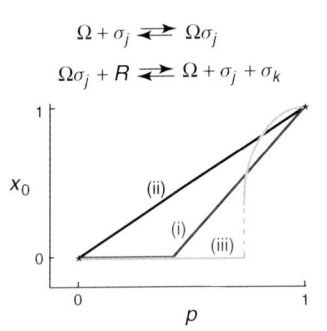

Figure 2.4 Quasispecies model with catalytic replication of different reaction orders. In this model, Ω is used as a general variable representing the material that binds with σ_j, a population of single-stranded sequences of some biopolymer (e.g., RNA or protein), to form the complex $\Omega\sigma_j$, and R is used for the raw materials. We considered three separate cases, characterized by the following replication schemes: (i) $\Omega = M$, describing the obligatory binding of σ_j to M, a nongenomic molecule (e.g., small molecule) before replication, (ii) $\Omega = \sigma_j$, reflecting the formation of a homodimer catalyst for replication, and (iii) $\Omega = \sigma_i$, for the formation of a heterodimer catalyst for replication. The following qualitative solutions for the three cases are shown: case (i) (dark gray) shows an error threshold where the viable population fraction is continuous, case (ii) (black) shows no error threshold, and case (iii) (light gray) yields a jump discontinuity in the viable population fraction at the error threshold.

RNA systems at the early stages of the evolution of ribozymes – key species in the postulated "RNA world" that can serve both as genetic material and catalysts for replication [68] – that gain or lose replication efficiency because of dimerization. The more practical aspect suggests studying experimental virology systems in light of the new findings. Especially relevant systems include viral RNA in which higher order interactions, due to the association of two or more RNA molecules, affect their replication.

2.5
Taking the Steps toward More Complex Arithmetic

We have shown recently how catalytic pathways can be wired together to perform more complex arithmetic units and network motifs, such as the half-adder and half-subtractor computational modules and the coherent feed-forward loop (FFL) network motifs under different sets of parameters [58]. As in previous studies of chemical networks, some of the systems described display behavior that would be difficult to predict without the numerical simulations. Furthermore, the simulations reveal trends and characteristics that should be useful as "recipes" for future design of experimental functional motifs and for potential integration into modular circuits and molecular computation devices.

2.5.1
Arithmetic Units

Logic gates may be joined together into simple arithmetic units such as the full adder, half adder, full subtractor, and half subtractor [2]. The integration of the gates can be achieved in chemical systems by two different methods. The first makes use of multifunctional molecules that exhibit different logic in response to different stimuli (e.g., different light wavelengths). In the second method, one follows two or more molecules in parallel. While each of the molecules performs a single Boolean logic, the overall product of the network is the expected arithmetic. A half adder is a simple logical circuit that performs an addition operation on two binary digits. It has two inputs and two outputs, the sum S and the carry C. S and C are the two-bit XOR and AND operations, respectively, of the same two inputs. A half subtractor circuit subtracts one bit from another and places the difference D and the borrow B. It thus implements INHIBIT and XOR logics that share the same two inputs.

Using the information in Section 2.4.1, together with the principles used for constructing the rather difficult XOR logic gate, it was possible to simulate an "asymmetric" network, in which the formation of each of the three nodes is controlled by the other two nodes and follows a different Boolean logic. We have termed this network as a system with *"maximal molecular information."* To that aim, we have manipulated the simulation parameters so that the heterodimer T_1T_2 catalyzes only T_2 and not T_1, while T_1 is produced autocatalytically [58]. As a result

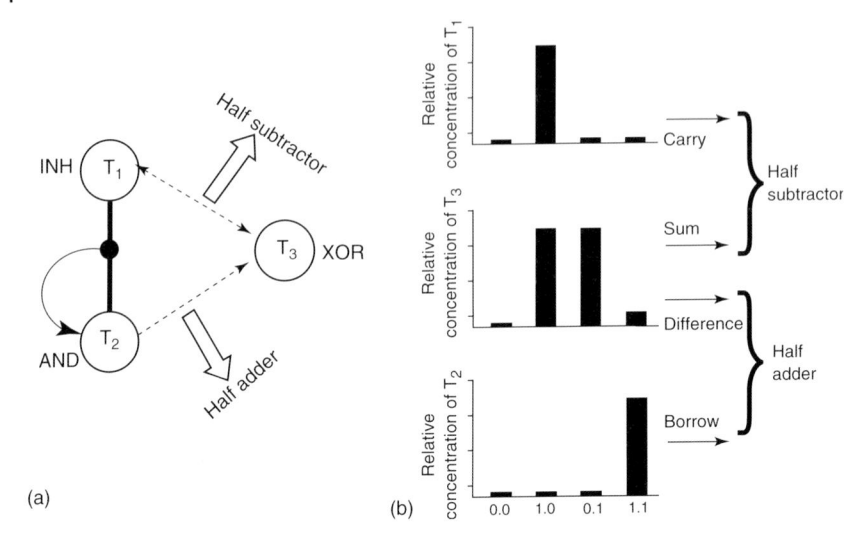

Figure 2.5 Graphical description (a) and simulation (b) of a network with "maximal molecular information." An XOR gate in T_3 is formed, and the network simultaneously produces an INHIBIT in T_1 and an AND in T_2, allowing the construction of a half adder and a half subtractor. The dashed lines in (a) represent especially weak catalytic pathways.

of constructing this specific configuration, the T_3 output is an XOR, while the T_1 output is an INHIBIT and the T_2 output is an AND (Figure 2.5). This design further allowed the simultaneous formation of the half subtractor (using T_1 and T_3) and half adder (T_2 and T_3) arithmetic units within a single network.

2.5.2
Network Motifs

Specific modules that are found highly frequently within complex networks were given the general term *"network motifs."* This concept was used to detect basic building blocks of cellular networks, with a special focus on gene regulatory networks [29, 69–71]. The studied network motifs, such as the FFL, include auto- or cross-regulation of small subnetworks that consist of three nodes, thus conceptually similar to the three-nodes networks used for constructing the above-described logic gates. The FFL represents a slightly higher level of complexity relative to the logic gates. No one had yet explored the network motifs on chemical systems, and to do so we extended our study to include motifs that are similar to the FFL motifs, using the same network building blocks, and simulated with the same equations and computations, as described earlier [58]. We explored the behavior of the FFL network motifs in comparison with two control functions: the direct IF gate and a simple "feed-forward" (FF) cascade logic, where T_1 indirectly catalyzes the formation of T_3 by first catalyzing the formation of T_2, which in turn catalyzes the formation of T_3. The FF was simulated as a concatenation of two YES gates, while the FFL OR motif is a concatenation of a YES with an OR, and the FFL AND motif is a concatenation of a YES with an AND.

2.6
Experimental Logic Gates

Ashkenasy and Ghadiri [72] have designed synthetic peptide networks that can mimic some of the basic functions of complex biological networks. The formation of each node within such a network is regulated by more than one catalytic pathway; thus the goal was to examine whether selected segments of the network could also perform basic logic operations [72]. The network shown in Figure 2.6a shows that a small network whose graph structure is composed of 5 nodes (out of 9 from the network shown in Figure 2.2) and 15 directed edges can express OR, NOR, and NOTIF logic. Nodes T_3 and T_7 were focused on separately as the outputs from two-input logic operations because of their prominent positions in the network and their patterns of network interconnectivity.

2.6.1
OR Logic

The linear nature of the $T_3 \leftrightarrow T_7 \leftrightarrow T_4$ subnetwork connectivity indicates that T_7 might function as the output of an OR gate with T_3 and T_4 as the inputs. The OR logic function was verified in isolated experiments in which the rate of T_7 production was monitored (Figure 2.6b, left). In the absence of added initial amounts of either input, the background autocatalytic rate of T_7 production was low. However, because T_3 and T_4 are efficient templates for the production of T_7, the presence of either or both inputs in the reaction mixture brought about significantly enhanced rates of T_7 production. These two peptides cannot catalyze each other's formation when operating as catalysts for T_7.

2.6.2
NOT, NOR, and NOTIF Logic

The simplest negative regulation is formed by the NOT inverter operation, in which the existence of the input ligand inhibits the normal activity of the catalyst. As an example, the autocatalytic production of T_3 in a reaction mixture composed of N, E_3, and T_3 can be negatively influenced by the addition of E_5 that sequesters N and T_3 for the production of T_5 via the more efficient $T_3 \leftrightarrow T_5$ pathway (Figure 2.6c, No. 6).

The NOT function was studied as part of the more intriguing two-input NOR and NOTIF logic operations. The NOR logic was examined in the context of the circular subnetwork formed by T_3, T_5, and T_7. The rate of T_3 production (output) was monitored in reaction mixtures composed of N, E_3, and T_3 fragments, in

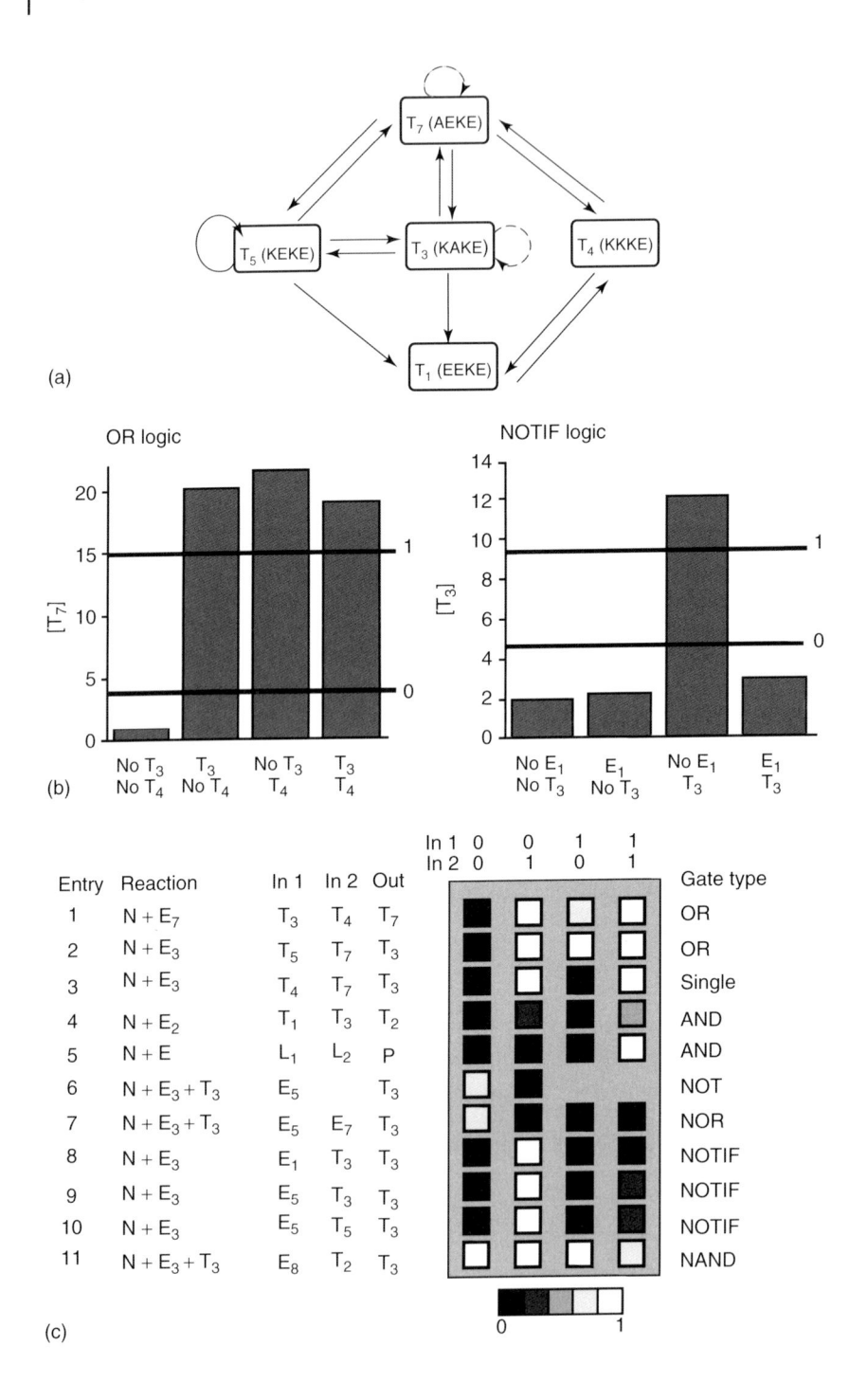

2 Peptide-Based Computation: Switches, Gates, and Simple Arithmetic

(a)

T_7 (AEKE)

T_5 (KEKE) T_3 (KAKE) T_4 (KKKE)

T_1 (EEKE)

(b) OR logic

| | No T_3 No T_4 | T_3 No T_4 | No T_3 T_4 | T_3 T_4 |

NOTIF logic

| | No E_1 No T_3 | E_1 No T_3 | No E_1 T_3 | E_1 T_3 |

(c)

Entry	Reaction	In 1	In 2	Out	Gate type
1	$N + E_7$	T_3	T_4	T_7	OR
2	$N + E_3$	T_5	T_7	T_3	OR
3	$N + E_3$	T_4	T_7	T_3	Single
4	$N + E_2$	T_1	T_3	T_2	AND
5	$N + E$	L_1	L_2	P	AND
6	$N + E_3 + T_3$	E_5		T_3	NOT
7	$N + E_3 + T_3$	E_5	E_7	T_3	NOR
8	$N + E_3$	E_1	T_3	T_3	NOTIF
9	$N + E_3$	E_5	T_3	T_3	NOTIF
10	$N + E_3$	E_5	T_5	T_3	NOTIF
11	$N + E_3 + T_3$	E_8	T_2	T_3	NAND

In 1: 0 0 1 1
In 2: 0 1 0 1

0 1

the absence or presence of the following inputs: E_5, E_7, or an equal mixture of E_5 and E_7. In the absence of any input, T_3 production proceeded aptly through its autocatalytic cycle. However, in the presence of the inputs E_5 and/or E_7, the autocatalytic rate of T_3 production was lessened because of its involvement in the more efficient $T_3 \rightarrow T_5$ and/or $T_3 \rightarrow T_7$ pathways, respectively (Figure 2.6c, No. 7).

The NOTIF logic function was first examined in the context of the $T_3 \rightarrow T_1$ pathway. The autocatalytic rates of T_3 production were monitored in reaction mixtures composed of N and E_3, in the absence or presence of either inputs or a mixture of E_1 and T_3 (Figure 2.6b, right). A T_3 output was observed in the presence of T_3 only when E_1 was absent. The observed NOTIF function is consistent with the inhibition of T_3 autocatalysis in the presence of competing substrate E_1 that redirects T_3 for the production of T_1, which is neither an autocatalyst nor a back catalyst for the formation of T_3. This can be seen as a "host–parasite" relationship in the subnetwork (Figure 2.6a). Only when the replicator "host" T_3 is present does the T_1 "parasite" replicate, while T_3 loses its replication ability in the process of producing T_1. Two additional NOTIF functions were performed from different subnetworks (Figure 2.6c, Nos. 9, 10).

2.6.3
Additional Logic Operations

On the basis of the above-described examples, it is possible to claim that the formation of a given node within the network can be regulated by more than one template-directed pathway, and certain nodes serve as outputs of logic processes in response to selected system inputs. Figure 2.6c summarizes various different gates obtained during the experiments (including the examples discussed earlier). It is arranged to show the required "gate hardware" for each gate, namely, the reaction that were followed, the chemical inputs that were used (either templates or substrates), and the network node that served as the output. Since the differences between "0" and "1" can be decided arbitrarily, the figure is calibrated to show differences in rates of up to fivefold or higher.

Figure 2.6 (a) A synthetic five-member network used as a media for studying logic gates. Each node in the graph represents a distinct peptide template or product, marked with a four-letter code for the type of amino acids at the mutatable recognition interface. The edges designate the experimentally observed template-assisted ligation pathways pointing from the template to the product. Circular arrows are shown for autocatalytic processes and dashed arrows for weak processes. (b) Logic gate operations expressed by selected peptide subnetworks. The graph shows the amounts of product formed (output), in the absence or presence of different combinations of input peptides, resulting in OR and NOTIF logic operations. (c) The relative amount of peptide output formed is shown for several studied systems, using a linear color scale as a function of the existence of the network inputs, where the maximum change from black to white corresponds to fivefold rate enhancement for a "1" output relative to that of the "0" output.

2.7 Adaptive Networks

2.7.1 Chemical Triggering

The ability of a system to selectively amplify the production of one or more molecules is an important and challenging task. Chmielewski and coworkers [45] described a four-component peptide-based network capable of autocatalysis and cross-catalysis that allows for selective amplification of product formation by changing the reaction conditions. Their design was based on coiled-coil fragments that contain either polylysine or polyglutamate residues in the e and g heptad positions that flank the hydrophobic core. The four product peptides were studied for their catalytic ability under conditions that minimize repulsion between the coiled-coil helices. Accordingly, it was found that the mixed Glu-Lys can form autocatalytically in physiological conditions and the poly-Lys and poly-Glu can promote each other's formation. The autocatalytic processes for each peptide were made available when the system was studied in high salt concentration (poly-K) or low pH (poly-E). From several experiments, it was shown that seeding the mixture with an individual template peptide did not promote the production of only a single product, but rather several unexpected template-assisted processes were affected.

In a larger artificial system, the 9-mer network described earlier and shown in Figure 2.2, the productive catalytic pathways revealing the network connectivity were assessed in neutral aqueous solutions (pH 7.2). Under "native" conditions, the rates of product formation follow the relative order of the predicted pathway efficiencies based on the ion-pairing interactions across the helices. The effects of different environmental conditions on the network connectivity were studied by Ashkenasy and Ghadiri [73] by reacting less than stoichiometric amounts of N with the nine electrophiles (E_i) in several different buffers. At very early stages of the experiment, the products T_i formed via template-free background reactions, all at about the same rate. Once formed, each product could serve as a catalyst in auto- or cross-catalysis reactions, featuring different rates of product formation attributed to the overall catalytic processes that lead to the formation of each of the products. Analyzing the template-assisted reactions led to the characterization of the connectivity and graph architectures of the new networks operating under the new conditions. This system shows that characterizing the most robust interactions in molecular networks under a single set of environmental conditions can provide only a snapshot of how these networks operate. The network can rewire to form alternate sets of connectivity that give rise to new central nodes (molecules that interact with large numbers of molecules) and functional subnetworks, as a response to assay conditions.

2.7.2 Light Triggering

The ability to test and control the response of nonenzymatic networks to external signals might be used to perform different logic gates. Light was suggested

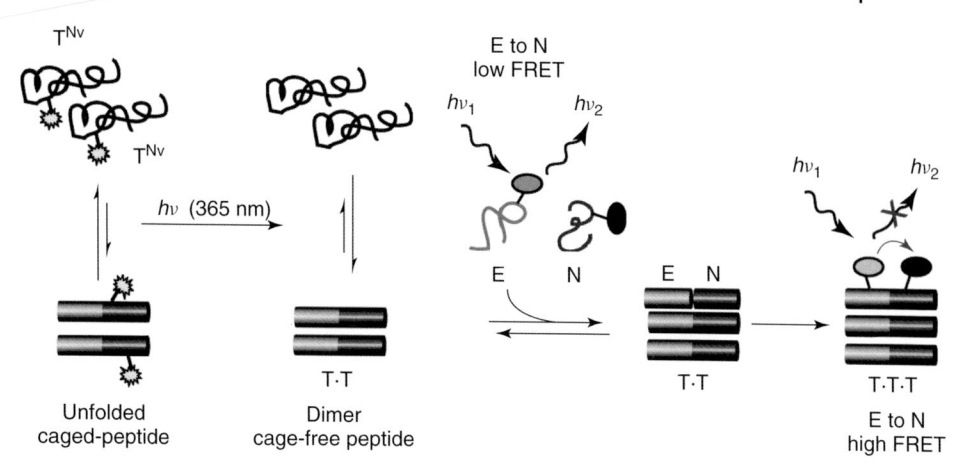

Figure 2.7 General scheme describing the initiation and detection of peptide replication using light. In the dark (left), the caged template is catalytically inactive since it cannot dimerize as a coiled coil or associate with the shorter fragments E and N. After illumination, it becomes active; the cage-free peptide T forms a stable dimeric coiled-coil complex that serves as a template for replication. The process is followed by measuring the FRET quenching of the fluorescence signal from a donor dye attached to the E fragment.

as an external trigger for quantitative control of peptide tertiary structures and, consequently, as a tool for controlling peptide-based replication-dependent processes. We have used light both to control and to follow *in situ* the replication reaction (Figure 2.7). Light-induced uncaging of coiled coils was practiced as a mechanism to initiate template-assisted replication processes. In order to achieve external control over replication, a caged peptide (T^{Nv}), having the photocleavable moiety 6-nitroveratryloxycarbonyl (Nvoc) attached to a Lys residue on an *e* heptad position, was synthesized. The caged molecule destabilizes the coiled-coil structure and abolishes its availability as a catalyst for replication.

2.7.3
Light-Induced Logic Operations

The light-induced replication process was further studied as a means to affect competition experiments within ternary networks (Figure 2.8) [74]. We expanded this system to a network by synthesizing two new nucleophilic fragments, N^{aa} and N^z, which compete together with N for ligation to E as a common reactant. The formation of each of the three products, T, T^{aa}, and T^z, was analyzed in the dark and after shining light. While in the dark, when T^{Nv} was present, all three compounds were formed with similar rates; after shining light and uncaging of T^{Nv}, T replicated at a faster rate, resulting in a higher concentration (Figure 2.8a). Introduction of T^{Nv} and shining light can be considered as two separate "inputs"

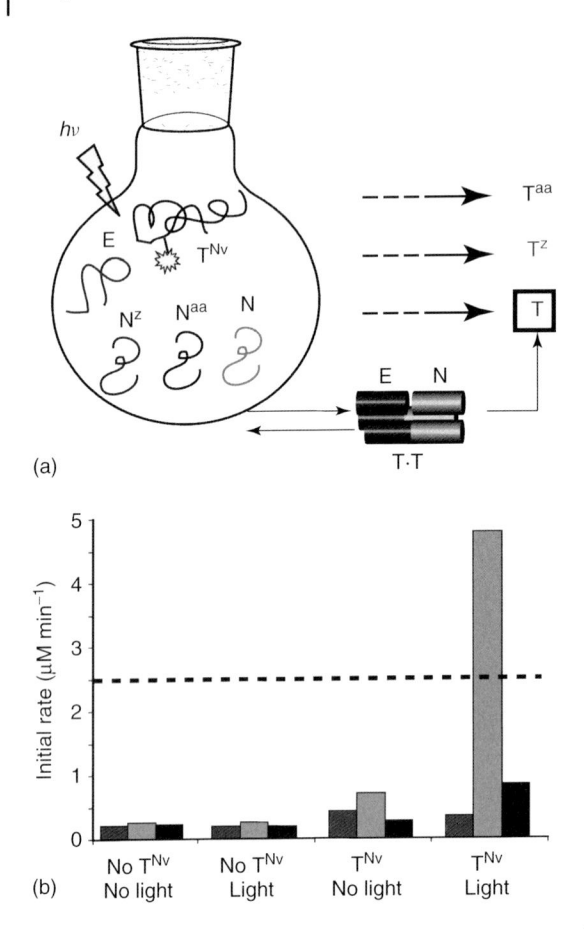

Figure 2.8 Light-induced selection in small networks. (a) A ternary network under illumination. The initial reaction mixtures included E, N, N^{aa}, N^z, and T^{Nv}. The free, uncaged template T can form stable complexes only with E and N, facilitating fast template-assisted reaction at the expense of production of the other two templates, T^{aa} and T^z. (b) AND logic gate observed for production of T within the network. The initial rates of products formation (output) – T (light gray), T^{aa} (black), and T^z (dark gray) – are shown for the various combinations of T^{Nv} and 365 nm light as inputs.

for the study of logic operations within networks. Thus, the competition experiment was studied four times, in which the rates of product formation in the absence or presence of T^{Nv}, and in the dark or after shining light, were characterized. The Boolean logic presentation of the data in Figure 2.8b clearly shows selective product formation of T obeying the AND gate operation. This data provided the first example of manipulating logic operations in replication systems by shining light; furthermore, this method facilitates the formation of the AND gate via an algorithm much simpler than the above-described operation using heterodimeric protein templates.

In order to construct the replication NAND gate [9, 75], we have coupled the förster resonance energy transfer (FRET) system with a light-induced replication protocol (Figure 2.7) [76]. This allowed fast *in situ* tracking of the replication processes (Figure 2.9a), obtaining results that are more accurate than before in terms of reproducibility in repeating experiments, and even yielding some direct insight into the replication mechanism. The NAND gate's output is "1" when neither or either of the inputs is present and "0" when both the inputs are present. The two following inputs were chosen: (i) a caged template molecule (T^{Nv}) and (ii) UV light at 365 nm. Introduction of T^{Nv} alone, in the dark, is not expected to enhance the ligation, since it interferes with the formation of tertiary structures, thus reducing significantly the concentrations of dimeric template and quaternary intermediate complexes. On shining light for a short time and removing the Nvoc group, a new peptide (T) is formed, which can fold and aggregate correctly and thus enhance the replication. The NAND logic behavior of this system is shown in Figure 2.9b. The ligation experiment was studied four times, in the absence or presence of T^{Nv} and in the dark or after shining light for 15 min. The donor fluorescence signal (420 nm), observed 60 min after initiating the reaction, was utilized as a negating signal of the product formation. Very little product formation

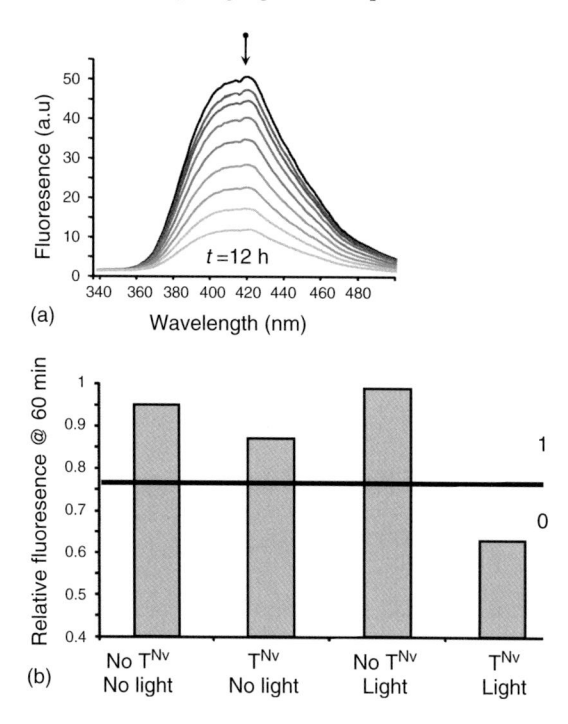

Figure 2.9 Monitoring peptide self-replication by FRET (see general scheme in Figure 2.7). (a) Fluorescence spectra of the donor dye attached to E, obtained at different times after initiating the reaction between E and N. (b) Fluorescence data of the replication NAND gate. The reaction mixture, made of E and N, was treated with four different combinations of T^{Nv} and UV light as inputs.

was observed for the template-free reactions in the dark (input 0, 0) or after shining light (0, 1), and thus interpreted as a "1" output. A slightly lower signal was observed for the reaction in the presence of T^{Nv} in the dark (input 1, 0), still regarded as a "1," while a much lower signal (50% quenching), reflecting a "0" output, was obtained when the latter reaction was carried out after shining light (input 1, 1).

2.8
Peptide-Based Switches and Gates for Molecular Electronics

One important challenge in studying chemical logic gates has to do with finding ways for interfacing such organic entities with the "real world" and furthermore identifying technological applications for which they are relevant or even crucial. An obvious area in which switches and logic gates can be attractive is molecular electronics, and thus we intend to incorporate the peptide logic gates described earlier or related systems into molecular electronic devices. We have noticed that achieving this target can be realized only in a stepwise manner; thus we have studied recently the possibility of exploiting the coiled-coil molecules toward switching and controlling "molecular bridges" conductivity.

Conformational changes of proteins are widely used in nature for controlling cellular functions, including ligand binding, oligomerization, and catalysis. Despite the fact that different proteins and artificial peptides have been utilized as electron-transfer mediators in electronic devices [77, 78], the unique propensity of proteins to switch between different conformations had not yet been used as a mechanism to control device properties and performance. We have designed and prepared dimeric coiled-coil proteins that adopt different conformations due to parallel or antiparallel relative orientations of their monomers (Figure 2.10a) [79]. Controlling the conformations of these proteins attached as monolayers to gold, which dictates the direction and magnitude of the internal molecular dipole relative to the surface, resulted in quantitative modulation of the gold work function. Furthermore, we have shown that charge transport through the proteins as molecular bridges is controlled by the different protein conformations, producing either rectifying or ohmic-like behavior (Figure 2.10b).

To the best of our knowledge, this system makes use of the first artificial coiled-coil proteins with variable conformations as components of molecular electronic devices. We have suggested that the ability of coiled-coil proteins, as well as other systems, to undergo major dynamic conformational changes might be used in the future for controlling the molecular junction electronic behavior *in situ*. While studying this system as is can provide new insights into unresolved issues in the phenomenon of electron transport through proteins, one should not ignore the fact that the structural versatility embedded in the coiled-coil structures, which is the reason for their diverse functionality as discussed in this chapter, offers ample options for future design of artificial molecular electronic devices with tailor-made properties. For example, the addition of appropriate solvent-exposed side chains can be used to get better conductivity.

(a)   Parallel (PC)   Antiparallel (AP)   Parallel (PN)

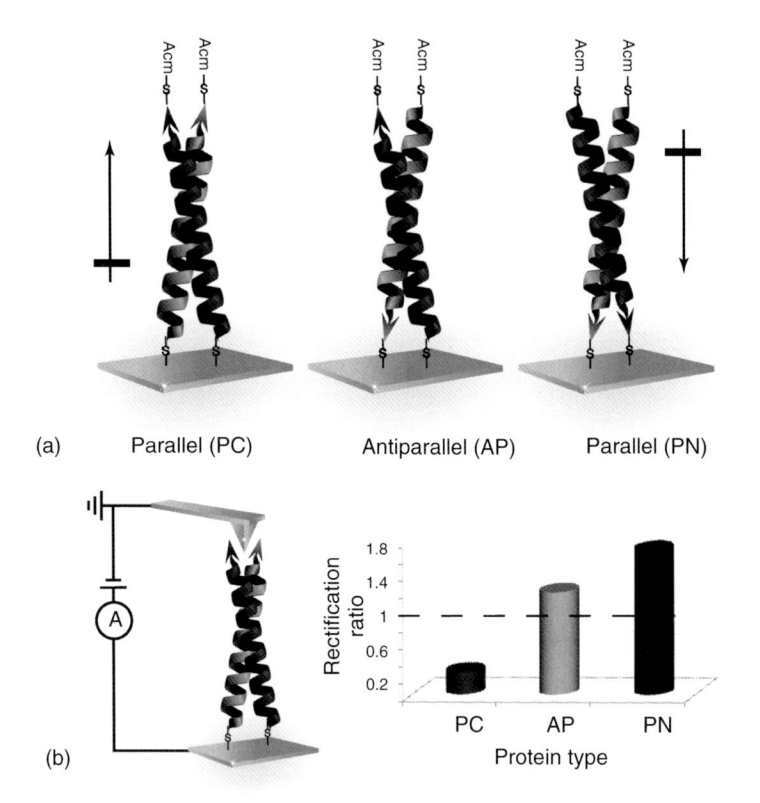

(b)

Figure 2.10 (a) Illustration of parallel and antiparallel coiled coils adsorbed onto gold surfaces. Black color is used for glutamic acid-rich peptides and gray for lysine-rich peptides. PN and PC are the structures obtained after adsorbing a monolayer of a parallel coiled-coil protein with the N-termini or C-termini thiols to the gold. The illustrative arrowheads mark the monomeric molecular dipoles, emphasizing that net molecular dipoles of opposite directions relative to the gold surface are present in the parallel dimeric structures (PC vs PN), while a negligible dipole is present in the antiparallel structure, AP. (b) Schematic illustration of the experimental conductive probe atomic force microscopy setup used to measure the I–V curves of coiled-coil molecular bridges (left). Rectification ratios (right) that were calculated from the I–V curves of PC, AP, and PN (defined as $R = I_{V=0.5\ V}/I_{V=-0.5\ V}$).

2.9 Summary and Conclusion

We suggest that our theoretical, simulation, and experimental studies can provide guidelines for the preparation of logic gates not only with peptides but also using other entities, including DNA and small organic molecules. As noted earlier, rapid progress has been achieved in the field using DNA and RNA molecules *in vitro* and within cells, mainly because the interaction rules for nucleic acid hybridization are very well understood. For the peptide-based computation, we had to define these interaction rules at the single residue level, as explained earlier (where we discuss the scoring algorithm). Consequently, we have shown that one can exploit the structural information and further probe how the stability of different intermolecular complexes is translated to the hierarchy in the kinetic space that is required for computation.

As a future direction, we suggest researching networks that operate via reversible replication, such as those discussed recently in the context of dynamic combinatorial chemistry [50, 80, 81]. It was postulated, and observed in our preliminary experiments, that the reversible replication can allow decomposition of undesired structures, thus leading to better selectivity and probably also to homogeneous concatenation of outputs and inputs in cascade reactions [82].

Finally, we have shown the first steps toward incorporating peptide-based switches (but not yet the gates) as active entities in molecular electronics. Since there is a large interest in controlling the performance of molecular devices using molecules, we expect rapid progress in this field, including major involvement of molecules that can provide high versatility in structure and function, such as peptides and proteins.

Acknowledgments

This research is supported by grants from the European Research Council (ERC 259204) and the Israeli Science Foundation (ISF 1291/08). G.A. and N.A. thank the Edmond J. Safra Foundation for support. We thank the students and coworkers who were involved in the described research, Eliya. Y. Safranchik, Clara Shlizerman, Alexander Atanassov, and Dr Rivka Cohen-Luria.

References

1. de Silva, A.P. and McClenaghan, N.D. (2000) *J. Am. Chem. Soc.*, **122**, 3965.
2. de Silva, A.P. and Uchiyama, S. (2007) *Nat. Nanotech.*, **2**, 399.
3. Credi, A. (2007) *Angew. Chem. Int. Ed.*, **46**, 5472.
4. Pischel, U. (2007) *Angew. Chem. Int. Ed.*, **46**, 4026.
5. Margulies, D., Melman, G., and Shanzer, A. (2005) *Nat. Mater.*, **4**, 768.
6. Margulies, D., Melman, G., and Shanzer, A. (2006) *J. Am. Chem. Soc.*, **128**, 4865.
7. Margulies, D. and Hamilton, A.D. (2009) *J. Am. Chem. Soc.*, **131**, 9142.
8. Benenson, Y., Paz-Elizur, T., Adar, R., Keinan, E., Livneh, Z., and Shapiro, E. (2001) *Nature*, **414**, 430.
9. Saghatelian, A., Voelcker, N.H., Guckian, K.M., Lin, V.S.Y., and Ghadiri, M.R. (2003) *J. Am. Chem. Soc.*, **125**, 346.
10. Seelig, G., Soloveichik, D., Zhang, D.Y., and Winfree, E. (2006) *Science*, **314**, 1585.
11. Shapiro, E. and Benenson, Y. (2006) *Sci. Am.*, **294**, 44.
12. Gianneschi, N.C. and Ghadiri, M.R. (2007) *Angew. Chem. Int. Ed.*, **46**, 3955.
13. Kossoy, E., Lavid, N., Soreni-Harari, M., Shoham, Y., and Keinan, E. (2007) *ChemBioChem*, **8**, 1255.
14. Macdonald, J., Stefanovic, D., and Stojanovic, M.N. (2008) *Sci. Am.*, **299**, 84.
15. Gupta, T. and van der Boom, M.E. (2008) *Angew. Chem. Int. Ed.*, **47**, 5322.

16. Willner, I., Shlyahovsky, B., Zayats, M., and Willner, B. (2008) *Chem. Soc. Rev.*, **37**, 1153.
17. de Ruiter, G., Tartakovsky, E., Oded, N., and van der Boom, M.E. (2010) *Angew. Chem. Int. Ed.*, **49**, 169.
18. Allen, V.C., Robertson, C.C., Turega, S.M., and Philp, D. (2010) *Org. Lett.*, **12**, 1920.
19. Katz, E. and Privman, V. (2010) *Chem. Soc. Rev.*, **39**, 1835.
20. Wang, J. and Katz, E. (2011) *Isr. J. Chem.*, **51**, 141.
21. Katz, E. (2011) *Isr. J. Chem.*, **51**, 132.
22. Shoshani, S., Ratner, T., Piran, R., and Keinan, E. (2011) *Isr. J. Chem.*, **51**, 67.
23. Benenson, Y. (2011) *Isr. J. Chem.*, **51**, 87.
24. Stojanovic, M.N. (2011) *Isr. J. Chem.*, **51**, 99.
25. Guet, C., Elowitz, M., Hsing, W., and Leibler, S. (2002) *Science*, **296**, 1466.
26. Dueber, J.E., Yeh, B.J., Bhattacharyya, R.P., and Lim, W.A. (2004) *Curr. Opin. Struct. Biol.*, **14**, 690.
27. McDaniel, R. and Weiss, R. (2005) *Curr. Opin. Biotech.*, **16**, 476.
28. Mayo, A.E., Setty, Y., Shavit, S., Zaslaver, A., and Alon, U. (2006) *PLoS Biol.*, **4**, 555.
29. Alon, U. (2007) *Nat. Rev. Genet.*, **8**, 450.
30. Carteret, H.A., Rose, K.J., and Kauffman, S.A. (2008) *Phys. Rev. Lett.*, **101**, 218702/1.
31. Zhu, R., Ribeiro, A.S., Salahub, D., and Kauffman, S.A. (2007) *J. Theor. Biol.*, **246**, 725.
32. Lehn, J.M. and Eliseev, A.V. (2001) *Science*, **291**, 2331.
33. Otto, S., Furlan, R.L.E., and Sanders, J.K.M. (2002) *Science*, **297**, 590.
34. Severin, K. (2004) *Chem. Eur. J.*, **10**, 2565.
35. Corbett, P.T., Leclaire, J., Vial, L., West, K.R., Wietor, J.-L., Sanders, J.K.M., and Otto, S. (2006) *Chem. Rev.*, **106**, 3652.
36. Otto, S. and Severin, K. (2007) *Top. Curr. Chem.*, **277**, 267.
37. Ludlow, R.F. and Otto, S. (2008) *Chem. Soc. Rev.*, **37**, 101.
38. Hunt, R.A.R. and Otto, S. (2011) *Chem. Commun.*, **47**, 847.
39. Sievers, D. and von Kiedrowski, G. (1994) *Nature*, **369**, 221.
40. Levy, M. and Ellington, A.D. (2003) *Proc. Natl. Acad. Sci. U.S.A.*, **100**, 6416.
41. Kim, D.-E. and Joyce, G.F. (2004) *Chem. Biol.*, **11**, 1505.
42. Lincoln, T.A. and Joyce, G.F. (2009) *Science*, **323**, 1229.
43. Severin, K., Lee, D.H., Martinez, J.A., and Ghadiri, M.R. (1997) *Chem. Eur. J.*, **3**, 1017.
44. Lee, D.H., Severin, K., and Ghadiri, M.R. (1997) *Curr. Opin. Chem. Biol.*, **1**, 491.
45. Yao, S., Ghosh, I., Zutshi, R., and Chmielewski, J. (1998) *Nature*, **396**, 447.
46. Ashkenasy, G., Jagasia, R., Yadav, M., and Ghadiri, M.R. (2004) *Proc. Natl. Acad. Sci. U.S.A.*, **101**, 11872.
47. Wintner, E.A. and Rebek, J. Jr. (1996) *Acta Chem. Scand.*, **50**, 469.
48. Kindermann, M., Stahl, I., Reimold, M., Pankau, W.M., and von Kiedrowski, G. (2005) *Angew. Chem. Int. Ed.*, **44**, 6750.
49. Kassianidis, E. and Philp, D. (2006) *Angew. Chem. Int. Ed.*, **45**, 6344.
50. Sadownik, J.W. and Philp, D. (2008) *Angew. Chem. Int. Ed.*, **47**, 9965.
51. Dadon, Z., Wagner, N., and Ashkenasy, G. (2008) *Angew. Chem. Int. Ed.*, **47**, 6128.
52. Lee, D.H., Granja, J.R., Martinez, J.A., Severin, K., and Ghadiri, M.R. (1996) *Nature*, **382**, 525.
53. Krylov, D., Mikhailenko, I., and Vinson, C. (1994) *EMBO J.*, **13**, 2849.
54. Krylov, D., Barchi, J., and Vinson, C. (1998) *J. Mol. Biol.*, **279**, 959.
55. Grigoryan, G. and Keating, A.E. (2008) *Curr. Opin. Chem. Biol.*, **18**, 477.
56. Steinkruger, J.D., Woolfson, D.N., and Gellman, S.H. (2010) *J. Am. Chem. Soc.*, **132**, 7586.
57. Rubinov, B., Wagner, N., Rapaport, H., and Ashkenasy, G. (2009) *Angew. Chem. Int. Ed.*, **48**, 6683.
58. Wagner, N. and Ashkenasy, G. (2009) *Chem. Eur. J.*, **15**, 1765.
59. Wagner, N. and Ashkenasy, G. (2009) *J. Chem. Phys.*, **130**, 164907.
60. Wagner, N., Alesebi, S., and Ashkenasy, G. (2011) *J. Comput. Theor. Nanosci.*, **8**, 471.
61. Frezza, B.M., Cockroft, S.L., and Ghadiri, M.R. (2007) *J. Am. Chem. Soc.*, **129**, 14875.

62. de Silva, A.P. (2005) *Nat. Mater.*, **4**, 15.
63. Biebricher, C.K. and Eigen, M. (2006) *Curr. Top. Microbiol. Immunol.*, **299**, 1.
64. Szathmary, E. and Smith, J.M. (1997) *J. Theor. Biol.*, **187**, 555.
65. Stadler, B.M.R., Stadler, P.F., and Schuster, P. (2000) *Bull. Math. Biol.*, **62**, 1061.
66. Pross, A. (2003) *J. Theor. Biol.*, **220**, 393.
67. Wagner, N., Tannenbaum, E., and Ashkenasy, G. (2010) *Phys. Rev. Lett.*, **104**, 188101/1.
68. Kruger, K., Grabowski, P.J., Zaug, A.J., Sands, J., Gottschling, D.E., and Cech, T.R. (1982) *Cell*, **31**, 147.
69. Milo, R., Shen-Orr, S., Itzkovitz, S., Kashtan, N., Chklovskii, D., and Alon, U. (2002) *Science*, **298**, 824.
70. Shen-Orr, S.S., Milo, R., Mangan, S., and Alon, U. (2002) *Nat. Genet.*, **31**, 64.
71. Alon, U. (2007) *An Introduction to Systems Biology: Design Principles of Biological Circuits*, Chapman and Hall.
72. Ashkenasy, G. and Ghadiri, M.R. (2004) *J. Am. Chem. Soc.*, **126**, 11140.
73. Ashkenasy, G. and Ghadiri, M.R. (2006) Understanding Biology Using Peptides: Proceedings of the 19th American Peptide Symposium, p. 645.
74. Dadon, Z., Samiappan, M., Safranchik, E.Y., and Ashkenasy, G. (2010) *Chem. Eur. J.*, **16**, 12096.
75. Zhou, J., Arugula, M.A., Halamek, J., Pita, M., and Katz, E. (2009) *J. Phys. Chem. B*, **113**, 16065.
76. Samiappan, M., Dadon, Z., and Ashkenasy, G. (2011) *Chem. Commun.*, **47**, 710.
77. Sek, S., Misicka, A., Swiatek, K., and Maicka, E. (2006) *J. Phys. Chem. B*, **110**, 19671.
78. Kimura, S. (2008) *Org. Biomol. Chem.*, **6**, 1143.
79. Shlizerman, C., Atanassov, A., Berkovich, I., Ashkenasy, G., and Ashkenasy, N. (2010) *J. Am. Chem. Soc.*, **132**, 5070.
80. Xu, S. and Giuseppone, N. (2008) *J. Am. Chem. Soc.*, **130**, 1826.
81. Carnall, J.M.A., Waudby, C.A., Belenguer, A.M., Stuart, M.C.A., Peyralans, J.J.P., and Otto, S. (2010) *Science*, **327**, 1502.
82. Mal, P. and Nitschke, J.R. (2010) *Chem. Commun.*, **46**, 2417.

3
Biomolecular Electronics and Protein-Based Optical Computing

Jordan A. Greco, Nicole L. Wagner, Matthew J. Ranaghan, Sanguthevar Rajasekaran, and Robert R. Birge

3.1
Introduction

Molecular electronics explores the encoding, manipulation, and retrieval of information at the molecular or macromolecular level. Biomolecular electronics (or bioelectronics) is a subfield of molecular electronics that investigates the use of both native and modified biological molecules as media, in place of molecules synthesized in the laboratory. Bioelectronics has shown significant promise because of the sophisticated control and manipulation that is obtainable through self-assembly mechanisms and genetic optimization of large macromolecules. The ability to explore new architectures unique to biomolecular-based systems enhances the value of bioelectronics beyond the materials realm and opens up new possibilities that reflect the creative processes inherent in nature and revealed through natural selection.

Semiconductor electronics are the cornerstone for a majority of the recent and developing computing and processing systems. We do not propose that biomolecular electronic systems can replace these semiconductor-based devices; rather, we believe that bioelectronics can provide new options that can ultimately yield improved energy efficiency or unique architectures. The continuous push toward miniaturization and high-speed computing will inevitably succumb to the limits of lithographic manipulation of bulk materials, as reflected in Moore's Law [1]. Experts in the industry have predicted that complimentary metal-oxide semiconductors will soon reach scalable limits and will require new paradigms (i.e., new transistors) and new architectures to circumvent current device limitations in power and scalability [2–4]. As smaller transistors are crowded onto silicon chips, both power and heating issues start to dominate [4]. Biomolecular devices have the potential to operate at higher efficiencies. This chapter begins with a comparison of biomolecular and semiconductor electronics and highlights the exploration of these new architectures as the molecular limit is approached.

A bioelectronics material that has shown great potential in photonic devices is the protein bacteriorhodopsin (BR). The native organism, *Halobacterium salinarum*,

Biomolecular Information Processing: From Logic Systems to Smart Sensors and Actuators,
First Edition. Edited by Evgeny Katz.
© 2012 Wiley-VCH Verlag GmbH & Co. KGaA. Published 2012 by Wiley-VCH Verlag GmbH & Co. KGaA.

uses this protein as a photosynthetic energy source that converts light energy into a proton gradient (chemical energy). The photochemistry involved in this conversion is inherent to its potential for biophotonic devices and facilitates complex optical data storage and information processing. Implementing BR as a medium for molecular computing has the potential to solve some of the problems unique to integrated circuits. In addition, the architecture of these protein-based devices may serve as a model for scientists and computer engineers who are investigating other molecular media. Optical memories and processors are discussed in this chapter, and prototypes successfully implementing the proposed architectures are described.

3.2
Biomolecular and Semiconductor Electronics

One of the best ways to introduce bioelectronic computing is to compare the potential advantages of these developing technologies with the current advantages of semiconductor-based electronics. The following sections elaborate on key characteristics of computing and serve as a sample of the technical advancements achieved by scientists and engineers over the past few decades. Similar discussions can be found in Refs. [5–7]. We note that many of the same challenges faced by molecular engineers to achieve the goals of miniaturization and speed will soon be faced by semiconductor engineers as consumer technology advances. As the molecular level is approached, the problems involved in developing novel computational systems will unify independent of the technology used.

3.2.1
Size and Speed

The strength of molecular electronics is the potential capacity for using synthesized molecules that are of the nanometer scale. Additive synthesis allows a certain flexibility in creating a feasible material for these devices "from the bottom-up," starting with readily available organic compounds. Semiconductor devices are generated "from the top-down" through the lithographic manipulation of bulk materials. A synthetic chemist can selectively add an oxygen atom to a chromophore with a precision that far exceeds a comparable lithographic step, in which electron beams or X-rays etch a semiconductor surface. Molecular-based gates approach a size that is hundreds of times smaller than lithographic components. Some investigations have even produced logic gates down to the atomic level [8], while others have been shown to exploit the spin states of quantum dots as one- and two-quantum-bit gates [9].

Figure 3.1 illustrates trends in bit size that have characterized the past few decades of computer memory development. The results indicate that the area per bit has decreased logarithmically for multiple technologies since the early 1970s. As a comparison, we also show the cross-sectional area per bit calculated for the

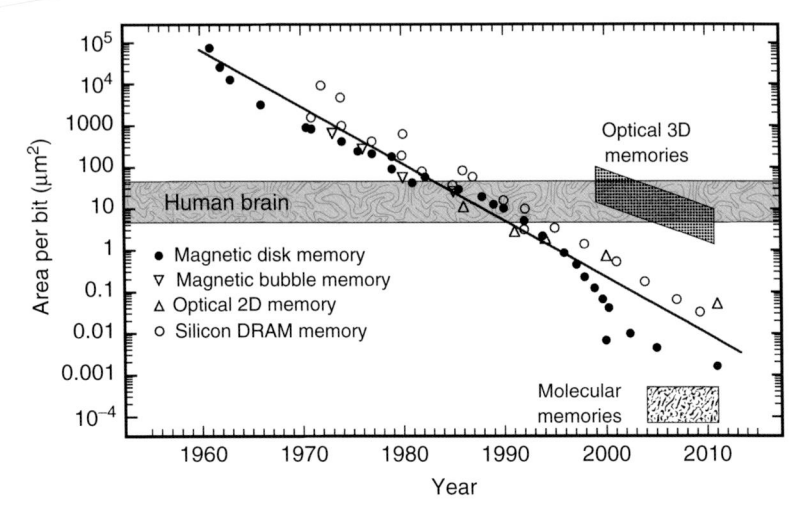

Figure 3.1 Analysis of the area in square microns required to store a single bit of information as a function of time (date in calendar years). The data for magnetic disk, magnetic bubble, optical 2D, and silicon dynamic random-access memory (DRAM) memories are shown. (Taken from Refs. [10–12] and the references therein.) These data are compared to the cross-sectional area per bit (neuron) for the human brain as well as anticipated areas and implementation times for optical three-dimensional memories and molecular memories [13]. Note that the optical 3D memory, the brain, and the molecular memories are three-dimensional and therefore the cross-sectional area (A) per bit is plotted for comparison. The area is calculated in terms of the volume per bit, V/bit, by the formula $A = (V)^{2/3}$.

human brain (assuming that one neuron is equivalent to 1 bit), for our proposed three-dimensional memory, and for generic molecular memories. Despite the fact that current technologies have surpassed the cross-sectional density of the human brain, the major advantage of the neural system is that information is stored in three dimensions and is capable of associative recall. Admittedly, it is a considerable stretch to compare the mind of a human being to digital technologies; however, the analogy emphasizes the fact that current computational technology is still far behind the complexity of the human brain. Figure 3.1 also demonstrates the rationale for and the potential of the development of three-dimensional memories based on BR, which is discussed further in this chapter. We can conclude that the trend of memory densities will soon force the bulk semiconductor industry to address the approaching limitations that proponents for molecular electronics are beginning to solve.

The capability of rapid switching of molecular logic gates can largely be attributed to size. In general terms, gates operating by electron transfer, electron tunneling, or conformational photochromism will increase in speed as there is a decrease in size of the material. This speed-versus-size tradeoff applies to all gates in use, whether they are already implemented or currently being envisioned. The mass of the carrier will limit how rapidly the conformational change can take place. In principle, one can argue that the generation of an excited electronic state in less than 1 fs (10^{-15} s) within a large chromophore-containing protein is an exception to this general statement. Nevertheless, the reaction of the system to the charge shift remains to be size dependent, and the direct relationship between the total size of the device and the response time remains valid.

The speed of molecular- and biomolecular-based devices is limited in part by the Heisenberg uncertainty principle. The maximum frequency of operation of a monoelectronic or monomolecular device, f_{max}, can be estimated by the following relationship [14]:

$$f_{max} \cong \frac{0.00800801 \cdot \tilde{v}_s \cdot \pi^2}{hN\left[2\pi + 2\tan^{-1}(-2) + \ln\left(\frac{\tilde{v}_s^2}{4}\right) - \ln\left(\frac{5\tilde{v}_s^2}{4}\right)\right]} \quad (3.1a)$$

$$f_{max}(GHz) \approx \frac{0.963\tilde{v}_s}{N} \quad (3.1b)$$

where $\tilde{v}_s$ is the energy separation of the two states of the device in wavenumbers and N is the number of state assignments that must be averaged to achieve a reliable state assignment. Equations (3.1a) and (3.1b) apply to only monoelectronic or monomolecular devices; however, Heisenberg's uncertainty principle permits higher frequencies for devices using ensemble averaging. For example, if a device requires 1000 state assignment averages to achieve reliability and $\tilde{v}_s = 1000$ cm^{-1}, the maximum operating frequency will be $\sim$960 MHz [14]. Almost all monomolecular or monoelectronic devices require $N > 500$ at ambient temperature; however, cryogenic devices ($\sim$1.2 K) can approach $N = 1$. While quantum mechanics places constraints on the operating frequency at ambient temperatures, molecular devices may have an inherent advantage of faster data transfer and calculation speeds compared to current technologies.

3.2.2
Architecture

The fundamental building block of current computer gates and signal processing circuitry is the three-terminal transistor. Inherent in the ability to exploit this element of modern devices, lithography offers one advantage that none of the techniques available to bioelectronics can duplicate. Lithographic materials can be used to build ultralarge-scale integrated devices that involve from 10^7 to 10^9 discrete components with complex interconnections. The true power of the integrated circuit does not come from the capability of fabricating a chip with millions of devices, but stems from the ability to interconnect the devices in a complex circuit. The replication of this particular architecture is near impossible using molecules, and it should be noted that it is becoming more difficult to implement with semiconductor materials with the need to keep up with the increasing modern demands of computational power and energy efficiency.

Biomolecular electronics offers significant potential as a catalyst to explore completely novel architectures and represents one of the defining sources of enthusiasm for researchers in the field. The need for new architectures as devices approach the limits of lithographic materials should encourage the investigation and development of new designs based on neural, associative, or parallel architectures and lead to hybrid systems with enhanced capabilities relative to current technology. For example, optical associative memories and three-dimensional memories can be implemented with unique capabilities based on photoactive proteins (Sections 3.4 and 3.5) [15, 16]. The incorporation of these architectures within hybrid systems is anticipated to have near-term applications. Furthermore, the human brain, a computer with capabilities that far exceed the most advanced supercomputer, is an elegant example of the potential of molecular electronics and a hint into the future of computation and artificial intelligence [13, 17–20]. Although the development of an artificial neural computer is beyond our current technology, it would be unreasonable to assume that such an accomplishment is impossible. Current research investigating molecular electronics should be viewed as an opening to new architectural opportunities that will lead to overall advances in computer and signal processing systems.

3.2.3
Nanoscale Engineering

During the evolution of computer technology, the feature size of high-speed semiconductor devices has decreased dramatically and still continues along this trend (Figure 3.1). The informational revolution has driven the demand for higher speeds and densities, which has lead to submicron feature sizes in the most commonplace devices of today. Extreme ultraviolet lithography can provide modest improvement over the projected densities of current materials; however, there is a need for new light sources for the development of nanoscale feature sizes [21]. Such lithographic methods are well understood, although several challenges remain to be solved before the technology is successfully implemented [22, 23].

As we have noted earlier, organic synthesis provides a "bottom-up" approach that offers a 100- to 1000-fold improvement in resolution relative to the best lithographic methods. Two alternatives to organic synthesis that have made an impact on current efforts in biomolecular electronics have been self-assembly and genetic engineering. The Langmuir-Blodgett technique that is used to prepare organized structures is the most known and best characterized example of self-assembly; however, self-assembly can also be used in the generation of membrane-based devices, microtubule-based devices, and liquid crystal holographic films [5, 24–29]. Genetic engineering offers a unique approach to the manipulation of large biological molecules and is the described technique in this chapter. Specifically, here we discuss the enhancement and application of BR using mutagenesis and directed evolution (DE) for various bioelectronic devices (Section 3.6). Genetic engineering is crucial to the early success of implementing large and complex molecules into these devices. We note, however, that recent advances in DNA architectures for computing and processing information have been made and the reader is directed to relevant references for further discussions of these technologies [17, 18, 30–32].

3.2.4
Stability

All electronic devices produce heat as a byproduct of their operation. The thermal stability of the biomolecular components must be considered because most organic-based molecules, including proteins, are irreversibly damaged at temperatures far lower than the operational temperature of the inorganic materials that are used in current semiconductors. However, there are examples of biological molecules that can operate at moderate to high temperatures for extended periods of time. Bacteriorhodopsin, for example, is stable at temperatures above 80 °C [33] and is often regarded as a viable material for optical computing (Section 3.3) [15, 34–37].

When many write–read–erase cycles are required for biophotonic applications, such as holographic associative memories, concerns of stability are extended to the photochemical degradation of molecules. Photochemical stability is measured in terms of cyclicity and approximates the ability to photochemically switch between functional states before becoming unreliable. There is no accepted definition of cyclicity, but it is often considered to be the number of events required to degrade (or denature) ~37% (1/e) of the material [15]. Bacteriorhodopsin is capable of $\geq 10^6$ conversions at ambient temperature [38–40] and, when paired with the thermal stability, is more stable than a majority of organic photochromic molecules [41]. Much of this stability is attributed to the protein structure [42], which sequesters the photoactive chromophore within the protein core and protects it from oxidation and free radicals while simultaneously providing steric constraints that enhance stability.

The reliability of the protein-based device can also be implemented to ensure that no data is corrupted or lost if any protein denatures via thermal or photochemical stress during operation. One such method is ensemble averaging, whereby many molecules are simultaneously used to represent a single bit of information. This technique is implemented as a fail-safe to reduce the risk of molecular degradation affecting a device. Other methods have been designed, and the reader is directed to articles by Lawrence et al. [43] and Lawrence and Birge [44] for further information about the topic. In summary, thermal and photochemical stability is an important issue to consider when implementing molecules into electronics; however, it has been shown that both native and redesigned biological molecules have stabilities sufficient for the operating conditions of the proposed technologies [15].

3.2.5
Reliability

Advocates of semiconductor-based devices have cited the issue of reliability repeatedly as a reason to view molecular and biomolecular electronics as impractical. Some believe that the need to use ensemble averaging in optically coupled molecular gates and switches demonstrates the inherent unreliability of these devices. As a counterpoint, one must acknowledge that transistors require more than one charge carrier to function efficiently. In fact, most ambient temperature molecular and bulk semiconductor devices use more than one molecule or charge carrier to represent a bit. This is simply because ensemble averaging improves reliability and permits higher speeds. The use of ensemble averaging does not, however, rule out the possibility of building reliable monomolecular or monoelectronic devices.

The probability of correctly assigning the state of a single molecule, p_1, is never unity. This seemingly flawed assignment capability is due to quantum effects and inherent limitations in the state assignment process for many technologies. The probability of an error occurring in a single state assignment, P_{error}, is a function of p_1 and the number of molecules, n, within the ensemble used to represent a single bit of information. P_{error} can be approximated by the following formula:

$$P_{error}(n, p_1) \cong -erf \left[\frac{(2p_1 + 1)\sqrt{n}}{4\sqrt{2p_1(1-p_1)}}; \frac{(2p_1 - 1)\sqrt{n}}{4\sqrt{2p_1(1-p_1)}} \right] \quad (3.2)$$

where $erf[Z_0; Z_1]$ is the differential error function defined by

$$erf[Z_0; Z_1] = Erf[Z_1] - Erf[Z_0] \quad (3.3)$$

where

$$erf[Z] = \frac{2}{(\pi)^{1/2}} \int_0^Z \exp(-t^2)dt \quad (3.4)$$

Equation (3.2) is an approximate representation of the error involved in these processes and neglects the statistical error associated with the probability that the number of molecules in the correct conformation can stray from their expectation values. Nevertheless, these equations are sufficient to demonstrate the issue of reliability and ensemble size.

Let us define a logarithmic reliability parameter, ξ, that is related to the probability of error in the measurement of the state of the ensemble by the function, $P_{error} = 10^{-\xi}$. A value of $\xi = 10$ is considered a minimal requirement for reliability without the need for error-correcting routines. If we assume that a single molecule can be correctly assigned with a probability of 90% ($p_1 = 0.9$), then 95 molecules are required to collectively represent a single bit to yield $\xi > 10$ (P_{error} (95, 0.9) $\approx 8 \times 10^{-11}$). A value of $p_1 = 0.9$ is larger than what is normally observed. Ensembles larger than 10^3 are typically required for sufficient reliability, although this value diminishes if fault-tolerant or fault-correcting routines are built within the architecture. Analyses on specific molecular-based devices are discussed further by Tallent et al. [14].

The question next becomes whether or not a reliable processor that uses a single molecule to represent 1 bit of information can be designed. The answer is yes, provided that one of two conditions is met. First, the device can have fault-tolerant architectures that either recover from digital errors or operate reliably with sporadic error due to analog or analog-type environments. One example of digital error correction that is commonly used in semiconductor memories is the use of additional bits beyond the number required to represent a number. These architectures lower the required value of ξ to less than 4. Analog error tolerance is present in many optical computer designs that implement holographic and/or Fourier architectures to carry out complex functions. Second, it is possible to design molecular architectures that can probe the assigned state of a bit without disturbing the state of the molecule. For example, an optically coupled device can be read by using a wavelength that is absorbed or diffracted, but that does not initiate state conversion. Under these conditions, the variable n in Eq. (3.2) can be defined as the number of read operations instead of ensemble size. In the previous example, 95 molecules must be included in the ensemble to achieve reliability. Using this architectural system, a single molecule can be used to assign a state provided that 95 nondestructive measurements are used to define the state. Multiple measurements are equivalent to integrated measurements, and a continuous read with digital or analog averaging can achieve the same level of reliability.

3.3
Bacteriorhodopsin as a Photonic and Holographic Material for Bioelectronics

Bacteriorhodopsin is the seven transmembrane alpha helical protein that exists in the cell membrane of the archaeon *Halobacterium salinarum* [42, 45–50]. Under anaerobic conditions, the protein is expressed and forms a semicrystalline two-dimensional matrix known as the *purple membrane*. This densely packed structure, which consists entirely of BR (75%) and lipid (25%), efficiently converts light energy into chemical energy via the absorption of a photon by retinal. This organic chromophore is linked to the protein by a covalent Schiff base linkage to lysine 216. A series of thermally driven intermediates, which are defined by distinct molecular conformations of the retinylidene protein, are observed after the photoexcitation of BR. This process is known as a *photocycle* and occurs every time a single photon is absorbed by the protein [51–53].

The basis of BR-based devices relies heavily on both the spectral separation of specific photointermediates and the significant change of refractive index observed throughout the photocycle of the protein [54, 55]. Bacteriorhodopsin as a holographic material is unique in the fact that this change in refractive index is coupled with an unprecedented quantum efficiency (Φ) of approximately 65% [56–58]. Evolution has assisted in modifying BR to serve as a photonic material for devices, one that is thermally and photochemically stable enough to withstand significant fluctuations in light, temperature, and a self-induced pH gradient [59, 60]. The robust and stable nature of the protein has yielded numerous applications within the bioelectronics field. This stability allows the protein to be customized via chemical and genetic manipulations without perturbing the overall integrity of protein structure and function.

3.3.1
The Light-Induced Photocycle

When the chromophore absorbs a photon, BR undergoes a complex photocycle that generates photointermediates with absorption maxima spanning the entire visible region of the spectrum (Figure 3.2). The primary photochemical event of

BR (bR → K) involves the excitation of the chromophore to the Franck-Condon excited state, which subsequently causes a photoisomerization from all-*trans* to 13-*cis* retinal [15, 40]. A series of thermally driven photointermediates follow (L, M, N, and O) as the protein resets to its resting state (bR) after a complete photocycle, which typically has a 10 ms duration (Figure 3.2). Most holographic devices involving BR operate at or near ambient temperature and utilize the refractive index differential observed in the initial green–red absorbing bR state and the long-lived blue absorbing M state [56, 58].

$$\text{bR} (\lambda_{max} \cong 570 \text{ nm}) \underset{\Phi_2 \leq 0.95}{\overset{\Phi_1 \sim 0.65}{\rightleftarrows}} M (\lambda_{max} \cong 410 \text{ nm}) \tag{3.5}$$

The forward reaction takes place only via light activation and is complete in less than 50 μs, whereas the reverse reaction can be either light activated or can occur thermally. The light-activated M → bR transition is a photochemical transformation induced by a sequential photon following light activation. This transition is highly sensitive to temperature, environment, genetic modification, and chromophore substitution. These specific photonic properties have often

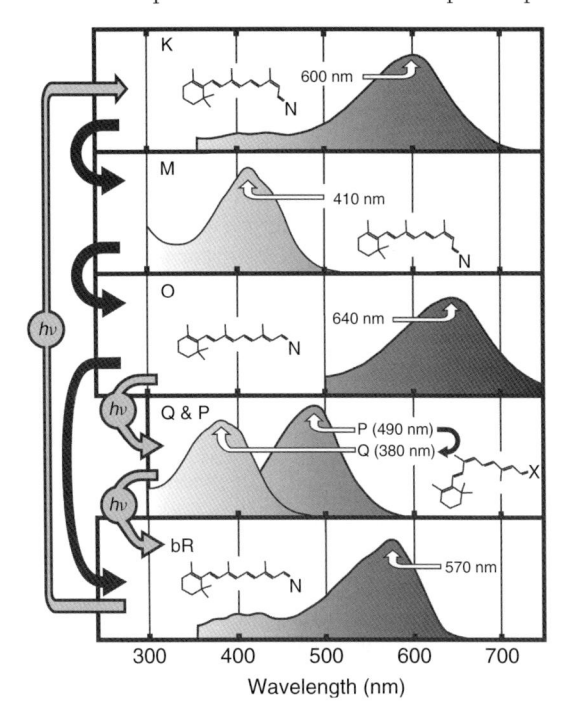

Figure 3.2 Absorption spectra of select intermediates of the bacteriorhodopsin photocycle. The outlined arrows indicate photochemical transitions, while the solid arrows represent thermal transitions between photointermediates. The insets indicate the corresponding configuration of retinal and the absorption maxima of each state. The nitrogen (N) of the Schiff base linkage is shown for each retinal model. The exception is the model for the P and Q states, where X represents the Schiff base nitrogen (for the P state) or carbonyl oxygen (for the Q state). (Adapted from Figure 6.4 of Ref. [7].)

been exploited for many BR-based optical devices, particularly those that involve holography. Although the M state is always transient, there are applications in which the bR resting state and the M state have been used to successfully implement BR in holographic systems. Norbert Hampp has pioneered technologies that apply real-time dynamic holography, including a holographic interferometer that uses a genetically and chemically modified BR mutant (D96N) with a long-lived M state [36, 61–64]. Researchers have still been searching for the conditions to create a permanent and reversible blueshifted M state photoproduct for holographic devices; however, they have yet to completely overcome the limitations of the transient M state lifetime. Holographic associative processors and the genetic modification of BR for bioelectronics are discussed further in Sections 3.4 and 3.6, respectively.

3.3.2
The Branched Photocycle

A branching reaction from the main photocycle of BR has been observed to produce a long-lived photoproduct known as the Q state. The presence of two stable photoproducts, bR and Q, facilitates the binary assignments of bits 0 and 1, respectively, and presents tremendous potential for long-term volumetric data storage. The branched photocycle consists of the P and Q states and was originally characterized by Popp *et al.* [65]. The branched photocycle is accessed via a sequential two-photon process by exposing the O state to red light. The P state is the direct product of this reaction and is characterized by photoisomerization from a protonated all-*trans* to 9-*cis* retinal [65]. The 9-*cis* chromophore in native BR is electrostatically unfavorable and subsequently hydrolyzes to form an unbound 9-*cis* retinal within the active site of the protein [66]. Reformation of this bond, and reversion to the bR state, is possible by two routes: thermally or photochemically. The former route is theoretically calculated to require several years ($\sim$190 kJ mol^{-1}) [15] and is ideal for long-term data storage. The latter route requires that a photon of UV light (380 nm) be absorbed by the Q state and is good for applications where the protein would undergo many read–write–erase cycles [67].

This branching reaction facilitates a dramatic increase of potential for BR in biomolecular devices. The absorption maximum of the Q state (380 nm) is more blueshifted than the M state (410 nm) and is ideal for probing and manipulating data via differential absorptivity. The stability of the Q state photoproduct enhances the potential of long-term data storage in these new architectural milieus. The technology behind two-photon volumetric memories is explained below in Section 3.5.

3.4
Fourier Transform Holographic Associative Processors

Associative processors based on holography present a unique advantage over conventional computers in the implementation of autocorrelation. These processors can be used to identify and optically compare real objects to an extensive holographic database and has shown potential for applications in optical computer architectures, robotic vision hardware, and generic pattern recognition systems [62, 68–70]. Unlike the architectures of current integrated circuits in semiconductor-based computers, holographic associative processors mimic the neural associative activity of the human brain. Neural association processes allow for the fast and efficient identification of objects in real time and the capability of distinguishing between multiple objects of a similar form. Conventional computers lack the necessary architecture to complete associative recall. They are capable, however, of compensating through complex algorithms that involve the digitization of the input image, the comparison of the input data to all digital images, and the exploitation of correlation subroutines that identify the closest match [71]. Holography allows us to surpass these limitations and to create a data processor that competes with the complexity and efficiency of the human brain.

Holographic associative loops were first described by Paek and Psaltis in 1987 and were revolutionary in their use of Fourier transformations to write and read reference planes using thin films [70]. In practice, these associative memories take an input data block and scan the entire memory for the data block that matches the input. In some implementations, the closest match will be found even if there is not a perfect match. The memory will then return the data block that satisfies the matching criteria. The most sophisticated associative memory hardware would permit data blocks of varying sizes for flexible association of input and output data. We have implemented the design proposed by Paek and Psaltis by using thin films made with BR as the photoactive material (Figure 3.3) [68]. Modifications to this system have been made to facilitate the introduction of both input and reference images using a single active matrix spatial light modulator (AMSLM) and output using a charge-coupled detector (CCD). Because BR thin films are dynamic holographic media, new databases can be actively fed in through the AMSLM as needed. Database images are focused via Fourier lenses (FLs) onto two hologram locations, H1 and H2, and then the input image is fed into the system. The image transverses the loop and is superimposed on H1 and H2, at which point an intensity enhancement is observed at the location of the closest match in the database array. The hologram stored at this location is preferentially selected and imaged at the CCD output. Figure 3.4 demonstrates the associative analysis of a data set using thresholding to obtain the closest match. Thresholding is handled electronically in this implementation; however, optical thresholding can improve the overall performance [68, 70, 72]. As shown in Figure 3.3, a partial input image is sufficient to generate a complete output image once a match is obtained within the database. For a more detailed discussion of associative loop technology and the architecture described above, the reader is referred to Refs. [15, 68, 71, 72], as well as further discussions by Paek and Psaltis [69, 70, 73–76].

A data search within the associative memory can be enhanced significantly if the holographic reference pattern is rapidly changed via the single optical input. For this to remain true, feedback and thresholding must be maintained. In

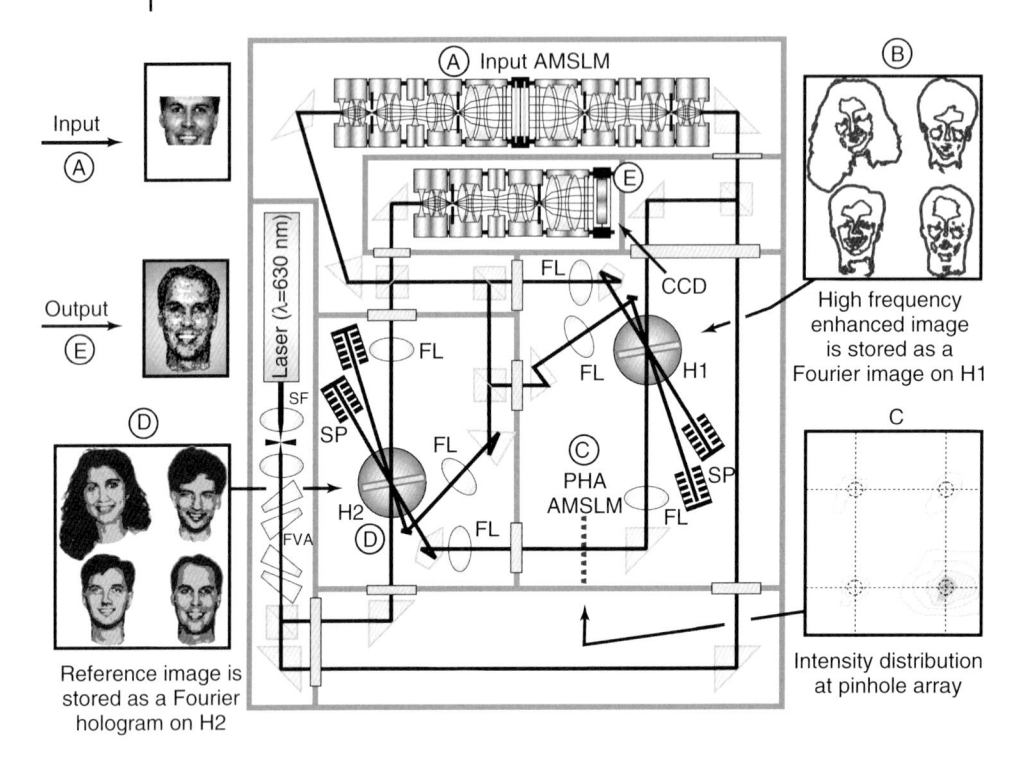

Figure 3.3 Schematic diagram of a Fourier transform holographic (FTH) associative memory. Thin polymer films of bacteriorhodopsin serve as media to create FTH reference planes for real-time storage of the holograms. A partial input image is enough to select and regenerate the entire associated image stored on the reference hologram. Although only four reference images are shown, the FTH associative memory can handle many hundreds or thousands of images simultaneously. This memory can also work on binary data by using redundant binary representation logic, and a small segment of data can be used to find which page has the largest association with the input segment. Selected components are labeled as follows: AMSLM, active matrix spatial light modulator; FL, Fourier lens; FVA, Fresnel variable attenuator; H1, H2, holographic films; PHA, pinhole array; SF, spatial filter; SP, beam stop. (Adapted from Figure 3.9 of Ref. [15].)

conjunction with solid-state hardware, a search engine of this type can be integrated into the hybrid computer architectures envisioned here. The diffraction limited performance and the high write and erase speeds associated with the excellent quantum efficiencies of these films represent key elements in the potential of this memory. The protein-based system described above provides the most efficient and cost-effective approach to developing large-scale associative processors for true artificial intelligence. Furthermore, the ability to genetically modify the protein (Section 3.6) or replace retinal with a modified chromophore analog provides tremendous flexibility and control in enhancing the photophysical properties of the protein for such devices [35, 62, 77].

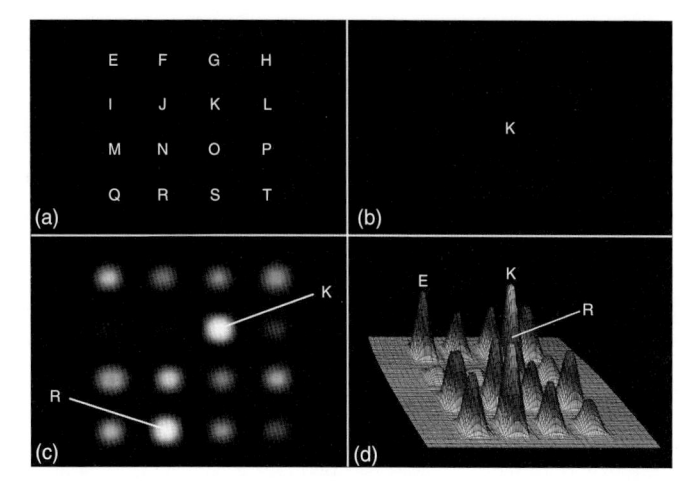

Figure 3.4 Demonstration of the importance of thresholding in the associative analysis of data sets with similar pages. A set of 16 pages each containing a single letter (a) is associated with a single letter, K, as input (b). The intensity of the Fourier association at each pinhole is shown in (c) and displayed as a profile at the CCD detector in (d). The letters R and K are so similar that either letter will strongly associate with the other, and without variable association, both letters would be displayed in the output. Associative memories work best when the pages contain more complex entities such as words, paragraphs, complex images, or combinations of words and images. The higher the complexity, the more accurate is the associative recall.

3.5
Three-Dimensional Optical Memories

Many scientists and computer engineers interested in nanotechnology believe that the most likely architecture that will be used for molecular electronics and computer hardware will involve volumetric memory. Several optical architectures based on BR have been proposed in the literature, including those involving holography [40, 78–80], simultaneous two-photon [81–83], and sequential two-photon [5, 15, 84] applications. Despite the variety of ways in which this architecture can be imagined, holographic media based on BR have not yet been successful because of the lack of a truly bistable bR/M system. Furthermore, simultaneous two-photon volumetric memories have shown to be problematic because of unwanted photochemistry induced by intense lasers and the subsequent need for cleanup operations [84]. The orthogonal sequential two-photon system has shown the greatest potential in recent years because the branched photocycle of BR and the presence of the long-lived Q state completely circumvent the major issues described earlier.

The sequential two-photon architecture minimizes unwanted photochemistry outside of the irradiated volume and provides a particularly straightforward parallel architecture. As discussed, the discovery of the P and Q states add a considerable advantage to all BR-based devices with a possibility of a long-lived photoproduct other than the bR resting state. These devices rely on the potential for binary

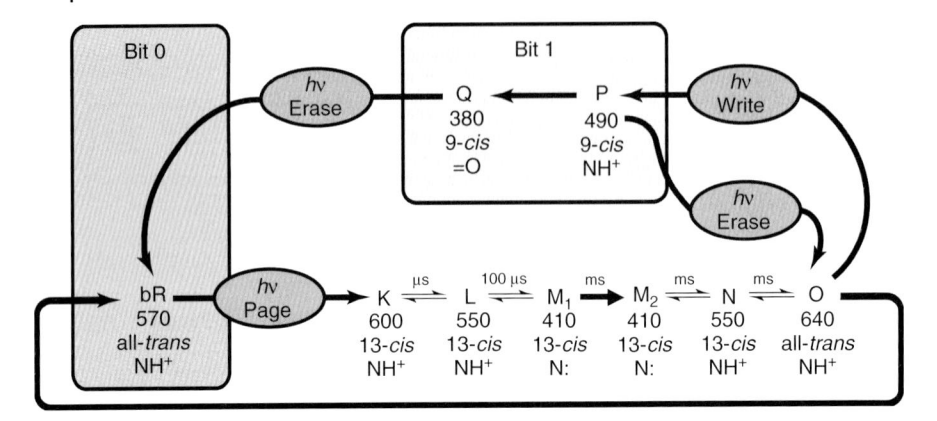

Figure 3.5 Data storage based on the branched photocycle reactions of bacteriorhodopsin. The photochemical transitions of the branched photocycle facilitate the successful assignment of bits 0 and 1 to the bR resting state and the Q state, respectively (see text). The absorption maxima (in nanometers), retinal configurations, and protonation states of the Schiff base linkage are provided below each photointermediate.

assignments of the two stable photoproducts, bR and Q, as bits 0 and 1, respectively. Because the P and Q states can be accessed only via a temporally separated dual pulse sequence, the protein operates like an optical AND gate in which data is written if and only if two conditions are satisfied. Under these terms, the two conditions refer to the activation by a green paging laser, followed by a delay and a red data laser initiation during the O state of the protein (Figure 3.5). The orthogonal laser excitation ensures that unwanted conversion to the Q state can be avoided in any given volume of the protein matrix. The exploitation of the photochemistry involved in driving BR to these desired photostates is based on the sequence shown in Figure 3.5, where the K, L, M, N, and O states are all transient intermediates within the main photocycle of the protein and the P and Q states are intermediates in the branched photocycle. It is important to note that the Q state is the only photoproduct that is defined by hydrolyzed 9-*cis* retinal (i.e., retinal unbound to the protein). Figure 3.6 provides schematic representations of the write, read, and erase operations required for implementing the branched photocycle architecture, and the components of the devices are described in greater detail below.

3.5.1
Write, Read, and Erase Operations

The writing and the reading processes both start by "paging" a very thin region (~15 μm) inside the memory cuvette, initiating the photocycle in a select volume. Paging lasers must have a wavelength between 550 and 640 nm to efficiently induce photoexcitation of native BR and most BR mutants. The photocycle of native BR is over after approximately 10 ms, and it is within this window of time that the second red data laser (680 nm) must be activated orthogonally to the page to successfully write or read data within the medium. This process is ideally accomplished once the

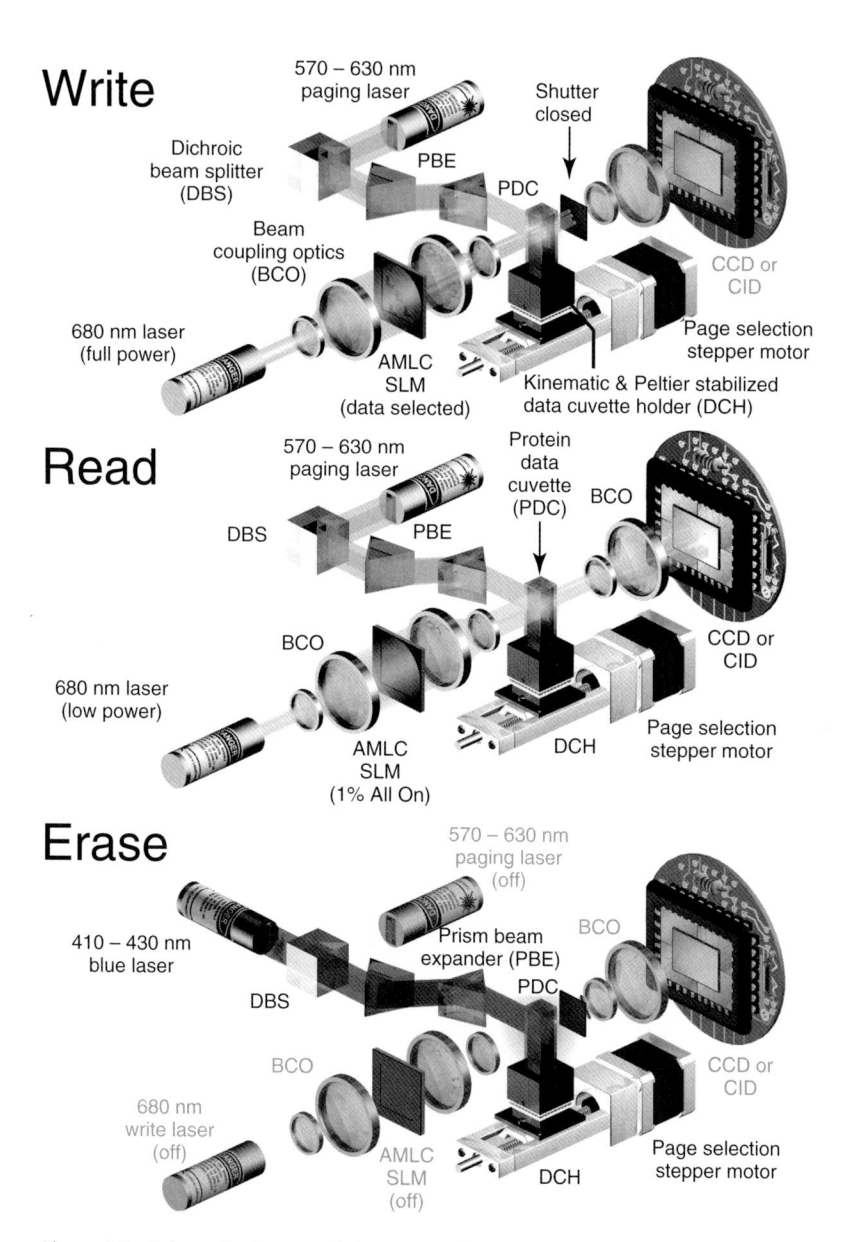

Figure 3.6 Schematic diagram of the branched photocycle three-dimensional memory. The write and read operations are illustrated in the top and center panels, respectively, and are based on a two-photon architecture in which the initiation of a paging laser is followed sequentially by an orthogonal write operation. The intensity of the data laser dictates whether bit assignments are designated or nondestructively probed. Data can then be erased, either globally or locally, by exposing the data cuvette to UV light (bottom). (Adapted from Figure 3.12 of Ref. [15].)

O state reaches a maximum concentration, at about 2–3 ms after initiation of the photocycle for native BR. If stimulation from the secondary laser is not present, the protein simply returns to the resting state in a nondestructive manner. To write data, information is spatially encoded in the incident red beam by a spatial light modulator (SLM), only orthogonally accessing your selection and completing the conversion from bit 0 to 1. As a result, volumetric pixels, or voxels, are encoded within the immobilized BR medium and remain until the erase operation is initiated.

In the center panel of Figure 3.6, the reading operation is illustrated. The first step is identical to the writing process via the paging action of the device. The difference here is that some volume in the page being probed is already written as the Q state and is therefore transparent to the orthogonal paging and data beams. The page is exposed to orthogonal red light during the O state through a data timing shutter (DTS) and a data read shutter (DRS); however, this time the laser is fired at a much lower intensity (roughly 1% of nominal write intensity) and no SLM is used to encode information. A CCD array then images all light passing through the data cuvette. Only elements in the paged region that began in the bit 0 state cycle into the O state and the CCD detector array observes differential absorptivity and the presence of bit 1. The selectivity described here allows for a reasonable signal to noise ratio, even with thick (1 cm) memory media containing more than 10^3 pages. The absorptivity of the O state within the paged region is more than 1000 times larger than the absorptivity of the remaining volume of the data cuvette, which allows for the relatively weak beam to generate a large differential signal. A protein with a photocycle of 10 ms completes one read cycle in that time, which leads to a rate of 10 MB/s. Each read operation must be monitored for each page in the cuvette, and a refresh operation is necessary after ~1000 reads. In practice, the three most recent pages are stored in semiconductor cache memory to reduce the number of refresh operations required.

The P and Q states can both be converted back to the bR resting state using blue light (~410 nm). Data erasure can be accomplished one of two ways. First, bit 1 can be converted to bit 0 by using a blue laser of coherent light with an output near 410 nm. Second, the cuvette can be erased globally by using incoherent light in the 360–450 nm range. Both options are possible now with the availability of blue diode lasers at 405 nm to selectively erase pages; however, all prototypes thus far have used the global erasure option because of a decrease in complexity and expense. Figure 3.7 provides two views of a branched photocycle volumetric memory prototype. Prototype development is in its third stage and currently implements the orthogonal two-photon architecture using paging and data lasers coupled to fiber optics.

3.5.2
Efficient Algorithms for Data Processing

While volumetric memories based on proteins offer significant advantages over traditional memories, a crucial issue remains with volumetric memories, namely, the diffraction problem. Protein-based memories require that the number of zeros

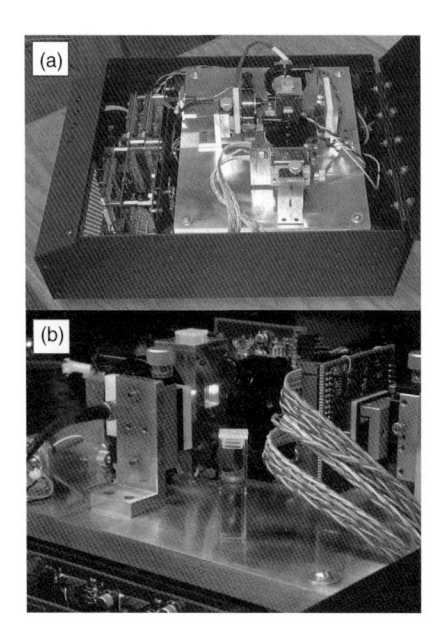

Figure 3.7 A prototype of the branched photocycle three-dimensional optical memory based on BR. This prototype currently implements the orthogonal two-photon architecture and contains paging and data lasers that are directed to the BR matrix via fiber optics (a). The data cuvette is placed into the cuvette holder on an x-y-z translation stage and high-speed linear actuators are employed to move the data cuvette relative to the fiber optic cables. The overall system measures 31 cm × 38 cm × 23 cm. A close-up view of the optical platform is shown (b), with a polymer-based BR data cuvette in the foreground.

and ones in the data stored be nearly the same. This issue presents an unreasonable demand for any user. To overcome this problem, many coding schemes have been proposed. One such option is to replace each zero in the data with 01 and every one with 10. In this case, the number of zeros and ones stored in the memory will be perfectly balanced. However, the memory utility (defined as the data size divided by the memory used) is only 50%. A series of coding techniques that can offer utility factors close to 100% have been proposed in Refs. [85–90].

A common coding scheme employs novel compression algorithms (see e.g., [86, 88–90]). The crucial insight here is that any data when compressed with the right algorithm tends to have nearly the same number of zeros and ones. Any efficient coding scheme should have not only good memory utility but also small coding and decoding times. Some of the coding algorithms presented in Ref. 85 are parallelizable (to run in $O(1)$ time, for example).

Volumetric protein-based memories also possess associative properties that make them ideal candidates for pattern recognition applications. In particular, we can think of protein-based volumetric memory as a device that is capable of performing convolutions in $O(1)$ time. A computational model based on this observation has been defined as convolution-based parallel machine (CONV-PAR) in Ref. [85].

CONV-PAR is a parallel model in which the convolution operation can be performed in $O(1)$ time. Through this model, it has been shown that various problems can be solved efficiently, including prefix computation, multiplying two polynomials, matrix multiplication, matrix inversion, string matching, sorting, motif search, and association rules mining. In particular, constant time algorithms have been devised for these problems. As an example, the sorting of n elements can be done in $O(1)$ time using n^2 processors, two given $n \times n$ matrices can be multiplied in $O(1)$ time using n^3 processors, and so on. Given that protein-based memories offer great parallelism, it is imperative that we develop efficient algorithms for various fundamental problems on the CONV-PAR model. Ideally, constant time algorithms will be preferred. Even when the run times are $O(1)$, it is essential to decrease the processor bounds as much as possible. For all the above problems, decreasing the processor bounds (or proving that these bounds are optimal) is an open problem. Devising better coding schemes to address the diffraction issue is also an open problem.

3.5.3
Multiplexing and Error Analysis

Polarization and gray-scale multiplexing of data in a three-dimensional memory can achieve higher storage densities of up to 10 bits per voxel (Figure 3.8) [15]. The importance of multiplexing for the proposed device lies in the ability to operate at or near the diffraction limit. The SLM and CCD in the prototypes are designed to provide 8-bit gray-scale capability, which is augmented to 16 bits into each voxel using polarization doubling. In practice, reliability limits and ensemble averaging permit that the writing and reading processes contain no more than 10 bits per voxel. Furthermore, the bits along the edges of the data cuvette must be allocated to alignment and checksum information (Figure 3.8). With all of these features fully implemented, a standard $1 \times 1 \times 3$ cm^3 data storage cuvette can store approximately 10 GB of information, which translates into 10^4 to 10^5 protein molecules per bit for the average protein concentrations used for experimentation. This value yields a statistically relevant ensemble to compensate for potential error in the write, read, and erase operations [14, 15].

Error analysis can be carried out by measuring read histograms, which measure the separation between bits 0 and 1 and any overlap between the two [15, 43]. A detailed discussion of error sources and error analysis for volumetric memories is beyond the scope of this chapter, and we direct the reader to Lawrence *et al.* [43]. A read operation is carried out using differential absorption, which is based on a normalized scale where 0 is low intensity reaching the detector (bit 0, O state absorption) and 1 is high intensity reaching the detector (bit 1, P and Q state absorption). The differential read process allows the user to normalize the signals across the detector array, so when no gray scaling is done, peaks are only observed at two locations (bits 0 and 1). When gray scaling is used, bit density increases and one observes peaks corresponding to each of the allowed levels [15]. Read errors increase the width of these peaks, which potentially creates a situation in which

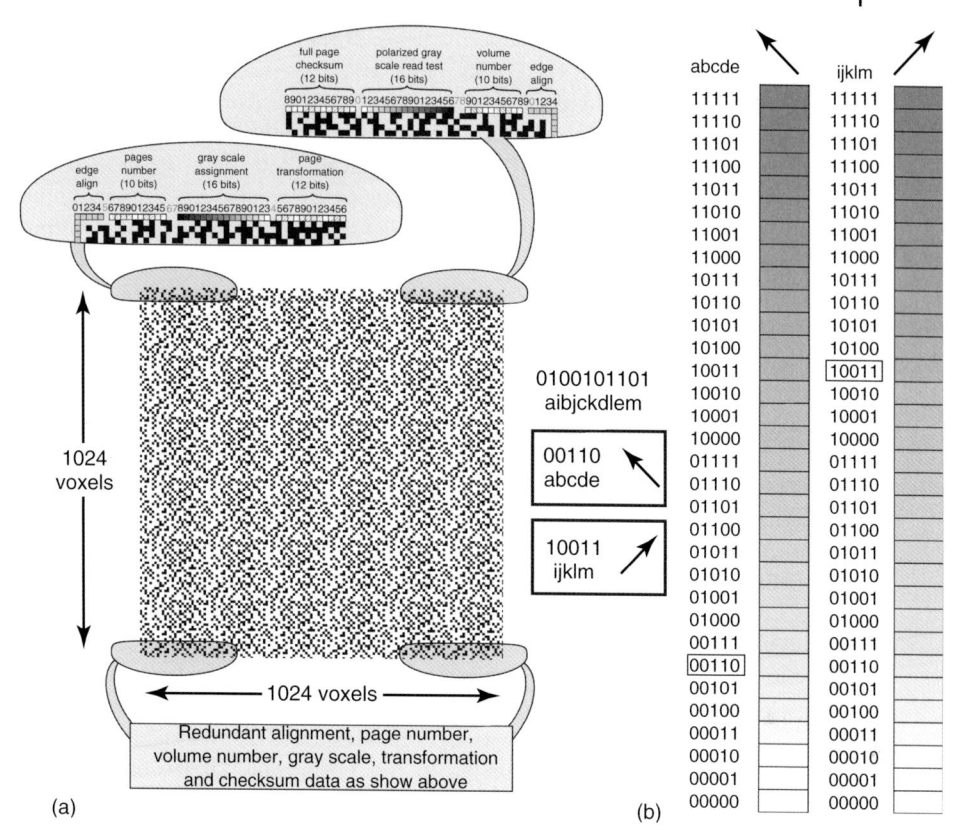

Figure 3.8 A polarization and gray-scale multiplexing scheme in which each voxel can be assigned to store either a single bit or many bits of information. The current design includes voxels devoted to optical alignment, page number, volume number, gray scale, transformation, and checksum data along the edges (a). The multiplexing scheme envisioned here can store 10 bits per voxel, provided that two orthogonal polarizations and 16 density levels are implemented into the system (b). (Adapted from Figure 3.15 of Ref. [15].)

the peaks overlap. The overall goal is to keep such error at a level that limits the necessity of error-correcting codes within a device.

3.6
Genetic Engineering of Bacteriorhodopsin for Device Applications

Protein engineering of BR, or the design and construction of new BR mutants, is an essential process for the successful implementation of BR in biomolecular electronics. Although nature has provided a robust photoactive protein, the use of BR in applied technologies requires further enhancement of the native protein. Because the native protein is inherently stable, mutagenesis of BR is often focused on the optimization of the photointermediates, primarily the M and O states, and the branched photoproduct, the Q state; however, mutagenesis of BR has also been employed to generate gold-binding BR mutants [91, 92], enhance the innate dipole moment of BR [93], and alter the photocycle lifetimes of the protein [38, 63].

Mutagenesis of the M and Q states is critical, particularly in holographic processors. Because the M state is not permanent, holograms made utilizing the bR to M state transition will last only as long as the M state lifetime, limiting the application of this state to real-time dynamic holography. Devices based on the Q state remove the time-sensitive nature of holography because the Q state is a photoproduct of BR. Manipulation of these photostates allows for the generation of more efficient bioelectronic memory systems. In order to improve or lengthen the lifetime of the M state, mutagenesis usually targets residues involved in proton transfer from the intracellular milieu to the Schiff base [35, 63, 77]. Mutagenesis of the Q state is more involved because the Q state does not exist in nature and is the result of a sequential two-photon event. Directed evolution has been used to improve the formation and reversion efficiencies of the Q state for BR for device applications.

Directed evolution, or the enhancement of a molecule toward a specific characteristic via repeated iterations of genetic mutation, screening, and differential selection, provides a cost-effective and time-efficient method for the genetic and chemical manipulation of biological molecules [94–97]. The use of DE to optimize proteins is most useful when full organism selection is possible; however, organism selection is difficult to implement when a complex photochemical reaction needs to be optimized, particularly when variables such as thermal stability and pH require simultaneous enhancement [94, 97, 98]. Historically, DE has been used to modify the properties of enzymes for industrial and pharmaceutical applications [99–102]; however, we implement DE to optimize the formation and reversion efficiencies of the Q state (Figure 3.9). In each case, a diversified genetic library is generated and screened to identify mutants with enhanced phenotypes, including efficient formation of the Q state. This molecular biological technique allows for the deliberate evolution of individual populations of molecules and proteins toward a specific trait.

The redesign and generation of new BR mutants is accomplished through site-directed mutagenesis (SDM), site-specific saturation mutagenesis (SSSM), semi-random mutagenesis (SRM), and random mutagenesis (RM). Site-directed mutagenesis alters the protein by strategically replacing or substituting single amino acid residues, thereby changing the structure or function of the protein. Semi-random mutagenesis and RM produce a large number of indiscriminant mutants through the use of doped oligonucleotides or error-prone polymerase chain reaction (PCR). Lastly, SSSM is a combination of SDM and SRM and allows for the deliberate examination of multiple amino acid substitutions via the saturation of a key residue or residues at specific loci on the bacterio-opsin (*bop*) gene.

We use type I DE to optimize the photochemical properties of BR by implementing automated screening and microgram protein characterization. In the first round of DE, mutants are generated via region-specific SRM and are screened and evaluated for altered photophysical properties, such as M state lifetime or

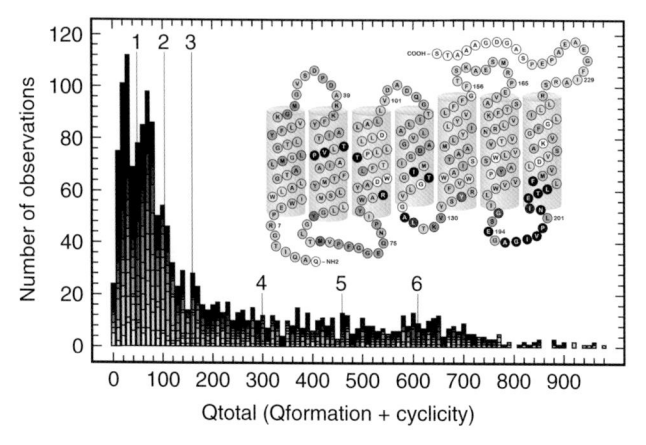

Figure 3.9 Histogram illustrating type I directed evolution of BR to enhance the formation of the Q state. The vertical bars of the histogram represent the 1604 unique BR mutants that were sequenced throughout the six stages of directed evolution. The shade of the vertical bars indicate the regions of the protein that were targeted for mutagenesis, where the white bars indicate regions near the N-terminus of the protein and the black bars indicate the last region near the C-terminus. The vertical lines mark the average value of Q_{total} for each stage. The inset is a two-dimensional map of the primary sequence of native BR. Mutagenesis of the darkest residues had the most significant impact toward the enhancement of Q_{total}, while the light gray residues were less important to this enhancement. It should be noted that the shade of gray on the two-dimensional map does not correspond to the shade of the vertical bars within the histogram.

Q state formation, using 96-well plates. Because the deliberate engineering of BR for devices requires several rounds of mutagenesis to identify a mutant that outperforms or exhibits enhanced photophysical properties, an *in-house* automated screening system was developed. At each stage in the DE process, the most efficient mutants are selected to serve as the parents to the next generation of genetic progeny, which are then produced via SDM, SSSM, and SRM. This process is reiterative and builds on successive improvements to the protein. After six stages of mutagenesis, over 10 000 mutants have been generated and screened for the ability to serve in protein-based devices (Figure 3.9).

3.7
Future Directions

Bacteriorhodopsin has captured the attention of bioelectronic engineers for nearly four decades and has served as the quintessential medium for both general biophysical investigations and biomolecular device applications. In this chapter, two types of optical computing systems have been presented based on the photonic properties of BR. These architectures have been successfully implemented into numerous prototypes and address the issues central to interfacing biological molecules with inorganic materials. We envision that the first step in the evolutionary development of these technologies will be the design of hybrid computers that incorporate the best features of semiconductors and optical and molecular architectures. Limitations of the native protein have notably discouraged the development of commercially viable systems in recent years; however, advancements in RM and DE have reignited interest in implementing BR into optical memories and processors. Upon discovery of BR mutants with enhanced photonic properties, the remaining issues lie in the development of materials capable of harnessing these proteins into electronic devices. We are currently in the third generation of prototyping and continue to improve on the designs presented in this chapter. The prototypes will combine past successes of architectural engineering with recent progress made using genetically engineered mutants, polarization and multiplexing schemes, and efficient algorithms to optimize these bioelectronic technologies. Molecular and semiconductor engineers have made dramatic advances in their respective fields using a wide variety of information-processing media. Bacteriorhodopsin and other biomolecules, however, offer a unique platform to investigate microelectronics, holography, memory storage, artificial intelligence, and sensing while exploiting elegant natural processes such as natural selection and evolution. The potential advantages of using biological materials in devices will continue to drive the development and optimization of these systems as the limitations of bulk material manipulation are approached.

Acknowledgments

Funding for this research carried out at the University of Connecticut was provided in part by grants from the National Science Foundation (EIA-0129731, BES-0412387, CCF-0432151, EMT-0829916), the National Institutes of Health (GM-34548), DARPA (HR0011-05-1-0027), and the Harold S. Schwenk, Sr. Distinguished Chair in Chemistry.

References

1. Moore, G.E. (1965) Cramming more components onto integrated circuits. *Electronics*, **38** (8), 477–490.
2. Meindl, J.D. (1993) Evolution of solid-state circuits: 1958-1992-20?? *IEEE ISSCC Commemorative Suppl.*, 23–26.
3. Horowitz, M., Alon, E., Patil, D., Naffziger, S., Kumar, R., and Bernstein, K. (2005) Scaling, power, and the future of CMOS. Electronic Devices Meeting, Washington, DC, December 5, 2005.
4. Esmaeilzadeh, H., Blem, E., St. Amant, R., Sankaralingam, K., and Burger, D. (2011) Dark silicon and the end of multicore scaling. Proceedings of the 38th International Symposium on Computer Architecture, San Jose, CA, June 4–8, 2011.
5. Birge, R.R. (1994) Molecular and biomolecular electronics. *Adv. Chem.*, **240**, 596.
6. Vought, B.W. and Birge, R.R. (1999) Molecular electronics and hybrid computers, in *Wiley Encyclopedia of Electrical and Electronics Engineering* (ed. J.G. Webster), Wiley-Interscience.
7. Marcy, D.L., Vought, B.W., and Birge, R.R. (2003) Bioelectronics and

protein-based optical memories and processors, in *Molecular Computing* (eds T. Sienko, A. Adamatzky, N.G. Rambidi, and M. Conrad), The MIT Press, Cambridge, MA, 191–220..
8. Khajetoorians, A.A., Wiebe, J., Chilian, B., and Wiesendanger, R. (2011) Realizing all-spin-based logic operations atom by atom. *Science*, **332**, 1062–1064.
9. Loss, D. and DiVencenzo, D.P. (1998) Quantum computation with quantum dots. *Phys. Rev. A: At., Mol., Opt. Phys.*, **57**, 120–126.
10. Keyes, R.W. (1992) Electronic devices in large systems. AIP Conference Proceedings.
11. Bandic, Z.Z. (2008) Advances in magnetic data storage technologies. *Proc. IEEE*, **96**, 1748–1753.
12. Eleftheriou, E., Haas, R., Jelitto, J., Lantz, M., Pozidis, H. (2008) Trends in storage technology. Bulletin of the IEEE Computer Society Technical Committee on Data Engineering, pp. 1–10.
13. Kandel, E.R., Schwartz, J.H., and Jessell, T. (1991) *Principles of Neuroscience*, Appleton & Lange, Norwalk, CT.
14. Birge, R.R., Lawrence, A.F., and Tallent, J.A. (1991) Quantum effects, thermal statistics, and reliability of nanoscale molecular and semiconductor devices. *Nanotechnology*, **2**, 73–87.
15. Birge, R.R., Gillespie, N.B., Izaguirre, E.W., Kunsnetzow, A., Lawrence, A.F., Singh, D., Song, Q.W., Schmidt, E., Stuart, J.A., Seetharaman, S., and Wise, K.J. (1999) Biomolecular electronics: protein-based associative processors and volumetric memories. *J. Phys. Chem B*, **103**, 10746–10766.
16. Hampp, N. (2000) Bacteriorhodopsin as a photochromic retinal protein for optical memories. *Chem. Rev.*, **100**, 1755–1776.
17. Qian, L., Winfree, E., and Bruck, J. (2011) Neural network computation with DNA strand displacement cascades. *Nature*, **475**, 368–372.
18. Baum, E.B. (1995) Building an associative memory vastly larger than the brain. *Science*, **268**, 583–585.
19. Laplante, J. and Pemberton, M. (1995) Experiments on pattern recognition by chemical kinetics. *J. Phys. Chem.*, **99**, 10063–10065.
20. Hjelmfelt, A., Weinberger, E.D., and Ross, J. (1991) Chemical implementation of neural networks and turing machines. *Proc. Natl. Aacd. Soc.*, **88**, 10983–10987.
21. Stamm, U. (2004) Extreme ultraviolet light sources for use in semiconductor lithography – state of the art and future development. *J. Phy. D: Appl. Phys.*, **37**, 3244.
22. Wu, B. and Kumar, A. (2007) Extreme ultraviolet lithography: a review. *J. Vac. Sci. Technol., B: Microelectron. Nanomet. Struct.--Process., Meas., Phenom.*, **25**, 1743.
23. Wu, B. and Kumar, A. (2009) *Extreme Ultraviolet Lithography*, McGraw-Hill Professionals.
24. Ratner, M.A. and Jortner, J. (1997) *Molecular Electronics*, Blackwell Science, Oxford.
25. Kuhn, H., Forsterling, H., and Waldeck, D.H. (2009) *Principles of Physical Chemistry*, Wiley-Interscience, Hoboken, NJ.
26. Takimoto, K., Kuroda, R., Shido, S., Yasuda, S., Matsuda, H., Eguchi, K., and Nakagiri, T. (1997) Writing and reading bit arrays for information storage using conductance change of a Langmuir-Blodgett film induced by scanning tunneling microscopy. *J. Vac. Sci. Technol., B: Microelectron. Nanomet. Struct.– Process., Meas., Phenom.*, **15**, 1429–1431.
27. Choi, J., Man, Y.S., and Fujihira, M. (2004) Nanoscale fabrication of biomolecular layer and its application to biodevices. *Biotechnol. Bioprocess Eng.*, **9**, 76–85.
28. Ye, J., Cui, H., Ottova, A., and Tien, H.T. (2006) Interfaces, bifaces, and nanotechnology. *Adv. Planar Lipid Bilayers Liposome*, **3**, 251–267.
29. Sponsler, M.B. (2001) Hologram switching and erasing strategies with liquid crystals. *Mol. Supramol. Photochem.*, **7**, 363–386.
30. Qian, L. and Winfree, E. (2011) Scaling up digital circuit computation with DNA strand displacement cascades. *Science*, **332**, 1196–1201.

31. Maley, C.C. (2003) DNA Computing and Its Frontiers, in *Molecular Computing*, 2nd edn (eds T. Sienko, A. Adamatzky, N.G. Rambidi, and M. Conrad), The MIT Press, Cambridge, MA, pp. 153–189.
32. Adleman, L.M. (1994) Molecular computation of solutions to combinatorial problems. *Science*, **266**, 1021–1024.
33. Jackson, M. and Sturtevant, J. (1978) Phase transitions of the purple membranes of Halobacterium halobium. *Biochemistry*, **17**, 911–915.
34. Ranaghan, M.J., Shima, S., Ramos, L., Poulin, D.S., Whited, G., Rajasekaran, S., Stuart, J.A., Albert, A.D., and Birge, R.R. (2010) Photochemical and thermal stability of green and blue proteorhodopsins: implications for protein-based bioelectronic devices. *J. Phys. Chem. B*, **114**, 14064–14070.
35. Hillebrecht, J.R., Koscielecki, J.F., Wise, K.J., Marcy, D.L., Tetley, W., Rangarajan, R., Sullivan, J., Brideau, M., Krebs, M.P., Stuart, J.A., and Birge, R.R. (2005) Optimization of protein-based volumetric optical memories and associative processors by using directed evolution. *Nanobiotechnology*, **1** (2), 141–151.
36. Hampp, N., Popp, A., and Bräuchle, C. (1992) Diffraction efficiency of bacteriorhodopsin films for holography containing bacteriorhodopsin wildtype BR and its variants D85E and D96N. *J. Phys. Chem.*, **96**, 4679–4685.
37. Oesterhelt, D., Bräuchle, C., and Hampp, N. (1991) Bacteriorhodopsin: a biological material for information processing. *Q. Rev. Biophys.*, **24** (4), 425–478.
38. Bräuchle, C., Hampp, N., and Oesterhelt, D. (1991) Optical applications of bacteriorhodopsin and its mutated variants. *Adv. Mater.*, **3**, 420–428.
39. Hampp, N., Bräuchle, C., and Oesterhelt, D. (1990) Bacteriorhodopsin wildtype and variant aspartate-96 to asparagine as reversible holographic media. *Annu. Rev. Phys. Chem.*, **41**, 83–93.
40. Birge, R.R. (1990) Photophysics and molecular electronic applications of the rhodopsins. *Ann. Rev. Phys. Chem.*, **41**, 683–733.
41. Vsevolodov, N.N. (1998) *Biomolecular Electronics: An Introduction via Photosensitive Proteins*, Birkhauser, Boston, MA.
42. Luecke, H., Schobert, B., Richter, H.T., Cartailler, J.P., and Lanyi, J.K. (1999) Structure of bacteriorhodopsin at 1.55 A resolution. *J. Mol. Biol.*, **291**, 899–911.
43. Lawrence, A.F., Stuart, J.A., Singh, D.L., and Birge, R.R. (1998) Bit-error sources in 3D optical memory: experiments and models. *Proc. SPIE*, **3468**, 258–268.
44. Lawrence, A.F. and Birge, R.R. (1994) Fundamentals of reliability calculations for molecular devices and photochromic memories. *Adv. Chem.*, **240**, 131–160.
45. Oesterhelt, D. and Stockenius, W. (1971) Rhodopsin-like protein from the purple membrane of Halobacterium halobium. *Nature*, **233**, 149–153.
46. Balashov, S.P., Litvin, F.F., and Sineshchekov, V.A. (1988) Photochemical processes of light energy transformation in bacteriorhodopsin. *Physiochem. Biol. Rev.*, **8**, 1–61.
47. Mathies, R.A., Lin, S.W., Ames, J.B., and Pollard, W.T. (1991) From femtoseconds to biology: mechanism of bacteriorhodopsin's light-driven proton pump. *Ann. Rev. Phys. Chem.*, **20**, 491–518.
48. El-Sayed, M.A. (1992) On the molecular mechanisms of the solar to electric energy conversion by the other photosynthetic system in nature, bacteriorhodopsin. *Acc. Chem. Res.*, **25**, 279–286.
49. Lanyi, J.K. (1992) Proton transfer and energy coupling in the bacteriorhodopsin photocycle. *J. Bioenerg. Biomembr.*, **24**, 169.
50. Ebrey, T.G. (1993) Light energy transduction in bacteriorhodopsin, in *Thermodynamics of Membrane Receptors and Channels* (ed. M. Jackson), CRC Press, Boca Raton, FL, 353–379.
51. Lanyi, J.K. (2004) Bacteriorhodopsin. *Annu. Rev. Physiol.*, **66**, 665–688.
52. Zimanyi, L., Varo, G., Chang, M., Ni, B., Needleman, R., and Lanyi, J.K. (1992)

Pathways of proton release in the bacteriorhodopsin photocycle. *Biochemistry*, **31**, 8535–8543.
53. Stuart, J.A. and Birge, R.R. (1996) Characterization of the primary photochemical events of bacteriorhodopsin and rhodopsin. *Biomembranes*, **2A**, 33–139.
54. Zimanyi, L. and Lanyi, J.K. (1993) Deriving the intermediate spectra and photocycle kinetics from time-resolved difference spectra of bacteriorhodopsin. The simpler case of the recombinant D96N protein. *Biophys. J.*, **64**, 240–251.
55. Lanyi, J.K. and Varo, G. (1995) The photocycles of bacteriorhodopsin. *Isr. J. Chem.*, **35**, 365–385.
56. Tittor, J. and Oesterhelt, D. (1990) The quantum yield of bacteriorhodopsin. *FEBS Lett.*, **263** (2), 269–273.
57. Birge, R.R., Cooper, T.M., Lawrence, A.F., Masthay, M.B., Vasilakis, C., Zhang, C., and Zidovetzki, R. (1989) A spectroscopic, photocalorimetric, and theoretical investigation of the quantum efficiency of the primary event in bacteriorhodopsin. *J. Am. Chem. Soc.*, **111**, 4063–4074.
58. Govindjee, R., Balashov, S.P., and Ebrey, T.G. (1990) Quantum efficiency of the photochemical cycle of bacteriorhodopsin. *Biophys. J.*, **58** (3), 597–608.
59. Lukashev, E.P. and Robertson, B. (1995) Bacteriorhodopsin retains its light-induced proton-pumping function after being heated to 140 °C. *Bioelectrochem. Bioenerg.*, **37**, 157–160.
60. Shen, Y., Safinya, C.R., Liang, K.S., Ruppert, A.F., and Rothschild, K.J. (1993) Stabilization of the membrane protein bacteriorhodopsin to 140 °C in two-dimensional films. *Nature*, **366**, 48–50.
61. Hampp, N. and Juchem, T. (2000) Fringemaker-the first technical system based on bacteriorhodopsin, in *Bioelectronic applications of photochromic pigments* (eds A. Der and L. Keszthelyi), IOS Press, Szeged, Hungary, pp. 44–53.
62. Hampp, N., Thoma, R., Zeisel, D., and Bräuchle, C. (1994) Bacteriorhodopsin variants for holographic pattern recognition. *Adv. Chem.*, **240**, 511–526.
63. Hampp, N. (2000) Bacteriorhodopsin: mutating a biomaterial into an optoelectronic material. *App. Microbiol. Biotechnol.*, **53** (6), 633–639.
64. Juchem, T. and Hampp, N. (2000) Interferometric system for non-destructive testing based on large diameter bacteriorhodopsin films. *Opt. Lasers Eng.*, **34**, 87–100.
65. Popp, A., Wolperdinger, M., Hampp, N., Bräuchle, C., and Oesterhelt, D. (1993) Photochemical conversion of the O-intermediate to 9-cis-retinal containing products in bacteriorhodopsin films. *Biophys. J.*, **65**, 1449–1459.
66. Tallent, J.R., Stuart, J.A., Song, Q.W., Schmidt, E.J., Martin, C.H., and Birge, R.R. (1998) Photochemistry in dried polymer films incorporating the deionized blue membrane form of bacteriorhodopsin. *Biophys. J.*, **75**, 1619–1634.
67. Gillespie, N.B., Wise, K.J., Ren, L., Stuart, J.A., Marcy, D.L., Hillebrecht, J., Li, Q., Ramos, L., Jordan, K., Fyvie, S., and Birge, R.R. (2002) Characterization of the branched-photocycle intermediates P and Q of bacteriorhodopsin. *J. Phys. Chem. B*, **106**, 13352–13361.
68. Birge, R.R., Parsons, B., Song, Q.W., and Tallent, J.R. (1997) Protein-based three-dimensional memories and associative processors, in *Molecular Electronics* (eds M.A. Ratner and J. Jortner), Blackwell Science Ltd, Oxford, 439–471.
69. Paek, E.G. and Jung, E.C. (1991) Simplified holographic associated memory using enhanced nonlinear processing with a thermoplastic plate. *Opt. Lett.*, **16**, 1034–1036.
70. Paek, E.G. and Psaltis, D. (1987) Optical associative memory using Fourier transform holograms. *Opt. Eng.*, **26**, 428–433.
71. Stuart, J.A., Marcy, D.L., Wise, K.J., and Birge, R.R. (2003) Biomolecular Electronic Device Applications of Bacteriorhodopsin, in *Molecular Electronics: Bio-Sensors and Bio-Computers* (ed. L. Barsanti), Kluwer Academic Publishers, The Netherlands, pp. 265–299.
72. Gross, R.B., Izgi, K.C., and Birge, R.R. (1992) Holographic thin films, spatial light modulators, and optical associative

memories based on bacteriorhodopsin. *Proc. SPIE*, **1662**, 186–196.
73. Psaltis, D., Brady, D., and Wagner, K. (1988) Adaptive optical networks using photorefractive crystals. *Appl. Opt.*, **27**, 1752–1759.
74. Psaltis, D., Neifeld, M.A., and Yamamura, A. (1989) Image correlators using optical memory disks. *Opt. Lett.*, **14**, 429–431.
75. Psaltis, D., Brady, D., Guang, X., and Lin, S. (1990) Holography in artificial neural networks. *Science*, **343**, 325–330.
76. Psaltis, D., Neifeld, M.A., Yamamura, A., and Kobayashi, S. (1990) Optical memory disks in optical information processing. *Appl. Opt.*, **29**, 2038–2057.
77. Wise, K.J., Gillespie, N.B., Stuart, J.A., Krebs, M.P., and Birge, R.R. (2002) Optimization of bacteriorhodopsin for bioelectronic devices. *Trends Biotechnol.*, **20**, 387–394.
78. d'Auria, L., Huignard, J.P., Slezak, C., and Spitz, C. (1994) Experimental holographic read-write memory using 3-D storage. *Appl. Opt.*, **13**, 808–818.
79. Heanue, J.F., Bashaw, M.C., and Hesselink, L. (1994) Volume holographic storage and retrieval of digital data. *Science*, **265**, 749–752.
80. Psaltis, D. and Pu, A. (1996) Holographic 3-D disks. *IEEE Nonvol. Mem. Tech. (INVMTC)*, **6**, 34–39.
81. Parthenopoulos, D.A. and Rentzepis, P.M. (1989) Three-dimensional optical storage memory. *Science*, **245**, 428–433.
82. Chen, Z., Govender, D., Gross, R., and Birge, R.R. (1995) Advances in protein-based three-dimensional memories and associative processors. *BioSystems*, **35**, 145–151.
83. Dvornikov, A.S. and Rentzepis, P.M. (1996) 3D optical memory devices: system and materials characteristics. *IEEE Nonvol. Mem. Tech. (INVMTC)*, **6**, 40–44.
84. Stuart, J.A., Marcy, D.L., Wise, K.J., and Birge, R.R. (2002) Volumetric optical memory based on bacteriorhodopsin. *Synth. Met.*, **127**, 3–15.
85. Rajasekaran, S., Kundeti, V., Birge, R.R., Kumar, V., and Sahni, S. (2011) Efficient algorithms for computing with protein-based volumetric memory processors. *IEEE Trans. Nanotech.*, **10**, 881–890.
86. Trinca, D. and Rajasekaran, S. (2010) Specialized compression for coping with diffraction effects in protein-based volumetric memories: solving some challenging instances. *J. Nanoelectron. Optoelectron.*, **5**, 290–294.
87. Trinca, D. and Rajasekaran, S. (2010) Coping with diffraction effects in protein-based computing through a specialized approximation algorithm with constant overhead. Proceedings IEEE NANO, Seoul, Korea, August 17–20, 2010.
88. Trinca, D. and Rajasekaran, S. (2010) Using a specialized compression scheme based on antidictionaries for coping with diffraction effects in protein-based computing. Proceedings of the 14th International Biotechnology Symposium and Exhibition, Rimini, Italy, September 14–18, 2010.
89. Trinca, D. and Rajasekaran, S. (2009) Optimized bzip2 compression for reducing diffraction effects in protein-based computing: a study of feasibility. Proceedings of the 31st Annual International IEEE EMBS Conference, Minneapolis, MN, September 2–6.
90. Kundeti, V. and Rajasekaran, S. (2009) Generalized algorithms for generating balanced modulation codes in protein-based volumetric memories. Proceedings of the 10th Annual IEEE Congress on Evolutionary Computation (IEEE CEC), Norway, May 18–21, 2009.
91. Brizzolara, R.A., Boyd, J.L., and Tate, A.E. (1997) Evidence for covalent attachment of purple membrane to a gold surface via genetic modification of bacteriorhodopsin. *J. Vac. Sci. Technol. A*, **15**, 773–778.
92. Schranz, M., Noll, F., and Hampp, N. (2007) Oriented purple membrane monolayers covalently attached to gold by multiple thiole linkages analyzed by single molecule force spectroscopy. *Langmuir*, **23**, 11134–11138.
93. Hsu, K.C., Rayfield, G.W., and Needleman, R. (1996) Reversal of the surface charge asymmetry in purple membrane due to single amino

acid substitutions. *Biophys. J.*, **70**, 2358–2365.

94. Arnold, F.H. (1998) Design by directed evolution. *Acc. Chem. Res.*, **31**, 125–131.

95. Arnold, F.H. and Volkov, A.A. (1999) Directed evolution of biocatalysts. *Curr. Opin. Chem. Biol.*, **3**, 54–59.

96. Schmidt-Dannert, C. and Arnold, F.H. (1999) Directed evolution of industrial enzyme. *Trends Biotechnol.*, **17**, 135–136.

97. Jäckel, C., Kast, P., and Hilvert, D. (2008) Protein design by directed evolution. *Annu. Rev. Biophys.*, **37**, 153–173.

98. Jäckel, C., Bloom, J.D., Kast, P., and Arnold, F.H. (2010) Consensus protein design without phylogenetic bias. *J. Mol. Biol.*, **399**, 541–546.

99. Yano, T., Oue, S., and Kagamiyama, H. (1998) Directed evolution of an aspartate aminotransferase with new substrate specificities. *Proc. Natl. Acad. Sci. U.S.A.*, **95**, 5511–5515.

100. Miyazaki, K., Wintrode, P.L., Grayling, R.A., Rubingh, D.N., and Arnold, F.H. (2000) Directed evolution study of temperature adaptation in a psychrophilic enzyme. *J. Mol. Biol.*, **297**, 1015–1026.

101. Moore, J.C. and Arnold, F.H. (1996) Directed evolution of a para-nitrobenzyl esterase for evolutionary divergence. *Nat. Biotechnol.*, **14**, 458–467.

102. Spiller, B., Gershenson, A., Arnold, F.H., and Stevens, R.C. (1999) A structural view of evolutionary divergence. *Proc. Natl. Acad. Sci. U.S.A.*, **96**, 12305–12310.

4
Bioelectronic Devices Controlled by Enzyme-Based Information Processing Systems
Evgeny Katz

4.1
Introduction

Molecular [1–7] and biomolecular [8–10] logic gates and their networks processing chemical input signals similar to computers received high attention and were rapidly developed in the last decade. Being a subarea of unconventional computing [11, 12], they can process chemical information mimicking Boolean logic operations using binary definitions (1, 0; in other words: YES/NO) for concentrations of reacting species. Using this approach, chemical reactions could be reformulated as information processing steps with built-in logic operations [13]. Then, the chemical processes could be programmed similar to computer programming [14, 15], yielding networks performing several logic operations. Despite the fact that chemical systems based on organic molecules [16–31] or biomolecules [9, 10], [32–41] achieved significant success in the formulation of single logic operations, and their short sequences mimicking natural biochemical pathways were successfully designed [42, 43], there is no clear opinion about their possible applications. The present complexity of the chemical information processing systems is far below that of electronic computers and the timescale of their operation (minutes to hours) is too long to be competitive with electronics. On the other hand, an obvious advantage of biocomputing systems over their electronic counterparts is their compatibility with biochemical systems and their ability to operate in a biochemical environment [44]. Therefore, one of the possible applications of chemical, and particularly biochemical, computing systems might be in the interfacing electronic (in a more narrow definition – electrochemical) and biological systems.

The recently developed concept of biocomputing with the use of enzyme-biocatalyzed reactions [8] allows logic processing of multiple biochemical signals before their transduction to a bioelectronic device. A new strategy applying digital multisignal biosensors became possible through this approach [45]. Novel biosensors based on enzyme-based concatenated logic gates have already been developed for the analysis of pathophysiological conditions related to different kinds of injuries [46–50].

Biomolecular Information Processing: From Logic Systems to Smart Sensors and Actuators, First Edition. Edited by Evgeny Katz.
© 2012 Wiley-VCH Verlag GmbH & Co. KGaA. Published 2012 by Wiley-VCH Verlag GmbH & Co. KGaA.

Signals generated by enzyme logic systems in the form of concentration changes of reactant species can be read out using various analytical techniques: optical [41, 51, 52], electrochemical [40], and so on. Sensitive analytical methods and instruments are required because the concentration changes produced during chemical reactions usually take place at low levels (micromolar or even nanomolar concentration ranges). Application of highly sensitive techniques requires electronic transduction and amplification of the output signals produced by enzyme logic systems. This approach was actually applied to most chemical computing systems based on nonbiological molecules [1–7, 16–31]. While most nonbiochemical computing systems are based on the application of noncatalytic reactions, enzyme logic systems have the advantage of catalyzing biochemical transformations when continuous production of the output species could potentially produce substantial chemical changes in the systems even when the concentrations of the reacting catalytic species are low. The chemical changes in enzyme biocatalytic systems can be coupled to signal-responsive materials resulting in changes in their bulk properties, thereby amplifying the chemical changes generated by enzymes [53]. Signal-responsive materials coupled with enzyme logic systems could be represented by polymers responding to external chemical signals by restructuring between swollen and shrunken states [54–56]. The structural reorganization of the polymers will substantially amplify the chemical changes generated by the enzyme reactions, thereby obviating the need for highly sensitive analytical techniques to observe the output signals from biocomputing systems. Application of signal-responsive polymers in a biochemical environment has already a well-established background, allowing their integration with biocomputing systems [57].

4.2
Enzyme-Based Logic Systems Producing pH Changes as Output Signals

Since many signal-responsive polymer systems are pH sensitive and switchable, we designed enzyme-based logic gates using enzymes as biocatalytic input signals, processing information according to the Boolean operations AND or OR and generating pH changes as the output signals from the gates [58–61]. The AND gate performed a sequence of biocatalytic reactions (Figure 4.1A): sucrose hydrolysis was biocatalyzed by invertase (Inv) yielding glucose, which was then oxidized by oxygen in the presence of glucose oxidase (GOx). The later reaction resulted in the formation of gluconic acid and therefore lowered the pH of the solution. The absence of the enzymes was considered as the input signals "0", while their presence under optimized concentrations was interpreted as the input signals "1". The biocatalytic reaction chain was activated only in the presence of both enzymes (Inv and GOx) (input signals 1,1) resulting in the decrease of the solution pH (Figure 4.1B). The absence of any of two enzymes (input signals 0,1 or 1,0) or both of them (input signals 0,0) resulted in the inhibition of gluconic acid formation and therefore no pH changes were produced. Thus, the biocatalytic chain mimics

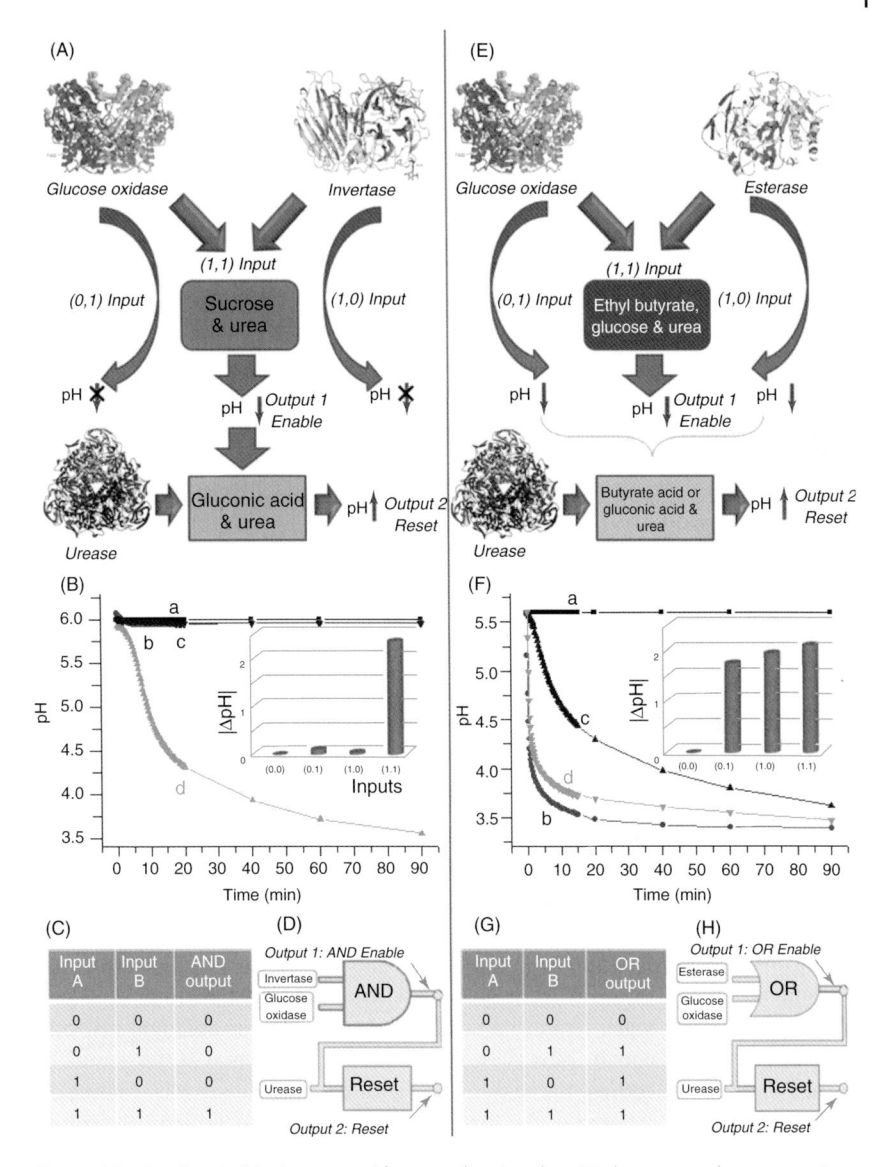

Figure 4.1 Biochemical logic gates with the enzymes used as input signals to activate the gate operation. The absence of the enzyme is considered as "0" and the presence as "1" input signals. The Reset function was catalyzed by urease. (A) The AND gate based on GOx- and Inv-catalyzed reactions. (B) pH-changes generated *in situ* by the AND gate upon different combinations of the input signals: (a) "0,0", (b) "0,1", (c) "1,0" and (d) "1,1". Inset: bar diagram showing the pH changes as the output signals of the AND gate. (C) The truth table of the AND gate showing the output signals in the form of pH changes generated upon different combinations of the input signals. (D) Equivalent electronic circuit for the biochemical AND–Reset logic operations. (E) The OR gate based on GOx- and Est-catalyzed reactions. (F–H) The same as (B–D) for the OR gate. (Adapted from [58] with permission. Copyright American Chemical Society, 2009.)

the AND logic operation expressed by the standard truth table corresponding to the AND Boolean operation (Figure 4.1C). After completion of the biocatalytic reactions and reaching the final pH value, the system might be reset to the original pH by using another biochemical reaction in the presence of urease and urea, resulting in the production of ammonia and the elevation of pH (Figure 4.1A). While urea is a part of the biomolecular "machinery" being included into the system from the very beginning, urease is a signal applied by an operator to reset the system after the AND logic operation is completed and the pH changes reach their final values. The performance of the biochemical system can be described in terms of a logic circuitry with an AND/Reset function (Figure 4.1D). Another gate operating as Boolean OR logic function was composed of two parallel reactions (Figure 4.1E): hydrolysis of ethyl butyrate and oxidation of glucose biocatalyzed by esterase (Est) and GOx, respectively, and resulting in the formation of butyric acid and gluconic acid. Any of the produced acids or both of them together resulted in the formation of the acidic solution (Figure 4.1F). Thus, in the absence of both enzymes (Est and GOx) (input signals 0,0), both reactions were inhibited and the pH value was unchanged. When either enzyme (Est or GOx; input signals 0,1 or 1,0) or both of them were present together (input signals 1,1), one or both of the reactions proceeded and resulted in the acidification of the solution. The features of the system correspond to the OR logic operation and can be expressed by the standard truth table (Figure 4.1G). Similarly to the previous example, the system can be reset to the initial pH by the urease-catalyzed reaction, allowing the sequence of OR/Reset functions (Figure 4.1H).

4.3
Interfacing of the Enzyme Logic Systems with Electrodes Modified with Signal-Responsive Polymers

pH-switchable materials immobilized on interfaces of electronic/electrochemical transducers (e.g., Si-based chips [62] or conducting electrodes [63, 64]) were coupled with enzyme logic systems producing pH changes in solutions as logic responses to input signals. This allowed electronic transduction of the generated output signals, converting the systems into multisignal biosensors chemically processing various patterns of the input signals using logic "programs" built-in in the enzyme systems. For example, enzyme logic systems mimicking Boolean AND/OR logic operations and producing the output signal in the form of solution pH changes were coupled with charging–discharging organic shells around Au nanoparticles associated with an EIS (electrolyte–insulator–semiconductor) Si-chip surface (Figure 4.2a) [62]. This resulted in capacitance changes at the modified interface (Figure 4.2b), allowing electronic transduction of the biochemical signals processed by the AND/OR enzyme logic systems (Figure 4.3). Another approach to the electrochemical transduction of the output signals generated by enzyme logic systems in the form of pH changes was based on the application of polyelectrolyte-modified electrode surfaces [63, 64]. Polyelectrolytes

4.3 Enzyme Logic Systems integrated with Electrodes Modified with Signal-Responsive Polymers

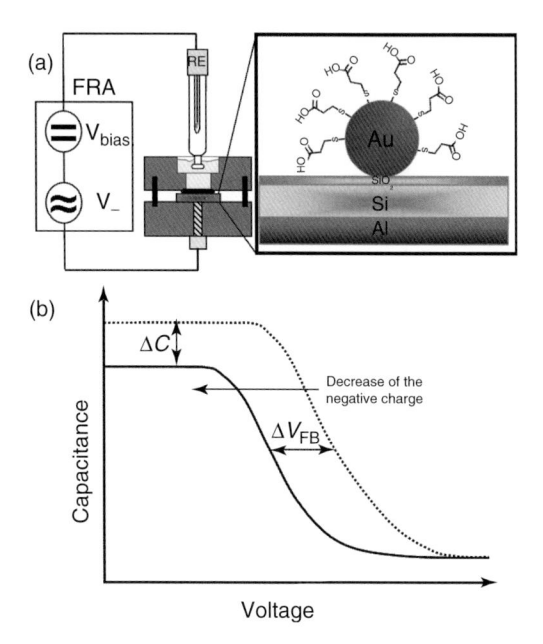

Figure 4.2 (a) Capacitive EIS (electrolyte–insulator–semiconductor) sensor chip functionalized with Au nanoparticles coated with a pH-responsive organic shell. (b) Schematic representation of capacitance versus voltage (C–V) curves obtained for the variable charges at the sensing interface upon protonation of the organic shell. (FRA = frequency response analyzer; RE = reference electrode; ΔC = change of the interfacial capacitance; ΔV_{FB} = change of the flat-band potential.) (Adapted from [62] with permission. Copyright American Chemical Society, 2009.)

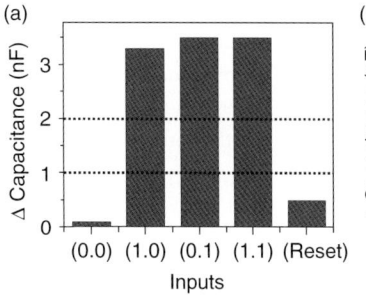

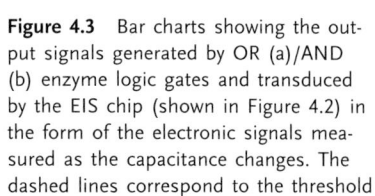

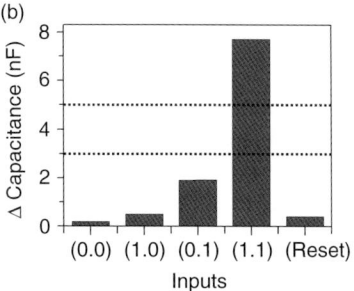

Figure 4.3 Bar charts showing the output signals generated by OR (a)/AND (b) enzyme logic gates and transduced by the EIS chip (shown in Figure 4.2) in the form of the electronic signals measured as the capacitance changes. The dashed lines correspond to the threshold values: the output signals located below the first threshold were considered as "0", while the signals higher than the second threshold were treated as "1". (Adapted from [62] with permission. Copyright American Chemical Society, 2009.)

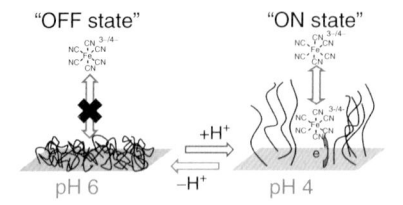

Figure 4.4 Electrode surface functionalized with a polymer brush switchable between shrunken and swollen states upon protonation–deprotonation of the polymer molecules, thus inhibiting and allowing penetration of redox active species to the conducting support and resulting in switching OFF/ON their electrochemical activity. (Adapted from [64] with permission. Copyright American Chemical Society, 2009.)

covalently bound to electrode surfaces as polymer brushes reveal pH sensitivity, allowing control of the electrode interfacial properties by varying the pH values. Charged states of the polymer brushes produce hydrophilic swollen thin films on the electrode surfaces, resulting in their high permeability for soluble redox probes to the conducting supports, thus yielding the electrochemically active "ON" states of the modified electrodes. Upon discharging the polymer chains, the produced hydrophobic shrunken states isolated the conducting supports, yielding the inactive "OFF" states of the modified electrodes. Switching between the ON and OFF states of the electrode modified with the polymer brush was achieved by varying the pH of the solution (Figure 4.4). This property of the polymer brush-functionalized electrodes was used to couple them with an enzyme logic system composed of several networked gates [64]. The logic network composed of three enzymes (alcohol dehydrogenase (ADH), glucose dehydrogenase (GDH), and GOx) (Figure 4.5a), operating in concert as four concatenated logic gates (AND/OR) (Figure 4.5b), was designed to process four different chemical input signals (NADH, acetaldehyde, glucose, and oxygen). The cascade of biochemical reactions resulted in pH changes controlled by the pattern of the applied biochemical input signals. The "successful" set of the inputs produced gluconic acid as the final product and yielded an acidic medium, lowering the pH of the solution from its initial value of 6–7 to the final value of about 4 (Table 4.1), thus switching ON the interface for the redox process of a diffusional redox probe, namely, $[Fe(CN)_6]^{3-/4-}$. The chemical signals processed by the enzyme logic system and transduced by the sensing interface were read out by electrochemical means using cyclic voltammetry (Figure 4.6A). Reversible activation–inactivation of the electrochemical interface was achieved upon logic processing of the biochemical input signals and then by the reset function activated in the presence of urease and urea (Figure 4.6A, inset). The whole set of the input signal combinations included 16 variants, while only 0,0,1,1; 0,1,1,1; 1,0,1,1; 1,1,1,0 and 1,1,1,1 combinations resulted in the ON state of the electrochemical interface (Figure 4.6B and Table 4.1). The present system exemplifies a multigate/multisignal processing enzyme logic ensemble associated with an electrochemical transduction readout of the output signal.

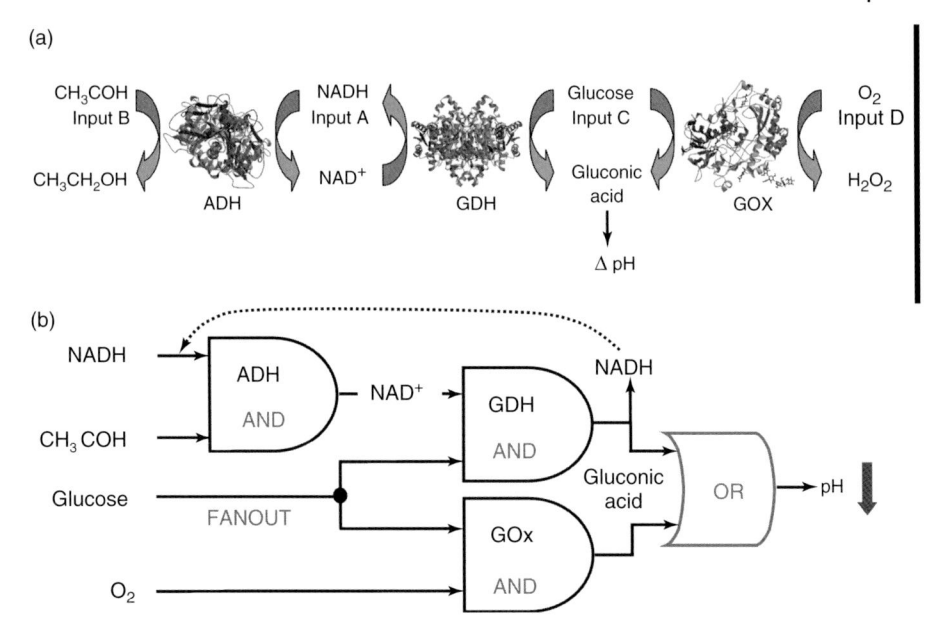

Figure 4.5 (a) Multigate/multisignal processing enzyme logic system producing *in situ* pH changes as the output signal. (b) Equivalent logic circuitry for the biocatalytic cascade. (Adapted from [64] with permission. Copyright American Chemical Society, 2009.)

More sophisticated electrochemical properties were found for the polymer brush bound to an electrode surface and functionalized with redox species [65]. Os(dmo-bpy)$_2$ redox groups (dmo-bpy = 4,4′-dimethoxy-2,2′-bipyridine) were covalently bound to poly(4-vinyl pyridine) (P4VP) polymer chains grafted on an indium–tin oxide (ITO) electrode. The polymer-bound redox species were found to be electrochemically active at pH < 5 when the polymer is protonated and swollen and the chains are flexible (Figure 4.7A). Upon changing the pH to values greater than 6, the polymer chains lose their charges and the produced shrunken state of the polymer does not show substantial electrochemical activity. This was explained by the poor mobility of the polymer chains in the shrunken state, restricting the direct contact between the Os-complex units and the conducting support. A low density of the Os-complex in the polymer film does not allow the electron hopping between the redox species, and their electrochemical activity can be achieved only upon quasidiffusional translocation of the polymer chains in the swollen state, bringing the redox species to a short distance from the conducting support. The pH-controlled reversible activation–inactivation of the interface-confined redox species (Figure 4.7B) was used to read out the pH signals produced *in situ* by enzyme systems performing AND/OR logic operations (Figure 4.8). The electrochemical signal produced by the modified interface in response to the pH changes generated by the enzyme reactions was further amplified by electrocatalytic oxidation of NADH being switched ON upon activation of the redox species on the electrode surface [63].

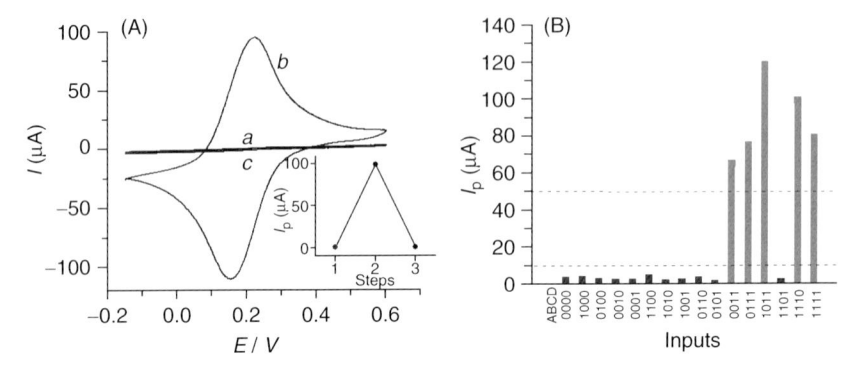

Figure 4.6 (A) Cyclic voltammograms obtained for the ITO electrode modified with the P4VP polymer brush in (a) the initial OFF state, pH about 6.7; (b) ON state enabled by the input combinations resulting in acidifying the solution to pH about 4.3; and (c) *in situ* reset to the OFF state, pH about 8.8. Inset: reversible current changes upon switching the electrode ON/OFF. Deoxygenated unbuffered solution of 0.1 M Na_2SO_4, 10 mM $K_3Fe(CN)_6$, and 10 mM $K_4Fe(CN)_6$ also contained ADH, GDH, and GOx, 10 units · ml^{-1} each. Input A was 0.5 mM NADH, input B was 5 mM acetaldehyde, input C was 12.5 mM glucose and input D was O_2 in equilibrium with air. Potential scan rate, 100 mV · s^{-1}. (B) Anodic peak currents I_p for the 16 possible input combinations. The dotted lines show threshold values separating logic 1, undefined, and logic 0 output signals. (Adapted from [64] with permission. Copyright American Chemical Society, 2009.)

4.4
Switchable Biofuel Cells Controlled by the Enzyme Logic Systems

The logically controlled Os-complex P4VP-modified electrode [63] was applied to mediate oxygen reduction in the presence of laccase – an enzyme reducing O_2 to water. This electrode was integrated into an enzyme-based biofuel cell serving as a switchable biocatalytic cathode (Figure 4.9) [66–68]. Glucose oxidation in the presence of soluble GOx and methylene blue (MB) mediating electron transport to a bare ITO electrode was used as an anodic process. The biofuel cell was switched ON on demand upon processing biochemical signals by enzyme logic gates, and reset back to the OFF state by another biocatalytic reaction in the presence of urea and urease. The coupling of the *in situ* enzyme reactions with the switchable electrochemical interface was achieved by pH changes generated in the course of the enzymatic reactions, resulting in swelling/shrinking of the redox polymer at the electrode surface as described earlier. At the beginning, single logic gates AND/OR were used to control the biofuel cell power production [67]. A higher complexity biocatalytic system based on the concerted operation of four enzymes activated by four chemical input signals was designed (Figure 4.10a) to mimic a logic network composed of three logic gates (AND/OR connected in parallel generating two intermediate signals for the final AND gate) (Figure 4.10b) [68]. The switchable biofuel cell was characterized by measuring polarization curves at its "mute" and active states. A low voltage–current production was characteristic

Table 4.1 Truth table for the Boolean logic function: O = {(A AND B) AND C} OR (C AND D), where O is the output, and A, B, C, and D are input signals

Input A: NADH	Input B: acetaldehyde	Input C: glucose	Input D: O_2	Output O: decrease in pH ($\Delta pH > 2$)
0	0	0	0	0
1	0	0	0	0
0	1	0	0	0
0	0	1	0	0
0	0	0	1	0
1	1	0	0	0
1	0	1	0	0
1	0	0	1	0
0	1	1	0	0
0	1	0	1	0
0	0	1	1	1
0	1	1	1	1
1	0	1	1	1
1	1	0	1	0
1	1	1	0	1
1	1	1	1	1

The chemical system was composed of three enzymes (glucose oxidase, glucose dehydrogenase, and alcohol dehydrogenase) performing four logic operations activated by four input signals.

of the initial inactive state of the biofuel cell at pH about 6 (Figure 4.11A, curve a). Upon receiving an output signal in the form of a pH decrease from the enzyme logic network, the voltage–current production by the biofuel cell was dramatically enhanced when the pH reached about 4.3 (Figure 4.11A, curve b). When the activation of the biofuel cell was achieved, another biochemical signal (urea in the presence of urease) resulted in the increase in pH, thereby resetting the cell to its inactive state with a small voltage–current production (Figure 4.11A, curve c). The cyclic operation of the biofuel cell upon receiving biochemical signals can be followed by the reversible changes of the current production (Figure 4.11A, inset). The biofuel cell switching from the "mute" state with a low activity to the active state was achieved upon appropriate combination of the input signals processed by the enzyme logic network. Only three combinations of the input signals, namely, 1,1,1,0; 1,1,0,1 and 1,1,1,1 from all 16 possible variants resulted in the change in the solution pH, thereby switching the biofuel cell to its active state (Figure 4.11B). The studied biofuel cells exemplify a new kind of bioelectronic devices where the bioelectronic function is controlled by a biocomputing system. Such devices will provide a new dimension in bioelectronics and biocomputing, benefiting from the integration of both concepts.

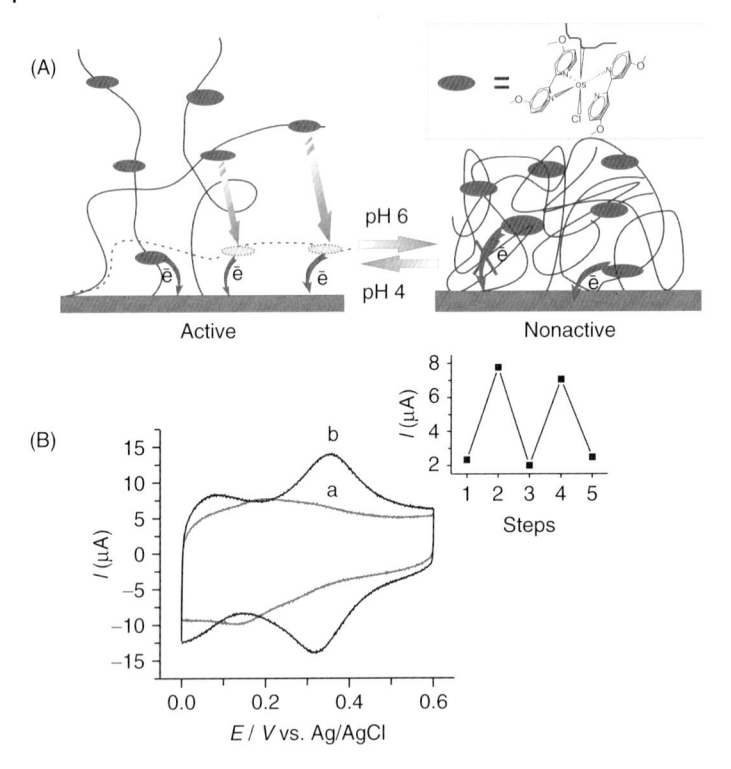

Figure 4.7 (A) Reversible pH-controlled transformation of the Os(dmo-bpy)$_2$-P4VP-polymer brush on the electrode surface between electrochemically active and inactive states. (B) Cyclic voltammograms of the modified electrode obtained upon measurements performed in neutral and acidic solutions: (a) pH 7.0 and (b) 3.0. Inset: reversible switching of the modified electrode activity. (Adapted from [65], with permission; Copyright American Chemical Society, 2008.)

4.5
Biomolecular Logic Systems Composed of Biocatalytic and Biorecognition Units and Their Integration with Biofuel Cells

Enzyme-based biochemical networks demonstrate robust, error-free processing of biochemical signals upon appropriate optimization of their components and interconnections [69–71]. However, the limit to the complexity of the biocomputing network is set by the cross-reactivity of the enzyme-catalyzed reactions. Only enzymes belonging to different biocatalytic classes (oxidases, dehydrogenases, peroxidises, hydrolases, etc.) could operate in a homogeneous system without significant cross-reactivity. If chemical reasons require the use of cross-reacting enzymes in the system, they must be compartmentalized using patterning on surfaces or applied in microfluidic devices. Application of more selective biomolecular interactions would be an advantage in making biocomputing systems more specific to various input signals and less cross-reactive in the chemical signal processing.

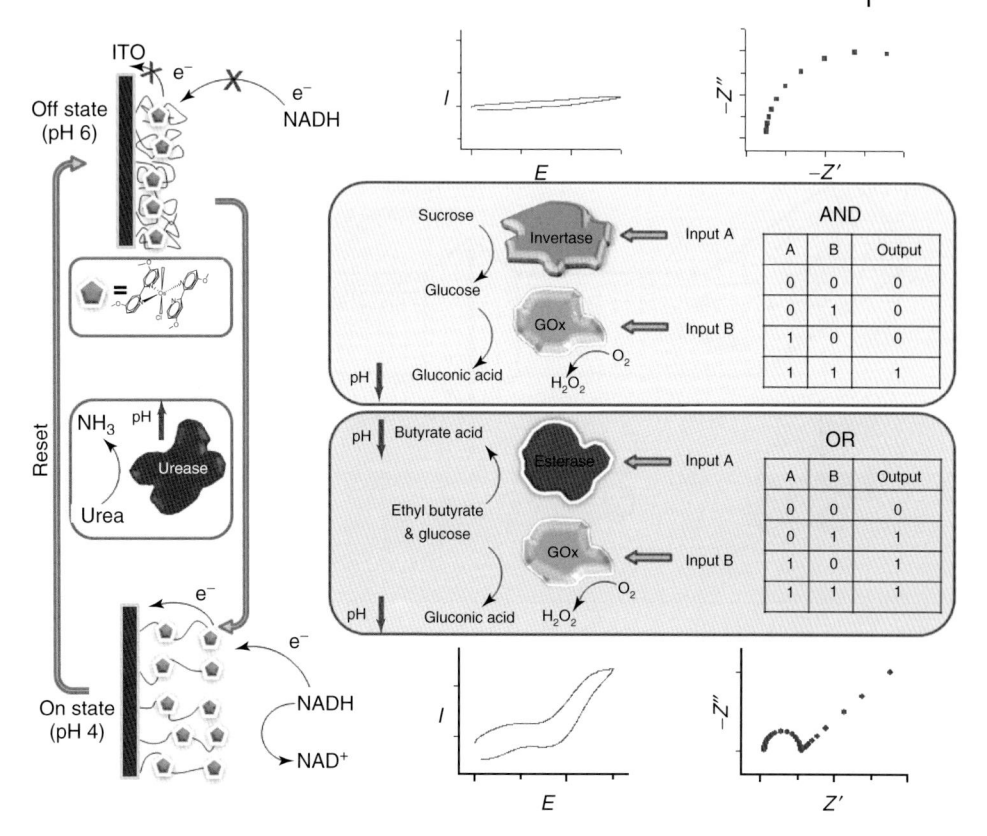

Figure 4.8 Logic operations AND/OR performed by the enzyme-based systems resulting in the ON and OFF states of the bioelectrocatalytic interface followed by the Reset function to complete the reversible cycle. Schematically shown cyclic voltammograms and impedance spectra correspond to the ON and OFF states of the bioelectrocatalytic electrode applied for the NADH oxidation. (Adapted from [63] with permission. Copyright American Chemical Society, 2009.)

This aim can be achieved by the application of highly selective biorecognition (e.g., immuno) interactions for biocomputing [72]. One of the novel immune-based biocomputing systems was already applied for switching the biofuel cell activity by the logically processed antibody signals [73].

A surface functionalized with a mixed monolayer of two different antigens, namely, 2,4-dinitrophenyl (DNP) and 3-nitro-L-tyrosine (NT) loaded on human serum albumin (HSA) and bovine serum albumin (BSA), respectively, was used to analyze the input signals of the corresponding antibodies, namely, anti-DNP (anti-dinitrophenyl IgG polyclonal from goat) and anti-NT (anti-nitrotyrosine IgG from rabbit) [73]. After binding to the surface, the primary antibodies were reacted with the secondary antibodies anti-goat-IgG-HRP and anti-rabbit-IgG-HRP (mouse origin IgG against goat immunoglobulin and mouse origin IgG against rabbit IgG,

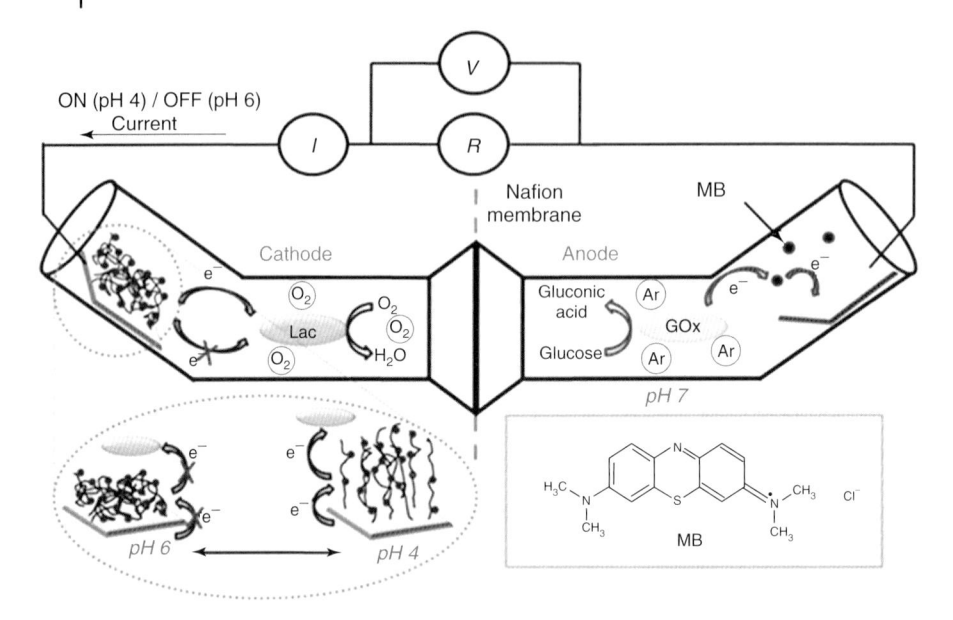

Figure 4.9 Biofuel cell composed of the pH-switchable logically controlled biocatalytic cathode and glucose oxidizing anode. (Adapted from [68] with permission.)

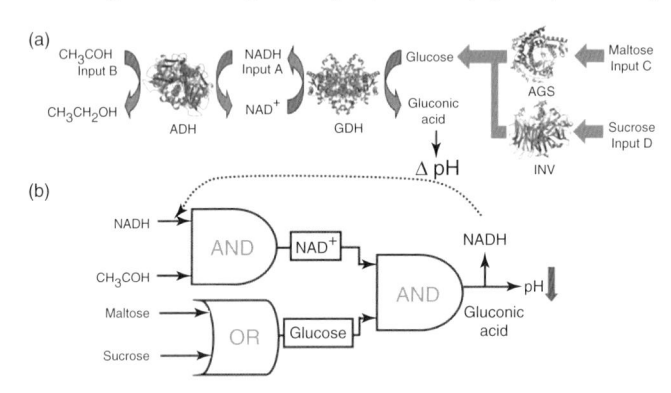

Figure 4.10 (a) Cascade of reactions biocatalyzed by alcohol dehydrogenase (ADH), amyloglucosidase (AGS), invertase (Inv), and glucose dehydrogenase (GDH) and triggered by chemical input signals: NADH, acetaldehyde, maltose, and sucrose added in different combinations. (b) Logic network composed of three concatenated gates and equivalent to the cascade of enzymatic reactions outlined in (a). (Adapted from [68] with permission.)

both labeled with horseradish peroxidise, (HRP)) to attach the biocatalytic HRP tag to the immune complexes generated on the surfaces (Figure 4.12a). The primary anti-DNP and anti-NT antibodies (signals *A* and *B*, respectively) were applied in four different combinations, 0,0; 0,1; 1,0 and 1,1, where the digital value 0 corresponded to the absence of the antibody and the value 1 corresponded to their presence under optimized concentrations. The secondary antibody labeled with the HRP

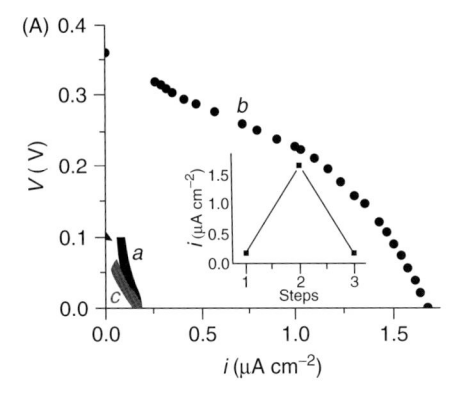

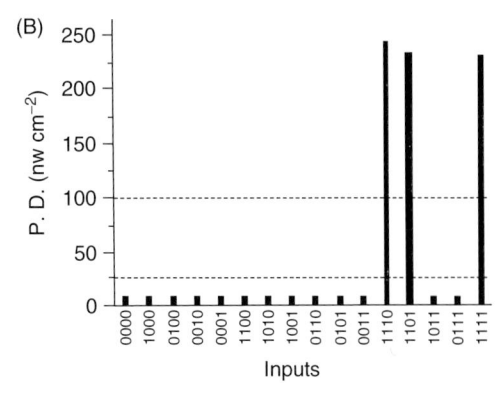

Figure 4.11 (A) $V - i$ polarization curves obtained for the biofuel cell with different load resistances: (a) in the inactive state before the addition of the biochemical input signals (pH value in the cathodic compartment about 6); (b) in the active state after the cathode was activated by changing pH to about 4.3 by the biochemical signals; and (c) after the Reset function activated by the addition of 5 mM urea. Inset: switchable i_{sc} upon transition of the biofuel cell from the mute state to the active state and back performed upon biochemical signals processed by the enzyme logic network. (B) Bar chart showing the power density produced by the biofuel cell in response to different patterns of the chemical input signals. Dash lines show thresholds separating digital 0, undefined, and 1 output signals produced by the system. (Adapted from [68] with permission.)

biocatalytic tag was bound to the surface only if the respective primary antibody was already present. Since both secondary antibodies were labeled with HRP, the biocatalytic entity appeared on the surface upon application of 0,1; 1,0 and 1,1 signal combinations. Only in the absence of both primary antibodies (signals 0,0) were the secondary antibodies not bound to the surface and the HRP biocatalyst did not appear there, thus resembling the OR logic operation. The assembled functional interface was reacted with 2,2'-azino-bis(3-ethylbenzothiazoline-6-sulfonic acid) (ABTS) and H_2O_2. The biocatalytic oxidation of ABTS and the concomitant reduction of H_2O_2 resulted in the increase of the solution pH only when the biocatalytic HRP tag was present on the surface (Figure 4.12b). This happened when the primary antibody signals were applied in the combinations 0,1; 1,0 and 1,1. The pH increase generated *in situ* by the enzyme reaction coupled with the immuno-recognition system yielded the inactive shrunken state of the polymer brush-modified electrode, thereby deactivating the entire biofuel cell. It should be noted that, for simplicity, the cathode was represented by a model redox system with a ferricyanide solution instead of the oxygen system (Figure 4.12c). The biofuel cell that was active at pH 4.5 (Figure 4.13A, B, curves *a*) was partially inactivated (curves *b*) by the pH increase up to 5.8 generated by the immuno-based logic system. Since the output signal 1 from the logic system resulted in the inactivation of the biofuel cell (operating as the inverter producing 0 output for input 1 and vice versa), the system modeled a NOR logic gate (Figure 4.13B, inset). After the biofuel cell inactivation, the next cycle was started by resetting to the initial pH value thus activating the switchable electrode again. To activate the biofuel cell,

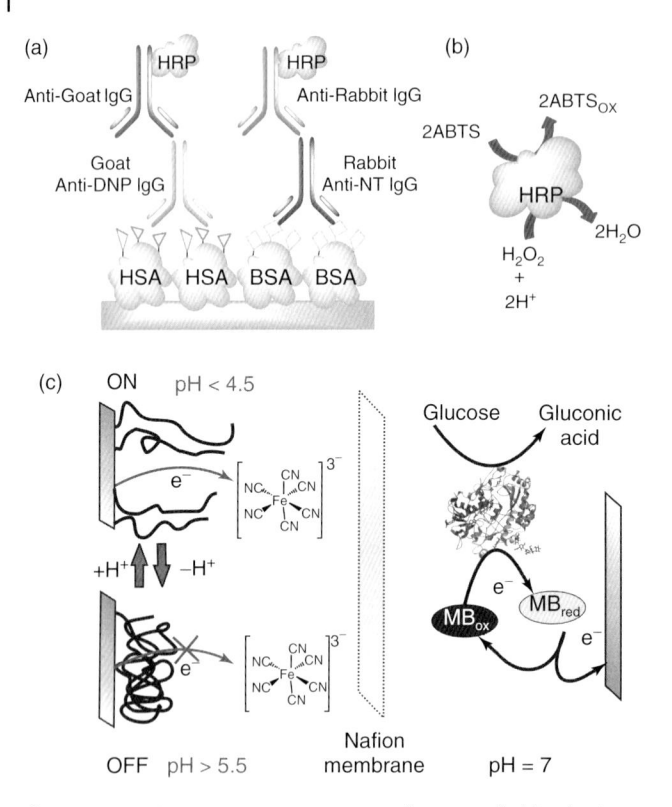

Figure 4.12 (a) Immune system composed of two antigens, two primary antibodies, and two secondary antibodies labeled with horseradish peroxidise (HRP) biocatalytic tag used for the OR logic gate. (b) Biocatalytic reaction producing pH changes to control the biofuel cell performance. (c) Biofuel cell controlled by the immuno-OR-logic gate due to the pH-switchable $[Fe(CN)_6]^{3-}$-reducing cathode. MB_{ox} and MB_{red} are oxidized and reduced states of the mediator methylene blue. (Adapted from [73] with permission; Copyright American Chemical Society, 2009.)

GOx and glucose were added to the cathodic compartment, resulting in the pH decrease to about 4.2 due to the biocatalytic oxidation of glucose and formation of gluconic acid.

4.6
Processing of Injury Biomarkers by Enzyme Logic Systems Associated with Switchable Electrodes

The logic gates and networks controlling states of switchable electrode interfaces were initially developed to demonstrate the concept of coupling between biomolecular computing systems and modified electrodes. However, their practical use is highly feasible in the area of "smart" multisignal processing biosensors and

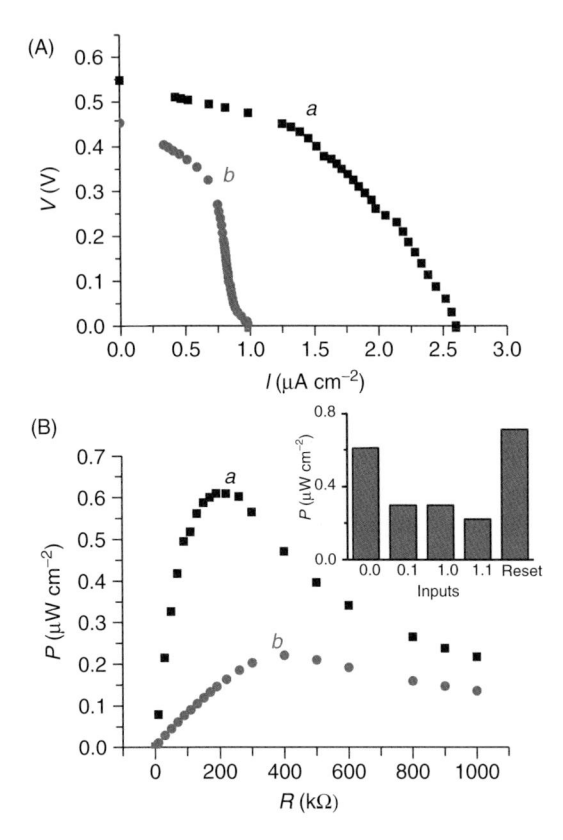

Figure 4.13 (A) Polarization curves of the biofuel cell with the pH-switchable cathode obtained at different pH values generated *in situ* by the immuno-OR-logic gate: (a) pH 4.5 and (b) pH 5.8. (B) Electrical power density generated by the biofuel cell on different load resistances at different pH values generated *in situ* by the immuno-OR-logic gate: (a) pH 4.5 and (b) pH 5.8. Inset: the maximum electrical power density produced by the biofuel cell upon different combinations of the immune input signals. (Adapted from [73] with permission; Copyright American Chemical Society, 2009.)

actuators, particularly for biomedical applications [45, 74]. Recently, logic gates were designed for the analysis of pathophysiological conditions corresponding to different injuries, which are activated by various injury biomarkers [46–50]. The logic analysis of many biomarker signals significantly increases fidelity of the biomedical conclusion, particularly when biomarkers with limited specificity are used in the analysis. It should be noted that the biggest challenge in the realization of these logic systems for biomedical applications is the relatively small difference between the logic 0 and logic 1 levels of the digitized input signals. While for the biocomputing concept demonstration convenient arbitrary concentrations of chemical inputs were used (usually logic 0 was represented by the absence of the corresponding species), in biomedical applications the logic levels of the biochemical input signals are defined on the basis of their medical meaning: logic 0 and logic 1 input levels correspond to normal physiological and elevated pathological concentrations of biomarkers, respectively, which may appear with a small difference.

A system representing the first example of an integrated sensing–actuating chemical device with the Boolean logic for processing natural biomarkers at their physiologically relevant concentrations has been developed [75]. Biomarkers characteristic of liver injury, alanine transaminase, (ALT) and lactate dehydrogenase (LDH) were processed by a biocatalytic system functioning as a logic AND gate (Figure 4.14A). The NAD^+ output signal produced by the system upon its activation

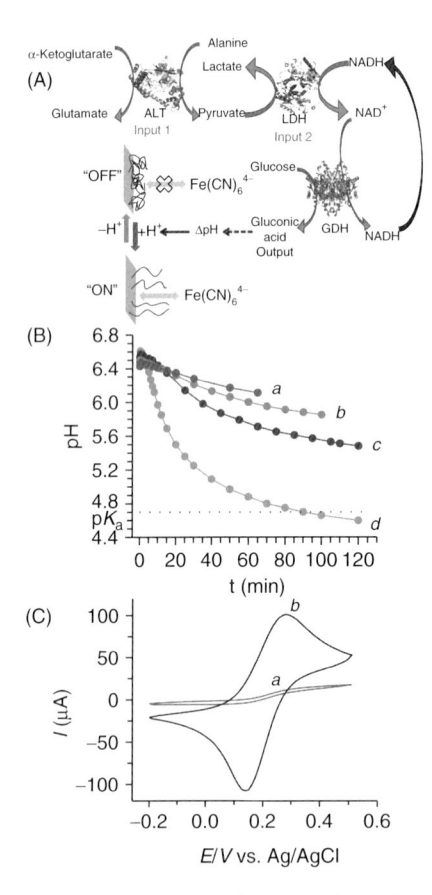

Figure 4.14 (A) Biocatalytic cascade used for the logic processing of the biomarkers characteristic of liver injury, resulting *in situ* pH changes and activation of the electrode interface. (B) pH changes generated *in situ* by the biocatalytic cascade activated with various combinations of the two biomarker input signals ALT and LDH: (a) 0,0; (b) 0,1; (c) 1,0; and (d) 1,1. The dotted line corresponds to the pK_a value of the P4VP-brush. (C) Cyclic voltammograms obtained for the ITO electrode modified with the P4VP-polymer brush in (a) the initial OFF state, pH 6.3 and (b) the ON state enabled by the ALT, LDH input combination 1,1, pH 4.75. (Adapted from [75], with permission; Copyright American Chemical Society, 2011.)

in the presence of both biomarkers was then biocatalytically converted to a pH decrease. The acidic pH value produced by the system as a response to the biomarkers (Figure 4.14B) triggered the restructuring of a polymer-modified electrode interface. It should be noted that the 1,1 signal combination (Figure 4.14B, curve d) was the only input producing enough pH changes to reach the pK_a value of the polymer brush, resulting in switching of the interfacial system from the original neutral hydrophobic state to the positively charged hydrophilic state. This allowed negatively charged soluble redox species to approach the electrode surface, thus switching the electrochemical reaction ON (Figure 4.14C). Small concentration changes of the NADH/NAD$^+$ system were converted into a large current corresponding to the electrochemical process of the [Fe(CN)$_6$]$^{4-}$ redox probe, thus amplifying the output signal generated by the enzyme logic system. Application of enzyme logic systems for processing medically relevant inputs applied at their physiological concentrations is highly important for future "smart" implantable bioelectronic devices controlled by physiological parameters in the body.

4.7
Summary and Outlook

The developed approach paves the way for the realization of novel digital biosensors and bioelectronic devices processing multiple biochemical input signals and producing a combination of output signals dependent on the whole pattern of various input signals. The biochemical signals are processed by chemical means based on the enzyme logic system before their electronic transduction, thereby obviating the need for computer analysis of the biosensing information. We anticipate that biochemical logic gates and networks connected with bioelectronic sensing and actuating devices will find numerous biomedical applications. They will facilitate decision-making in connection with an autonomous feedback-loop drug delivery system and will revolutionize the monitoring and treatment of patients. The designed systems exemplify the novel approach to multisignal processing biosensors mimicking natural biochemical pathways and operating according to the biocomputing concept [45, 74]. Further studies will be needed to transfer this approach from a conceptual demonstration to real-life biosensor applications. This will require much scientific and engineering work to integrate multienzyme systems in a rational design with modified electrodes so that a real practically applicable biosensor becomes possible. Particularly important will be the operation of the switchable bioelectronic interfaces upon local pH changes without affecting the composition of the bulk solution [76]. The broadening of the possible applications of this concept will result in the design of various bioelectronic devices and bioactuators [77] controlled by complex patterns of multiple inputs. Microrobotics and bioimplantable computing systems are among the most likely applications to benefit from advances in biomolecular computing. Future progress in these areas will depend on the development of novel computing concepts and design of new signal-responsive and information processing materials contributing to molecular information technology [78, 79]. An additional challenging aim in the future integration of biomolecular logic systems with signal-responsive materials and electronics will be their scaling down to micro and nano sizes [80].

Acknowledgments

This research was supported by the National Science Foundation (Grants DMR-0706209, CCF-1015983), by ONR (Grant N00014-08-1-1202), and by the Semiconductor Research Corporation (Award 2008-RJ-1839G).

References

1. de Silva, A.P., Uchiyama, S., Vance, T.P., and Wannalerse, B. (2007) *Coord. Chem. Rev.*, **251**, 1623–1632.
2. de Silva, A.P. and Uchiyama, S. (2007) *Nat. Nanotechnol.*, **2**, 399–410.
3. Szacilowski, K. (2008) *Chem. Rev.*, **108**, 3481–3548.
4. Credi, A. (2007) *Angew. Chem. Int. Ed.*, **46**, 5472–5475.
5. Pischel, U. (2007) *Angew. Chem. Int. Ed.*, **46**, 4026–4040.
6. Pischel, U. (2010) *Aust. J. Chem.*, **63**, 148–164.
7. Andreasson, J. and Pischel, U. (2010) *Chem. Soc. Rev.*, **39**, 174–188.
8. Katz, E. and Privman, V. (2010) *Chem. Soc. Rev.*, **39**, 1835–1857.
9. Saghatelian, A., Volcker, N.H., Guckian, K.M., Lin, V.S.Y., and Ghadiri, M.R. (2003) *J. Am. Chem. Soc.*, **125**, 346–347.
10. Ashkenasy, G. and Ghadiri, M.R. (2004) *J. Am. Chem. Soc.*, **126**, 11140–11141.
11. Calude, C.S., Costa, J.F., Dershowitz, N., Freire, E., and Rozenberg, G. (eds) (2009) *Unconventional Computation. Lecture Notes in Computer Science*, vol. **5715**, Springer, Berlin.
12. Adamatzky, A., De Lacy Costello, B., Bull, L., Stepney, S., and Teuscher, C. (eds) (2007) *Unconventional Computing*, Luniver Press.
13. Arugula, M.A., Halámek, J., Katz, E., Melnikov, D., Pita, M., Privman, V., and Strack, G. (2009) *IEEE Comp. Soc. Publ.* (Los Alamitos, CA), 1–7.
14. Van Noort, D. (2005) *DNA Computing, Lecture Notes in Computer Science*, Carbone, A. and Pierce, N.A. (eds) Springer, vol. **3384**, pp. 365–374.
15. Reif, J.H. and Sahu, S. (2009) *Theor. Comput. Sci.*, **410**, 1428–1439.
16. de Silva, A.P., Gunaratne, H.Q.N., and McCoy, C.P. (1993) *Nature*, **364**, 42–44.
17. de Silva, A.P., Gunaratne, H.Q.N., and McCoy, C.P. (1997) *J. Am. Chem. Soc.*, **119**, 7891–7892.
18. de Silva, A.P., Gunaratne, H.Q.N., and Maguire, G.E.M. (1994) *J. Chem. Soc., Chem. Commun.*, 1213–1214.
19. de Silva, A.P. and McClenaghan, N.D. (2002) *Chem. Eur. J.*, **8**, 4935–4945.
20. de Silva, A.P., Dixon, I.M., Gunaratne, H.Q.N., Gunnlaugsson, T., Maxwell, P.R.S., and Rice, T.E. (1999) *J. Am. Chem. Soc.*, **121**, 1393–1394.
21. Straight, S.D., Liddell, P.A., Terazono, Y., Moore, T.A., Moore, A.L., and Gust, D. (2007) *Adv. Funct. Mater.*, **17**, 777–785.
22. Turfan, B. and Akkaya, E.U. (2002) *Org. Lett.*, **4**, 2857–2859.
23. Wang, Z., Zheng, G., and Lu, P. (2005) *Org. Lett.*, **7**, 3669–3672.
24. Baytekin, H.T. and Akkaya, E.U. (2000) *Org. Lett.*, **2**, 1725–1727.
25. Zong, G., Xiana, L., and Lua, G. (2007) *Tetrahedron Lett.*, **48**, 3891–3894.
26. Gunnlaugsson, T., Mac Dónaill, D.A., and Parker, D. (2001) *J. Am. Chem. Soc.*, **123**, 12866–12876.
27. Gunnlaugsson, T., MacDónaill, D.A., and Parker, D. (2000) *Chem. Commun.*, 93–94.

28. de Sousa, M., de Castro, B., Abad, S., Miranda, M.A., and Pischel, U. (2006) *Chem. Commun.*, 2051–2053.
29. Li, L., Yu, M.-X., Li, F.Y., Yi, T., and Huang, C.H. (2007) *Colloids Surf., A*, **304**, 49–53.
30. Luxami, V. and Kumar, S. (2008) *New J. Chem.*, **32**, 2074–2079.
31. Qian, J.H., Qian, X.H., Xu, Y.F., and Zhang, S.Y. (2008) *Chem. Commun.*, 4141–4143.
32. Sivan, S. and Lotan, N. (1999) *Biotechnol. Prog.*, **15**, 964–970.
33. Sivan, S., Tuchman, S., and Lotan, N. (2003) *Biosystems*, **70**, 21–33.
34. Deonarine, A.S., Clark, S.M., and Konermann, L. (2003) *Future Generat. Comput. Syst.*, **19**, 87–97.
35. Ashkenazi, G., Ripoll, D.R., Lotan, N., and Scheraga, H.A. (1997) *Biosens. Bioelectron.*, **12**, 85–95.
36. Unger, R. and Moult, J. (2006) *Proteins*, **63**, 53–64.
37. Win, M.N. and Smolke, C.D. (2008) *Science*, **322**, 456–460.
38. Ogawa, A. and Maeda, M. (2009) *Chem. Commun.*, 4666–4668.
39. Stojanovic, M.N., Mitchell, T.E., and Stefanovic, D. (2002) *J. Am. Chem. Soc.*, **124**, 3555–3561.
40. Pita, M. and Katz, E. (2008) *J. Am. Chem. Soc.*, **130**, 36–37.
41. Baron, R., Lioubashevski, O., Katz, E., Niazov, T., and Willner, I. (2006) *J. Phys. Chem. A*, **110**, 8548–8553.
42. Wagner, N. and Ashkenasy, G. (2009) *Chem. Eur. J.*, **15**, 1765–1775.
43. Stojanovic, M.N. and Stefanovic, D. (2003) *Nat. Biotechnol.*, **21**, 1069–1074.
44. Kahan, M., Gil, B., Adar, R., and Shapiro, E. (2008) *Phisica D*, **237**, 1165–1172.
45. Wang, J. and Katz, E. (2010) *Anal. Bioanal. Chem.*, **398**, 1591–1603.
46. Manesh, K.M., Halámek, J., Pita, M., Zhou, J., Tam, T.K., Santhosh, P., Chuang, M.-C., Windmiller, J.R., Abidin, D., Katz, E., and Wang, J. (2009) *Biosens. Bioelectron.*, **24**, 3569–3574.
47. Pita, M., Zhou, J., Manesh, K.M., Halámek, J., Katz, E., and Wang, J. (2009) *Sens. Actuators, B*, **139**, 631–636.
48. Melnikov, D., Strack, G., Zhou, J., Windmiller, J.R., Halámek, J., Bocharova, V., Chuang, M.-C., Santhosh, P., Privman, V., Wang, J., and Katz, E. (2010) *J. Phys. Chem. B*, **114**, 12166–12174.
49. Halámek, J., Windmiller, J.R., Zhou, J., Chuang, M.-C., Santhosh, P., Strack, G., Arugula, M.A., Chinnapareddy, S., Bocharova, V., Wang, J., and Katz, E. (2010) *Analyst*, **135**, 2249–2259.
50. Windmiller, J.R., Strack, G., Chuan, M.-C., Halámek, J., Santhosh, P., Bocharova, V., Zhou, J., Katz, E., and Wang, J. (2010) *Sens. Actuators, B*, **150**, 285–290.
51. Baron, R., Lioubashevski, O., Katz, E., Niazov, T., and Willner, I. (2006) *Org. Biomol. Chem.*, **4**, 989–991.
52. Baron, R., Lioubashevski, O., Katz, E., Niazov, T., and Willner, I. (2006) *Angew. Chem. Int. Ed.*, **45**, 1572–1576.
53. Pita, M., Minko, S., and Katz, E. (2009) *J. Mater. Sci. Mater. Med.*, **20**, 457–462.
54. Tokarev, I. and Minko, S. (2009) *Soft Matter*, **5**, 511–524.
55. Ahn, S.K., Kasi, R.M., Kim, S.C., Sharma, N., and Zhou, Y.X. (2008) *Soft Matter*, **4**, 1151–1157.
56. Glinel, K., Dejugnat, C., Prevot, M., Scholer, B., Schonhoff, M., and Klitzing, R.V. (2007) *Colloids Surf., A*, **303**, 3–13.
57. Mendes, P.M. (2008) *Chem. Soc. Rev.*, **37**, 2512–2529.
58. Tokarev, I., Gopishetty, V., Zhou, J., Pita, M., Motornov, M., Katz, E., and Minko, S. (2009) *ACS Appl. Mater. Interfaces*, **1**, 532–536.
59. Motornov, M., Zhou, J., Pita, M., Gopishetty, V., Tokarev, I., Katz, E., and Minko, S. (2008) *Nano Lett.*, **8**, 2993–2997.
60. Pita, M., Krämer, M., Zhou, J., Poghossian, A., Schöning, M.J., Fernández, V.M., and Katz, E. (2008) *ACS Nano*, **2**, 2160–2166.
61. Motornov, M., Zhou, J., Pita, M., Tokarev, I., Gopishetty, V., Katz, E., and Minko, S. (2009) *Small*, **5**, 817–820.
62. Krämer, M., Pita, M., Zhou, J., Ornatska, M., Poghossian, A., Schöning, M.J., and Katz, E. (2009) *J. Phys. Chem. B*, **113**, 2573–2579.
63. Zhou, J., Tam, T.K., Pita, M., Ornatska, M., Minko, S., and Katz, E. (2009) *ACS Appl. Mater. Interfaces*, **1**, 144–149.

64. Privman, M., Tam, T.K., Pita, M., and Katz, E. (2009) *J. Am. Chem. Soc.*, **131**, 1314–1321.
65. Tam, T.K., Ornatska, M., Pita, M., Minko, S., and Katz, E. (2008) *J. Phys. Chem. C*, **112**, 8438–8445.
66. Katz, E. and Pita, M. (2009) *Chem. Eur. J.*, **15**, 12554–12564.
67. Amir, L., Tam, T.K., Pita, M., Meijler, M.M., Alfonta, L., and Katz, E. (2009) *J. Am. Chem. Soc.*, **131**, 826–832.
68. Tam, T.K., Pita, M., Ornatska, M., and Katz, E. (2009) *Bioelectrochemistry*, **76**, 4–9.
69. Melnikov, D., Strack, G., Pita, M., Privman, V., and Katz, E. (2009) *J. Phys. Chem. B*, **113**, 10472–10479.
70. Privman, V., Arugula, M.A., Halámek, J., Pita, M., and Katz, E. (2009) *J. Phys. Chem. B*, **113**, 5301–5310.
71. Privman, V., Strack, G., Solenov, D., Pita, M., and Katz, E. (2008) *J. Phys. Chem. B*, **112**, 11777–11784.
72. Strack, G., Chinnapareddy, S., Volkov, D., Halámek, J., Pita, M., Sokolov, I., and Katz, E. (2009) *J. Phys. Chem. B*, **113**, 12154–12159.
73. Tam, T.K., Strack, G., Pita, M., and Katz, E. (2009) *J. Am. Chem. Soc.*, **131**, 11670–11671.
74. Wang, J. and Katz, E. (2011) *Isr. J. Chem.*, **51**, 141–150.
75. Privman, M., Tam, T.K., Bocharova, V., Halámek, J., Wang, J., and Katz, E. (2011) *ACS Appl. Mater. Interfaces*, **3**, 1620–1623.
76. Bocharova, V., Tam, T.K., Halámek, J., Pita, M., and Katz, E. (2010) *Chem. Commun.*, **46**, 2088–2090.
77. Strack, G., Bocharova, V., Arugula, M.A., Pita, M., Halámek, J., and Katz, E. (2010) *J. Phys. Chem. Lett.*, **1**, 839–843.
78. Zauner, K.P. (2005) *Crit. Rev. Solid State Mater. Sci.*, **30**, 33–69.
79. Fu, L., Cao, L.C., Liu, Y.Q., and Zhu, D.B. (2004) *Adv. Colloid Interface Sci.*, **111**, 133–157.
80. Bychkova, V., Shvarev, A., Zhou, J., Pita, M., and Katz, E. (2010) *Chem. Commun.*, **46**, 94–96.

5
Enzyme Logic Digital Biosensors for Biomedical Applications
Evgeny Katz and Joseph Wang

5.1
Introduction

Despite the great success and rapid development of chemical [1–7] and biochemical [8–13] computing systems, most of them represent only the proof of the concept demonstrating the possibility of performing computing/logic operations with the use of molecular systems. They are not ready yet for any practical application. Indeed, these unconventional chemical computing systems are hardly organized in small circuitries, and are capable of solving only basic arithmetic/logic operations on the timescale of minutes or even hours. On the other hand, application of molecular logic systems for analytical purposes could yield a novel class of sensors that are able to accept many input signals and produce binary outputs in the form of "YES"/"NO," which are particularly important for biomedical applications. This approach has been already successfully applied to analyze protein libraries associated with multiple sclerosis [14]. Logically processed feedback between drug delivery application and physiological conditions can significantly improve drug targeting and efficiency [15]. The well-developed field of DNA biocomputing [16] spins out from solving combinatorial problems [17] to analyzing biomedical multiparameter physiological conditions [18]. Programmable and autonomous DNA computing systems operating *in vitro* have demonstrated logic multiparameter analysis of disease-related biomolecular markers, and can be applied in future for *in situ* medical diagnosis and cure [18, 19]. For example, biosensor systems for detecting genetic modifications in avian influenza [20] were developed on the basis of the DNA computing principles, in which various oligonucleotide signals were logically processed by a DNA logic network. Coupling enzyme logic systems with controlled self-assembly of nanoparticles allowed logic AND/OR responses to cancer markers matrix-metalloproteinases: MMP2 and MMP7) [21].The logically controlled aggregation of the superparamagnetic Fe_3O_4 nanoparticles was detected by magnetic resonance imaging (MRI), thus promising easy adaptation of the method to future *in vivo* medical applications. The results of the logically processed biomolecular signals can be stored in enzyme-controlled set–reset flip-flop memory units [22]. The terminal memory units can be connected to various biocatalytic pathways processing multiple biochemical signals. Programmable memory units based on two different protein kinases as information input signals allowed AND, OR, and NOR logic operations before the information storage, thus providing information processing diversity [23].

It should be noted that most of the presently developed biocomputing systems are commonly activated by signals that are not relevant to biomedical applications [24]. Even when the signals have biomedical significance, they are usually used at nonphysiologically relevant concentrations [20, 21]. This limits the immediate use of the developed biocomputing systems for biomedical applications. Extensive research, combining efforts in biochemistry, computer science, medicine, and engineering, is necessary to develop real biocomputing systems for biomedical applications.

This chapter overviews recent advances in biomedical applications of enzyme-based logic systems, particularly emphasizing the attractive use of real physiological concentrations of biomarker input signals for providing reliable assessment of common battlefield injuries.

5.2
Enzyme-Based Logic Systems for Identification of Injury Conditions

Common biosensing devices are based on a single input (analyte), while multisignal logic gate devices rely on multiple target analytes (inputs) to yield YES/NO responses. Such digital biosensors are expected to offer high-fidelity biosensing compared to single-input traditional biosensors. By using various biomarkers as inputs for the enzyme gates and automatically processing physiological information, such biochemical logic systems can provide rapid and reliable assessment of the overall physiological condition and would initiate optimal, timely therapeutic intervention. Such signal processing systems based on biocatalytic reactions mimic computing networks with the input/output signals represented by biochemical means.

Recently developed enzyme logic systems [8] were applied for the development of novel digital multianalyte biosensors [25], which were particularly used for the analysis of different physiological situations associated with various injuries [26–28]. Different types of injury result in distinct pathophysiological changes reflected by changes in the concentrations of many biochemical substances in a body. Some of these biomolecules undergo substantial concentration changes during a given injury and could be selected as biological signaling markers useful for biochemical processing, operating as input signals for enzyme logic gates/circuitries. For example, a biocomputing system composed of a combination of AND–IDENTITY logic gates based on the concerted operation of three enzymes, namely, lactate oxidase (LOx), horseradish peroxidase (HRP), and glucose dehydrogenase (GDH), was designed to process biochemical information related to pathophysiological conditions originating from traumatic brain injury (TBI) and hemorrhagic shock (HS) (Figure 5.1) [27]. Three biochemical substances, namely, glucose, lactate,

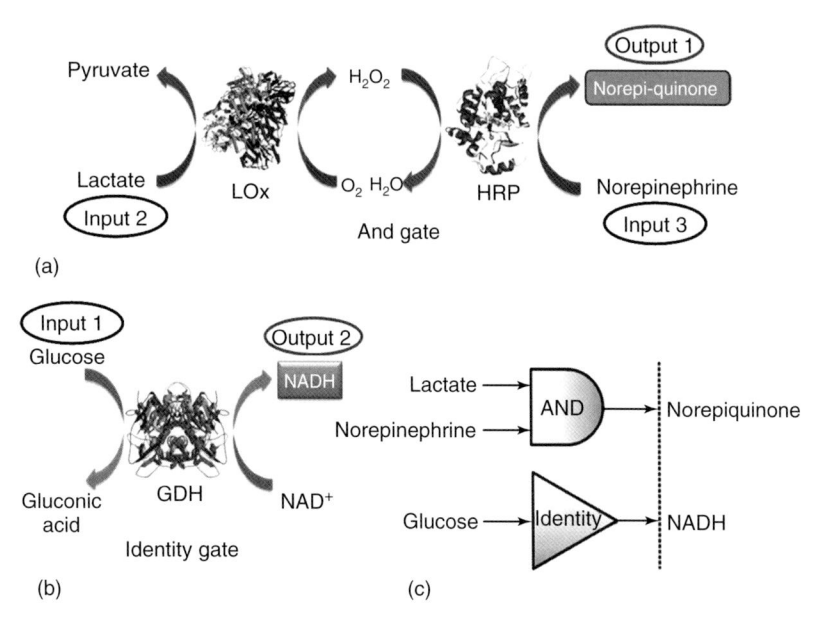

Figure 5.1 Biocomputing system for analysis of traumatic brain injury (TBI) and hemorrhagic shock (HS): biochemical reactions catalyzed by (a) LOx and HRP and (b) GDH being used to perform AND/IDENTITY logic operation and (c) the equivalent logic system used for processing the lactate, norepinephrine, and glucose input signals. (Adapted from [26] with permission.)

and norepinephrine (NE) were selected as physiological biomarkers signaling the injuries and were applied for demonstrating the concepts of biochemical signal processing and autonomous decision making. Specific concentration patterns of these biomarkers can provide sufficient information to identify the type of injury occurring in a body. Abnormal increase in the concentration of glucose might originate from HS [29, 30], while higher than normal physiological lactate concentration could be caused also by HS or/and TBI [30, 31]. A high concentration of NE can be indicative of any traumatic injury [31, 32]. Thus, glucose, lactate, and NE were applied as chemical input signals for the enzyme logic circuitry. The digitized signals were considered as logic 0 when the inputs were applied at their physiologically normal concentrations: 4 mM glucose, 2 mM lactate, and 2.2 nM NE. Abnormally high concentrations of glucose (30 mM) characteristic of HS [29, 30], of lactate (13 mM) observed in case of both TBI and HS [30, 31], and of NE (3.5 μM) typical for any traumatic injury [31, 32] were considered as logic 1 values for the input signals. Biochemical processing of different patterns of the biomarkers resulted in the formation of norepi-quinone (NQ) and NADH defined as the output signals (Figures 5.2 and 5.3), respectively. Optical and electrochemical means were used to follow the formation of the output signals for eight different combinations of three input signals. The enzymatically processed biochemical information, presented in the form of a logic truth

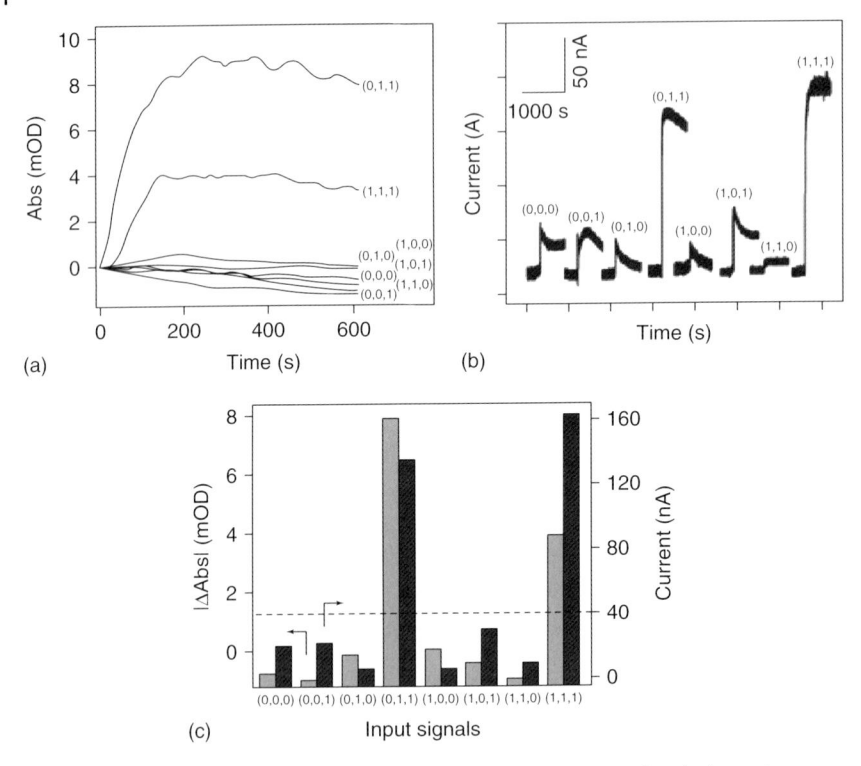

Figure 5.2 Time-dependent signals corresponding to the formation of NQ generated by the combined AND-IDENTITY logic system upon application of various combinations of the input signals (glucose, lactate, and NE) measured by (a) optical means and (b) amperometrically. (c) Bar diagram featuring the combined AND-IDENTITY logic operation of the optical and electrochemical systems. Absorbance measurement were performed at $\lambda = 465$ nm. Electrochemical measurements were performed at -0.25 V. Dashed line shows the threshold values separating digital 0 and 1 output signals produced by both systems. (Adapted from [26] with permission.)

table, allowed distinguishing the difference between normal physiological conditions, pathophysiological conditions corresponding to TBI and HS, as well as abnormal situations not corresponding to injury. Specifically, the input signals (glucose–lactate–NE) appearing with the digital pattern 0,1,1 resulted in the output signal combination (NQ–NADH) 1,0, which corresponds to TBI symptoms, while the input pattern 1,1,1 yielded the output combination 1,1, meaning HS. It is interesting to note that the characteristic output patterns can be easily represented by decimal numbers corresponding to the binary output code. For example, TBI will be represented by 2 (binary 10), while HS will have 3 (binary 11). This way of presenting the analytical results will be particularly convenient when the number of the used input/output signals is large, as described in the next section.

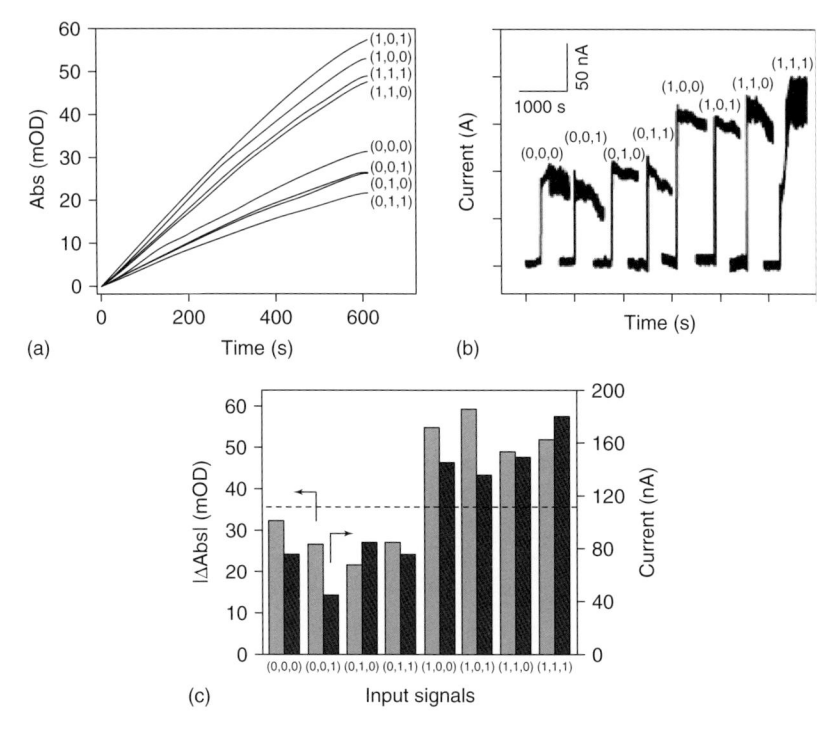

Figure 5.3 Time-dependent signals corresponding to the formation of NADH generated by the combined AND-IDENTITY logic system upon application of various combinations of the input signals (glucose, lactate, and NE) measured by (a) optical means and (b) amperometrically. (c) Bar diagram featuring the combined AND-IDENTITY logic operation of the optical and electrochemical systems. Absorbance measurement were performed at $\lambda = 340$ nm. Electrochemical measurements were performed at $+0.75$ V. Dashed line shows the threshold values separating digital 0 and 1 output signals produced by both systems. (Adapted from [26] with permission.)

5.3
Multiplexing of Injury Codes for the Parallel Operation of Enzyme Logic Gates

On further development of the biocomputing approach for the analysis of various injuries, we introduced a new "injury coding" diagnostic method [33] that could multiplex multiple injuries and assign to each pathophysiological state a distinct "injury code," thereby enabling highly parallelized operation in the digital domain while minimizing the complexity of the analog electronic integration required for multiple-potentiostat electrochemical devices. Owing to the Boolean nature of the enzyme logic concept [8], all normal physiological states on implementation of the AND logic operation can be ascribed a logical "0" value in the biochemical domain, which accounts for most of the injury combinations, before the transduction of the signals to the electrical domain. Only pathological conditions causing a change in the outputs relative to a predefined threshold level would result in a logical "1" value, thereby alleviating the complex decision routines that must be performed

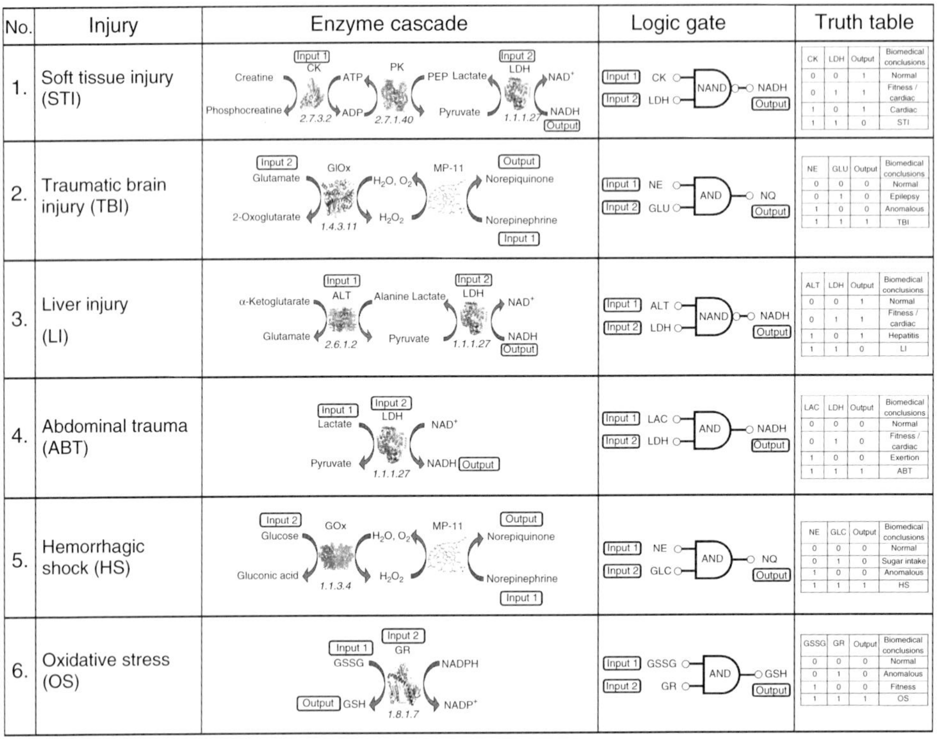

Figure 5.4 Enzyme cascades, equivalent logic gates, and truth tables corresponding to six unique injuries: STI, TBI, LI, ABT, HS, and OS. The abbreviations are detailed in the Appendix. (Adapted from [33] with permission.)

in the electronic domain. It should be noted that the inverted logic values "1" and "0" will be applied for normal and pathological conditions, respectively, on application of the NAND (Not-AND) logic operation. As such, n outputs can be multiplexed into an n-bit word or "injury code" for a comprehensive assessment of health conditions. A unique "injury code" can thus be ascribed to a specific pathophysiological state in accordance with a truth table. A simple lookup table (Figure 5.4) in digital logic circuitry could thus be employed to determine which injuries, if any, have been sustained in accordance with this distinct sequence of bits. In this manner, an array of n individual dual-input enzyme logic gates (each evaluating a separate injury) can assess 2^n possible pathological conditions among 2^{2n} possible physiological states. This new concept represents the first example of the *parallelization* of enzyme logic gates applied to diagnostic use, as well as the simultaneous *multiplexing* of the outputs of multiple logic gates in the *biochemical* domain into a binary injury code.

To illustrate the new concept, and in accordance with the goal of rapid and reliable diagnosis of multiple injury states, an array of two NAND and four AND enzyme logic gates was assembled, as illustrated in Figure 5.4. Six different

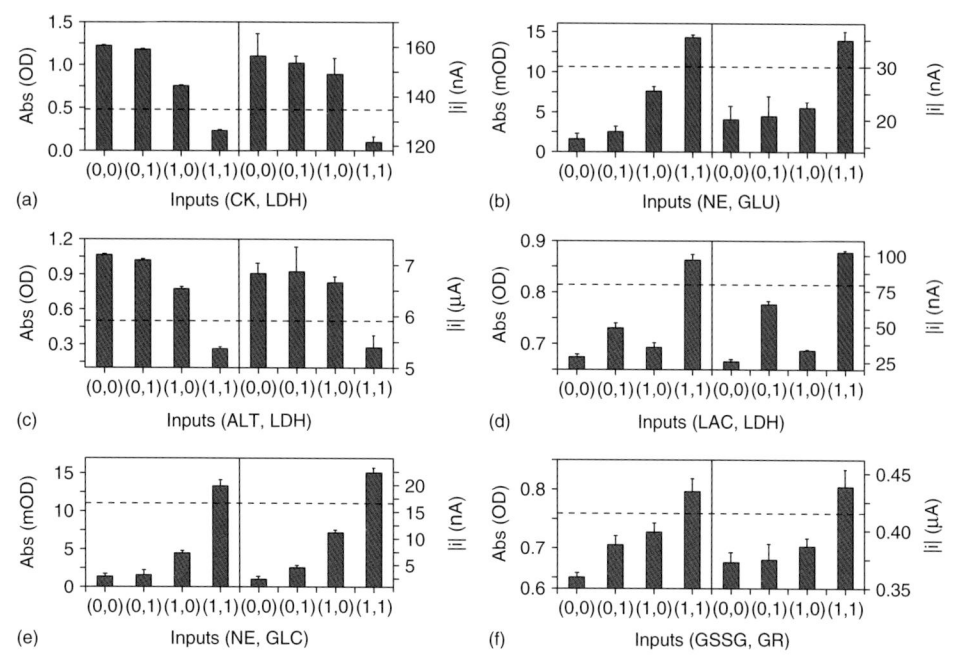

Figure 5.5 Optical (left) and electrochemical (right) bar charts obtained by sampling the output of the (a) STI NAND, (b) TBI AND, (c) LI NAND, (d) ABT AND, (e) HS AND, and (f) OS AND logic gates on various combinations of the input biomarkers (0,0), (0,1), (1,0), and (1,1). Optical absorbance measurements were extracted at $t = 300, 200, 150, 200, 100,$ and 60 s for the STI, TBI, LI, ABT, HS, and OS gates, respectively. Electrochemical chronoamperograms were sampled at $t = 60, 60, 60, 10, 60,$ and 30 s for the STI, TBI, LI, ABT, HS, and OS gates, respectively. Dashed lines indicate the decision threshold for the realization of Boolean logic gate operation. (Adapted from [33] with permission.)

pathological conditions were assessed: soft tissue injury (STI), TBI, liver injury (LI), abdominal trauma (ABT), HS, and oxidative stress (OS), using 12 biomarker inputs. Optical absorbance and amperometric characterization of the six-gate system were conducted (Figure 5.5) in order to verify compliance with a truth table as well as to ensure proper differentiation between the logical "0" and "1" output levels. The outputs of the six logic gates were subsequently multiplexed to yield a distinct 6-bit injury code representing 64 unique pathological conditions among 4096 possible physiological scenarios (Figure 5.6). This led to an additional (comprehensive) level of information on the overall nature of the injury, beyond the assessment of individual injuries performed by the individual gates. The system integration of clinically relevant enzyme logic gates and the subsequent multiplexing of their outputs into a cohesive injury code thus offer considerable promise for the rapid, reliable, and decentralized assessment of multiple injuries and polytrauma conditions. It should be noted that the system is comprised of separate channels individually tailored for specific biomedical needs. The modularity of the system

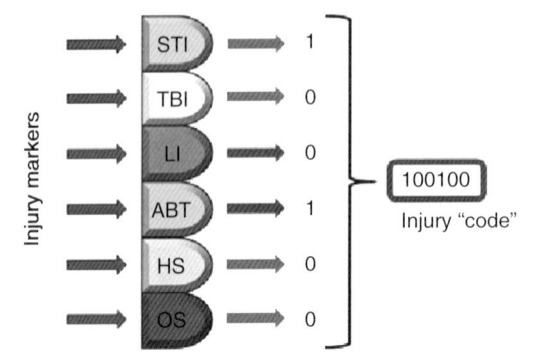

88 | 5 Enzyme Logic Digital Biosensors for Biomedical Applications

Figure 5.6 Scheme showing parallel operation of the logic gates analyzing complex patterns of biomarkers corresponding to different injury conditions. The injury "code" can be derived from the digital outputs generated by the gates (the specific shown example corresponds to the combination of LI and ABT).

allows straightforward reconfiguration of the constituents to enable the device to adapt, expand, and meet new requirements and applications. The potential of the new modular biocomputing coding concept extends beyond the diagnosis of multiple injuries, as the concept could be readily extended for assessing reliably a wide range of other practical real-world scenarios involving multitude changes to offer a distinct "situation code".

The enzyme logic systems designed for the injury biomarker analysis were originally optimized for their performance in buffer solutions containing biocomputing "machinery" and spiked with different concentrations of biomarker inputs [33]. However, the real operation of biosensors based on these logic systems is expected in biological fluids containing many interferants. In order to mimic future *in vivo* operation of biosensors based on biocomputing enzyme cascades, digital biosensor systems analyzing biomarkers characteristic of LI, STI, and ABT were developed and optimized for their performance in serum solutions spiked with injury biomarkers to mimic real medical samples [34].

Numerous compounds present in the serum samples can potentially interfere with the enzymatic "machinery" of the analytical systems described above. For example, pyruvate is an intermediate product in the LI and STI detection systems [33, 34]. In the LI system, pyruvate is generated as an intermediate product by the first system input, namely, the alanine transaminase (ALT) enzyme. In the STI system, it is also produced by the creatine kinase/pyruvate kinase (CK/PK) biocatalytic cascade. In both systems, the pyruvate production corresponds to the presence of one of the biomarker inputs, ALT or CK, respectively. However, pyruvate is present in blood serum (at around 40 μM) [35] and it could provide a false positive signal even in the absence of the first inputs. Similar complication may occur for adenine 5′-diphosphate (ADP) in the STI system. Since lactate is a common blood constituent, with its concentration significantly elevated by most traumatic injuries [30, 31], it can prevent the conversion of pyruvate to lactate and the readout of the output signal corresponding to the decrease of the NADH concentration. In order to suppress the lactate dehydrogenase (LDH) induced biocatalytic conversion of lactate to pyruvate for the proper performance of the STI system, the solution pH value should be optimized. Another potential complication can arise from the presence of various ions in human serum samples. The CK enzyme, which is a key component for the STI system [33, 34], can be inhibited by many bivalent cations (Ca^{2+}, Zn^{2+}, and Cu^{2+}) and anions (Cl^- and PO_4^{3-}) [36]. In order to achieve an adequate performance of the enzymes for the STI gate, Mg^{2+} ions were added to the solution for preventing the CK inhibition [33, 34]. Additionally, K^+ ions were added to the reaction mixture to enhance the PK and CK activities [33, 34].

The operation of all three logic systems presented in this study [34] was examined using different serum samples. Some minor sample-to-sample absorbance variations were observed, mostly because of the difference in the transparency of the serum samples. However, the robust operation of the bioanalytical systems always allowed convenient discrimination of 0 and 1 output signals, thus providing reliable diagnostics of the injury conditions.

5.4
Scaling Up the Complexity of the Biocomputing Systems for Biomedical Applications – Mimicking Biochemical Pathways

Enzyme-based logic networks, upon appropriate optimization [37–40], allow robust, low-noise operation up to 10 concatenated gates [41]. In order to evaluate the performance of complex multicomponent biocatalytic systems for digital biosensing, we assembled a system mimicking the biochemical pathways [42].

The assessment of two common battlefield injuries, namely, STI and TBI, was performed by a multienzyme biocatalytic cascade (Figure 5.7), which can be described as a comprehensive concatenated logic system (Figure 5.8). The logic network represented in Figure 5.8 is actually identical to the biocatalytic cascade depicted in Figure 5.7, but it offers another more convenient way of its representation that is more useful for the discussion of the system operation. The system architecture includes eight networked logic gates of the AND type. The biocatalytic system operated in two different modes: one for the analysis of TBI and the other for the analysis of STI with the possibility to switch between them. The system was designed to analyze five different biomarkers appearing in different combinations and to lead to a logical conclusion about the presence or absence of TBI or STI conditions. Three biomarkers, CK, LDH, and lactate (Lac), corresponded to the physiological conditions characteristic of STI, while two other biomarkers, enolase (EN) and glutamate (Glu), were reporting on the TBI diagnosis. All five biomarkers were applied as biochemical input signals activating the logic network at two different concentration levels: logic 0 corresponded to the normal physiological concentrations of the biomarkers, while logic 1 was selected at the

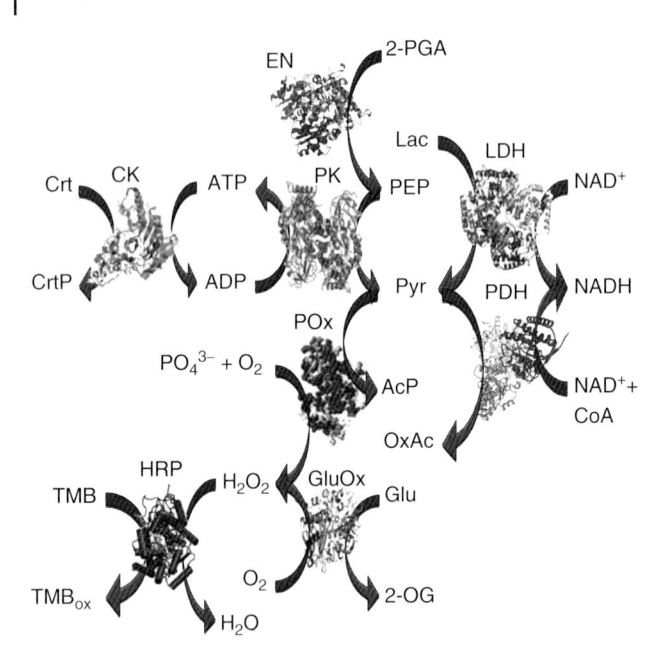

Figure 5.7 Multienzyme biocatalytic cascade for the analysis of STI and TBI. Biomarker inputs were used for STI (CK, Lac, and LDH) and for TBI (EN and Glu). Output signals for STI and for TBI are NADH and TMB$_{ox}$, respectively. Note that for simplicity the scheme does not include some reacting cofactors, promoters, and by-products. The abbreviations are detailed in the Appendix. (Adapted from [42] with permission.)

elevated pathophysiological concentrations corresponding to the respective injuries. Simultaneous processing of many biomarkers through the complex biocatalytic cascade required optimization of the biocatalytic reactions by tuning the reaction rates in order to have comparable output signals for various combinations of the biomarker inputs. The optimization was achieved by careful selection of the auxiliary inputs: adenine 5′-triphosphate (ATP), 2-phosphoglyceric acid (2-PGA), glutamate oxidase (GluOx), and 3,3′,5,5′-tetramethylbenzidine (TMB). In order to digitize the logic network operation, the auxiliary inputs were applied at two levels: logic 0 corresponded to the physical zero concentration, while logic 1 was selected experimentally on optimization of the system. The auxiliary inputs did not provide any information about the physiological conditions related to the injuries, but they were needed for optimal processing of the biomarker inputs. When they were applied at "0" levels, the system was mute and insensitive to the biomarker inputs, while their application at "1" levels provided optimized processing of the biomarker inputs applied in different combinations of "0" and "1" logic levels. Four additional inputs, namely, phosphoenolpyruvic acid (PEP), ADP, phosphate, and coenzyme A (CoA), were used to switch the system operation between the TBI and STI modes. These inputs were applied at logic 0 levels corresponding to the physical zero concentrations and logic 1 levels being experimentally optimized.

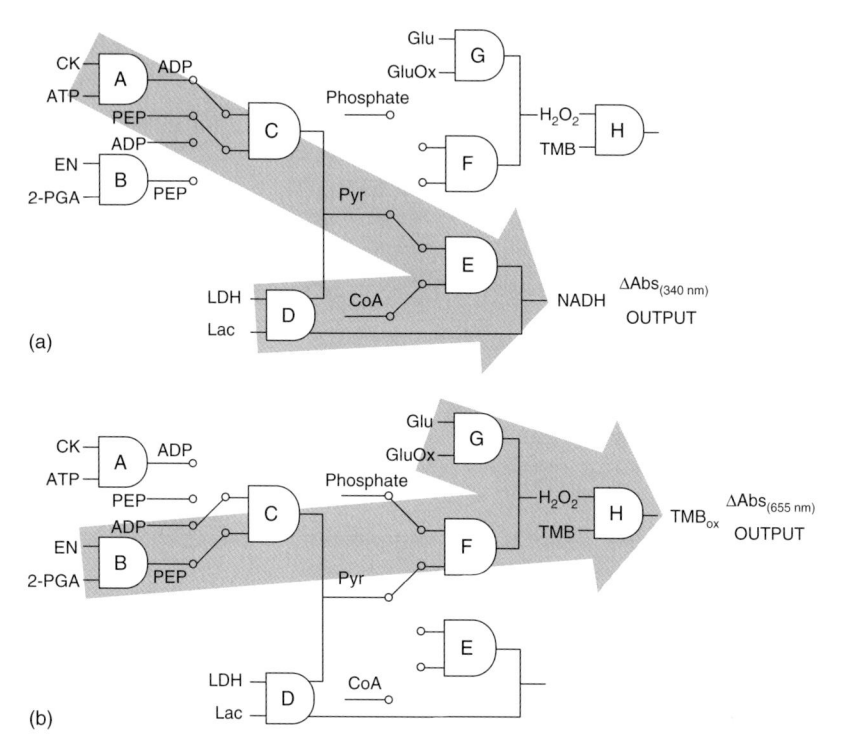

Figure 5.8 Equivalent logic schemes for the concatenated logic gates analyzing STI and TBI corresponding to the biocatalytic cascade shown in Figure 5.7. (a) The system switched to the STI analysis mode. (b) The system switched to the TBI analysis mode. The pathways for the STI and TBI operational modes are indicated by arrows. All abbreviations for the chemicals used in the system are explained in the Appendix. (Adapted from [42] with permission.)

Since none of the used biomarkers is specific enough for the STI diagnosis, only the simultaneous presence of all three STI-related biomarkers (CK, LDH, and Lac) at logic 1 values would provide reliable conclusion about the STI conditions. The system was operating in the following way (see Figure 5.7 for the biochemical representation and Figure 5.8a for the logic equivalent circuitry – the STI pathway is highlighted): the biocatalytic reaction of CK (STI biomarker) and ATP (gate A) resulted in the production of ADP. Further reaction of ADP with PEP biocatalyzed by PK (gate C) resulted in the formation of pyruvic acid (Pyr), which then reacted in the presence of CoA and pyruvate dehydrogenase (PDH) (gate E) to yield the reduced NADH considered as the output signal. Note that the pathway composed of A-C-E gates was activated only when the biomarker input CK, auxiliary input ATP, and switching inputs PEP and CoA appeared at logic 1 values. Simultaneous application of the LDH and Lac inputs (STI biomarkers) at logic 1 (gate D) resulted in the reduction of NAD^+ and further increase of the NADH output. It should be noted that this pathway resulted in the concomitant production of Pyr which was passing through gate E producing one more equivalent of NADH, thus further amplifying

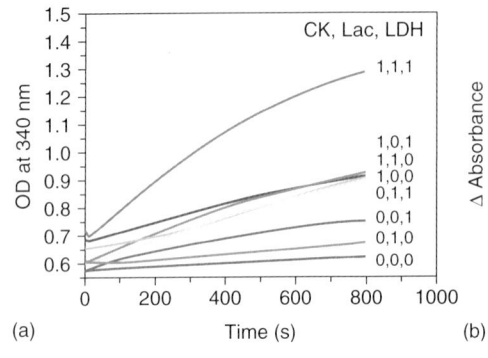

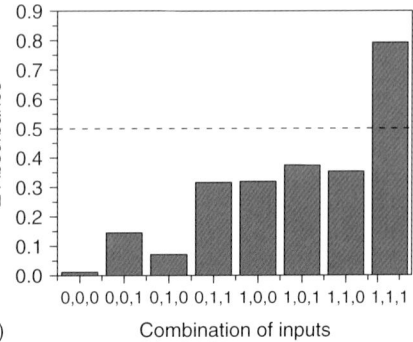

Figure 5.9 (a) Optical detection of the output signal (NADH) generated by the logic system operating for the STI analysis obtained for different combinations of the injury biomarker input signals (CK, Lac, and LDH). (b) Bar chart for the output signals generated by the enzyme logic system for the analysis of STI at 800 s. The STI diagnosis corresponds to the output signal above the decision threshold (dashed line). The auxiliary (ATP) and switching (PEP and CoA) inputs were at logic 1 for all combinations of the biomarker inputs. (Adapted from [42] with permission.)

the output signal. Figure 5.9a shows the optical changes in the system measured at $\lambda = 340$ nm (NADH absorbance) for different combinations of the biomarker inputs. Only the simultaneous application of all three biomarker inputs at logic 1 values (input combination 1,1,1) resulted in high optical absorbance changes allowing an unambiguous conclusion about the STI condition. The experimentally derived threshold level of 0.5 OD (optical density) allowed perfect separation of the logic 0 and 1 levels for the output signal, being different at least by twofold (Figure 5.9b). It should be noted that, in all the measurements shown in Figure 5.9, the auxiliary (ATP) and switching (PEP and CoA) inputs were applied at logic 1 values to allow optimal performance of the analytical pathway. At the same time, the switching inputs ADP and phosphate were applied at logic 0 levels to inhibit the alternative TBI pathway.

In contrast to STI, the TBI biomarkers (EN and Glu) are rather specific and can report on the presence of injury even when appearing alone. The system was operating in the following way (see Figure 5.7 for the traditional biochemical representation and Figure 5.8b for the logic equivalent circuitry – the TBI pathway is highlighted): the biocatalytic reaction of EN (TBI biomarker) and 2-PGA (gate B) resulted in the formation of PEP. The next step included the reaction of PEP with ADP (switching input) biocatalyzed by PK (gate C) and resulted in the production of Pyr. Further reaction of Pyr with phosphate (switching input) biocatalyzed by pyruvate oxidase (POx) (gate F) yielded H_2O_2 which reacted with TMB in the presence of HRP (gate H). This reaction resulted in the oxidation of TMB and produced the absorbance increase at $\lambda = 655$ nm characteristic of TMB_{ox}, being considered as the final output signal from the pathway composed of B-C-F-H gates. The same signal was generated through another pathway composed of G and H gates: The biocatalytic reaction of Glu (TBI biomarker) and GluOx resulted

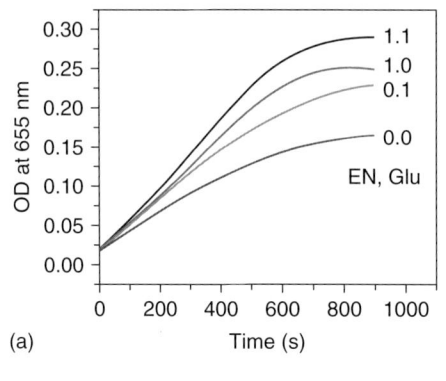

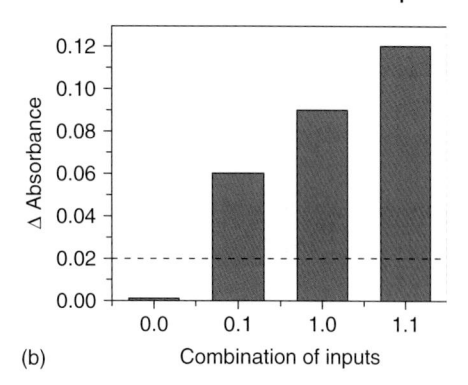

Figure 5.10 (a) Optical detection of the output signal (TMB_{ox}) generated by the logic system operating for the TBI analysis obtained for different combinations of the biomarker input signals (EN, Glu). (b) Bar chart for the output signals generated by the enzyme logic system for the analysis of TBI at 800 s. The TBI diagnosis corresponds to the output signals above the decision threshold (dashed line). The auxiliary (2-PGA, GluOx, and TMB) and switching (ADP and phosphate) inputs were at logic 1 for all combinations of the biomarker inputs. (Adapted from [42] with permission.)

in O_2 reduction and yielded H_2O_2 (G gate). Then H_2O_2 reacted with TMB in the presence of HRP to yield TMB_{ox} with the respective absorbance changes (H gate). Careful optimization of the system (involving tuning the concentrations of the auxiliary inputs) allowed comparable output signals produced in the both pathways. Figure 5.10a shows the optical changes in the system measured at $\lambda = 655$ nm (TMB_{ox} absorbance) for different combinations of the biomarker inputs. Any or both biomarkers appearing at logic 1 level resulted in high absorbance changes signaling about TBI conditions. The experimentally derived threshold level of 0.02 OD allowed perfect separation of the logic 0 and 1 levels for the output signal, being significantly different for the input combination 0,0 and all other combinations (0,1; 1,0; and 1,1) (Figure 5.10b). It should be noted that, in all the measurements shown in Figure 5.10, the auxiliary (2-PGA, TMB, and GluOx) and switching (ADP and phosphate) inputs were applied at logic 1 values to allow optimized performance of the analytical pathway. At the same time, the switching inputs PEP and CoA were applied at logic 0 levels to inhibit the alternative STI pathway.

The complex biocomputing–bioanalytic system operates in two different modes upon activation of the appropriate pathways. Operation of the logic system in the STI mode results in production of ADP as a product of the biocatalytic reaction at gate A. Further reaction of ADP in gate C requires the presence of PEP which is not produced by the system when it operates in the STI mode (note that the gate B is mute because of the absence of EN input). Therefore, PEP should be added artificially in order to activate gate C. In contrast, when the system operates in the TBI mode, PEP is produced *in situ* in gate B, whereas ADP, which is also needed for the operation of gate C, is missing (note that gate A is mute because of the absence of CK). Therefore, in this case ADP should be added artificially in order to activate gate C. Finally, for activation of the network in the STI mode, the switching inputs PEP and ADP should be applied at the logic 1 and 0 values, respectively, while for the TBI mode they should be at the opposite 0 and 1 values. This switch allowed the use of gates C and F in two different modes of operation. Similarly, in order to switch between two operational modes, gates F and E should be selectively activated by the correct concentrations of phosphate and CoA. Specifically, phosphate and CoA were applied at logic 0 and 1 levels in the STI mode and at logic 1 and 0 levels in the TBI mode.

The above study demonstrated that even a very sophisticated multienzyme/multistep biocatalytic cascade can provide reliable diagnostic of physiological conditions upon logical analysis of the complex patterns of various biomarkers. The designed system exemplifies the novel approach to multisignal-processing biosensors mimicking natural biochemical pathways and operating according to the biocomputing concept.

5.5
Application of Filter Systems for Improving Digitalization of the Output Signals Generated by Enzyme Logic Systems for Injury Analysis

It should be noted that the well-defined difference between 0 and 1 output signals generated by the enzyme logic systems described above required very careful optimization of the readout time. The problem of the readout timing and the solution of this problem with the help of chemical "filters" [43] are illustrated below for the LI system [33, 34].

The two enzyme inputs ALT and LDH are biomarkers characteristic of LI [44]. It should be noted that each of them is not specific enough to indicate LI, but their simultaneous increase in concentration, from normal to pathophysiological levels, provides an unambiguous evidence of LI [33, 34]. On the basis of the sequence of the biochemical reactions illustrated in Figure 5.11a, the final result – NADH oxidation followed by absorbance change (Figure 5.12a) – should happen only upon the cooperative work of both enzyme biomarkers. However, it should be remembered that logic 0 values in their present definition are not the absence of the enzymes but rather their presence at normal physiological levels. Therefore, the reactions proceed not only at the 1,1 combination of the input signals but also, albeit at slower rates, when the inputs are supplied at the 0,0, 0,1, and 1,0 combinations. When the readout time interval is limited to 50–200 s, the absorbance decrease is significantly larger for the 1,1 input combination (implying output signal 1) as

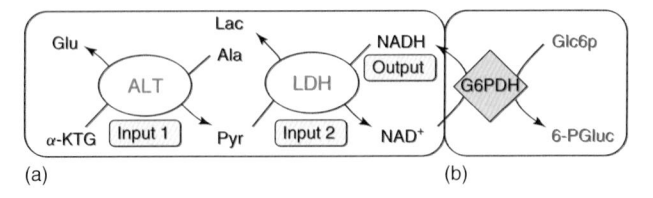

Figure 5.11 Biocatalytic cascade operating as AND logic gate for analysis of LI (a) without and (b) with the added biocatalytic filter reaction. The abbreviations are detailed in the Appendix. (Adapted from [46] with permission.)

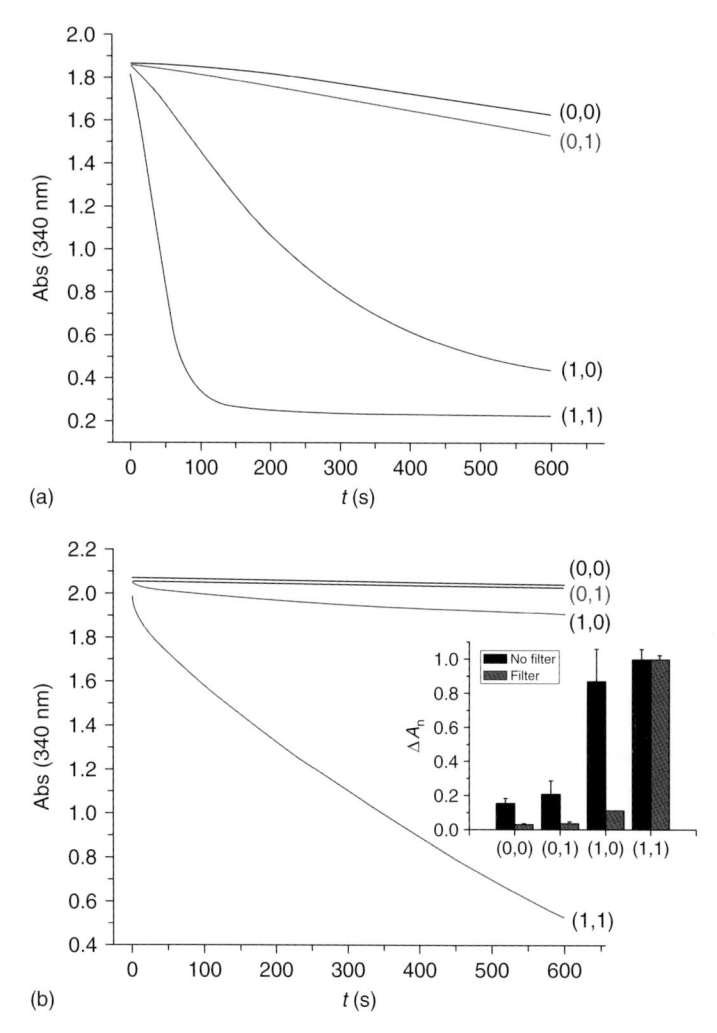

Figure 5.12 Time-dependent absorbance changes generated by the system outlined in Figure 5.11 (a) without and (b) with the added biocatalytic filter reaction upon application of various combinations of the ALT and LDH input signals. The inset shows the output signals measured at 600 s, obtained without and with the filter process, for different combinations of the digitalized inputs. The vertical axis in the inset gives the normalized values of the averaged absorbance decrease, $\Delta A_n = \Delta A / \Delta A_{max}$, ranging between 0 and 1, where A_{max} is the experimental mean value for the logic 1 output (the error bars shown are also normalized). (Adapted from [46] with permission.)

compared to three other input combinations that yield small absorbance changes (output signals 0) (Figure 5.12a). However, when the experiment extends to times longer than 200 s, which is important for actuation applications [45], the absorbance decrease for the 1,0 input combination becomes comparable with that for 1,1 inputs (Figure 5.12a). For even longer times, the results for the 1,0 and 1,1 input combinations will not be distinguishable, and the logic function can no longer be regarded as an AND gate (Figure 5.12b, inset: black bars).

In order to increase the gap separating 0 and 1 output signals, we have added a "filter" process [43] consuming the chemical product NAD^+ and converting it back to NADH for small input concentrations (Figure 5.11a,b) [46]. We note that such processes can likely be implemented for any so-called NAD^+-dependent dehydrogenase [47, 48], for example, GDH activated by physiological amounts of glucose. However, aiming at the ultimate application of our system in physiological environment, we selected glucose-6-phosphate dehydrogenase (G6PDH) activated by glucose-6-phosphate (Glc6P), which does not interfere with glucose naturally existing in blood. The filter system works in the following way: in the presence of G6PDH and Glc6P, the biocatalytically produced NAD^+ is converted back to NADH, thus preventing the absorbance changes, until Glc6P is totally consumed; then NAD^+ starts to accumulate resulting in the absorbance decrease. The delay in the biocatalytic formation of NAD^+ is controlled by the amount of Glc6P added to the system and should be optimized. A comprehensive approach to the filter performance optimization could include detailed analysis of the reactions kinetics [43]. However, a simple experimental optimization might suffice [46].

Application of the G6PDH–Glc6P filter following the biocatalytic cascade activated by ALT–LDH biomarker inputs (Figure 5.11a,b) has allowed a much better separation of the output signals generated by the system for the 1,1 versus all the other combinations of the inputs (Figure 5.12b). However, while improving the binary-signal separation, such filtering can decrease the overall signal strength, which could be an added source of relative noise [49]. Thus, this approach is useful for larger times when the decrease in the absorbance reaches its saturation, which is of relevance in actuation applications [45]. When the output signals were measured at 600 s, the desired system operation corresponding to high-tolerance AND logic realization was obtained in the presence of the filter (Figure 5.12b, inset: gray bars). Perfect separation of the 0 and 1 output signals was found to persist at significantly longer times as well, up to 3 h. The robustness of this analytical system has allowed its practical use in human serum solutions [46]. The "filter" approach is actually much broader than it is exemplified above and will be beneficial for many biocatalytic cascades processing digitized input signals [49].

5.6
Conclusions and Perspectives

The use of biomolecular tools, such as enzymes, allows novel biosensors with integrated biomolecular logic components. The developed systems represent biocomputing logic ensembles applied for the analysis of biomedical conditions related to various injuries. The approach can be easily extended to other biomedical applications: for example, biological dosimeters digitally indicating levels of tissue damage after exposure to radioactive irradiation [50]. One can anticipate that such biochemical logic gates will facilitate timely automated decision making in combination with an integrated therapeutic feedback-loop system and hence will revolutionize the diagnosis and treatment of injured patients. Biocatalytic systems mimicking logic gates can produce pH or redox potential changes which can be used to modify the properties of signal-responsive materials [51] and produce the mechanical actuation essential for such therapeutic intervention [52].

The major challenge in developing this kind of the digital multisignal biosensor system is obtaining a significant difference between the logic 0 and 1 output values. One should remember that the input signals appear at their physiological levels and the input signals with the logic value 0 do not have any more physically zero concentrations [25, 33, 34]. In contrast, the physical concentrations of the input signals corresponding to their logic values of 0 and 1 may have a relatively small difference. In order to obtain significant difference in the output signals, while they are induced by the inputs with a small difference, the response function should be sigmoid rather than linear [43, 49]. In other words, the system should demonstrate a nonlinearity with a sharp transition between the states. The first steps in this direction have been already demonstrated experimentally and analyzed theoretically [43, 53]. However, extensive research in this direction to design chemical "filter" systems similar to the electronic counterparts is needed. The thresholds separating logic 0 and 1 values could be personally tailored for a given patient by following circulating biomarkers in physiological liquids. This will be an important step toward future personalized medicine.

Biochemical computing and logic gate systems based on biomolecules have the potential to revolutionize the field of biosensors, particularly upon moving the signal transduction method from optical [34, 46] to electrochemical [33, 54] techniques. Interfacing biocomputing elements with sensing processes would allow multisignal analysis followed by biochemical processing of the data, giving a final digital ("YES" or "NO") analytical answer. Such "Yes/No" information allows also direct coupling of the signal processing with signal-responsive materials [51, 55–57] and chemical actuators [52] to offer a closed-loop "Sense/Act" operation. Biochemical networks can offer robust, error-free operation upon appropriate optimization of their components and interconnections. Chemical stability of the biomolecular components will be improved upon their immobilization in signal-responsive materials or functional interfaces [58]. Further development of this research area would require the cooperative work of engineers, biochemists, and computer specialists. The ultimate goal of this work will be the design of a microfluidic lab-on-a-chip performing multienzyme-catalyzed cascades and operating similar to an electronic chip by being able to integrate large networks for processing biochemical signals. Special hardware will be required for effective communication of the biomolecular logic systems with biological fluids on one side and electronic transducers on the other.

Minimally invasive microneedle devices for the electrochemical monitoring of biomarkers [59] and potentiostats logically processing digitized output signals [60] will be essential parts of future biosensor systems with a built-in biomolecular logic.

Sensing devices based on biochemical logic systems require a fundamentally new approach to the sensor design and operation, and hence careful attention must be paid to interface biocomputing systems and electronic or optical transducers. As common in conventional biosensors, the success of the enzyme logic biosensor would depend, in part, on the immobilization of the signal-processing reagent layer. The goal is to combine efficiently the individual logic gate elements (to ensure effective coupling of the enzyme cascade) and to address potentially largely different input levels (i.e., potential stoichiometric limitations in the cascade reaction). Such efficient coupling should be accomplished while maintaining high enzymatic stability and retaining the individual reagents. The resulting digital biosensors would benefit diverse and important fields that require immediate intervention or corrective action on the basis of reliable analytical data, ranging from biomedical applications [25, 61] and environmental monitoring to homeland security [62].

Acknowledgment

This research was supported by the National Science Foundation (Grants DMR-0706209, CCF-1015983), by ONR (Grant N00014-08-1-1202), and by the Semiconductor Research Corporation (Award 2008-RJ-1839G).

Appendix

Enzymes and Biocatalysts

ALT	alanine transaminase from porcine heart (E.C. 2.6.1.2)
CK	creatine kinase from rabbit muscle (E.C. 2.7.3.2)
EN	enolase from bacterial yeast (E.C. 4.2.1.11)
G6PDH	glucose-6-phosphate dehydrogenase from *Leuconostoc mesenteroides* (E.C. 1.1.1.49)
GDH	glucose dehydrogenase from *Pseudomonas* sp. (E.C. 1.1.1.47)
GluOx	glutamate oxidase from *Streptomyces* sp. (E.C.1.4.3.11)
GOx	glucose oxidase type X-S from *Aspergillus niger* (E.C. 1.1.3.4)
GR	glutathione reductase from *S. cerevisiae* (E.C. 1.6.4.2)
HRP	peroxidase from horseradish type VI (E.C.1.11.1.7)
LDH	lactate dehydrogenase from porcine heart (E.C. 1.1.1.27)
LOx	lactate oxidase from *Aerococcus viridans* (E.C. 1.1.3.2)
MP-11	microperoxidase-11
PDH	pyruvate dehydrogenase from porcine heart (E.C. 1.2.4.1)
PK	pyruvate kinase from rabbit muscle (E.C. 2.7.1.40)
POx	pyruvate oxidase from *Aerococcus* sp. (E.C. 1.2.3.3)

Other Chemicals (Substrates, Cosubstrates, and Products of Enzyme Reactions)

2-OG	2-oxoglutarate
2-PGA	D(+)2-phosphoglyceric acid
6-PGluc	6-phospho-gluconic acid
AcP	acetyl phosphate
ADP	adenine 5′-diphosphate
Ala	L-alanine
ATP	adenine 5′-triphosphate
CoA	coenzyme A
Crt	creatine
CrtP	creatine-phosphate
Glc6P	D-glucose-6 phosphate
Glu	L-glutamic acid
GSH	glutathione reduced (thiol) form
GSSG	glutathione oxidized (disulfide) form
Lac	L(+)-lactic acid
α-KTG	α-ketoglutaric acid
NAD^+	β-nicotinamide adenine dinucleotide
NADH	reduced form of NAD^+
NE	norepinephrine
NQ	norepiquinone (oxidized form of NE)
OxAc	oxalacetate
PEP	phosphoenolpyruvic acid
Pyr	pyruvic acid
TMB	3,3′,5,5′-tetramethylbenzidine dihydrochloride (redox mediator)
TMB_{ox}	oxidized form of TMB

Injuries

ABT	abdominal trauma
HS	hemorrhagic shock
LI	liver injury
OS	oxidative stress
STI	soft tissue injury
TBI	traumatic brain injury

References

1. de Silva, A.P., Uchiyama, S., Vance, T.P., and Wannalerse, B. (2007) *Coord. Chem. Rev.*, **251**, 1623–1632.
2. de Silva, A.P. and Uchiyama, S. (2007) *Nat. Nanotechnol.*, **2**, 399–410.
3. Szacilowski, K. (2008) *Chem. Rev.*, **108**, 3481–3548.
4. Credi, A. (2007) *Angew. Chem. Int. Ed.*, **46**, 5472–5475.
5. Pischel, U. (2007) *Angew. Chem. Int. Ed.*, **46**, 4026–4040.
6. Pischel, U. (2010) *Aust. J. Chem.*, **63**, 148–164.
7. Andreasson, J. and Pischel, U. (2010) *Chem. Soc. Rev.*, **39**, 174–188.
8. Katz, E. and Privman, V. (2010) *Chem. Soc. Rev.*, **39**, 1835–1857.
9. Saghatelian, A., Volcker, N.H., Guckian, K.M., Lin, V.S.Y., and Ghadiri, M.R. (2003) *J. Am. Chem. Soc.*, **125**, 346–347.
10. Ashkenasy, G. and Ghadiri, M.R. (2004) *J. Am. Chem. Soc.*, **126**, 11140–11141.

11. Win, M.N. and Smolke, C.D. (2008) *Science*, **322**, 456–460.
12. Ogawa, A. and Maeda, M. (2009) *Chem. Commun.*, 4666–4668.
13. Stojanovic, M.N., Mitchell, T.E., and Stefanovic, D. (2002) *J. Am. Chem. Soc.*, **124**, 3555–3561.
14. Margulies, D. and Hamilton, A.D. (2009) *J. Am. Chem. Soc.*, **131**, 9142–9143.
15. Szacilowski, K. (2007) *Biosystems*, **90**, 738–749.
16. Stojanovic, M.N., Stefanovic, D., LaBean, T., and Yan, H. (2005) in *Bioelectronics: from Theory to Applications* (eds I. Willner and E. Katz), Wiley-VCH Verlag GmbH, Weinheim, pp. 427–455.
17. Ezziane, Z. (2006) *Nanotechnology*, **17**, R27–R39.
18. Adar, R., Benenson, Y., Linshiz, G., Rosner, A., Tishby, N., and Shapiro, E. (2004) *Proc. Natl. Acad. U.S.A.*, **101**, 9960–9965.
19. Simmel, F.C. (2007) *Nanomedicine*, **2**, 817–830.
20. May, E.E., Dolan, P.L., Crozier, P.S., Brozik, S., and Manginell, M. (2008) *IEEE Sens. J.*, **8**, 1011–1019.
21. von Maltzahn, G., Harris, T.J., Park, J.H., Min, D.H., Schmidt, A.J., Sailor, M.J., and Bhatia, S.N. (2007) *J. Am. Chem. Soc.*, **129**, 6064–6065.
22. Pita, M., Strack, G., MacVittie, K., Zhou, J., and Katz, E. (2009) *J. Phys. Chem. B*, **113**, 16071–16076.
23. Tomizaki, K. and Mihara, H. (2007) *J. Am. Chem. Soc.*, **129**, 8345–8352.
24. Xia, F., Zuo, X., Yang, R., White, R.J., Xiao, Y., Kang, D., Gong, X., Lubin, A.A., Vallée-Bélisle, A., Yuen, J.D., Hsu, B.Y.B., and Plaxco, K.W. (2010) *J. Am. Chem. Soc.*, **132**, 8557–8559.
25. Wang, J. and Katz, E. (2010) *Anal. Bioanal. Chem.*, **398**, 1591–1603.
26. Pita, M., Zhou, J., Manesh, K.M., Halámek, J., Katz, E., and Wang, J. (2009) *Sens. Actuators, B*, **139**, 631–636.
27. Manesh, K.M., Halámek, J., Pita, M., Zhou, J., Tam, T.K., Santhosh, P., Chuang, M.C., Windmiller, J.R., Abidin, D., Katz, E., and Wang, J. (2009) *Biosens. Bioelectron.*, **24**, 3569–3574.
28. Windmiller, J.R., Strack, G., Chuan, M.C., Halámek, J., Santhosh, P., Bocharova, V., Zhou, J., Katz, E., and Wang, J. (2010) *Sens. Actuators, B*, **150**, 285–290.
29. Kline, J.A., Maiorano, P.C., Schroeder, J.D., Grattan, R.M., Vary, T.C., and Watts, J.A. (1997) *J. Mol. Cell. Cardiol.*, **29**, 2465–2474.
30. Zink, B.J., Schultz, C.H., Wang, X., Mertz, M., Stern, S.A., and Betz, A.L. (1999) *Brain Res.*, **837**, 1–7.
31. Prasad, M.R., Ramaiah, C., McIntosh, T.K., Dempsey, R.J., Hipkeos, S., and Yurek, D. (1994) *J. Neurochem.*, **63**, 1086–1094.
32. Rosenberg, J.C., Lillehei, R.C., Longerbean, J., and Zini-Nierinann, B. (1961) *Ann. Surg.*, **154**, 611–627.
33. Halámek, J., Windmiller, J.R., Zhou, J., Chuang, M.C., Santhosh, P., Strack, G., Arugula, M.A., Chinnapareddy, S., Bocharova, V., Wang, J., and Katz, E. (2010) *Analyst*, **135**, 2249–2259.
34. Zhou, J., Halámek, J., Bocharova, V., Wang, J., and Katz, E. (2011) *Talanta*, **83**, 955–959.
35. Gajovic, N., Binyamin, G., Warsinke, A., Scheller, F.W., and Heller, A. (2000) *Anal. Chem.*, **72**, 2963–2968.
36. Chen, L.H., Borders, C.L. Jr., Vásquez, J.R., and Kenyon, G.L. (1996) *Biochemistry*, **35**, 7895–7902.
37. Melnikov, D., Strack, G., Zhou, J., Windmiller, J.R., Halámek, J., Bocharova, V., Chuang, M.C., Santhosh, P., Privman, V., Wang, J., and Katz, E. (2010) *J. Phys. Chem. B*, **114**, 12166–12174.
38. Privman, V., Pedrosa, V., Melnikov, D., Pita, M., Simonian, A., and Katz, E. (2009) *Biosens. Bioelectron.*, **25**, 695–701.
39. Melnikov, D., Strack, G., Pita, M., Privman, V., and Katz, E. (2009) *J. Phys. Chem. B*, **113**, 10472–10479.
40. Privman, V., Arugula, M.A., Halámek, J., Pita, M., and Katz, E. (2009) *J. Phys. Chem. B*, **113**, 5301–5310.
41. Privman, V., Strack, G., Solenov, D., Pita, M., and Katz, E. (2008) *J. Phys. Chem. B*, **112**, 11777–11784.
42. Halámek, J., Bocharova, V., Chinnapareddy, S., Windmiller, J.R., Strack, G., Chuang, M.C., Zhou, J., Santhosh, P., Ramirez, G.V., Arugula,

M.A., Wang, J., and Katz, E. (2010) *Mol. BioSyst.*, **6**, 2554–2560.
43. Privman, V., Halámek, J., Arugula, M.A., Melnikov, D., Bocharova, V., and Katz, E. (2010) *J. Phys. Chem. B*, **114**, 14103–14109.
44. Tan, K.K., Bang, S.L., Vijayan, A., and Chiu, M.T. (2009) *Injury*, **40**, 978–983.
45. Privman, M., Tam, T.K., Bocharova, V., Halámek, J., Wang, J., and Katz, E. (2011) *ACS Appl. Mater. Interfaces*, **3**, 1620–1623.
46. Halámek, J., Zhou, J., Halámková, L., Bocharova, V., Privman, V., Wang, J., and Katz, E. (2011) *Anal. Chem.*, **83**, 8383-8386.
47. Rosemeyer, M.A. (1987) *Cell Biochem. Funct.*, **5**, 79–95.
48. Beutler, E. (2008) *Blood*, **111**, 16–24.
49. Privman, V. (2011) *Isr. J. Chem.*, **51**, 118–131.
50. Bocharova, V., Halámek, J., Zhou, J., Strack, G., Wang, J., and Katz, E. (2011) *Talanta*, **85**, 800-803.
51. Pita, M., Minko, S., and Katz, E. (2009) *J. Mater. Sci. - Mater. Med.*, **20**, 457–462.
52. Strack, G., Bocharova, V., Arugula, M.A., Pita, M., Halámek, J., and Katz, E. (2010) *J. Phys. Chem. Lett.*, **1**, 839–843.
53. Pita, M., Privman, V., Arugula, M.A., Melnikov, D., Bocharova, V., and Katz, E. (2011) *Phys. Chem. Chem. Phys.*, **13**, 4507–4513.
54. Zhou, N., Windmiller, J.R., Valdés Ramírez, G., Zhou, M., Halámek, J., Katz, E., and Wang, J. (2011) *Anal. Chim. Acta*, **703**, 94–100.
55. Tokarev, I., Gopishetty, V., Zhou, J., Pita, M., Motornov, M., Katz, E., and Minko, S. (2009) *ACS Appl. Mater. Interfaces*, **1**, 532–536.
56. Motornov, M., Zhou, J., Pita, M., Tokarev, I., Gopishetty, V., Katz, E., and Minko, S. (2009) *Small*, **5**, 817–820.
57. Motornov, M., Zhou, J., Pita, M., Gopishetty, V., Tokarev, I., Katz, E., and Minko, S. (2008) *Nano Lett.*, **8**, 2993–2997.
58. Strack, G., Luckarift, H.R., Nichols, R., Cozart, K., Katz, E., and Johnson, G.R. (2011) *Chem. Commun.*, **47**, 7662–7664.
59. Windmiller, J.R., Valdés-Ramírez, G., Zhou, N., Zhou, M., Miller, P.R., Jin, C., Brozik, S.M., Polsky, R., Katz, E., Narayan, R., and Wang, J. (2011) *Electroanalysis*, **23**, 2302-2309.
60. Windmiller, J.R., Santhosh, P., Katz, E., and Wang, J. (2011) *Sens. Actuators, B*, **155**, 206–213.
61. Wang, J. and Katz, E. (2011) *Isr. J. Chem.*, **51**, 141–150.
62. Chuang, M.C., Windmiller, J.R., Santhosh, P., Valdés-Ramírez, G., Katz, E., and Wang, J. (2011) *Chem. Commun.*, **47**, 3087–3089.

6
Information Security Applications Based on Biomolecular Systems

Guinevere Strack, Heather R. Luckarift, Glenn R. Johnson, and Evgeny Katz

6.1
Introduction

Modern computers rely on networked logic gates performing Boolean operations to carry out binary computational functions. By extending the same digital paradigms to molecular [1–4] and biomolecular computing systems [5, 6], chemical processes can mimic electronic counterparts and perform logic operations. For example, Boolean logic gates such as AND, OR, XOR, NOR, NAND, and INHIB, have been demonstrated based on switchable molecular systems [1–4]. In addition, a wide range of biomolecules have been used for information processing [5, 6]. The simplicity of biomolecular processing is attributed to the inherent substrate specificity and versatility of biomolecules; thus the application of biomolecular systems for processing chemical and biochemical information has reached high levels of complexity, despite employing only simple chemical signals.

Compared to the advanced power and complexity of computing performed *in silico*, molecular and biomolecular computing is primitive; however, biomolecular information processing has a niche application in small portable devices that rely on a specific bioelectronic interface. In this area of technological development, portable biomedical sensors [7] with built-in diagnostic capabilities and controlled by simple biomolecular logic operations can be realized [8–11].

Despite significant advances in DNA-based computing [12–14], examples of enzyme-based computing are sparse, but the inherent advantage that enzymes exhibit over DNA is catalytic specificity. To date, enzyme-based logic gates using chemical inputs have been demonstrated for XOR, INHIBIT A, INHIBIT B, AND, OR, NOR, Identity, and Inverter Boolean operators [15, 16]. These single logic gates all operate as independent systems, that is, the output does not act as an input for a subsequent logic operation. In contrast, biocatalytic logic systems with increased complexity can be designed by defining specific reaction cascades [17]. In fact, biocatalytic cascades found in Nature provide insight into engineering such systems. Thus, by combining enzymes with interconnected and compatible reactions, Baron *et al.* [16] have demonstrated half-adder and

Biomolecular Information Processing: From Logic Systems to Smart Sensors and Actuators, First Edition. Edited by Evgeny Katz.
© 2012 Wiley-VCH Verlag GmbH & Co. KGaA. Published 2012 by Wiley-VCH Verlag GmbH & Co. KGaA.

half-subtractor biocatalytic computing systems that perform one-digit addition (or subtraction) of two inputs. Extending the principle to multiple enzyme-based logic gates allows assembly of concatenated strings of programming that can rapidly escalate the computational functionality [7–9, 15–23].

With a "toolbox" of digital paradigms based on molecular and biomolecular systems in hand, a natural exploration toward security systems, cryptography, steganography, and code relay is progressing. An interesting example of molecular cryptography can be found in using programmable DNA to carry strands of encrypted messages [24, 25]. As noted previously, however, few examples of unconventional encryption that use biomolecules other than DNA have been explored. Herein, we focus on various methodologies that demonstrate the breadth of unconventional bimolecular encryption that can be realized using enzymes and antibodies.

6.2
Molecular and Bio-molecular Keypad Locks

One of the simplest modes of information security is a lock, with a defined security mechanism, such as a keypad code. Using conventional electronic circuits, a three-input AND gate can serve as a simple electronic lock (Figure 6.1a). When all three inputs, A, B, and C, are equal to 1, the output produced will also be equal to 1. Thus, the correct three inputs can be applied in any order (Figure 6.1b) to activate the lock. Alternatively, a keypad lock that uses multiple interconnected logic gates may be designed such that the lock is activated by the appropriate input signals – but only when applied in the correct order – by employing a three-input electronic keypad circuit composed of three concatenated two-input AND gates (Figure 6.1c). The three inputs, A, B, and C, can be applied in six possible combinations: ABC, ACB, BAC, BCA, CAB, and CBA; however, only one combination will produce a high output (Figure 6.1d). The first gate is enabled by pin 1 and the addition of a high input (1); the high output of that gate then serves as the activation pin for the second gate, which is enabled by a subsequent input (2). Gate 3 is in turn activated in the same manner by the high output from the preceding gate along with a third input. The output from the third gate is the final output and activates a light-emitting diode (LED) or other electronic system component, which may trigger a switch, alarm, or light. This digital design can inspire molecular and biomolecular security systems operating with unique computational "machinery" while conserving the information processing outcome.

A molecular keypad lock (or a molecular device capable of distinguishing input sequences) was also demonstrated by utilizing an output-specific fluorophore; Fluorescein–LInker–Pyrene (FLIP) [26]. The presence of the two fluorophores, pyrene and fluorescein, enables the FLIP molecule to fluoresce in either blue or green regions, depending on the binding of specific ferric ions. Taking advantage of the complex chemistry of FLIP allowed for the design of one- and two-input logic gates, such as YES, NOT, NOR, OR, AND, XOR, and INHIBIT. To design

6.2 Molecular and Bio-molecular Keypad Locks

(b)

Input A	Input B	Input C	Output
0	0	0	0
0	0	1	0
0	1	0	0
0	1	1	0
1	0	0	0
1	0	1	0
1	1	0	0
1	1	1	1

(d)

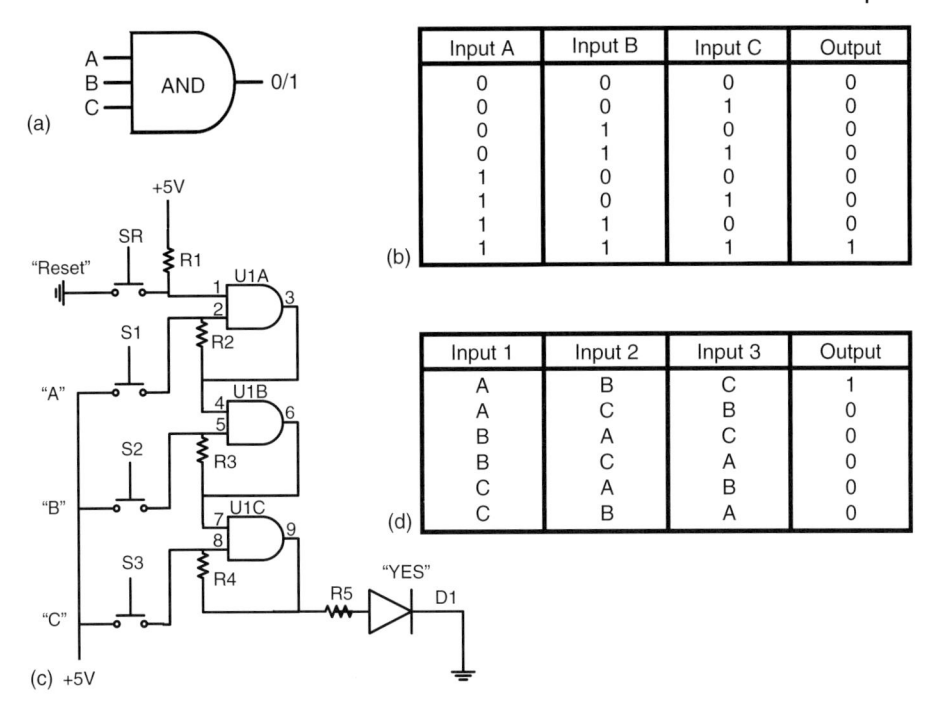

Input 1	Input 2	Input 3	Output
A	B	C	1
A	C	B	0
B	A	C	0
B	C	A	0
C	A	B	0
C	B	A	0

Figure 6.1 Three-input AND gate (a) and the corresponding truth table (b). Three-input electronic keypad circuit (c) and the table corresponding to the six input combinations when A, B, and C are equal to 1 (d).

a security system, the FLIP logic system was reconfigured to distinguish between different input sequences and serve as a three-input keypad lock. The keypad lock is activated by the insertion of the correct "key" or three-input sequence, that is, ethylenediaminetetraacetic acid (E), base (B), and UV light (U). Only the sequence E, B, and U applied in the appropriate order will successfully result in a high output that is monitored optically. For an additional level of security, the lock could be inactivated if the incorrect character key is applied, by integrating the release of a fluorescence quencher [26].

A similar fluorescence-based security system was developed utilizing DNA strands immobilized on silica nanoparticles (SiNPs) [27]. *Inputs* were defined as the presence or absence of DNA fragments applied to the solution containing the DNA-modified SiNPs. The output was detected optically as an increase in fluorescent intensity of a nucleic acid intercalator, Genefinder, which binds to double-stranded DNA. The addition of each DNA strand resulted in a four-fragment hybridization scheme, which starts with an immobilized strand attached to the SiNP. The first input formed a duplex, with the immobilized strand leaving additional bases to hybridize with the next input, and so on. When the DNA strands are introduced into the DNA-modified SiNP solution in the correct order, the fluorescent output intensity increases, and the lock is activated. The DNA-modified SiNP hybridization scheme is only one example of the versatility of DNA-based computing systems that offer a unique technology developing in parallel to security and encryption systems based on enzymes and antibodies.

This first example of an enzyme-based keypad lock drew inspiration from biochemical pathways found in Nature [28]. In such pathways, specific inhibition of a biocatalyst, or lack of an activating substrate, renders the biocascade inactive. This principle was exploited to design a model biochemical reaction chain in which three catalytic steps were incorporated, specifically the hydrolysis of sucrose to glucose, reduction of oxygen by glucose to yield hydrogen peroxide, and oxidation by H_2O_2 of a synthetic dye (2,2′-azino-*bis*(3-ethylbenzothiazoline-6-sulfonic acid; ABTS), which results in the formation of a colored product (Figure 6.2a). These reaction steps were catalyzed specifically and exclusively by invertase (Inv), glucose oxidase (GOx), and microperoxidase-11 (MP-11), respectively. In a manner similar to the FLIP-based molecular keypad lock, the cascade uses three concatenated AND gates, the output from the first gate serving as the input for the second, whose output to the third gate elicits the signal that "opens" the lock [28]. The enzymes were immobilized on glass beads to serve as reusable catalytic inputs.

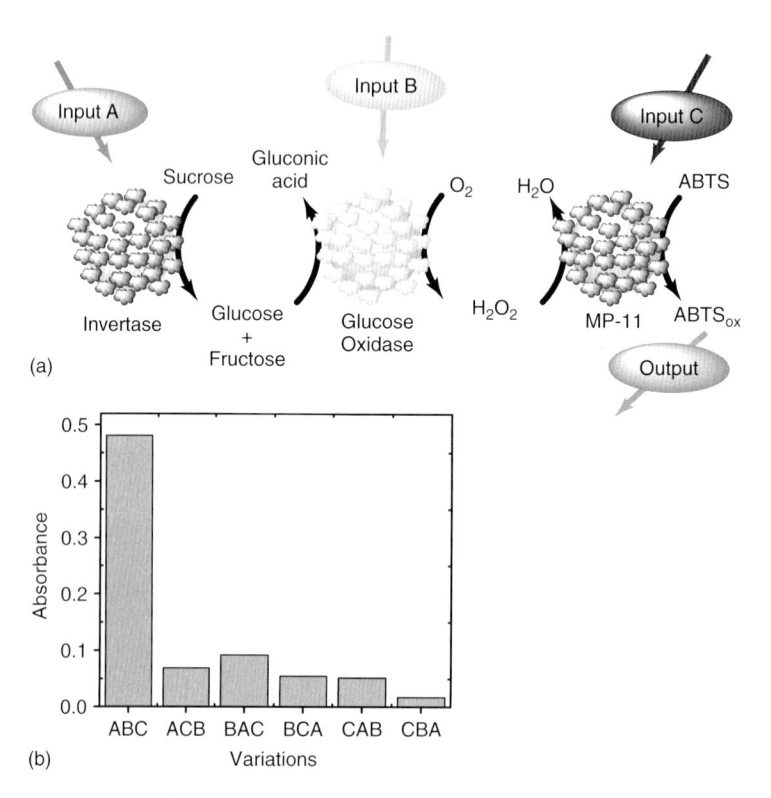

Figure 6.2 (a) Biocatalytic cascade operating with invertase, glucose oxidase, and microperoxidase (MP-11) as inputs A, B, and C, respectively. With inputs applied in the correct order, 2,2′-azino-*bis*(3-ethylbenzothiazoline-6-sulfonic acid) (ABTS) is oxidized and visualized as an absorbance change at $\lambda = 415$ nm (b) that activates the lock.

As the "buttons" are pushed on the keypad lock, the immobilized enzymes are introduced into a chemical system containing sucrose, oxygen, and ABTS. Each biocatalyst is exposed to the chemical solution until the reaction product has accumulated to a concentration high enough to trigger the next logic gate. When threshold is reached, the biocatalyst is removed from the chemical system and the next is introduced. As a result, the sucrose input signal for the first AND gate is 1, and the subsequent glucose and H_2O_2 input signals for the second and third AND gates are generated *in situ* by consecutive biocatalytic reactions. The resulting oxidized ABTS product exhibits a distinctive green color that can be measured as absorbance change and is actually visible to the naked eye. Using this simple three-enzyme system, experiments with six different input permutations were performed, confirming that only the correct input order activated the lock (Figure 6.2b). As an extension to the system described, additional buttons on the keypad lock that release inhibitors that deactivate specific biocatalysts and enhance the level of security inherent to the system could be added. Lastly, one of the specific advantages of this system is that the lock can be reset by simply flushing the chemical system.

The FLIP-based and enzyme-based keypad lock systems described were all performed in solution, and the outputs were detected optically. To advance the technology and integrate the chemical/biochemical security system with an electronic transducer, one must combine the initial concept with a bioelectronic interface. Thus far, biomolecular information processing systems have been integrated with switchable bioelectronic devices, including modified electrodes [21, 29–32] and biofuel cells [23, 33]. One such bioelectronic interface is an indium tin oxide (ITO) electrode modified with a pH-responsive polymer that, when functionalized with poly(4-vinylpyridine), allows the penetration of an electrochemically active anionic redox species at specific pH ranges [29, 33–35]. The resulting polymer brush exhibits a hydrophobic "collapsed" state at pH > 5.5, which is impermeable to soluble redox species and inhibits electrochemical reactions at the electrode surface (Figure 6.3). In contrast, at pH < 4.5, the pyridine groups in the polymer brush are protonated, resulting in a positively charged, swollen, hydrophilic state that is permeable to anionic redox species and allows the electrochemical reactions at the surface of the electrode to take place. As a result, the electrode response is specifically controlled by pH changes and the biocatalytic cascade provides built-in information processing capabilities [21, 32].

A similar pH-responsive biocatalytic keypad lock was designed in which the final output results in a decrease in the solution pH (Figure 6.4). In the first reaction, β-amylase (βAm; enzyme – input A) catalyzes the hydrolysis of starch to yield maltose. In the second step, maltose phosphorylase (MPh), coupled with acid phosphatase (AP) as input B, produces glucose from maltose. In the final step, GOx, (enzyme – input C) catalyzes the oxidation of glucose to gluconic acid, which decreases the pH [36].

By combining prior methodologies, a biological fuel cell was developed in which the fuel cell's power output was controlled by a keypad lock biochemical cascade (Figure 6.4). When the biocatalysts were applied to the cathode compartment

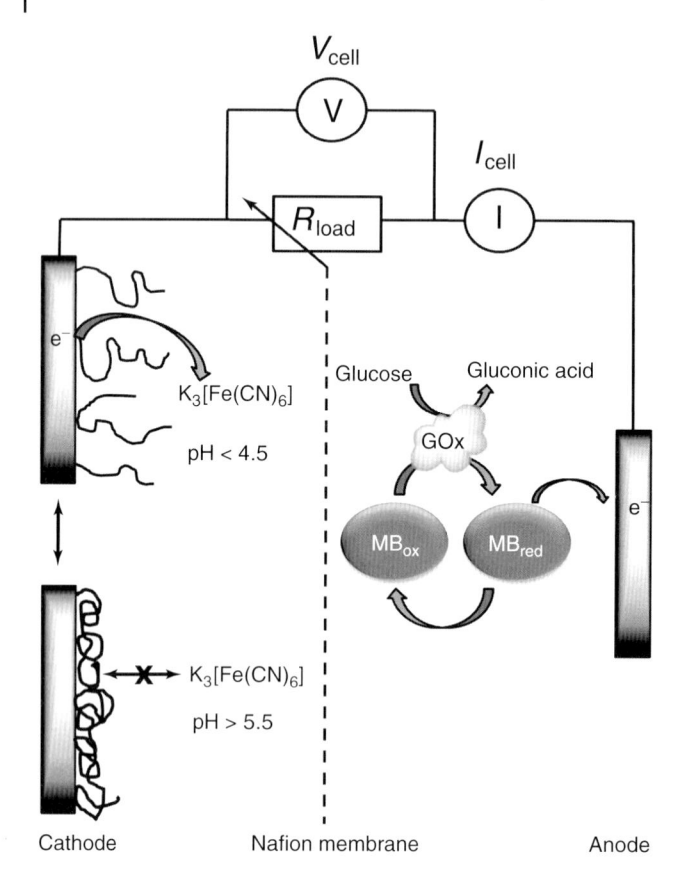

Figure 6.3 Schematic of a pH-responsive biofuel cell; a pH-responsive, poly(4-vinylpyridine)-modified cathode collapses at pH > 5.5 and extends at pH < 4.5, respectively, blocking and allowing diffusion of $K_3[Fe(CN)_6]$ to the electrode. Electron transfer at the anode is mediated by methylene blue (MB), which is reduced by glucose oxidase (GOx) in the presence of glucose.

in the correct order (A, B, and C), the resultant output – decreased pH – led to protonation of the pH-responsive polymer brush, which allowed ferricyanide (a redox mediator) to diffuse to the surface of the electrode. In principle, increasing the concentration of the mediator at the electrode surface resulted in an increased power output (Figure 6.5). All other permutations of the enzyme inputs resulted in no pH change and, therefore, the power output remained low [36].

6.3
Antibody Encryption and Steganography

Although biocatalytic cascades can be reconfigured to mimic keypad locks, the design and operation can be restricted because of cross-reactivity of parallel

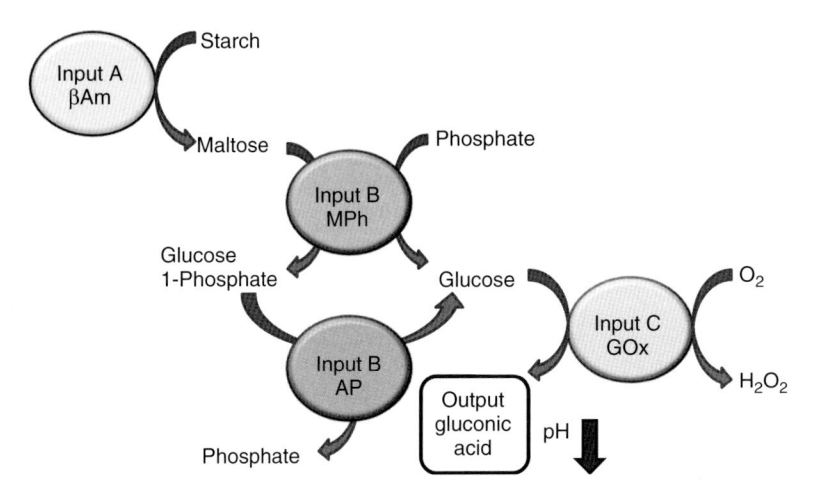

Figure 6.4 Biocatalytic cascade operating with β-amylase (βAm) (input A), maltose phosphorylase (MPh) and acid phosphatase (AP) (input B), and glucose oxidase (GOx) (input C). On the application of the three inputs in the correct order, the pH of the solution decreases, resulting in activation of the lock [36]. (Reproduced by permission of The Royal Society of Chemistry.)

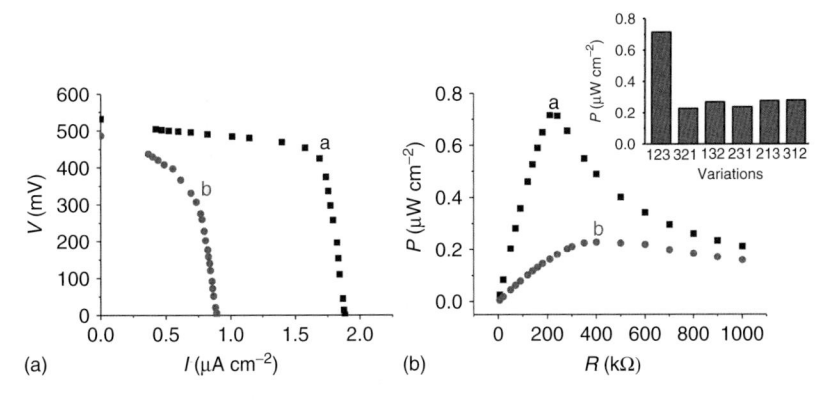

Figure 6.5 Galvanostatic polarization curve (a) and power output (b) of the biofuel cell obtained at (b, circles) initial pH 6.7 and (a, squares) pH 4.2 generated after exposure to the correct A–B–C sequence of the enzyme inputs [36]. (Reproduced by permission of The Royal Society of Chemistry.)

biocatalytic cascades, specific operational circumstances, and a limited set of reaction types. It is therefore prudent to explore other biomolecular modes of unconventional security, including antibody-based systems that can be engineered to recognize an abundance of biomolecules and chemicals. In a similar fuel cell design, an antibody-based keypad lock that relies on specific antibody biorecognition properties [37] was envisioned (Figure 6.6). For demonstrative purposes, an antigen, 3-nitro-L-tyrosine (NT), bound to a carrier protein (bovine

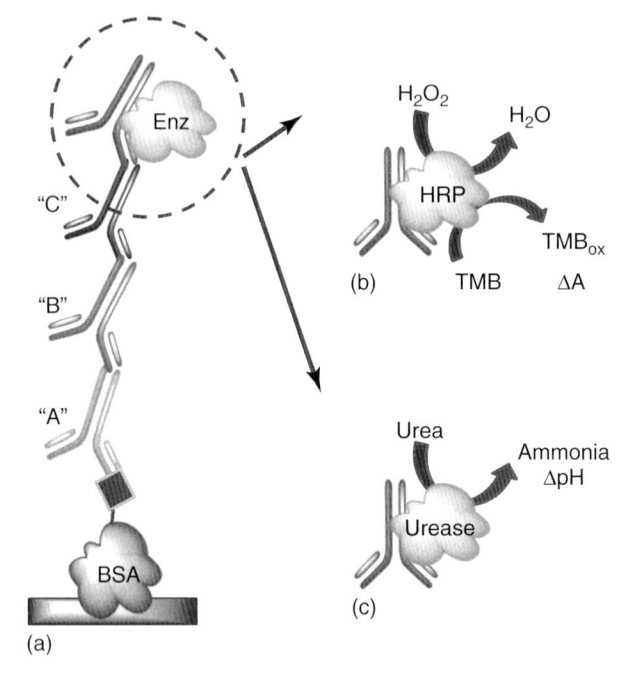

Figure 6.6 Multicomponent A,B, and C antibody stack including a fourth reporter antibody with a linked enzyme (Enz) (a). An immobilized antigen is bound to bovine serum albumin (BSA), which is recognized by antibody input A. Inputs B and C are specific for antibody input A and B, respectively. Confirmation of the assembly is monitored either optically via oxidation of 3,3′,5,5′-tetramethylbenzidine (TMB) by horseradish peroxidase (HRP) (b) or by pH changes induced by urease conjugated to the multicomponent antibody system (c).

serum albumin; BSA), was physically adsorbed to polystyrene wells of an ELISA plate. A three-antibody "stack" was then captured on the functionalized surface: rabbit-origin IgG anti-nitrotyrosine (input A), goat-origin IgG specific against rabbit IgG (input B), and mouse-origin IgG specific against goat IgG (input C). The antibody stack was monitored by enzyme-labeled goat-origin IgG specific against mouse IgG. This fourth antibody was labeled with horseradish peroxidase (HRP), which provides a colorimetric response (by biocatalytic oxidation of 3,3′,5,5′-tetramethylbenzidine (TMB)) to indicate correct assembly of the antibody sequence. The security of the keypad lock was confirmed when only one antibody permutation out of six possibilities resulted in a colorimetric response.

The antibody assembly was incorporated into a biological fuel cell analogous to the one previously described, using a Nafion® membrane to separate ITO electrodes at the cathode and anode [37]. In the cathodic compartment, the functionalized, pH-responsive polymer brush was signaled by the antibody stack to change conformation. Urease was conjugated to the reporter antibody (goat-origin IgG specific against mouse IgG) and, in the presence of urea, catalyzed ammonia production, which caused the solution pH to rise from 4.2 to >5.5. The responsive polymer brush was deprotonated at the higher pH, became hydrophobic, and collapsed onto the surface of the electrode, thus blocking diffusion of the mediator to the electrode surface and decreasing the power output of the fuel cell. In contrast (or complement) to the enzyme-based activation of a biological fuel cell lock, the correct antibody assembly effectively shuts off the fuel cell output.

Given the tailored specificity of antibodies, the biorecognition process could be exploited not only for security purposes but also for encryption and steganography. This process was demonstrated by using a mixture of IgG antibodies adsorbed to a polystyrene ELISA plate [38]. Mouse anti-human IgG (target), goat anti-cat IgG (masking), and donkey anti-chicken IgG (masking) were mixed in a ratio of 1 : 1 : 1, with the target antibody applied in a specific pattern that encoded a hidden message. Application of enzyme-labeled antibodies to empty spots, which are not encoded with corresponding bioaffinity units, does not result in antibody binding and does not yield any color-producing reaction (Figure 6.7a). The only way to expose the prepatterned antibodies was to add the correct labeled anti-mouse IgG, along with a redox dye specific for the enzyme. When the correct enzyme-labeled antibody is applied, the enzyme will oxidize the dye, resulting in color development (Figure 6.7b), thus exposing the pattern (Figure 6.8) [38]. The described encryption technique relies on a digital pattern on a surface. When the target antibody is exposed, the oxidized dye will reveal the hidden digital information (digital 1), whereas areas without the oxidized dye are considered as digital 0. The level of encryption can be enhanced by controlling the antibody ratio used such that the quantity of oxidized dye will correspond to the amount of target antibody applied at the surface [38]. Thus, the assigned value of digital 1 or 0 can also be controlled and

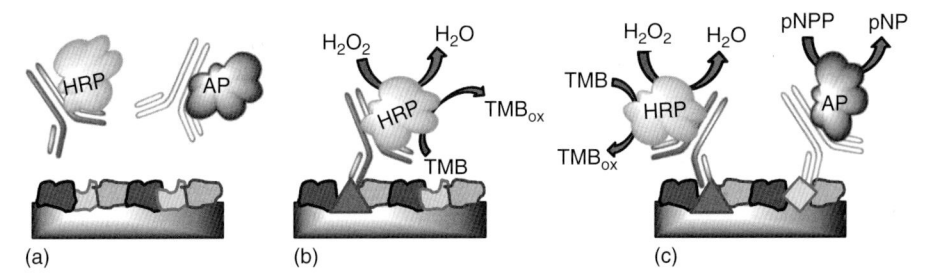

Figure 6.7 Steganography based on immune-specific interactions that result in color development by enzyme-labeled complementary antibodies. Binding does not result between the secondary enzyme-labeled antibodies and masking antibodies (a). The secondary enzyme-labeled antibody (horseradish peroxidase (HRP)-labeled anti-mouse IgG) binds to the corresponding target antibody (mouse anti-human IgG), resulting in the biocatalytic oxidation of 3,3′,5,5′-tetramethylbenzidine (TMB), (b) and two secondary enzyme-labeled antibodies (HRP-labeled anti-mouse IgG and acid phosphatase (AP)-labeled anti-rabbit IgG) bind to the corresponding target antibodies (mouse anti-human IgG and rabbit anti-rat IgG), resulting in the biocatalytic oxidation of TMB and formation of p-nitrophenol (pNP).

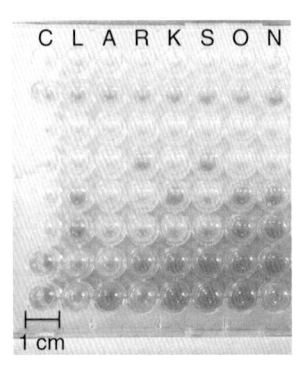

Figure 6.8 Photo of the ELISA plate with text "CLARKSON" (ASCII 12-bit code 39) encoded with mouse anti-human IgG in the presence of two masking proteins (goat anti-cat IgG and donkey anti-chicken IgG) after the treatment with HRP-labeled anti-mouse IgG and the color-developing solution for 30 min.

introduces an additional level of information security (or a hierarchy); the correct secondary antibody along with knowledge of the digital threshold will decode the hidden text.

Another variation on the ELISA-inspired encryption is the conjugation of various enzyme labels to the secondary antibodies [38]. In this way, multiple codes can be encrypted on the same ELISA plate, the target is revealed only when the enzyme-linked secondary antibody adheres, and only if the appropriate assay is performed to develop the secondary antibody signal. An ELISA plate modified with several IgG antibodies, for example, may include two different target antibodies placed in specific patterns that encode two overlapping patterns (Figure 6.7c). After the secondary enzyme-linked antibodies are added, both patterns can be visualized only if subjected to two separate colorimetric assays, each tailored for the catalytic specificity of the enzyme tag. Therefore, multiple lines of text can be encoded in one device as information, or to provide disinformation that distracts from the hidden text.

To advance the antibody encryption technique toward the practical realm, protein spots (mouse anti-human IgG) were deposited on nitrocellulose paper as an "invisible ink" to pattern code onto seemingly blank paper [38]. In this way, steganographic applications can be combined with cryptography. The antibody spots can be applied with an inkjet printer for miniaturization and precision [38]. Scaling down the technique to microsize can also be extended to the detection mode [39]. Recently, force–volume atomic force microscopy (AFM) studies demonstrated that the spot size needed to detect a specifically bound antibody layer can be reduced to approximately 3 μm^2, thus scaling down the required spot size by a factor of $\sim 10^6$.

The molecular and biomolecular systems described are essentially conceptual demonstrations that have inherent limitations but nevertheless exhibit specificity that may prove valuable in specific-niche operational conditions.

6.4
Bio-barcode

Most conventional encryption uses logic and binary functions for essential processing; an example is barcoding. Using biocatalysts to generate barcodes provides an interesting mechanism of encryption processing that is based entirely on the catalytic mechanism of an enzyme. Not all examples of unconventional bar codes employ biogenerated digital sequences [40–47]; encryption tags, for example, are fabricated from specific alignment of metallic nanowires and quantum dots [40, 41]. These types of unconventional barcodes include medical diagnostic models and multiplexed bioanalytical assays that can detect single proteins or single molecules of DNA [41–44, 46].

Generating a barcode in time domain, however, creates significant amounts of binary code, which implies a complementary, reversible on–off toggle. In biological terms, tailoring this kind of response requires the integration of a catalyst with a reliable and sustainable bioelectronic interface. Given that the addition of the inputs into chemical information processing systems is essentially a fluctuating environment, the response integrity, sustainability, and longevity of the bioelectronic interface must be addressed. In other words, an electrode capable of processing relatively large amounts of information must withstand constant input flow and flux over a considerable period of time.

As a demonstration, a laccase-functionalized carbon electrode was fabricated, which exhibited sustainable bioelectrocatalytic functionality for 20 days under continuous input fluctuation [48]. The catalytic response can be programmed in such a way that output is dependent on the catalysis of the enzyme. By varying the substrate input under continuous conditions, the first unconventional barcode generated was demonstrated by continuous relay of digital information through a bioelectronic interface (Figure 6.9). The electrode longevity was tested by programing an ON/OFF change in the substrate range [48]. In this manner, the enzyme interface responds to an abrupt transition between N_2 (OFF) and O_2 (ON) substrate feeds and forms a series of "bars" observed as an amperometric output

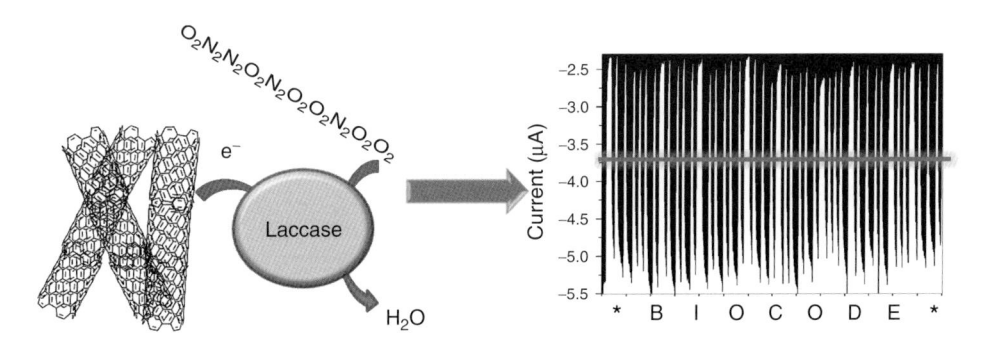

Figure 6.9 Barcode (reading BIOCODE) generated from the direct biocatalytic activity of laccase immobilized on carbon nanotubes, in response to inputs of oxygen and nitrogen.

that are read immediately without further treatment or manipulation. The barcode was generated by implementing a series of binary codes. In this case, ASCII 12-bit code 39 was chosen to generate a readable code currently available on barcode reader cell phone applications.

The bioelectronic barcode method was demonstrated using simple input sequences and electrode fabrication techniques. Several levels of encryption can be generated, including traditional electronic encryption of the binary sequence before it is translated by the electrode. Furthermore, the bioelectronic interface can be prepared with other biomolecules capable of biocatalytically responding to other chemical inputs and biological cascades; in this manner, the electrode can serve as a tailored biological key used to unlock a series of chemical inputs.

6.5
Conclusion

In conclusion, the inherent selectivity of DNA, biocatalysts, and antibodies allows for tailored reactions, in which unique functionalities, unobtainable using existing *in silico* computing techniques, can be realized. Security systems, cryptography, steganography, and code relay have now been demonstrated using unconventional molecular and biomolecular methods. Biomolecule-based information processing will not only aid development of medical sensors capable of injury or disease diagnosis, but also advance encryption, security systems, and code relay. The design and optimization of such systems could lead to a wealth of tools and technologies, such as logically controlled biofuel cells, programmable signal-responsive transducers, biocompatible security apparatuses, and high-fidelity biosensors.

Acknowledgment

The research at Clarkson University was supported by the National Science Foundation (Grants DMR-0706209, CCF-1015983), Office of Naval Research (Grant N00014-08-1-1202), and the Semiconductor Research Corporation (Award 2008-RJ-1839G). The authors acknowledge funding from the Air Force Research Laboratory Materials and Manufacturing Directorate and the Air Force Office of Scientific Research (Walt Kozumbo, Program Manager).

References

1. de Silva, A.P. and Uchiyama, S. (2007) *Nat. Nanotechnol.*, **2**, 399–410.
2. de Silva, A.P., Uchiyama, S., Vance, T.P., and Wannalerse, B. (2007) *Coord. Chem. Rev.*, **251**, 1623–1632.
3. Szacilowski, K. (2008) *Chem. Rev.*, **108**, 3481–3548.
4. Credi, A. (2007) *Angew. Chem. Int. Ed.*, **46**, 5472–5475.
5. Saghatelian, A., Volcker, N.H., Guckian, K.M., Lin, V.S., and Ghadiri, M.R. (2003) *J. Am. Chem. Soc.*, **125**, 346–347.
6. Ashkenasy, G. and Ghadiri, M.R. (2004) *J. Am. Chem. Soc.*, **126**, 11140–11141.

7. Chuang, M.C., Windmiller, J.R., Santhosh, P., Valdes Ramirez, G., Katz, E., and Wang, J. (2011) *Chem. Commun.*, **47**, 3087–3089.
8. Bocharova, V., Halamek, J., Zhou, J., Strack, G., Wang, J., and Katz, E. (2011) *Talanta*, **85**, 800–803.
9. Halamek, J., Bocharova, V., Chinnapareddy, S., Windmiller, J.R., Strack, G., Chuang, M.C., Zhou, J., Santhosh, P., Ramirez, G.V., Arugula, M.A., Wang, J., and Katz, E. (2010) *Mol. BioSyst.*, **6**, 2554–2560.
10. Halamek, J., Windmiller, J.R., Zhou, J., Chuang, M.C., Santhosh, P., Strack, G., Arugula, M.A., Chinnapareddy, S., Bocharova, V., Wang, J., and Katz, E. (2010) *Analyst*, **135**, 2249–2259.
11. Manesh, K.M., Halamek, J., Pita, M., Zhou, J., Tam, T.K., Santhosh, P., Chuang, M.C., Windmiller, J.R., Abidin, D., Katz, E., and Wang, J. (2009) *Biosens. Bioelectron.*, **24**, 3569–3574.
12. Fu, P. (2007) *Biotechnol. J.*, **2**, 91–101.
13. Parker, J. (2003) *EMBO Rep.*, **4**, 7–10.
14. Stojanovic, M.N. (2008) *Prog. Nucleic Acid Res.*, **82**, 199–217.
15. Baron, R., Lioubashevski, O., Katz, E., Niazov, T., and Willner, I. (2006) *J. Phys. Chem. B*, **110**, 8548–8553.
16. Baron, R., Lioubashevski, O., Katz, E., Niazov, T., and Willner, I. (2006) *Angew. Chem. Int. Ed.*, **45**, 1572–1576.
17. Niazov, T., Baron, R., Katz, E., Lioubashevski, O., and Willner, I. (2006) *Proc. Natl. Acad. Sci. U.S.A.*, **103**, 17160–17163.
18. Arugula, M.A., Bocharova, V., Halamek, J., Pita, M., and Katz, E. (2010) *J. Phys. Chem. B*, **114**, 5222–5226.
19. Katz, E. and Privman, V. (2009) *Chem. Soc. Rev.*, **39**, 1835–1857.
20. Pita, M., Strack, G., MacVittie, K., Zhou, J., and Katz, E. (2009) *J. Phys. Chem. B*, **113**, 16071–16076.
21. Privman, M., Tam, T.K., Pita, M., and Katz, E. (2009) *J. Am. Chem. Soc.*, **131**, 1314–1321.
22. Strack, G., Chinnapareddy, S., Volkov, D., Halamek, J., Pita, M., Sokolov, I., and Katz, E. (2009) *J. Phys. Chem. B*, **113**, 12154–12159.
23. Tam, T.K., Pita, M., Ornatska, M., and Katz, E. (2009) *Bioelectrochemistry*, **76**, 4–9.
24. Clelland, C.T., Risca, V., and Bancroft, C. (1999) *Nature*, **399**, 533–534.
25. Mao, C., LaBean, T.H., Relf, J.H., and Seeman, N.C. (2000) *Nature*, **407**, 493–496.
26. Margulies, D., Felder, C.E., Melman, G., and Shanzer, A. (2007) *J. Am. Chem. Soc.*, **129**, 347–354.
27. Zhen, F.L.P., Xinjian, Y., Jinsong, R., and Xiaogang, Q. (2011) *Chem. Commun.*, **47**, 6024–6026.
28. Strack, G., Ornatska, M., Pita, M., and Katz, E. (2008) *J. Am. Chem. Soc.*, **130**, 4234–4235.
29. Bocharova, V., Tam, T.K., Halamek, J., Pita, M., and Katz, E. (2010) *Chem. Commun.*, **46**, 2088–2090.
30. Pita, M., Minko, S., and Katz, E. (2009) *J. Mater. Sci.*, **20**, 457–462.
31. Tam, T.K., Zhou, J., Pita, M., Ornatska, M., Minko, S., and Katz, E. (2008) *J. Am. Chem. Soc.*, **130**, 10888–10889.
32. Zhou, J., Tam, T.K., Pita, M., Ornatska, M., Minko, S., and Katz, E. (2009) *ACS Appl. Mater. Interfaces*, **1**, 144–149.
33. Katz, E. and Pita, M. (2009) *Chem. Eur. J.*, **15**, 12554–12564.
34. Pita, M., Tam, T.K., Minko, S., and Katz, E. (2009) *ACS Appl. Mater. Interfaces*, **1**, 1166–1168.
35. Tam, T.K., Pita, M., Trotsenko, O., Motornov, M., Tokarev, I., Halamek, J., Minko, S., and Katz, E. (2010) *Langmuir*, **26**, 4506–4513.
36. Halamek, J., Tam, T.K., Strack, G., Bocharova, V., Pita, M., and Katz, E. (2010) *Chem. Commun.*, **46**, 2405–2407.
37. Halamek, J., Tam, T.K., Chinnapareddy, S., Bocharova, V., and Katz, E. (2010) *J. Phys. Chem. Lett.*, **1**, 973–977.
38. Kim, K.W., Bocharova, V., Halamek, J., Oh, M.K., and Katz, E. (2011) *Biotechnol. Bioeng.*, **108**, 1100–1107.
39. Volkov, D., Strack, G., Halamek, J., Katz, E., and Sokolov, I. (2010) *Nanotechnology*, **21** (article #145503).
40. Xiang, Y., Zhang, Y., Chang, Y., Chai, Y., Wang, J., and Yuan, R. (2010) *Anal. Chem.*, **82**, 1138–1141.

41. Kim, J., Seo, K., and Wang, J. (2004) *Conf. Proc. IEEE Eng. Med. Biol. Soc.*, **1**, 137–140.
42. Demirok, U.K., Burdick, J., and Wang, J. (2009) *J. Am. Chem. Soc.*, **131**, 22–23.
43. Thaxton, C.S., Elghanian, R., Thomas, A.D., Stoeva, S.I., Lee, J.S., Smith, N.D., Schaeffer, A.J., Klocker, H., Horninger, W., Bartsch, G., and Mirkin, C.A. (2009) *Proc. Natl. Acad. Sci. U.S.A.*, **106**, 18437–18442.
44. Appleyard, D.C., Chapin, S.C., and Doyle, P.S. (2011) *Anal. Chem.*, **83**, 193–199.
45. White, K.A., Chengelis, D.A., Gogick, K.A., Stehman, J., Rosi, N.L., and Petoud, S. (2009) *J. Am. Chem. Soc.*, **131**, 18069–18071.
46. Gunnarsson, A., Sjovall, P., and Hook, F. (2010) *Nano Lett.*, **10**, 732–737.
47. Dejneka, M.J., Streltsov, A., Pal, S., Frutos, A.G., Powell, C.L., Yost, K., Yuen, P.K., Muller, U., and Lahiri, J. (2003) *Proc. Natl. Acad. Sci. U.S.A.*, **100**, 389–393.
48. Strack, G., Luckarift, H.R., Nichols, R., Cozart, K., Katz, E., and Johnson, G.R. (2011) *Chem. Commun.*, **47**, 7662–7664.

7
Biocomputing: Explore Its Realization and Intelligent Logic Detection

Ming Zhou and Shaojun Dong

7.1
Introduction

Logic lies at the heart of modern computers, and the components that carry out its operations are logic gates [1]. In electronic computers, logic gates are sculpted on the surface of silicon wafers. The gates have names such as AND, OR, and IMP to describe the kind of output that is generated in response to different inputs. Their inputs and outputs are electrical voltages, but these are not the only systems that can form logic circuits. In recent years, much research interest was directed to unconventional chemical computing [1–4], performing Boolean logic operations without the involvement of electronic computers and responding to a large variety of activating input signals in homogeneous solutions or at interfaces functionalized with molecular/supramolecular moieties [4]. Despite the fact that a very promising future is expected for unconventional chemical computing systems [1], the development of these systems is limited by their synthetic complexity and difficulty to scale them up for assembling large networking systems. The latter problem originates from incompatibility of most chemical systems performing individual computing operations [5]. This limitation can be solved by the application of biochemical computing (biocomputing) systems designed by Nature to perform highly specific catalytic or recognition reactions where the individual steps are complementary and the reacting components are compatible [6, 7].

Biocomputing, belonging to a subarea of unconventional chemical computing and performed by living organisms (e.g., DNA [8, 9], proteins/enzymes [10, 11], and whole biological cells [12]), is aiming at the information processing using biochemical means without the involvement of electronic computers. The fundamental concepts for using living organisms in logic gates were developed by the pioneering work of Rosen in 1967 [13, 14]. He first introduced a "two-factor model" based on the idea that the dynamic behavior of physiological and biochemical systems is regulated by the combined action of two factors, an excitatory one and an inhibitory one. After this work, lots of exciting biocomputing concepts and results

Biomolecular Information Processing: From Logic Systems to Smart Sensors and Actuators,
First Edition. Edited by Evgeny Katz.
© 2012 Wiley-VCH Verlag GmbH & Co. KGaA. Published 2012 by Wiley-VCH Verlag GmbH & Co. KGaA.

based on different living organisms have been reported [8–12], which exhibited unique and interesting applications in such field.

DNA biocomputing is a form of computing that uses DNA, biochemistry, and molecular biology, instead of the traditional silicon-based computer technologies and is thought to be the future of computer technologies, while DNA molecules that perform logic operations are a prerequisite for digital information processing and computing [1, 6]. The DNA biocomputing field was initially developed by Leonard Adleman of the University of Southern California in 1994 [15], who demonstrated a proof-of-concept use of DNA as a form of computation, which solved the seven-point Hamiltonian path problem. Since the initial Adleman experiments, advances have been made and various "Turing machines" have been proved to be constructible [16, 17]. Ultimately, such gates could serve as dissolved "doctors" – sensing molecules such as markers on cells and jointly choosing how to respond.

Aptamers, which are novel artificially selected functional DNA or RNA oligonucleotides, possess high recognition ability to a broad range of specific targets from proteins to peptides, amino acids, drugs, metal ions, and even whole cells [18–21]. Owing to their unique characteristics (e.g., versatility, specificity, and synthetic nature), aptamers exhibit many unprecedented advantages compared with antibodies or other biomimetic receptors, including a wide range of targets, comparable or even better target affinity, easy and cost-effective synthesis with high reproducibility and purity, and simple and straightforward chemical modification [18, 19]. Thus, aptamers can be considered as a potential alternative to antibodies or other biomimetic receptors, for the development of biotechnology, diagnostics, and therapy [19–21]. In 2005, Stojanovic's group reported the first proof-of-concept aptamer logic system [22]. Such molecular computation performed by aptamers-based logic gates can be used to control the functional state of aptamers, switching them on or off based on the outcome of such computation.

Since the concept of enzyme-based device was presented by Clark in 1962 [23], enzyme has been used in conjunction with various electrodes for constructing electrochemical enzyme biodevices because of the inherent selectivity shown by the enzymes to promote selective reaction of the enzyme substrates [24–26]. One of the early biocomputing attempts to utilize enzymes was presented by Sivan and Lotan [14]. In that work, an enzyme chymotrypsin was chemically modified to render its activity sensitive to certain irradiation wavelengths (with one wavelength capable of activation and another, of inactivation). In addition, they employed a small molecule inhibitor that could be rendered inactive by a reducing chemical environment. Taken together, only the combination of active enzyme form and inactive inhibitor would trigger an output from the gate. In a related line of research, purely molecular enzymatic systems were developed by Willner's group [27] and later by Katz's group [7].

With the development of biocomputing, two different branches of the biocomputing system are being reported in different directions. One is for competing with traditional electronic computation, taking advantage of parallel computing performed by numerous biomolecules [28]. However, so far, most of these devices still cannot compete with the classic semiconductor-based processors, in terms of computational power, efficiency, and user-friendliness [29]. The other direction is not aiming at any complex computation but rather creating a "smart" information processing interface between biological and electronic systems [30], which would be further used as logic biosensors for the potential intelligent medical diagnostics [31–34] and may also lead to a better understanding of nature [5, 35]. Being different from the traditional biosensors, logic biosensors based on biocomputing are smart and able to intelligently analyze the relationship between different targets in complex samples according to the Boolean logic operations "programmed" into biocomputing systems. In this chapter, we review our recent exploration on both biocomputing realization and intelligent logic biocomputing detection. Finally, future challenges and perspectives toward biocomputing field are described.

7.2 DNA Biocomputing

DNA molecules can act as elementary logic gates analogous to the silicon-based gates of ordinary computers. Adleman's seminal 1994 insight that computation could be encoded in DNA set off a raft of speculation on such topics as whether massively parallel DNA computations might one day break the data encryption standard [15]. In the past two decades, such a field has attracted many efforts focused on problems in Boolean logic at the molecular scale to find ideal candidates that satisfy logic operations [8, 36–39]. Toward this goal, diverse DNA logic gates have been designed and constructed, most of which are, however, based on base pairing or DNAzyme-catalyzed DNA cleavage [1, 6].

In 2010, we utilized a cation-driven allosteric G-quadruplex DNAzyme to devise a DNA logic gate based on cation-tuned ligand binding and release [40]. Three monomolecular G-quadruplex DNAzymes (i.e., PS2.M, PW17, and T30695) [41] were chosen as the models to be investigated. With no input, PW17 was in the random coil state, whereas it folded, respectively, into a parallel or antiparallel quadruplex structure on input of K^+ or Pb^{2+} (Figure 7.1). It was found that Pb^{2+} can induce K^+-stabilized PW17 to undergo a parallel-to-antiparallel conformation transition as a result of its unusually high efficiency at stabilizing G-quadruplexes. The conformational transition is accompanied with a remarkable decrease in the DNAzyme activity. Since the enzyme activity can be straightforwardly represented in the form of absorbance at 420 nm (i.e., the characteristic absorption of the product ABTS•$^+$), A420 here serves as the output (1 or 0) with a threshold that is three times above the background imparted by hemin catalysis. It was observed that when K^+ was input alone, the output was 1; otherwise it was 0. These observations demonstrate that K^+-Pb^{2+}-switched PW17 really functions as a two-input Inhibit logic gate. Some common cations (e.g., Mg^{2+} and Ca^{2+}) have no obvious influence on the DNAzyme activity and the output, and thus, this logic gate can operate in the presence of these cations without any interference. Unlike most previous counterparts based on base pairing or DNA cleavage, which cannot be reversibly

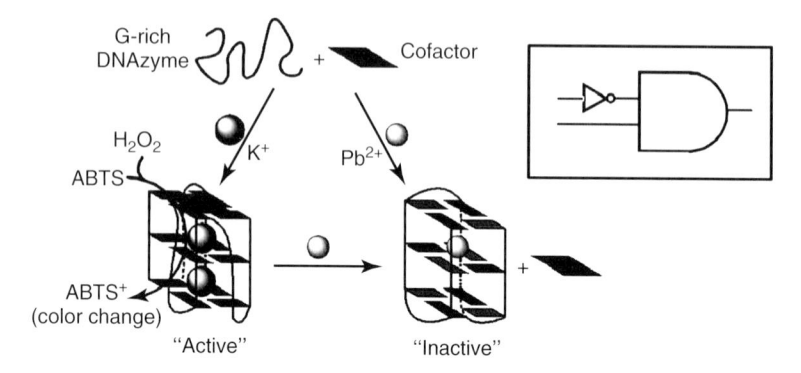

Figure 7.1 Construction of an INHIBIT logic gate (represented in the box) based on the G-rich DNAzyme PW17, with K^+ and Pb^{2+} as two inputs and absorbance as an output [40].

operated, this logic gate employs different cations to control ligand binding to or release from the G-quadruplex, thereby modulating the output reversibly. Such a logic process includes two unique features different from the conventional: (i) it provides the first demonstration of the transformation from the parallel to the antiparallel G-quadruplex induced by cations, in contrast to the usually reported antiparallel-to-parallel conformation transition and (ii) it indicates that a stable G-quadruplex does not always favor ligand binding (hemin), in contrast to the conventional opinion. Our study not only introduces a new concept for devising DNA logic gates but also provides new insight into the structures and functions of DNA.

Later, we devised a new kind of molecular logic device utilizing the ion-tuned DNA/Ag fluorescent nanoclusters [42]. Figure 7.2 depicts the concept for the construction of a versatile logic device utilizing a hairpin DNA template (HP26)-stabilized fluorescent Ag nanoclusters, with K^+ and H^+ as two inputs. The ion-tuned DNA/Ag fluorescent nanoclusters, in fact, behave as a novel kind of versatile logic device. HP26 consists of G-rich, poly-C, and C-rich regions, and first it is utilized as the stabilizer for synthesizing two species of Ag nanoclusters. With no input, HP26 forms a hairpin structure with a poly-C loop, where highly fluorescent Ag nanoclusters can be synthesized. On addition of K^+, the G-tracts of HP26 fold into a bimolecular G-quadruplex (G4) structure stabilized by K^+. This will induce the hairpin structure to unwind, namely, HP26 is subject to a hairpin-to-G4 structure conversion. On addition of H^+, the C-tracts of HP26 fold into the i-motif structure, suggesting that HP26 will undergo a conformational change from the hairpin to i-motif structure. In the presence of K^+ and H^+, the G4 and i-motif structures will be formed together. These structural changes are expected to remarkably influence the specific interaction between C residues and Ag nanoclusters, thereby modulating the fluorescence behaviors of Ag nanoclusters. This enables two or more logic operations to be performed together via multichannel fluorescence output. In contrast to previous counterparts performing only one logic operation [2, 8, 36, 37, 40], such designed logic gates show their versatility: (i) by changing the excitation wavelength, the DNA-tuned

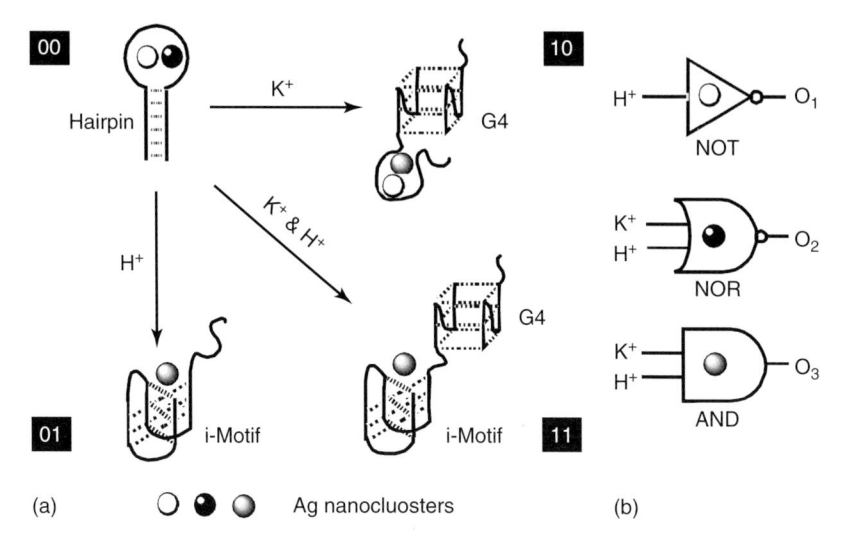

Figure 7.2 (a) Schematic diagram of logic operations based on HP26-tuned fluorescent Ag nanoclusters, and (b) the corresponding symbols of logic gates. K^+ and H^+ serve as two inputs to trigger the allosterism of HP26 and modulate the fluorescence output [42].

HP26: 5'-GGGTTAGGGTCCCCCCACCCTTACCC-3'

Ag nanoclusters are able to perform multiple logic operations together via multichannel fluorescence output and (ii) simply altering the specific sequence of the DNA stabilizer, more logic gates can be constructed utilizing Ag nanoclusters. The developed versatile logic device may find applications in further development of DNA circuit.

7.3 Aptamer Biocomputing

Aptamers, which are important members of the artificial functional nucleic acids family [43], possess high recognition ability to specific targets [20, 44]. Owing to the inherent selectivity, affinity, and multifarious advantages over the traditional recognition elements, aptamers have attracted more and more attention and have been widely developed in many research fields [20, 44]. In 2005, Stojanovic's group reported the first proof-of-concept aptamer logic system [22]. Such molecular computation performed by aptamers-based logic gates can be used to control the functional state of aptamers, switching them on or off based on the outcome of such computation.

Recently, we described the first model of controlled power release of biofuel cells (BFCs) by aptamer-based biochemical signals processed according to the Boolean logic operations "programed" into biocomputing systems [31]. As shown in Figure 7.3, the aptamer-based biochemical signals-controlled BFC-mimicking

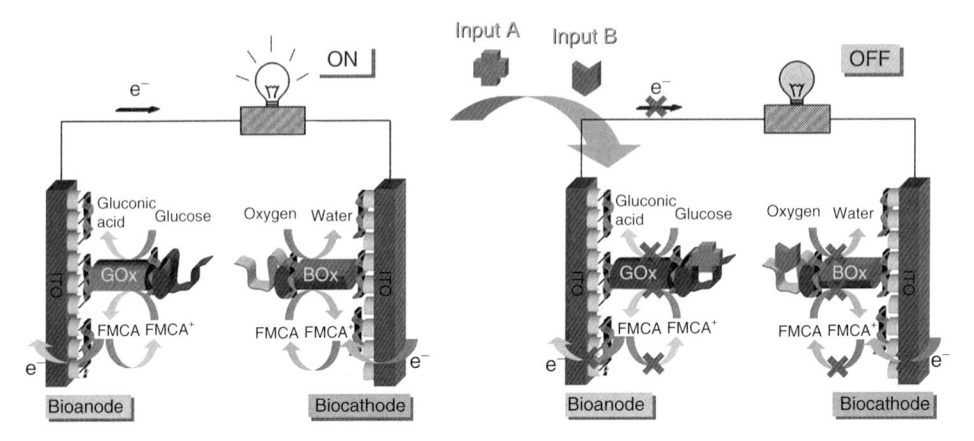

Figure 7.3 Schematic illustration of the assembled aptamer-based BFC logically controlled by biochemical signals that mimic a Boolean NAND logic gate [31].

Boolean logic gate was composed of on-chip patterned glucose oxidase (GOx)/thrombin-binding aptamer-based bioanode and bilirubin oxidase (BOx)/lysozyme-binding aptamer-based biocathode operating. Glucose was used as the fuel, and ferrocene monocarboxylic acid (FMCA) was applied as the diffusional redox mediator helping GOx oxidize glucose fuel and BOx reduce O_2 oxidizer. In the presence of only thrombin (input (1,0)) (or only lysozyme (input (0,1)), thrombin-binding (or lysozyme-binding) aptamer on the outermost layer of the bioanode (or biocathode) would catch thrombin (or lysozyme) and block the electrode interface, resulting in an increased (or decreased) onset potential for glucose oxidation (or O_2 reduction) at the bioanode (or biocathode); accordingly, the open circuit potential of the BFC decreased, but the open circuit potential was still above the threshold (output 1). In the presence of both substrates (input (1,1)), aptamer-target recognition occurred at both the bioanode (between thrombin-binding aptamer and thrombin) and biocathode (between lysozyme-binding aptamer and lysozyme), and the onset potential for bioanode and biocathode increased and decreased, respectively; consequently, the open circuit potential decreased (output 0). Furthermore, application of (0,0) input did not obviously change the open circuit potential of BFC (output 0). Therefore, the features of the system correspond to the equivalent circuitry of NAND logic gate performing the Boolean logic operations of $A' + B'$. NAND logic is represented by the situation where the output of the gate is inhibited only when both inputs are present. So if the presence of both thrombin and lysozyme inputs in a sample (1,1), the open circuit potential greatly reduces (output 0) and NAND gate switches OFF. If one (input (0,1) or (1,0)) or neither (input (0,0)) of the analytes is present in the sample, the open circuit potential is still "activated" (output 1). Thus, NAND logic gate presented here queries two target biological species in a sample to determine in a single test whether they are both present.

The integration of BFCs with aptamer logic may not only give us an avenue to control BFC power release by aptamer-based biocomputing systems but also indicate an interesting "mutual benefits" concept between aptamer and BFCs, that is, utilizing aptamer-based biochemical signals as logic operation for controlling BFC power release and applying BFCs as self-powered and intelligent biosensors for logic aptasensing. Being different from the traditional biosensors, the fabricated logic aptasensors are self-powered, smart, and amenable to whether both specific targets are present by using the built-in Boolean logic. If one applies the pathologically related target-binding aptamers into the aptamer logic systems-controlled BFCs [45], one possible application of such logic systems could be potentially used for performing self-powered smart diagnostics, which would be direct screening of various medical conditions that are dependent on combinations of diagnostic aptamers [45]. However, such a system was unreusable and can only realize the NAND logic gate, which may limit its potential application.

So we further demonstrated an IMP-Reset logic-based reusable, self-powered, intelligent, and microfluidic aptasensor [32]. The IMP-Reset gate-based reusable and intelligent on-chip aptasensor was constructed on the basis of micro-BFC, which was composed of two microchip-based Au electrodes (Figure 7.4). The cathode was modified with a part DNA duplex containing a 5'-thiolated partly complementary strand and a mixed aptamer, operating in the presence of $K_3(Fe(CN)_6$ used as an oxidizer. The anode was an unmodified Au electrode operating in the presence of soluble GOx, which oxidized glucose fuel with the help of a diffusional redox mediator FMCA. The cathode and anode were separated with a Nafion membrane. The switchable aptamer-based cathode mimicking Boolean IMP-Reset logic operation was composed of a Au electrode support modified with a part DNA duplex containing a 5'-thiolated partly complementary strand and a mixed aptamer. After adding adenosine triphosphate (ATP) (input (0,1)), ATP would interact with the extended ATP-binding aptamer embedded in the mixed aptamer and draw it away from the electrode surface (input (0,1)), which increased the onset potential and plateau of current density for $Fe(CN)_6^{3-}$ reduction (input (0,1)); accordingly, the maximum power density increased (output 1). When only thrombin was present (input (1,0)), the thrombin-binding aptamer in mixed aptamer on Au electrode would catch the thrombin, resulting in a blocked electrode interface. Accordingly, the onset potential and the plateau of current density for $Fe(CN)_6^{3-}$ reduction both obviously decreased, which decreased the maximum power density (output 0). However, when thrombin and ATP were both present (input (1,1)), ATP in the mixture may also interact with the extended ATP-binding aptamer embedded in the mixed aptamer and draw the mixed aptamer away from the electrode surface, greatly increasing the onset potential and plateau of current density for $Fe(CN)_6^{3-}$ reduction; accordingly the maximum power density of the BFC increased (output 1).

After logic operation, ATP and mixed aptamer were consecutively added, resulting in the regeneration of the electrode interface, thus re-activating the electrode. Therefore, the features of the system correspond to the equivalent circuitry of an IMP logic gate performing the Boolean logic operation of $A' + B$. Owing to the unique function of IMP logic, the aptamer IMP-Reset logic system proposed can be used to "smartly" determine the presence of one specific target in the absence of another target in human serum in a single test.

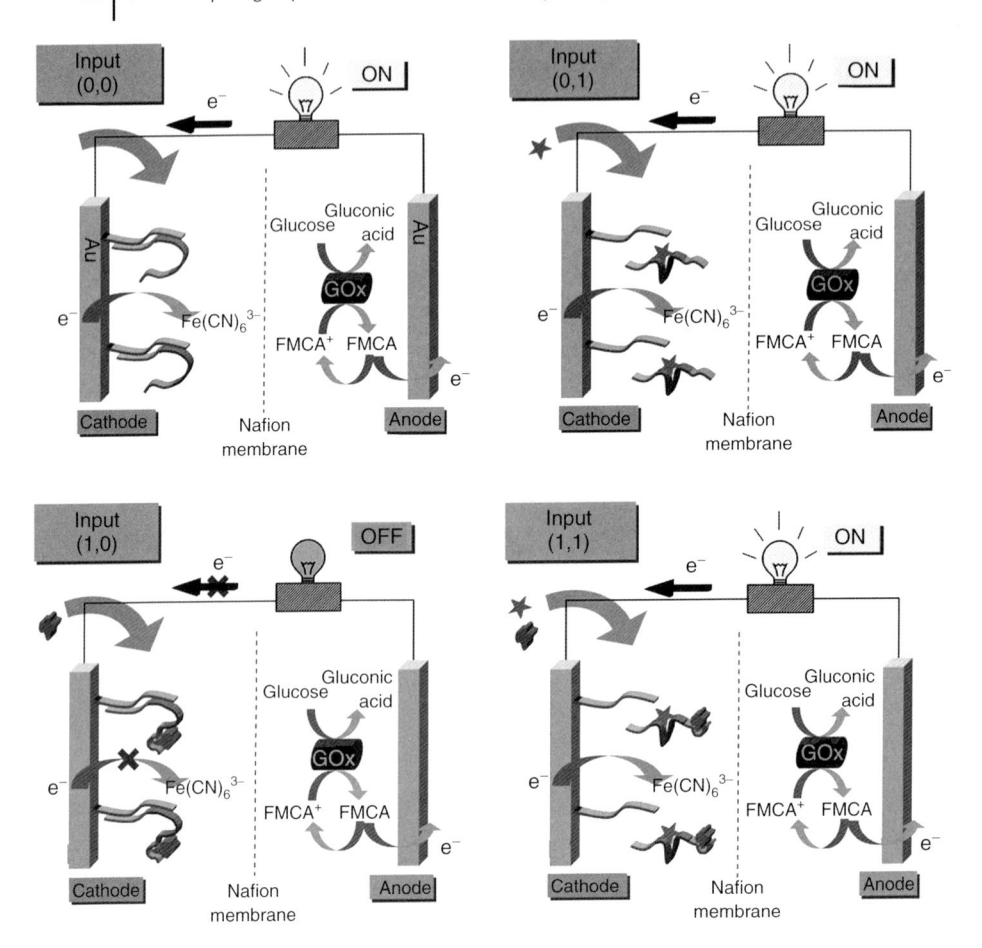

Figure 7.4 Schematic illustration of the switchable BFC for logically controlling the power release by different combinations of aptamer-target recognition-based input signals [32].

7.4
Enzyme Biocomputing

Since the first example of enzyme logic presented by Sivan and Lotan in 1999 [14], different enzyme logic gates are reported [10, 27, 46–55]. The mechanisms for early enzymatic logic gates are mainly based on enzyme-catalyzed reactions with the optical read of the output signals [10, 27, 46–48]. Recently, on the basis of pH-switchable polymer brushes/membranes [51, 56], different Boolean logic gates were realized [49–52]. By achieving different logic functions, enzymatic logic biosensing systems were developed [53–55, 57, 58].

In 2011, we demonstrated the first "undestructive" enzymatic logic gates system based on O_2-controlled BFC to realize multiple logic operations [33]. We first modified FMCA and 2,2′-azinobis(3-ethylbenzthiazoline-6-sulfonic acid) diammonium salt (ABTS) on ordered mesoporous carbons (OMCs), which were used as the mediators for glucose electrooxidation at GOx-bioanode and O_2 electroreduction at BOx-biocathode, respectively. On the basis of the interplay between bioanode and biocathode, such a system permits AND and XNOR Boolean logic gates readily achieved, reset, and interconverted into each other in "one pot." Compared with the reported enzymatic logic gates, especially the BFC-based enzymatic logic gates, the present O_2-controlled enzymatic logic gates exhibit several advantages: (i) This is the first strategy that constructs enzymatic logic gates based on O_2 control for regulating both the bioanode and biocathode containing O_2-sensitive materials. This approach is more convenient and versatile compared with the traditional enzymatic logic gates based on the liquid input [10, 27, 46–48], especially controlling only biocathode/one electrode by liquid input [49, 50, 54, 59]. (ii) The unique O_2-controlled strategy permits different Boolean logic gates (with two-input and one-output) readily to be realized, reused, and interconverted with each other in "one pot." This would be extremely advantageous for the development of "all in one" bioelectronic devices that can fulfill multiple logic functions in one component. (iii) This is also the first compartmentless BFC-based enzymatic logic gate being integrated with biocatalytic electrodes rather than enzyme-catalyzed reactions operating in solutions. So in such system, smaller quantities of enzymes are needed and the logic gates are easier to recover and reuse; more value for our money [49, 60]. This would not only be of advantage to fabricating BFC with higher power output but also be of extreme benefit to manufacturing more simple and efficient enzymatic logic gates for the versatile logic applications as well as the self-powered and "smart" implantable medical diagnosis aim.

Later, by introducing the gas control concept into biocomputing security keypad lock field, we fabricated the first reusable BFC-based biocomputing security system mimicking a keypad lock device [33]. Owing to the unique "RESET" reagent of N_2 applied in this work, the prepared biocomputing security system can be reset and cycled for a large number of times with no "RESET" reagent-based "waste," which would be advantageous for the potential practical applications of such a keypad lock as well as the development of biocomputing security devices (Figure 7.5). Coupling BFCs with keypad locks might not only significantly increase the versatility of BFCs resulting in the BFC-based biocomputing security system but also greatly enhance the adaptability of keypad lock to future self-powered and reusable biocomputing-security-system-based bioelectronic devices. In order to validate the universality of the system and also to harvest energy directly from biofuels with enhanced power output, glucose was replaced with orange juice as the biofuel to operate a BFC-based biocomputing system, which also possesses the function of keypad lock. In addition, by introducing BFCs into the biocomputing security system, the adaptive behavior of the BFCs self-regulating the power release would be an immense advantage of such security keypad lock devices in potential self-powered implantable medical systems. The principle of the influence of gas injection and electrolyte agitation on BFC performance is based on dissolved O_2 and liquid flow, respectively. Thus, based on the relationship between the dissolved O_2 concentration in blood and ischemic state/heart attack [61], as well

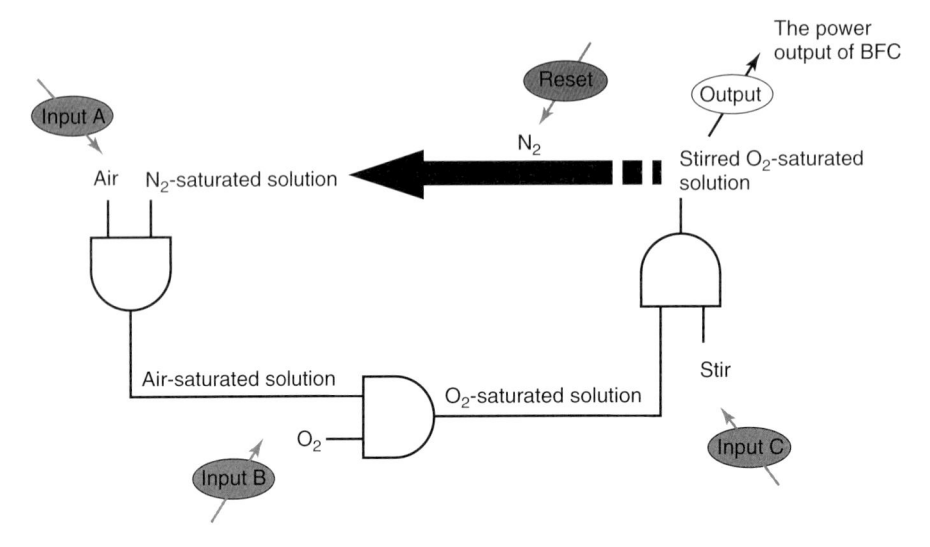

Figure 7.5 Representation of the BFC-based nondestructive "SET-RESET" biocomputing keypad lock system as a network of three concatenated AND gates and RESET function [33].

as the relationship between the blood flow rate in blood vessel and the cardiac output/mean blood pressure [62, 63], the current biocomputing security system can be used as the potential implant device model for assessing physiological conditions in living organisms and for making autonomous decisions on the use of specific tools/drugs in various implantable medical systems. Nearly all living organisms need O_2 to carry on life processes and blood flow to maintain life. Thus, on the other hand, on the conceptual level, this work might help us understand how living organisms manage to control extremely complex and coupled biochemical reactions; that is, the hope is to cast biochemical (metabolic) pathways in the language of information theory.

Furthermore, we report another reusable BFC-based biocomputing security system that mimics a keypad lock device, depending on the enzyme-based parameters as readin and the open circuit potential of BFC as readout [35], based on the OMCs-BFC we reported before [64], which permits the biocomputing security system to be self-powered (Figure 7.6a). The study may also shed light on mimicking and designing natural signal transduction and metabolic and gene regulation systems [65, 66]. One can think of Figure 7.6a as composed of three motifs. The three inputs act as connectors to the reaction modules, so that the three reactions can cascade down normally to produce the positive output responses. This system is analogous to a common signal transduction subnetwork mitogen-activated protein (MAP) kinase with three motifs of enzyme reactions of phosphorylation–dephosphorylation connected together [67, 68]. These motifs mimic the "logic gates." The function of MAP kinase subnetwork is to transmit information from upstream to downstream in a fast and reliable way. The specific sequence combination of the ambition MAP kinase motif generates the overall funneled landscape [67, 68], which guarantees the unique direction of information

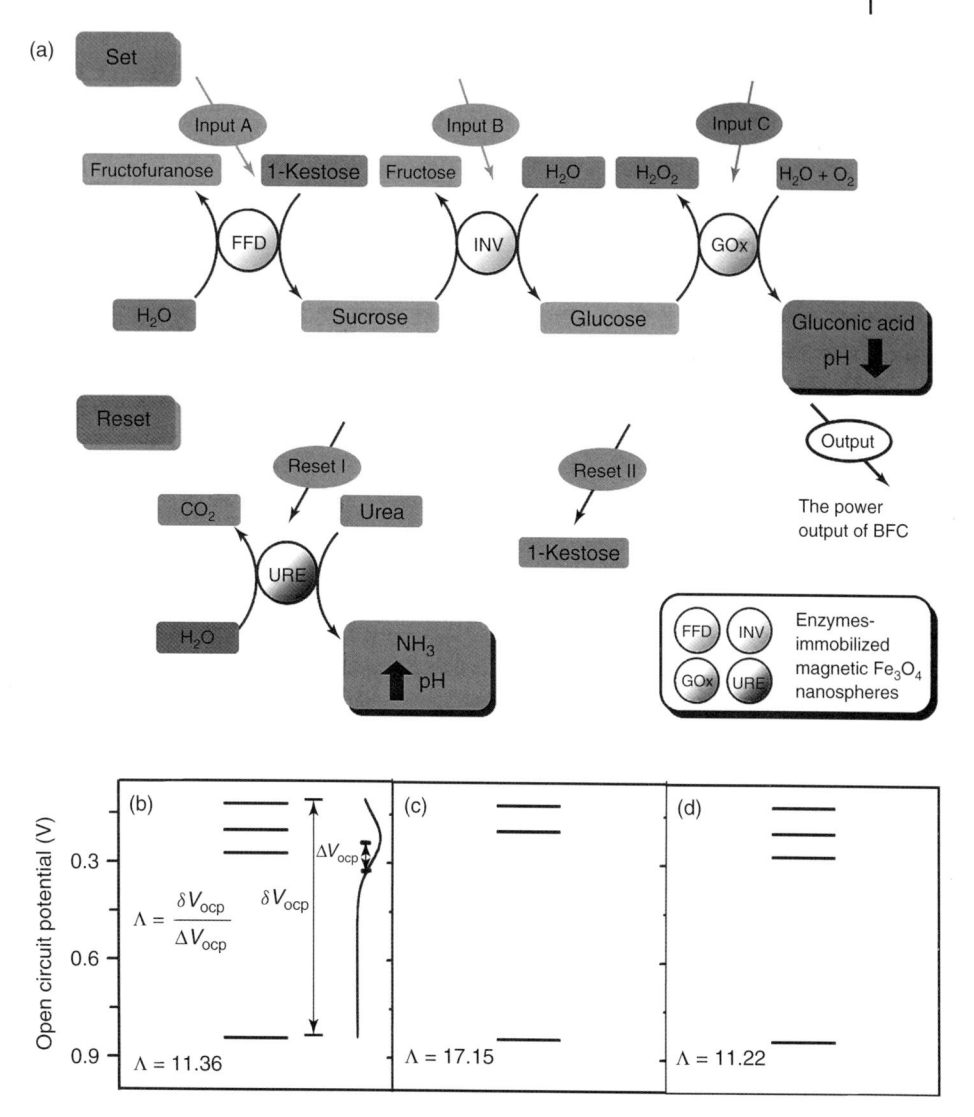

Figure 7.6 (a) Biocatalytic reactions in the biocathode chamber of the BFC-based self-powered and reusable biocomputing keypad lock system [35]. FFD = fructan beta-fructosidase, INV = invertase, GOx = glucose oxidase, URE = urease. Funneled landscape for the open circuit potential of the BFC [35]; (b) open circuit potential spectrum for eight different patterns of biochemical input signals; (c) open circuit potential spectrum for six different order of biochemical input signals; and (d) open circuit potential spectrum for eight different patterns of biochemical input signals and six different order of biochemical input signals. δV_{OCP} is the gap between the maximum and average open circuit potential, ΔV is the standard deviation measuring the variance of open circuit potentials from the mean. The ratio $\Lambda = \delta V_{OCP}/\Delta V_{OCP}$ gives an absolute and dimensionless measure of the degree of funnel and discrimination. The open circuit potential spectrum forms a funnel-like landscape with a large gap discriminating the best sequence from the others.

flow and stability of function. In this work, we built up a system to perform logic gate functions. Only with specific sequences of three AND gates, it can guarantee the resulting functions, therefore effectively generating an overall funneled landscape. The system with other combinations of sequences will effectively generate a rough landscape and will not guarantee the required function (Figure 7.6b–d). Such an approach provides a new way of mimicking the biological systems and is therefore important for design in systems and synthetic biology [65, 66]. It also bridges the gaps between theoretical concepts, such as information/landscape theory, and experiments as well as practical realizations of chemical and biological systems.

7.5
Conclusions and Perspectives

Biocomputers are manmade biological networks whose goal is to probe and control biological host cells and organisms in which they operate. Their key design features, informed by computer science and engineering, are programmability, modularity, and versatility. While still a work in progress, biocomputers will eventually enable disease diagnosis and treatment with single-cell precision, lead to "designer" cell functions for biotechnology, and bring about a new generation of biological measurement tools [1, 69].

DNA computing is thought to be the future of computer technologies, while DNA molecules that perform logic operations are a prerequisite for digital information processing and computing. A conceptually new class of DNA logic gate based on cation-tuned ligand binding and release was devised by utilizing a cation-driven allosteric G-quadruplex DNAzyme, which can function as a two-input INHIBIT logic gate. With the introduction of another input ethylenediaminetetraacetic acid (EDTA), this G-quadruplex can be further utilized to construct a reversibly operated IMPLICATION gate. On the basis of DNA/Ag fluorescent nanoclusters, another novel kind of versatile logic device utilizing the ion tuning was constructed. Aptamers, which are novel artificially selected functional DNA or RNA oligonucleotides, exhibit many unprecedented advantages in comparison with antibodies (i.e., immune system) or other biomimetic receptors, including a wide range of targets (ranging from proteins to peptides, amino acids, drugs, metal ions, and even whole cells), comparable or even better target affinity, easy and cost-effective synthesis with high reproducibility and purity, simple and straightforward chemical modification, and so on. For the first time, we demonstrate the controlled power release of BFCs by aptamer logic systems processed according to the Boolean logic operations "programed" into the biocomputing systems. On the basis of the built-in Boolean NAND logic, the fabricated aptamer-based BFCs logically controlled by biochemical signals enabled us to construct self-powered and intelligent logic aptasensors that can determine whether the two specific targets are both present in a sample. In addition, by combining the adaptive behavior of microfluidic BFCs self-regulating the power release with aptamer IMP logic, we further constructed a novel IMP-Reset gate-based reusable and self-powered on-chip aptasensor, which can be used to logically determine the presence of one specific target in the absence of another target in complex physiological samples (such as human serum) in a single test. The inherent selectivity of enzymes to promote selective reaction of enzymatic substrates as well as their wide existence in human bodies make enzymes suitable for both logic realization and potential biocomputing logic detection for various diseases. By introducing the gas control concept into the enzymatic biocomputing field, we not only realize the enzymatic logic gates for the versatile logic applications but also make enzymatic biocomputing systems for the self-powered and "smart" implantable medical diagnosis aim.

Although a great deal of progress has already been reported for biocomputing, several challenges and obstacles still keep them far away from real-world applications. Most biocomputing systems reported until now represent only the proof of the concept, demonstrating the possibility of performing computing/logic operations with the use of biomolecular systems. They are not ready yet for any practical application in terms of computational power, efficiency, and user-friendliness. An expert opinion of Stojanovic can be cited supporting this conclusion: "after 10 years of intensive efforts, and large investment, we have to admit that DNA computation is unlikely to make modern silicon computers obsolete, or, indeed, ever to solve any useful computational problem much faster than the average human can" [30, 70]. If biocomputing could replace silicon-based electronic devices in the future, lots of work is needed. The other direction of biocomputing is not aiming at any complex computation but rather creating a "smart" information processing interface between biological and electronic systems, which would be further used as logic biosensors for the potential intelligent medical diagnostics. However, most logic biosensors developed are tested in buffer systems in the laboratory. It should be deliberated on the sample matrix effects as well as the biocomputing system stabilities for the potential commercial medical diagnosis and real-time environmental monitoring. With scaling up, the complexity and diversity of biocomputing systems, data analysis for optimization, and noise reduction are beneficial to their real applications. The preceding decade has seen great progress in biocomputing. The exploration of novel biocomputing applications as well as the creation of new concepts and methods also merits further attention.

References

1. Szacilowski, K. (2008) *Chem. Rev.*, **108**, 3481–3548.
2. de Silva, P.A., Gunaratne, N.H.Q., and McCoy, C.P. (1993) *Nature*, **364**, 42–44.
3. Kompa, K.L. and Levine, R.D. (2001) *Proc. Natl. Acad. Sci. U.S.A.*, **98**, 410–414.
4. de Silva, A.P. and Uchiyama, S. (2007) *Nat. Nanotechnol.*, **2**, 399–410.
5. Nurse, P. (2008) *Nature*, **454**, 424–426.
6. Willner, I., Shlyahovsky, B., Zayats, M., and Willner, B. (2008) *Chem. Soc. Rev.*, **37**, 1153–1165.
7. Katz, E. and Privman, V. (2010) *Chem. Soc. Rev.*, **39**, 1835–1857.
8. Stojanovic, M.N. and Stefanovic, D. (2003) *J. Am. Chem. Soc.*, **125**, 6673–6676.

9. Saghatelian, A., Volcker, N.H., Guckian, K.M., Lin, V.S.Y., and Ghadiri, M.R. (2003) *J. Am. Chem. Soc.*, **125**, 346–347.
10. Niazov, T., Baron, R., Katz, E., Lioubashevski, O., and Willner, I. (2006) *Proc. Natl. Acad. Sci. U.S.A.*, **103**, 17160–17163.
11. Muramatsu, S., Kinbara, K., Taguchi, H., Ishii, N., and Aida, T. (2006) *J. Am. Chem. Soc.*, **128**, 3764–3769.
12. Simpson, M.L., Sayler, G.S., Fleming, J.T., and Applegate, B. (2001) *Trends Biotechnol.*, **19**, 317–323.
13. Rosen, R. (1967) *J. Theor. Biol.*, **15**, 282–297.
14. Sivan, S. and Lotan, N. (1999) *Biotechnol. Prog.*, **15**, 964–970.
15. Adleman, L.M. (1994) *Science*, **266**, 1021–1024.
16. Qian, L. and Winfree, E. (2011) *Science*, **332**, 1196–1201.
17. Kari, L., Gloor, G., and Yu, S. (2000) *Theor. Comput. Sci.*, **231**, 193–203.
18. Famulok, M., Hartig, J.S., and Mayer, G. (2007) *Chem. Rev.*, **107**, 3715–3743.
19. Tombelli, S. and Mascini, M. (2009) *Curr. Opin. Mol. Ther.*, **11**, 179–188.
20. Liu, J., Cao, Z., and Lu, Y. (2009) *Chem. Rev.*, **109**, 1948–1998.
21. Sefah, K., Phillips, J.A., Xiong, X., Meng, L., Simaeys, D.V., Chen, H., Martin, J., and Tan, W. (2009) *Analyst*, **134**, 1765–1775.
22. Kolpashchikov, D.M. and Stojanovic, M.N. (2005) *J. Am. Chem. Soc.*, **127**, 11348–11351.
23. Clark, L.C. and Lyons, C. (1962) *Ann. N. Y. Acad. Sci.*, **102**, 29–45.
24. Barton, S.C., Gallaway, J., and Atanassov, P. (2004) *Chem. Rev.*, **104**, 4867–4886.
25. Cracknell, J.A., Vincent, K.A., and Armstrong, F.A. (2008) *Chem. Rev.*, **108**, 2439–2461.
26. Tsujimura, S., Tatsumi, B., Ogawa, J., Shimizu, S., Kano, K., and Ikeda, T. (2001) *J. Electroanal. Chem.*, **496**, 69–75.
27. Baron, R., Lioubashevski, O., Katz, E., Niazov, T., and Willner, I. (2006) *Angew. Chem. Int. Ed.*, **45**, 1572–1576.
28. Benenson, Y., Paz-Elizur, T., Adar, R., Keinan, E., Livneh, Z., and Shapiro, E. (2001) *Nature*, **414**, 430–434.
29. Xu, J. and Tan, G.J. (2007) *J. Comput. Theor. Nanosci.*, **4**, 1219–1230.
30. Wang, J. and Katz, E. (2010) *Anal. Bioanal. Chem.*, **398**, 1591–1603.
31. Zhou, M., Du, Y., Chen, C., Li, B., Wen, D., Dong, S., and Wang, E. (2010) *J. Am. Chem. Soc.*, **132**, 2172–2174.
32. Zhou, M., Chen, C., Du, Y., Li, B., Wen, D., Dong, S., and Wang, E. (2010) *Lab Chip*, **10**, 2932–2936.
33. Zhou, M., Zheng, X., Wang, J., and Dong, S. (2011) *Bioinformatics*, **27**, 399–404.
34. Zhou, M., Wang, F., and Dong, S. (2011) *Electrochim. Acta*, **56**, 4112–4118.
35. Zhou, M., Zheng, X., Wang, J., and Dong, S. (2010) *Chem. Eur. J.*, **16**, 7719–7724.
36. Freeman, R., Finder, T., and Willner, I. (2009) *Angew. Chem. Int. Ed.*, **48**, 7818–7821.
37. Frezza, B.M., Cockroft, S.L., and Ghadiri, M.R. (2007) *J. Am. Chem. Soc.*, **129**, 14875–14879.
38. Elbaz, J., Lioubashevski, O., Wang, F.A., Remacle, F., Lioubashevski, R.D., and Willner, I. (2010) *Nat. Nanotechnol.*, **5**, 417–422.
39. Qian, L.L. and Winfree, E. (2011) *Science*, **332**, 1196–1201.
40. Li, T., Wang, E., and Dong, S. (2009) *J. Am. Chem. Soc.*, **131**, 15082–15083.
41. Li, T., Dong, S., and Wang, E. (2009) *Chem. Asian J.*, **4**, 918–922.
42. Li, T., Zhang, L., Ai, J., Dong, S., and Wang, E. (2011) *ACS Nano*, **5**, 6334–6338.
43. Ellington, A.D. and Szostak, J.W. (1990) *Nature*, **346**, 818–822.
44. Tombelli, S., Minunni, A., and Mascini, A. (2005) *Biosens. Bioelectron.*, **20**, 2424–2434.
45. Konry, T. and Walt, D.R. (2009) *J. Am. Chem. Soc.*, **131**, 13232–13233.
46. Baron, R., Lioubashevski, O., Katz, E., Niazov, T., and Willner, I. (2006) *J. Phys. Chem. A*, **110**, 8548–8553.
47. Baron, R., Lioubashevski, O., Katz, E., Niazov, T., and Willner, I. (2006) *Org. Biomol. Chem.*, **4**, 989–991.
48. Strack, G., Pita, M., Ornatska, M., and Katz, E. (2008) *ChemBioChem*, **9**, 1260–1266.

49. Amir, L., Tam, T.K., Pita, M., Meijler, M.M., Alfonta, L., and Katz, E. (2009) *J. Am. Chem. Soc.*, **131**, 826–832.
50. Privman, M., Tam, T.K., Pita, M., and Katz, E. (2009) *J. Am. Chem. Soc.*, **131**, 1314–1321.
51. Motornov, M., Zhou, J., Pita, M., Gopishetty, V., Tokarev, I., Katz, E., and Minko, S. (2008) *Nano Lett.*, **8**, 2993–2997.
52. Motornov, M., Marcos, J.Z., Ihor, P., Venkateshwarlu, T., Gopishetty, V., Katz, E., and Minko, S. (2009) *Small*, **5**, 817–820.
53. Manesh, K.M., Halámek, J., Pita, M., Zhou, J., Tam, T.K., Santhosh, P., Chuang, M.-C., Windmiller, J.R., Abidin, D., Katz, E., and Wang, J. (2009) *Biosens. Bioelectron.*, **24**, 3569–3574.
54. Pita, M., Zhou, J., Manesh, K.M., Halámek, J., Katz, E., and Wang, J. (2009) *Sens. Actuator. B: Chem.*, **139**, 631–636.
55. Tam, T.K., Pita, M., and Katz, E. (2009) *Sens. Actuator. B: Chem.*, **140**, 1–4.
56. Tokarev, I., Orlov, M., Katz, E., and Minko, S. (2007) *J. Phys. Chem. B*, **111**, 12141–12145.
57. Halamek, J., Windmiller, J.R., Zhou, J., Chuang, M.-C., Santhosh, P., Strack, G., Arugula, M.A., Chinnapareddy, S., Bocharova, V., Wang, J., and Katz, E. (2010) *Analyst*, **135**, 2249–2259.
58. Chuang, M.-C., Windmiller, J.R., Santhosh, P., Ramirez, G.V., Katz, E., and Wang, J. (2011) *Chem. Commun.*, **47**, 3087–3089.
59. Pita, M. and Katz, E. (2008) *J. Am. Chem. Soc.*, **130**, 36–37.
60. de Silva, A.P. (2008) *Nature*, **454**, 417–418.
61. Martin, L. (1999) *All You Really Need to Know to Interpret Arterial Blood Gases*, 2nd edn, Lippincott Williams and Wilkins, New York.
62. Guyton, A.C. (1981) *Textbook of Medical Physiology*, 6th edn, W.B. Saunders Company.
63. Taylor, G.L., Patel, B., and Sullivan, A.T. (2007) *J. Pharmacol. Toxicol. Methods*, **56**, 212–217.
64. Zhou, M., Deng, L., Wen, D., Shang, L., Jin, L., and Dong, S. (2009) *Biosens. Bioelectron.*, **24**, 2904–2908.
65. Davidson, E.H. and Erwin, D.H. (2006) *Science*, **311**, 796–797.
66. Buchler, N.E., Gerland, U., and Hwa, T. (2003) *Proc. Natl. Acad. Sci. U.S.A.*, **100**, 5136–5141.
67. Wang, J., Huang, B., Xia, X., and Sun, Z. (2006) *Biophys. J.*, **91**, L54–L56.
68. Wang, J., Zhang, K., and Wang, E. (2008) *J. Chem. Phys.*, **129**, 135101–1351019.
69. Benenson, Y. (2009) *Mol. BioSyst.*, **5**, 675–685.
70. Stojanovic, M., Stefanovic, D., LaBean, T., and Yan, H. (2005) in *Bioelectronics: from Theory to Applications* (eds I. Willner and E. Katz), Wiley-VCH Verlag GmbH, Weinheim, pp. 427–455.

8
Some Experiments and Models in Molecular Computing and Robotics

Milan N. Stojanovic and Darko Stefanovic

8.1
Introduction

Molecular computing lies at the interface of experimental chemistry and computer science. The field is still in such a raw state that any attempt to define it as anything more precise than a *mingling of concepts from computer science and (bio)chemistry* would be counterproductive and may forestall further intriguing and completely original developments. Almost every process in chemistry contains some aspect of information processing and some properties that can be mapped to concepts from computer science and reinterpreted as "molecular computing." The question for chemists considering working at this interface is whether recognizing such a mapping will do them any good – for instance, facilitate their understanding of natural biochemical processes or of the chemical origins of life, or perhaps lead them to some practical applications. It is very rewarding to see that even the straightforward approaches of recognizing logic gate-like information processing in enzymatic reactions can become a treasure trove of bioanalytical breakthroughs [1].

In this chapter, we describe projects in which computational thinking helped steer our work into new directions in chemistry. It is a review of our quest for unusual and complex behaviors of molecules, which can be seen as the present point on the historical line of challenges in synthetic chemistry and as our answer to a synthetic chemist's favorite question: Can this molecule do something interesting?

The first section is on what is by now almost considered a traditional topic, using molecular logic gates to build complex circuits, but we focus on our game-playing molecular automata. In the second section, we describe our progress in molecular robotics.

8.2
From Gates to Programmable Automata

In our work, information processing is performed by deoxyribozyme phosphodiesterases [2], all-DNA enzymes that are allosterically regulated by other oligonucleotides and that cleave oligonucleotide substrates. We can monitor the

Biomolecular Information Processing: From Logic Systems to Smart Sensors and Actuators,
First Edition. Edited by Evgeny Katz.
© 2012 Wiley-VCH Verlag GmbH & Co. KGaA. Published 2012 by Wiley-VCH Verlag GmbH & Co. KGaA.

activity of deoxyribozymes by means of fluorogenic labeling of their substrates (Figure 8.1a). Allosteric regulation [3] by oligonucleotides is accomplished by combining deoxyribozyme modules with stem–loop modules inspired by molecular beacons [4] (Figure 8.1b). The engineering principle is simple: the recognition module blocks the access of the substrate to the enzyme; the addition of an effector oligonucleotide complementary to the stem activates the enzyme by removing this blockade (Figure 8.1c). Two or more such modules can be added to a deoxyribozyme, allowing allosteric regulation by more than one oligonucleotide. We can describe the relationship ⟨presence/absence of allosteric effectors⟩ → ⟨enzymatic activity⟩ through correlation or lookup tables. Ideally, the presence of effectors and the presence of output could be assigned a value of 1, and their absence a value of 0 [5]. The correlation tables then have the form of truth tables of Boolean logic, as in the logic gates that form the basis of digital computing. This point of view allows us to directly relate our approach to the existing digital computing paradigm and to organize molecules into systems that perform more complex logic computation.

With a single stem–loop hindering the access of the deoxyribozyme substrate to the substrate recognition region (Figure 8.1c, YESi$_1$) we obtain a YES gate [6]. The binding of the complementary oligonucleotide to the loop opens the stem, allowing the substrate to bind and be cleaved. In contrast, the binding of an oligonucleotide to a stem–loop region embedded within the catalytic core of some deoxyribozymes turns its enzymatic activity off, leading to a NOT-gate-like behavior [5] (Figure 8.1d). The placement of two stem–loop regions controlling the two substrate recognition regions of an enzyme requires the presence of both inputs for enzymatic activity, resulting in an AND gate (Figure 8.1e). Finally, we can combine motifs leading to YES and NOT gate behaviors, and obtain ANDNOT and ANDANDNOT gates (Figure 8.1f,g) [6, 7]. Once we have the basic molecular logic units, we can arrange them together in solution by simply mixing them and allowing them to cleave the same substrate (so-called implicit-OR arrangement because the integrated truth table shows that any of the gates can cleave the substrate to produce output 1) or different substrates. For example, an XOR gate is obtained by combining two ANDNOT gates (Figure 8.1g) [5]. From a chemistry perspective, this circuit is also interesting because it allows an effector to function at the same time as an inhibitory or promotory element in a biochemical circuit.

Our work on more complex circuits has mostly focused on molecular circuits or automata known as Molecular Array of YES and AND gates (MAYAs). To date, we have constructed three generations of game-playing MAYA's; now MAYA-I) [7–9]. The task of game playing provides us with an objective test bed for the ability of molecules to perform more complex processing of information that is exchanged with an environment (here represented by a human player adding oligonucleotides). The rules of the game, once they are set, are beyond our control. And, success is judged objectively – it must be demonstrated by playing the game successfully. Our first two automata, MAYA-I and MAYA-II, played versions of the tic-tac-toe game; interestingly, the first game ever played by an electronic computer was tic-tac-toe, so this seemed a proper choice for our computing devices as well.

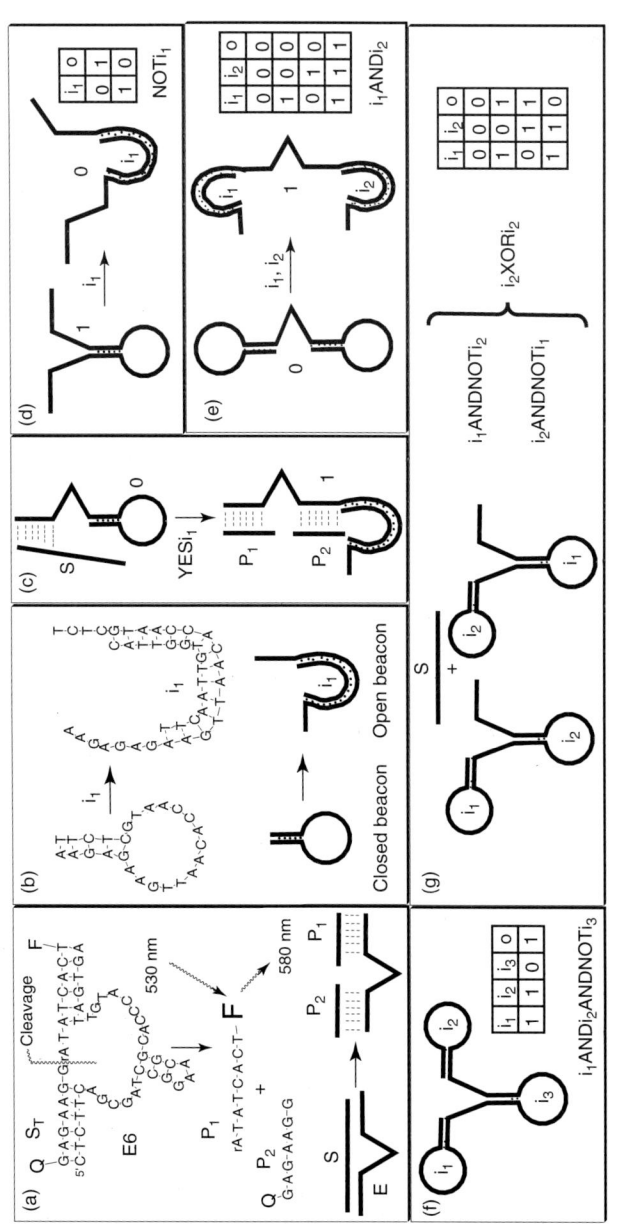

Figure 8.1 (a) Deoxyribozyme (E) in a complex with its oligonucleotide substrate (S). The cleavage reaction produces the two shorter oligonucleotide products (P_1 and P_2) as output (O). (b) The molecular beacon stem–loop is our preferred recognition module for oligonucleotides: a closed beacon has a stem–loop conformation, but the addition of oligonucleotides complementary to the loop opens the stem and can be interpreted as an input i_1. (c) A catalytic molecular beacon or $YESi_1$ or sensor gate is constructed by attaching a beacon module to one of the substrate recognition regions of the deoxyribozyme module. On the addition of an input (i_1), the gate switches into its active form. The reaction can be monitored fluorogenically. (d) A $NOTi_1$ gate is constructed when a stem–loop is added to the catalytic core of enzyme E6. Opening of the stem distorts the catalytic core and inhibits the enzymatic reaction. We also show the input/output correlation table, which is the truth table of NOT logic. (e) An i_1ANDi_2 gate is constructed when two stem–loops are added to both substrate recognition regions of the deoxyribozyme module. Both stems have to open for a substrate to be cleaved, thus both inputs have to be present. (f) A three-input gate $i_1ANDi_2ANDNOTi_3$ is active when two inputs (i_1 and i_2) are present, but not the third (i_3); it is a combination of an i_1ANDi_2 gate and a $NOTi_3$ gate. We show a partial truth table corresponding only to the unique combination of inputs resulting in an active gate. (g) A molecular XOR gate is constructed by mixing two gates, $i_1ANDNOTi_2$ and $i_2ANDNOTi_1$, that cleave the same substrate and are thus in an implicit-OR arrangement.

MAYA-III, in contrast, is a programmable protoautomaton that can be trained to play according to any of the strategies of an invented retributive game. In all automata, we introduce human moves of the game to the automaton through a "language" that molecular circuits can accept – input oligonucleotides – to form a sequence of inputs, and we obtain a sequence of responses interpretable as game moves that form a successful game play by the automaton.

Tic-tac-toe is played against a set of solutions distributed in the nine (3 × 3) wells of a 384-well plate mimicking the nine squares of a tic-tac-toe game board [7]. Individual wells are sequentially numbered 1–9. In MAYA-I (Figure 8.2), the game was symmetry-pruned: the automaton always goes first in well #5, and the human player is restricted to responding in either the corner square #1 or the side square #4. These two simplifications limit the number of legal game plays to 19, with 18 resulting in a win for the automaton after any mistake by a human, and one ending in a draw, with both players playing perfectly.

The automaton is activated by adding Mg^{2+} to all wells; this starts a reaction by a constitutively active enzyme in well #5. The input oligonucleotides keyed to human moves are added to all the wells, triggering a specific response by the automaton in only one of the wells. In MAYA-I, eight inputs are used, keyed to the human move into wells #1–4 or #6–9 (with #5 already used by the automaton in the opening move). In order to move into well #1, the human player will add input i_1;

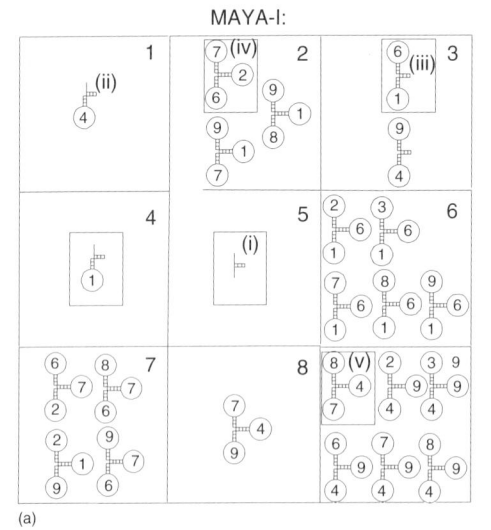

Figure 8.2 The first MAYA, a molecular automaton that plays a symmetry-pruned tic-tac-toe game. (a) Distribution of gates in wells. The central well (#5) contains a constitutively active deoxyribozyme, while the other wells contain logic gates. Gates used in the example game play (as in b) are boxed. (b) An example of a game play in which the human plays perfectly and draws. There are total of 19 individual game plays encoded in this distribution of logic gates; the remaining 18 end in MAYA's forced victory.

A game-play example (draw):

(i) By design, the automaton moves first in the middle well.

(ii) Human chooses well 1 and signals this choice by adding i_1 to all wells; this activates only the $YESi_1$ gate in well 4.

(iii) Human blocks three in a row, adding i_6 to all wells; this, together with previously added i_1, activates only one of the gates, $i_1 ANDi_6$ in well 3.

(iv) Human blocks three in a row, adding i_7 to all wells; this, together with previously added i_6 activates only one of the gates, $i_6 ANDi_7 ANDNOTi_2$ in well 2.

(v) Human blocks three in a row, adding i_8 to all wells; this, together with previously added i_7 activates one of the gates, $i_7 ANDi_8 ANDNOTi_4$ in well 9.

to move into well #2, the human player will add input i_2, and so on. The automaton was initially constructed by deducing the required Boolean logic in each well, with human moves as inputs and automaton moves as outputs, and then mixing up individual molecular gates to obtain such logic in each of the wells. As a result, an array of 23 logic gates distributed in eight wells calculates a response to the human player's input. The cycle of human player input followed by automaton response continues until there is a draw (if the human plays perfectly) or a victory for the automaton. The perfect game-playing ability of MAYA-I was demonstrated in over 100 test game plays.

After MAYA-I, we addressed a purely engineering challenge of building a larger automaton with MAYA-II and its 129 enzymes distributed over nine wells [8]. This time, the human player was free to choose the first move into any of the remaining wells, and the automaton could play all 76 games provided for by its hardwired game strategy. MAYA-II uses an array of 96 logic gates to respond to human moves, analyzing 32 input oligonucleotides that encode both the well position and the order of the human move (the first, second, third, or fourth move). MAYA-II is also considered "user friendly" in the sense that it echoes the human moves via 32 YES gates that cleave a differently colored fluorogenic substrate. In MAYA-II, human moves were displayed in the green channel (fluorescein) and automaton moves were displayed in the red channel (TAMRA). An important aspect of MAYA-II was that we had to go through an optimization and selection process to get a set of gates that behaved satisfactorily. As a result, we now have a "parts library" and we can predict relatively easily how a gate will behave in a larger construct.

Yet, except quantitatively (particularly regarding the time invested in its construction), even MAYA-II was not that conceptually different from the first circuit that we ever constructed, the XOR. In essence, by mixing the necessary gates, we hardwired the automaton as a realization of a particular Boolean formula that itself is a realization of a particular (carefully chosen) strategy for tic-tac-toe. (A strategy is a self-consistent subset of the complete game tree, providing unique responses to all permissible moves by the opponent.) But, is there a method to *program* or *teach* molecules to play a particular strategy of a game, instead of hardwiring it? And then perhaps *reprogram* it to play another strategy (following different paths in the game tree) or even to play a completely different game? This would require us to address the issue of reconfiguring a molecular mixture, and the difficulty is that most of what we know about programing electronic computers is not really applicable to molecular computing. This led us to introduce two concepts into the exploration of molecular automata, field-programmable (reconfigurable) molecular logic arrays and "teaching by example," and one new game, *tit for tat*. They are embodied in the automaton MAYA-III [9].

The tit-for-tat game is an example of trivial two-player two-move games played on a board split into four fields, invented explicitly for the purpose of studying the ability to train/program molecular automata. These two-player two-move games are indeed trivial, but have one great advantage: after some restrictions in the way we observe moves (focusing only on the remaining fields for the second move),

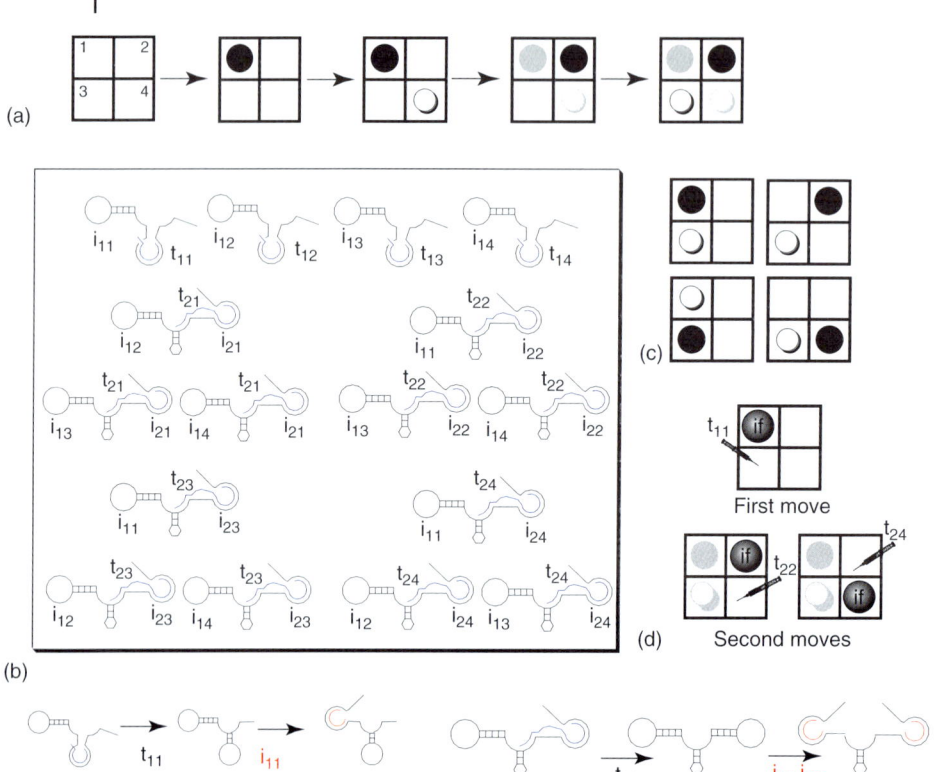

Figure 8.3 MAYA-III is an automaton that plays all strategies of two-player two-move games, with one example being tit for tat. (a) An example of a tit-for-tat game play, with human moves shown as filled circles and automaton moves as hollow circles. Past moves are shown in gray, current moves in black. (b) The distribution of gates before training is identical in all the wells; during the training the t inputs are evaluated, differentially changing the Boolean logic in individual wells. (c) An example of the strategy or a set of possible responses to all human first moves (in winning strategies the automaton has no choice on its second move). (d) Part of a training session, with training for one of four game plays shown. Training consists of injecting training inputs in an intuitive way, mimicking the actual game play. The complete training consists of covering all four possible game plays in a single strategy in four sessions such as this, for example, for strategy under (c). (e) Changes in the gates during the training sessions and during the game.

their complete action space can be represented with a field array with a total of 16 gates: a set of 4 YES gates, responding to the inputs keyed to the first moves, and 12 AND gates, responding to all legal combinations of inputs keyed to the first and second moves. This allows us to select any function within this action space, without having to optimize and coordinate an impractical number of gates. Before we explain how the function is selected, we describe the game itself (Figure 8.3).

In the game, the automaton's goal is simply to match each human move into one field with a move into another field, tit for tat. The game has only 81 *winning game strategies*, defined as comprehensive sets of responses to all possible moves by the human leading to automaton fulfilling its goal (each strategy has 4 × 2 possible game plays). This gives us the opportunity to select each of these strategies for tit for tat within the field-programmable array representing an action space. But in order to perform the selection, we need to introduce one more set of inputs, so-called training inputs. The training inputs allow us to activate the required gates in individual wells of the automaton in training sessions. Individual training sessions resemble playing individual games; the result is that the automaton learns how to play all possible game plays within a single strategy. Effectively, the training sessions turn a fully symmetric distribution of gates in the four wells into MAYA-I-like specialized distributions that play the one chosen strategy. Importantly, the human trainer (and likewise the human player) does not need to understand any molecular logic in order to select a strategy and teach it to the automaton. Teaching is by example, and the procedure requires one only to have a key for using the training inputs.

The importance of MAYA-III is not in the complexity of its logic and certainly not in the excitement of the game itself. It is in the way its function is programed or selected. MAYA-III is a molecular mixture that is molded into different functions by exposure to the instructional inputs in a very intuitive way. In future, these inputs could be environmental, and thus, MAYA-III hints at the possibility of harnessing molecular networks capable of adaptive behaviors for some good. MAYA-III also signifies a departure from the mechanical challenge of pursuing increasingly large networks of molecules. Eventually, we want to pursue complex and adaptive autonomous networks that will serve as standalone molecular devices in a two-way information exchange with the environment (the idea is pursued by other groups as well, e.g., [10, 11]). But, in order to fulfill that goal, we have to start somewhere, and, in MAYA-III, we probably took a step in the right direction, going "up" in the hierarchy of molecular behaviors.

8.3
From Random Walker to Molecular Robotics

Our starting position has been that robots, whether macroscopic or molecular, ought to continuously process information they sense in their environment and then affect this environment (locally) in a manner that we can program. Thus, the most interesting part of the field of molecular robotics would be to learn how to design molecules that will execute dynamic and realistic sets of local rules of interaction – namely, interaction among themselves and interaction with their ever-changing environment, with the latter feeding back continuously into this process [12, 13]. Many systems within a cell would fit this point of view of molecular robotics perfectly (e.g., the protein production machinery). Unlike in macroscopic robotics, where rules of interactions can often be reduced to mechanics and, thus, readily translated to different macroscopic environments, with molecules we are talking about Brownian movement and various affinity/catalysis interactions, the latter often being very specific for a given environment. Thus, we propose to focus on generic systems in which we can study the impact of rules for local interactions on global behaviors, on deviations from completely random behavior, and on accomplishment of trivial tasks. Nucleic-acid-based systems seem particularly suitable for such studies because of their modularity; the interface between nucleic-acid-based systems executing such rules and their environment may consist of aptamers (nucleic-acid-based receptors for proteins and small molecules). By changing the sets of aptamers employed, we may be able to adapt nucleic-acid-based robotic systems to a variety of environments and a variety of tasks within them.

In our work, we focused on harnessing the interactions of nucleic-acid-based random walkers with a well-defined environment, as a first step toward building the foundations of molecular robotics. We used nanoassemblies known as *molecular spiders* or nanoassembly incorporating catalytic kinesis (NICK[14]) (Figure 8.1a). The spiders have legs made of catalytic nucleic acids that cleave other oligonucleotides, either in solution on or surfaces. Spider legs bind to both the substrates that are cleaved and the products that result from this cleavage. When a leg binds to a product on the surface, it dissociates at some rate that determines the average residency time spent at that (or similar) position, for example, $t = 1$ arbitrary unit (a.u.) (Figure 8.4a). When the same leg binds to a substrate, it dissociates more slowly and mostly only after cleaving the substrate, leading to a longer residency time ($t > 1$ a.u., Figure 8.4a). One-legged spiders can diffuse (i.e., undergo random walk) over the surface covered with either products or uncleavable substrates, and the diffusion should be slower on the substrate-covered surface (assuming that a single leg has no directionality – biochemically interpreted, this assumes that k_{on} rates of binding to products and substrates are the same; in reality, for the only substrate we used, when tested in solution, k_{on} rates for the substrate were actually lower than for the product, because of additional secondary structures in the substrate).

If we introduce legs that cleave substrates (Figure 8.4b) Brownian diffusion on a surface changes to a previously uncharacterized type of random walk – a random walk with memory [15]. We can say "memory" because sites already visited once are recognized through a change of residency time on subsequent visits. There is an important consequence of memory on the rate of diffusion, in multilegged spiders (Figure 8.4c,d): at a boundary between products and substrates, the probability will be higher for the spider to move toward the substrates, because its legs that are bound to substrates will dissociate less often (the spider will more likely flip forward); at the same time, the boundary will be continuously moving, because of catalysis. As a result of these two factors, in spiders with multiple legs, the diffusion will stop being purely Brownian, and the walk will be directional toward new substrates. This directional walk was mathematically derived by Krapivsky and Antal from a single "memory" rule [15]: *if a site is not previously visited stick harder*, combined with the assumption that spiders have *multiple* legs that can reach neighboring substrates or products. This is not an obvious result: one might think

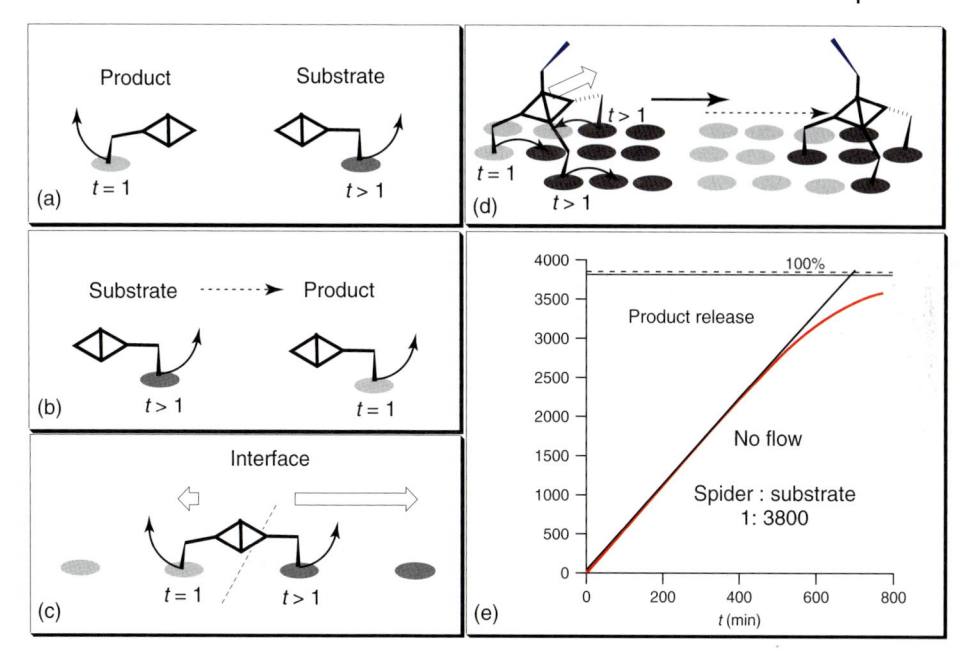

Figure 8.4 Rule-based movement of spiders [14]. (a) A leg binds to a product displayed on a surface and spends an average time on it that can be expressed as 1 a.u. Residency time is then >1 a.u. on substrates. (b) We can introduce a "memory" of past visits by means of catalytic events on the surface. A catalytic event changes the behavior of a leg on an already visited site. (c) Multilegged species on the interface will preferentially move toward substrates, because legs on products have shorter residency times. (d) Implementation of this rule leads to rapid diffusion that is faster than on product-covered and on uncleavable-substrate-covered surfaces, that is, to the movement in the direction of new substrates. (e) Release profile (no. of substrates vs time) of products by a four-legged spider from a matrix displaying substrates. The measurement was performed by surface plasmon resonance. One can observe nearly linear release for up to 400 min (at 1 : 3800 ratio of spiders to substrates).

that such a multivalent species will immediately capture itself in the field of grazed products, moving very slowly toward new substrates, at the rate that is limited by the rate of diffusion on the product surface. The model indeed predicts this in the asymptotic limit of very long times, in most cases beyond the experimental horizon (interestingly, we did observe such behavior with species with shorter legs). We later refined the theoretical model of spider behavior [16], interpreting the superdiffusive transient and its eventual decay in terms of an alternation between two metastates, sticking to the boundary (B) and diffusing in the field of grazed products (D). The B periods are Markovian, but the D periods are not, and are of increasing duration.

This theoretical treatment provides a rule-based explanation for a multivalent species behavior that was observed earlier in bulk [14]. Figure 8.4e shows time-dependent release of products. Using surface plasmon resonance, we monitored the substrate cleavage from a matrix displaying substrates by a four-legged spider. The observed release of products was nearly constant over prolonged periods of time, at a rate that was between two and four times the rate at which a single leg cleaved substrates in solution. This indicates that at 100, 200, or 300 min into the experiment, spiders were finding themselves in almost identical environments, with fully replenished substrates (otherwise the rate would be dropping with time). Yet, substrates had been attached to the matrix, so "replenishing" of the substrates could have happened only via spiders moving toward new substrates, consistent with the theoretical explanation. Next, with a large group of collaborators we expanded these experiments to three-substrate-wide tracks deposited on planar DNA objects (origami, [17]). The behavior of spiders was monitored on origami at the single-molecule level, which was on its own a significant contribution to the field of DNA nanotechnology [18]. These results were also the first single-molecule demonstration of interactions between moving autonomous molecular robots (i.e., molecules continuously implementing local rules) and engineered environments guiding them directionally.

These initial experiments demonstrated that the spiders are molecules that process information displayed on the surface; one point of view is that the spider's body integrates information from multiple sensors and does one of two things: if all sensors touch identical elements, the spider moves randomly ("decides" to move randomly), while if they touch a mixture of products and substrate, the spider moves with higher probability toward new substrates. The advantage of the rule-based approach is that we can now start thinking about the ways to add additional layers of sensing and decision making to the initially randomly moving molecules, and observe further deviations from nonrandom behavior that these new layers induce.

8.4
Conclusions

We need not defend the reasons for pursuing molecular computing and molecular robotics. On his first visit to the United States in 1924, Archibald V. Hill was asked to give examples of basic science concepts unexpectedly metamorphosing into practical devices with broad impact on human life; he retorted, instead, *"to tell the truth, sir, we don't do it because it's useful; we do it because it's amusing"* [19]. And, undeniably, there is a hint of that in our field today. However, one cannot resist hoping that one day "beads of water" may be injected into a human organism in support of failing metabolic networks, as in this science-fiction image, "...[the] system software looked like an old-fashioned canvas water bag, a sort of canteen... worn, and spectacularly organic, with tiny beads of water bulging through the tight weave of fabric[1]."

1) Description of bead-based bio-organic computers, "Sandbenders," taken from the science-fiction novel Idoru by William Gibson.

Acknowledgments

Our work in this field has been supported by NSF since 2002. Numerous colleagues contributed significantly to this research, and their names are listed on joint papers.

References

1. Katz, E. and Privman, V. (2010) Enzyme-based logic systems for information processing. *Chem. Soc. Rev.*, **39**, 1835–1857.
2. Breaker, R.R. and Joyce, G.F. (1995) A DNA enzyme with Mg(2+)-dependent RNA phosphoesterase activity. *Chem. Biol.*, **2** (10), 655–660.
3. Breaker, R.R. (2002) Engineered allosteric ribozymes as biosensor components. *Curr. Opin. Biotechnol.*, **13**, 31–39.
4. Tyagi, S. and Krammer, F.R. (1996) Molecular beacons: probes that fluoresce upon hybridization. *Nat. Biotechnol.*, **14**, 303–309.
5. de Silva, A.P. and McClenaghan, N.D. (2000) Proof-of-principle of molecular-scale arithmetic. *J. Am. Chem. Soc.*, **122**, 3965–3966.
6. Stojanovic, M.N., Mitchell, T.E., and Stefanovic, D. (2002) Deoxyribozyme-based logic gates. *J. Am. Chem. Soc.*, **124** (14), 3555–3561.
7. Stojanovic, M.N. and Stefanovic, D. (2003) Deoxyribozyme-based automaton. *Nat. Biotechnol.*, **21**, 1069–1073.
8. Macdonald, J., Li, Y., Sutovic, M., Lederman, H., Pendri, K., Lu, W., Andrews, B., Stefanovic, D., and Stojanovic, M.N. (2006) Medium Scale integration of molecular logic gates in an automaton. *Nano Lett.*, **6**, 2598–2603.
9. Pei, R., Matamoros, E., Li, M., Stefanovic, D., and Stojanovic, M.N. (2010) Training a molecular automaton to play a game. *Nat. Nanotechnol.*, **5**, 773–777.
10. Ashkenasy, G. and Ghadiri, M.R. (2004) Boolean logic functions of a synthetic peptide network. *J. Am. Chem. Soc.*, **126**, 11140–11141.
11. Wang, Z.G., Elbaz, J., Remacle, F., Levine, R.D., and Willner, I. (2010) All-DNA finite-state automata with finite memory. *Proc. Natl. Aacd. Sci.*, **107** (51), 21996–22001.
12. Jones, J.L. (2004) Robot Programming: A Practical Guide to Behavior-Based Robotics.
13. Brooks, R.A. (1991) Intelligence without Reason, web document.
14. Pei, R., Taylor, S., Rudchenko, S., Stefanovic, D., Mitchell, T.E., and Stojanovic, M.N. (2006) Behavior of polycatalytic nanoassemblies on substrate-displaying matrices. *J. Am. Chem. Soc.*, **128**, 12693–12697.
15. Antal, T. and Krapivsky, P.L. (2007) Molecular spiders with memory. *Phys. Rev. E*, **76**, 021121.
16. Semenov, O., Olah, M.J., and Stefanovic, D. (2011) Mechanism of diffusive transport in molecular spider models. *Phys. Rev. E*, **83**, 021117.
17. Rothemund, P.W. (2006) Folding DNA to create nanoscale shapes and patterns. *Nature*, **440**, 297–302.
18. Lund, K., Manzo, A.J., Dabby, N., Michelotti, N., Johnson-Buck, A., Nangreave, J., Taylor, S., Pei, R., Stojanovic, M.N., Walter, N.G., Winfree, E., and Yan, H. (2010) Molecular robots guided by prescriptive landscapes. *Nature*, **465** (7295), 206–210.
19. Kandel, E.R. (2007) *In Search of Memory: The Emergence of a New Science of Mind*, W.W. Norton & Company, New York.

9
Biomolecular Finite Automata
Tamar Ratner, Sivan Shoshani, Ron Piran, and Ehud Keinan

9.1
Introduction

The growing interest in the design of new computing systems is attributed to the notion that, although the silicon-based computing provides outstanding speed, it cannot meet some of the challenges posed by the developing world of biotechnology and synthetic biology. New abilities, such as direct interface between computation processes and biological environment, are necessary. In addition, the challenges of parallelism [1] and miniaturization are still driving forces for developing innovative computing technologies. Moreover, the growth in speed for silicon-based computers, as described by Moor's law, may be nearing its limit [2]. The design of new computing architectures involves two main challenges: reduction of computation time and solving intractable problems. Most of the celebrated computationally intractable problems can be solved with electronic computers by an exhaustive search through all possible solutions. However, an insurmountable difficulty lies in the fact that such a search is too vast to be carried out using the currently available technology.

In his visionary talk "There's plenty of room at the bottom" in 1959, Richard Feynman suggested the use of atomic and molecular scale components for building machinery [3]. This idea has stimulated several research studies, but it was not until 1994 that an active use of DNA molecules was presented in the form of computation [4]. Biomolecular computing (BMC) has rapidly evolved since then as an independent field at the interface of computer science, mathematics, chemistry, and biology [5–7]. Living organisms carry out complex physical processes dictated by molecular information. For example, biochemical reactions, and ultimately the entire organism's operations, are ruled by instructions stored in its genome, encoded in sequences of nucleic acids. It is tempting to draw an analogy between the intracellular processing of DNA and RNA and the processing of information stored in the tape of the Turing machine. Both systems process information stored in a string of symbols built on a fixed alphabet, and both operate by moving step by step along those strings, modifying or adding symbols according to a given set of rules. These parallels have inspired the idea that biological molecules could become the raw material of new computer species.

DNA molecules enjoy many advantages over other biological molecules as building blocks for the construction of biocomputers. These highly stable molecules can be translated to proteins and thereby create biological structure and function without being consumed. Furthermore, DNA strands can store very high densities of information; 1 g of DNA in a volume of $1\,cm^3$ can store as much information as a trillion compact discs, approximately 750 TB [3, 8, 9]. The combination of the remarkable information storage capacity with the base-pairing rules, which can serve as a programming language, offers attractive opportunities in BMC. In addition, the DNA molecules can be easily copied and amplified to countless copies and can be easily manipulated with high fidelity using readily available enzymes. These advantages can be conveniently used not only for the design of biomolecular computing devices but also for dictating the computation rules.

Biological computers would not necessarily offer greater performance and speed in traditional computing tasks. Since the natural molecular machines depend on the catalytic rates of enzymes, they are obviously slower than the electronic computers. Moreover, the dependence on operations such as gel electrophoresis and polymerase chain reaction (PCR) renders biological computers slower than the conventional computers by orders of magnitude [10].

However, DNA-based computing devices have numerous advantages, including the direct interaction with biological systems, the miniaturization of the computing devices to a molecular scale, and the potential for massive parallelism which allows for a large number of operations per second. Transistor-based computers typically handle operations in a sequential manner, with the basic von Neumann architecture repeating the same "fetch and execute cycle" over and over again. In contrast, the DNA computers are stochastic machines that approach computation in a different way and therefore can be attractive tools for solving different classes of problems [11]. The combination of parallelism and miniaturization may offer orders of magnitude more operations per second than current supercomputers [12].

Over the years, numerous architectures for autonomous molecular computing devices have been developed on the basis of opportunities offered by molecular biology techniques [13–21]. Several of these have been explored experimentally and proven feasible [13, 22–25]. This chapter focuses on the realization of programmable DNA-based finite-state automata that can compute autonomously on mixing all their components in solution.

9.2
Biomolecular Finite Automata

9.2.1
Theoretical Models of a Molecular Turing Machine

The Turing machine is a theoretical device that manipulates symbols on a strip of tape according to a table of rules. Despite its simplicity, this machine can be

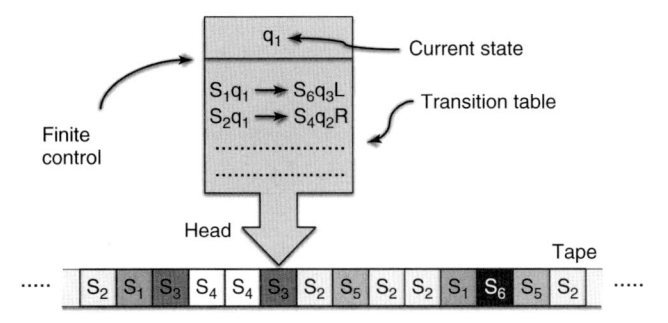

Figure 9.1 Basic structure of the Turing machine. The tape comprises an infinite number of cells, each containing one symbol. Following the instruction by the finite control, the head reads a symbol, replaces it by another symbol, changes its internal state, and moves to the next cell, either to the right or to the left.

adapted to simulate the logic of any computer algorithm. Consequently, the Turing machine can reach universal computing power, and is one of the most advanced computers ever conceived [26]. The simplicity of the machine combined with its power renders it an attractive target for molecular computing.

The *machine* is defined as a device that is composed of an infinite sequence of cells, known as the tape, and a head that can move back and forth along the tape, governed by a finite control (Figure 9.1) [27]. Each cell in the tape can store a single symbol from the set $S = \{s_0, s_1, \ldots, s_N\}$. The finite control contains two parts: one is the state of the machine within the set $Q = \{q_0, \ldots, q_P\}$ and the second is the table of transition rules. In each step, the head reads the symbol and then writes another symbol: that is, replaces the original symbol by a new one, as instructed by the relevant transition rule. Then the head moves to the next cell, either to the right or to the left (R or L) or halts (H). When the transition rule directs to H, the machine enters a special state and the computation terminates.

The analogy between the sequence of symbols stored in the Turing tape and the genetic information stored in the DNA has triggered the imagination of many scientists for many decades, and ideas of using biomolecules for constructing a Turing machine were expressed already in the early 1970s. For example, the use of RNA polymerase for this purpose was suggested as early as 1972 [28]. A schematic description of a Turing machine was later formulated on the basis of DNA with imaginary enzymes capable of recognizing and changing a single base pair (bp) of DNA [29]. Another idea was the construction of chemical neural networks, which, in turn, could be relevant to other general computers such as Turing machines [30].

In 1995, Rothemund delineated a more practical Turing machine on the basis of DNA and restriction enzymes [31]. The basic idea was to use circular double-stranded DNA (dsDNA) to represent the tape of symbols, while all enzyme-catalyzed manipulations, including restriction, insertion, and ligation, represented the computational steps, the position of the head, and its internal state. Small dsDNA fragments where designed to represent the transition rules.

The specific example proposed by Rothemund referred to the well-known Busy Beaver problem of a three-state Turing machine (BB-3). The more general BB-N

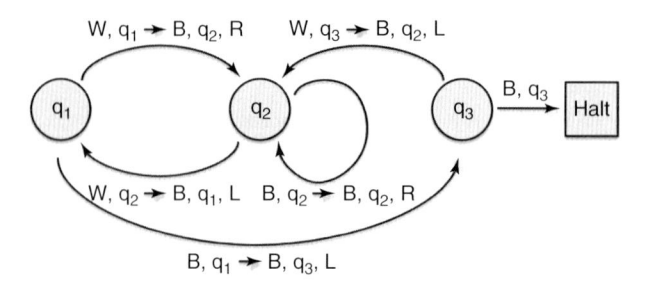

Figure 9.2 State transition diagram designed to solve the Busy Beaver problem for three states (BB-3) q_1, q_2, and q_3. This machine has to write the maximum number of B symbols before halting. In each step, the head reads the symbol, either W or B, changes it to either W or B, changes the state from q_i to q_j, and then moves to the next cell, either to the right (R) or to the left (L).

problem relates to the design of an *N*-state Turing machine that has two symbols, black and white (B and W). This machine is designed to print the greatest number of black symbols before halting [32]. Figure 9.2 shows graphically the entire set of the transition rules needed to solve the problem of a BB-3 Turing machine. A schematic demonstration of computing steps that solve the BB-3 problem (Figure 9.3) shows that in each step the tape undergoes a change in one cell according to the intrinsic state of the head (q_i) and the relevant transition rule, where the final outcome is six printed black symbols.

Rothemund represented the B and W symbols by two distinct dsDNA sequences, each subdivided into two halves (Figure 9.4). Both B and W sequences began and ended with two invariant short segments: L on the left side and R on the right side. The head of the Turing machine was represented by two recognition sites of restriction enzymes, labeled Inv, which point at the adjacent invariant sequence, and q_n (or Sta, standing for state cutter), which points at the current symbol. There were two locations in a given symbol that the head could point at: if it pointed at the right side of the symbol, it meant that the last move of the machine on the tape was to the right, and vice versa. The frame in which the Sta enzyme cut within the first half of the symbol determined the state of the machine. Cutting in three different modes defined three different states. The appropriate choice of a sequence for each symbol ensured that any of the resultant sticky ends was unique.

The transition molecules (TMs) encoded for the new (written) symbol, new state, and direction of the head toward the next symbol. The unique sticky end produced by the restriction enzyme enabled the insertion of the new symbol into the tape. Each TM was comprised of several parts (Figure 9.4): Coh was the sticky end complementary to the sticky end formed by a state cutter; Sta was a recognition site of the restriction enzyme employed as a state cutter; Res was a sequence encoding for the new symbol; Cap, X (symbol excision site), and Inv were class II recognitions sites of restriction enzymes that cut the L or R regions. Although a two-symbol three-state machine required six transition rules, the molecular model had 12 TMs because the tape could be read from both directions, depending on the direction of the last move.

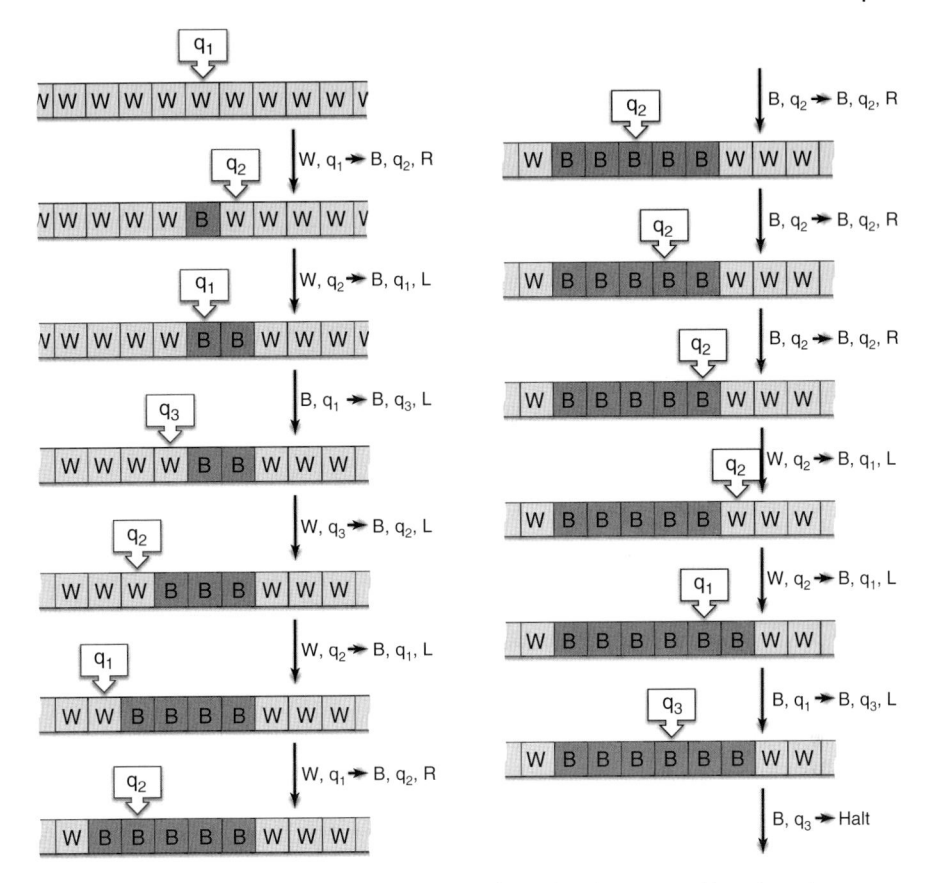

Figure 9.3 Schematic computing sequence designed to solve the BB-3 problem. In each step, the tape undergoes a change in one cell according to the intrinsic state of the head (q_i) and the relevant transition rule (shown on the arrow). Overall, this machine performs 13 steps in which it writes 6 B symbols before it halts.

Each step of the computation (reading, writing, and moving to the next step) was comprised of six chemical events (specifically illustrated in Figure 9.5): (i) cutting the initial tape using Sta and Inv enzyme with Sta, creating a sticky end unique to the current state and symbol, whereas Inv cleaves either the R or L regions; (ii) mixing all 12 TMs with the dsDNA tape, allowing the Coh sticky end to hybridize and ligate with the complementary one; (iii) cutting by the Cap enzyme to reveal the invariant sticky end sequence, either R or L; (iv) ligation of the two cohesive ends of the plasmid to one another in order to create a circular plasmid; (v) cutting by the X restriction enzyme in order to remove the previous symbol; and (vi) ligation to reproduce the circular plasmid.

The entire set of events 1–6 was repeated until a Halt symbol was incorporated into the tape. In order to detect whether a halt step was reached, Rothemund proposed to apply PCR amplification every time after performing step 6. If the Halt

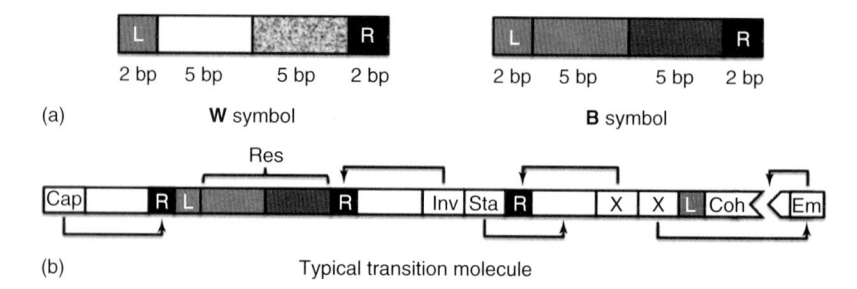

Figure 9.4 Schematic representation of (a) the two symbols W and B and (b) a typical transition molecule. The regions labeled Cap, Inv, Sta, X, and Em represent the recognition sites of restriction enzymes that cut the dsDNA away from the recognition site at the direction of the relevant arrow. The enzymes proposed are BsrDI, BseRI, FokI, BpmI, and BbvI, respectively. Coh is the cohesive end (sticky end), which is complementary to the sticky end produced by restriction (reading) of the last symbol according to the current state.

symbol was detected, sequencing could be performed to perceive the final output. This model was presented also as a small universal Turing machine (UTM). For example, a way to solve Minsky's four-symbol seven-state machine was proposed on the same general concept of the BB-3.

Although being a purely theoretical model, the Rothemund proposal discussed some potential problems that could occur in its experimental implementation, such as problems arising from failed or incorrect restrictions, failed or incorrect ligations, undesired dimerization, and so on. Less attention was given to other practical issues, such as the material balance, low yields, and accumulation of side products over the long sequence of multistep reactions. It was proposed that harnessing the advantages of immobilization on a solid support could technically simplify the computation process and make it more efficient. A realistic prediction was made that, with immobilized input, each transition from one state to another would take approximately 4.5 h. Overall, the seminal proposal of Rothemund has been sufficiently innovative and detailed to inspire future work in the field.

9.2.2
The First Realization of an Autonomous DNA-Based Finite Automaton

Although the experimental realization of the Turing machine represents a highly attractive goal, it represents a formidable task that has not been achieved yet. In contrast, the finite automaton, in which the head can only read and can move to only one direction, represents an achievable target for laboratory realization. Conceivably, realization of a molecular finite automaton, although being a rather limited computing model, could serve as an essential step on the way to the ultimate goal of a full-fledged molecular Turing machine.

Programmable finite automata that solve computational problems autonomously were prepared using dsDNA and DNA-manipulating enzymes. The automaton hardware consisted of a restriction nuclease and ligase, the software and input

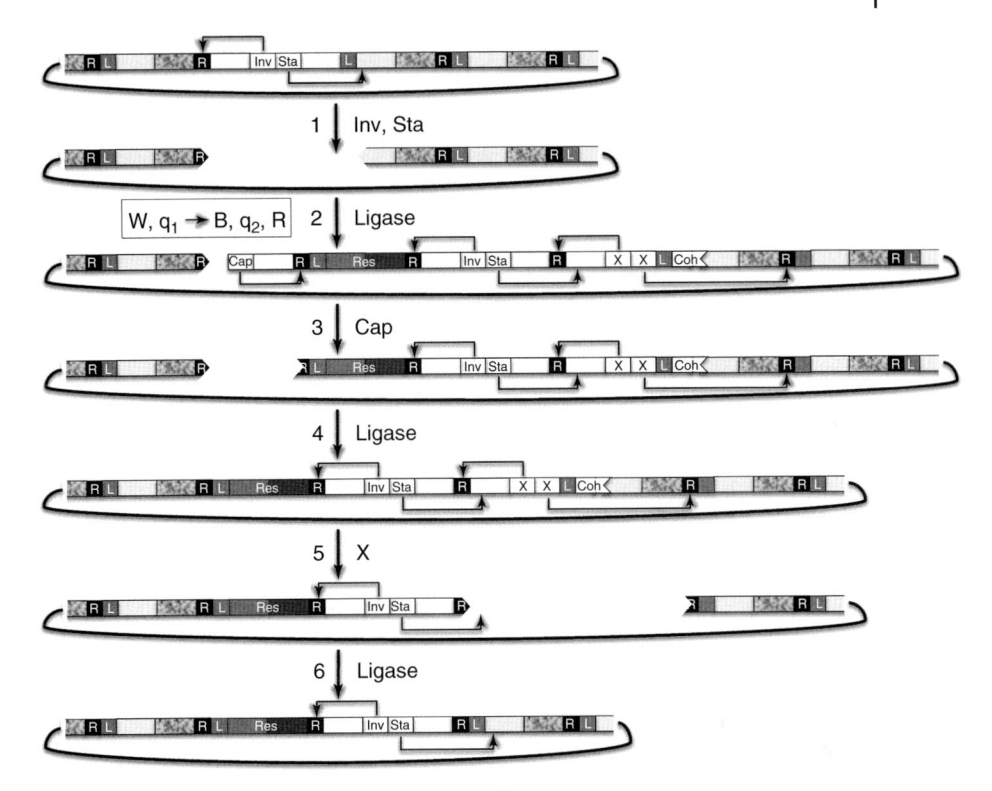

Figure 9.5 Schematic representation of the first step of the BB-3 machine simulated on the blank tape. In step 1, the plasmid input is restricted by Inv and Sta to produce a linear dsDNA. Sta cuts the plasmid according to the initial state (q_1 in this case) of the Turing machine depending on the spacing between the recognition site and the symbol. Inv is designed to cut an invariant sequence. In step 2, the appropriate transition molecule undergoes ligation to one side of the input molecule. In step 3, the protected end of the original transition molecule is restricted by Cap to produce a sticky end, which is ligated in step 4 to create a circular plasmid. In step 5, the plasmid is restricted twice, now by X (symbol excision enzyme) and thereby removes a symbol that was read. In step 6, the linear dsDNA ligates again to form a plasmid, which is now ready for the following sequence of events.

were encoded by dsDNA, and the program was defined by the appropriate choice of software molecules. On mixing solutions containing these components, the automaton processed the input molecule via a cascade of restriction, hybridization, and ligation cycles, producing a detectable output molecule that encoded for the automaton final state, and thus the computational result.

For example, a soluble mixture of molecules was designed to represent a two-symbol two-state finite automaton (Figure 9.6) [22]. The hardware comprised of a type-II endonuclease (*Fok*I), the T4 DNA ligase, and adenosine triphosphate (ATP). The software included a set of transition rules, represented by an appropriate set of TMs, all in the form of short dsDNA oligomers. A dsDNA molecule encoded for the input, with each input symbol being coded by a 6-bp dsDNA sequence

9 Biomolecular Finite Automata

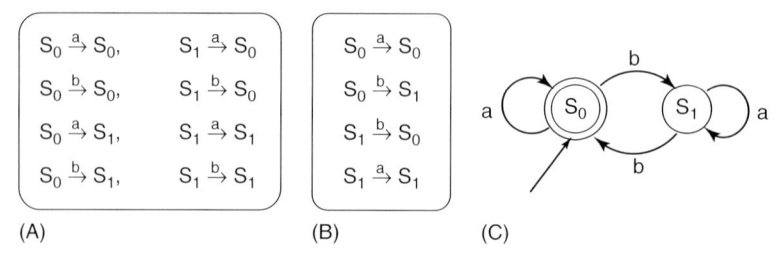

$S_0 \xrightarrow{a} S_0,$ $S_1 \xrightarrow{a} S_0$ $S_0 \xrightarrow{a} S_0$
$S_0 \xrightarrow{b} S_0,$ $S_1 \xrightarrow{b} S_0$ $S_0 \xrightarrow{b} S_1$
$S_0 \xrightarrow{a} S_1,$ $S_1 \xrightarrow{a} S_1$ $S_1 \xrightarrow{b} S_0$
$S_0 \xrightarrow{b} S_1,$ $S_1 \xrightarrow{b} S_1$ $S_1 \xrightarrow{a} S_1$

(A)   (B)   (C)

Figure 9.6 Two-symbol two-state automaton. (A) All eight possible transition rules of a finite automaton that has two internal states (S_0 and S_1) and two input symbols (a and b). (B) Selected subset of four transition rules that represent a specific finite automaton. (C) Graphic description of this automaton that includes two states: S_0 and S_1 (indicated by circles), two symbols: a and b, an initial state, S_0 (indicated by a straight arrow), and four transition rules (indicated by curved arrows). This automaton answers the question whether the number of b's in a given input is even. A positive answer to this question will result in the accepting state S_0 (indicated by a double circle). On the other hand, an even number of b's in the input will result in a final state S_1.

(Figure 9.7). The input molecule included the initial state of the automaton as well. The computing mixture contained the required "peripherals," namely, two output detection molecules (DMs) of different lengths. Each of these could hybridize and ligate selectively to a different output molecule, thus forming an output reporting molecule. The latter indicated a final state and could be readily detected by gel electrophoresis.

The two different internal states, either S_0 or S_1, were represented by two distinguishable restriction modes of any 6-bp symbol, either at the beginning of the symbol domain or 2-bp deeper into that domain, respectively (Figure 9.8). The different cleavage site was achieved by using a type-II, four-cutter endonuclease, *Fok*I, which cut 9 and 13 bases away from its recognition site. In this system, there were six unique four-nucleotide 5′-prime sticky end sequences that could be obtained by two restriction modes of two symbols and a terminator.

The automaton processed the input, first, by cleaving it with *Fok*I, thereby exposing a four-nucleotide sticky end. This result reflected the initial state and the first input symbol. The computation continued via a cascade of transition cycles (Figure 9.9). In each cycle, the sticky end of an appropriate TM ligated to the sticky end of the input molecule. This operation indicated the response of the system to the current state and symbol. At each restriction step, the most upstream symbol in the input molecule was cleaved by *Fok*I, exposing a new four-nucleotide sticky end. The design of the TMs (Figure 9.7) ensured that the 6-bp-long encodings of the input symbols a and b were cleaved by *Fok*I at the appropriate mode, either at the leftmost part encoding the state S_1 or at the rightmost part encoding S_0 (Figure 9.8). The exact next restriction site, and thus the next internal state, was determined by the current state and size of the spacers in an applicable TM. The computation proceeded until no TM matched the exposed sticky end of the input or until the special terminator symbol was cleaved, forming an output molecule that had a sticky end encoding the final state. This sticky end ligated to one of two output

9.2 Biomolecular Finite Automata

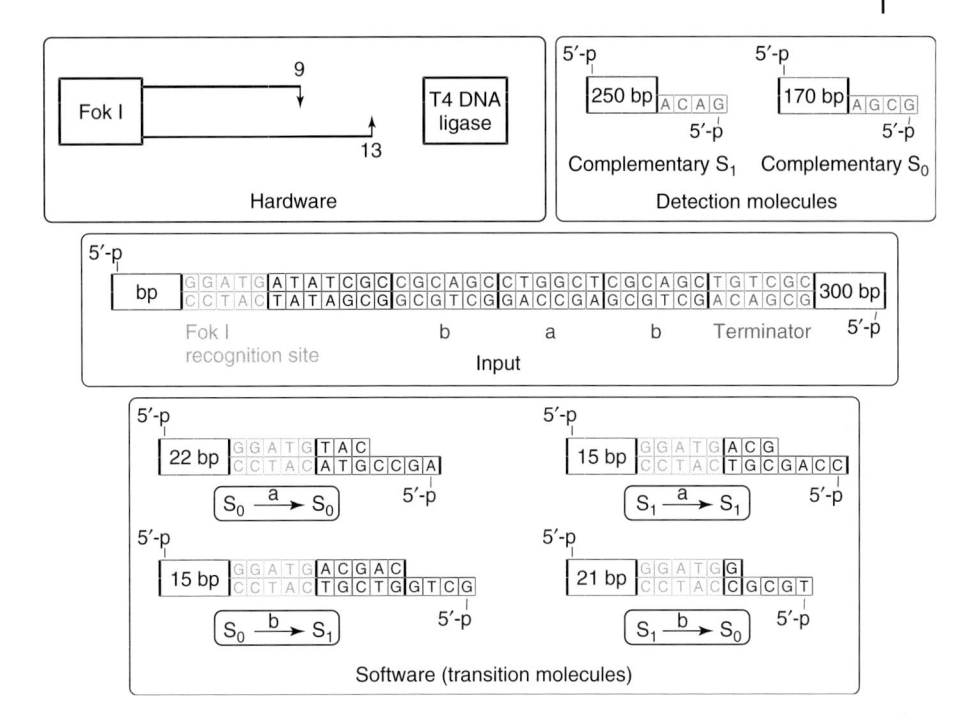

Figure 9.7 Molecular design of a two-symbol two-state finite automaton. The 10 components of the automaton included an input molecule, 2 enzymes, ATP, 4 transition molecules, and 2 detection molecules. The transition molecules shown here construct the automaton presented in Figure 9.6. Bases marked in light gray represent the recognition site of FokI.

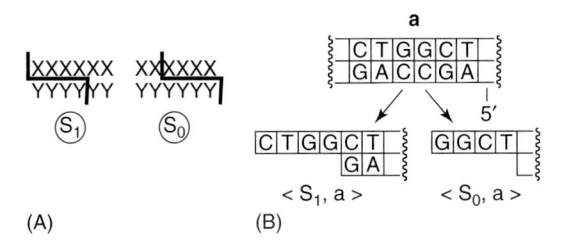

Figure 9.8 Two different internal states. (A) Two restriction modes of a 6-bp domain by a four-cutter enzyme. (B) The symbols a and its two restriction products produced by a four-cutter endonuclease.

detecting molecules (Figure 9.7) and the resultant output reporter was identified after purification by gel electrophoresis. The operation of several automata was tested on various inputs followed by detection of the outputs on gels (Figure 9.10).

Based on this design, a ligase-free system was also demonstrated. In this manner, molecules were not consumed but the input consumption drove the computation process to completion without the need to invest external energy in the form of ATP [33]. Furthermore, these principles allowed the design of

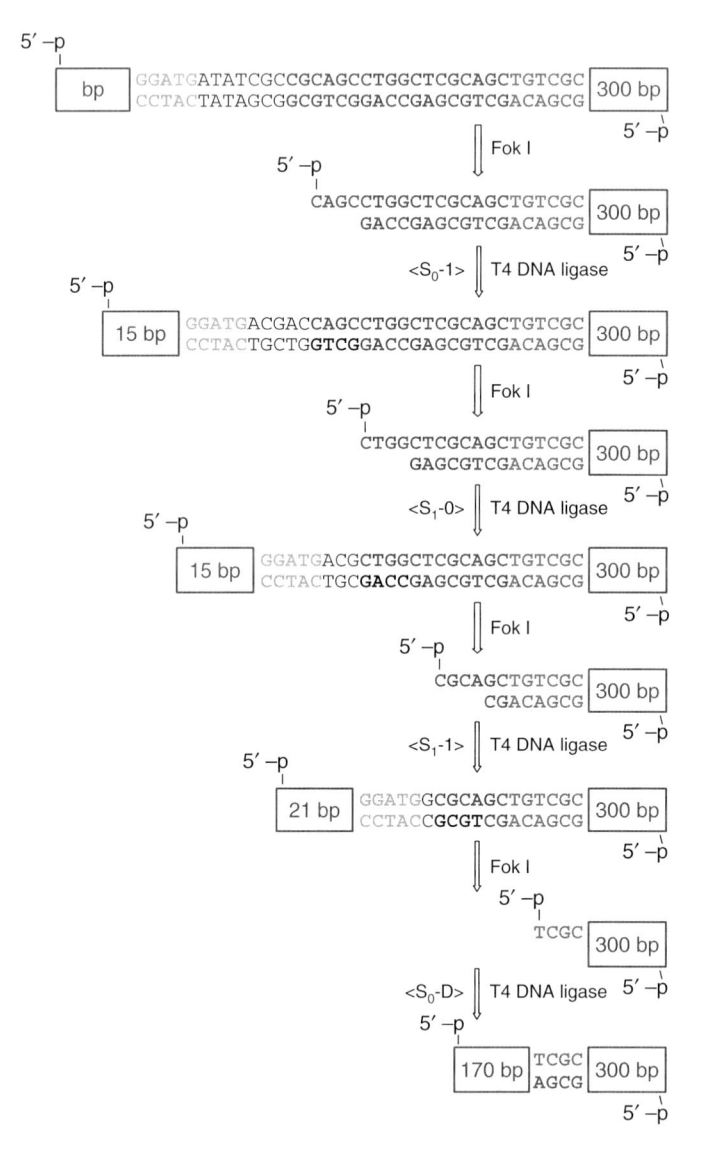

Figure 9.9 Description of a computing process with input bab. The input is cleaved by *Fok*I, forming the initial state. The complementary transition molecule ($<S_0\text{-}1>$) is hybridized, and in the presence of T4 DNA ligase it is ligated to the restricted input. Similarly, the input is processed via repetitive cycles of restriction, hybridization, and ligation with complementary transition molecules. In the final step, the output is ligated to S_0 detection molecule ($<S_0\text{-}D>$).

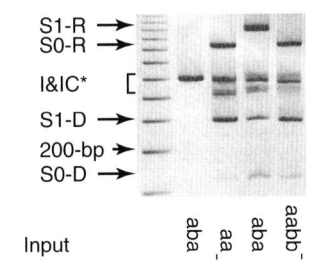

Figure 9.10 Output detection by gel electrophoresis. The processed mixtures of three inputs, aa, aba, and aabb, after ligation with the detection molecules S_0-D and S_1-D were analyzed by gel electrophoresis. The bands marked by I&IC represent the inputs and various other dsDNA molecules in the mixture. The relevant bands are the ligation products, marked S_0-R or S_1-R, indicating the final state, either S_0 or S_1, respectively. The left lane shows a molecular weight ladder and the second lane from left shows an intact input, aba, for reference. As expected, the outputs were S_0 for input aa, S_1 for input aba, and S_0 for input aabb.

biomolecular automata that could perform stochastic computing, reminiscent of natural biological processes [34].

9.2.3
Three-Symbol-Three-State DNA-Based Automata

Significant expansion of the complexity of the above-described automata was achieved by the demonstration that a type-II four-cutter endonuclease could restrict a 6-bp symbol in three distinguishable modes [35]. Three internal states, S_0, S_1, or S_2, could thus be represented by restriction at the beginning of the symbol domain, 1-bp deep or 2-bp deep into that domain, respectively (Figure 9.11). A three-state automaton with two symbols would have a total library of 18 transition rules and a much broader spectrum of possible programs as compared with the previous case of the two-symbol two-state automata [22].

Furthermore, the increased number of states and symbols resulted in significant enhancement of the computing power. For example, a three-symbol three-state device has a library of 27 possible transition rules and 134 217 727 possible selections of transition rule subsets from that library. Since there are seven possible selections of the accepting states (S_0, S_1, S_2, any combination of two and a combination of all three), the outcome becomes a remarkably large number of 939 524 089 syntactically distinct programs [36]. This number is considerably larger than the corresponding number of 765 possible programs offered by the above-described two-symbol two-state device [22].

Figure 9.11 Three restriction modes of a 6-bp symbol by a four-cutter endonuclease, representing three different internal states.

The applicability of the three-symbol three-state automaton was further enhanced by employing surface-anchored input molecules, using the surface plasmon resonance (SPR) technology to monitor the computation steps in real time. The realization of this computational design was achieved in a stepwise manner, with automatic real-time detection of each step, all carried out on a Biacore® chip.

While in the previously reported homogeneous system all components were placed in a single mixture [22], the heterogeneous design, which was based on an immobilized input, offered the advantages of applying the automaton in the form of two separate mixtures. These mixtures were added sequentially, allowing for separation between the restriction and ligation events. The first and most significant advantage was the ability to monitor the individual computation steps while carrying out parallel computation with multiple input molecules, all bound to a single chip. The second advantage resulted from the fact that, although none of the TMs was sufficiently long to become a substrate of *Bbv*I, they were all reversible inhibitors of this enzyme. Therefore, placing the enzyme and TMs in separate mixtures prevented this inhibition. The third advantage was related to the fact that *Bbv*I and T4 DNA ligase require different conditions for their optimal efficiency. In the previous case, the employment of both restriction enzyme and ligase in a single mixture required compromises in choosing the reaction conditions in order to partially satisfy each enzyme. Here, the use of immobilized input molecules allowed for convenient separation of the two enzymes, permitting each to operate under its optimal conditions.

Computation was performed by alternating the feed solutions between the endonuclease solution and a mixture containing the ligase, ATP, and the appropriate TMs. The binding and computing events were monitored while taking place at the sensor surface. The flow cell was first fed with a solution of *Bbv*I, then fed with a mixture of TMs and ligase, and so forth. The computation was terminated after executing a sufficient number of such cycles. Detection of the final state was carried out by sequential feed of three mixtures, each containing one of the DMs, D-S_0, D-S_1, or D-S_2, together with T4-DNA ligase and ATP. As the Biacore chip contained four independent sectors, it was possible to perform parallel computing with four different input molecules. This advantage was demonstrated by stepwise monitoring of the computation using one automaton and four different inputs: bc, a, ac, and acb (Figures 9.12 and 9.13).

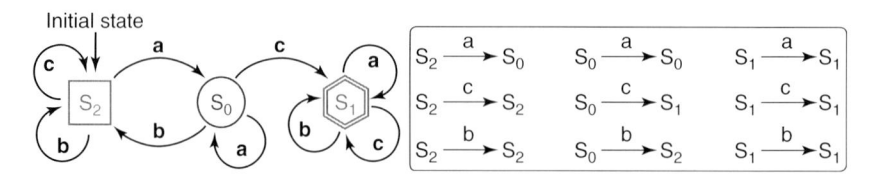

Figure 9.12 Molecular design of a three-symbol three-state finite automaton shown both graphically (left) and in the form of a table of nine transition rules (right).

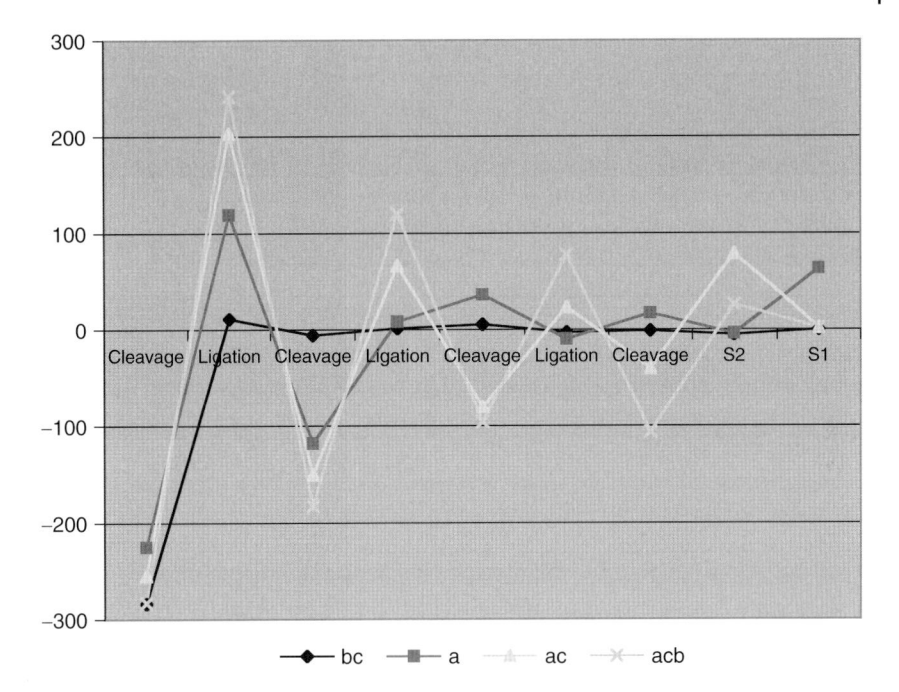

Figure 9.13 Monitoring the computation process by SPR. Stepwise computing with one automaton shown in Figure 9.12a and four input molecules: bc, a, ac, and acb. The transition molecules were supplied only to satisfy the computation needs of the latter three inputs, while no transition molecules were available for computation with the bc input. The differential RU values represent the changes in the SPR response between two consecutive steps. The computation was followed by detection with the soluble detection molecules D-S_0 and D-S_1.

This work presented significant enhancement of the computational power in comparison with the initially reported two-symbol two-state automata. The major improvements included (i) increase of the number of internal states, (ii) increase of the number of symbols, (iii) real-time detection of the output signal as well as real-time monitoring of stepwise computing by the use of the SPR technology, and (iv) separate parallel computing on different inputs, due to the ability to immobilize different input molecules on different sectors on the chip. The first two improvements increased the overall number of syntactically distinct programs from 48 to 137 781. Theoretically, this strategy can lead to automata with up to 37 different 6-bp symbols.

9.2.4
Molecular Cryptosystem for Images by DNA Computing

Besides tackling hard mathematical problems and taking advantage of the direct interaction molecular computing can have with living cells, molecular finite automata can be employed in new directions, for example, information encryption.

While the science of information encryption has a very long history, the use of DNA for this task is new. DNA molecules hold many advantages such as vast parallelism, immense information density, high chemical stability, and energy efficiency, thus serving as a promising tool for image encryption. Although DNA was demonstrated to be useful for steganography and encryption of text [37–42], no molecular encryption of images has yet been tested.

In 2011, a new molecular cryptosystem for images [43] based on two-symbol two-state finite automata [22, 35, 44, 45] was demonstrated. The automata representing the software were programmed by the choice of several molecules from a library of eight TMs, representing eight transition rules. A restriction enzyme, a DNA ligase, as well as ATP, represented the hardware. Computation was carried out by mixing all components in a solution, leading to processing of the input molecules via repetitive cycles of digestion, hybridization, and ligation. During computation, any symbol or terminator could be cut by the restriction enzyme in one of two different modes, each representing a different intrinsic state of the automaton, either S_0 or S_1. Thus, digestion of the input with the restriction enzyme resulted in the production of a sticky end that could hybridize and ligate with the sticky end of an appropriate TM, creating a new recognition site, and so forth. Finally, when digestion occurred inside a terminator, the resulting sticky end could not hybridize with any of the TMs. Instead, it hybridized with an appropriate DM which was either fluorescently tagged or not depending on the final state of the specific computation process.

This method employed parallel computing for deciphering two different images encrypted on a single DNA microarray chip. Two separate images, either the logo of the Technion or the logo of The Scripps Research Institute, were encrypted on a single DNA chip. Each pixel was designed to eventually contain either a fluorescent output or a nonfluorescent output, so when applying one automaton on the chip a unique combination of pixels became fluorescent, thus creating one image (Figure 9.14). Alternatively, applying the second automaton revealed the

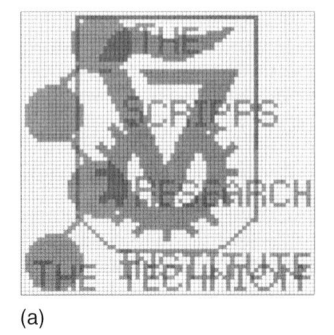

Figure 9.14 (a) Planned co-encryption of two images on a single DNA chip. Pixels show the Scripps logo in gray, the Technion logo in diagonal parallel lines, and those representing both are in dotted areas. Actual outputs from deciphering with (b) automaton 1 and (c) automaton 2. The round black defects are the result of air bubbles on the chip.

second image. In other words, mixtures of input molecules were processed with the appropriate molecular finite automata for deciphering each of the images by fluorescent visualization of the surface-bound output molecules.

This work presented the first molecular cryptosystem based on DNA computing. In order to decipher the visual information encrypted on the chip, one has to know the sequences of the input and DMs, the chosen automata, and the information on the chip. Also, the reaction conditions and the enzymes used add another dimension to the complexity and security of the system. In addition, many irrelevant input molecules can be added for increased security. Furthermore, the advanced DNA microarray technology with millions of printed pixels on a chip offers an immense diversity of potentially encrypted images. The enormous number of possible sequences, combined with the addition of nonrelevant DNA, renders the system very hard to break. Taking into consideration that the chip itself will not survive more than a few attacks, it becomes impossible to decipher the information encrypted on the chip without the appropriate knowledge.

9.2.5
Molecular Computing Device for Medical Diagnosis and Treatment *In Vitro*

An *in vitro* molecular computing device was constructed on the basis of a finite automaton model, which was capable of detecting and analyzing the levels of RNA disease indicators [46]. On proper detection of these indicators, the device could induce the release of an active drug. The automaton was programmed to identify and analyze mRNA of disease-related genes associated with small-cell lung cancer and prostate cancer. It was also designed to produce ssDNA molecules modeled after an anticancer drug, which can affect levels of gene expression via antisense activity.

The operation of this computing device was divided to three parts: a computing module that performed stochastic calculations; an input module in which specific mRNA levels or point mutations regulated the concentrations of active TMs; and an output module that controlled the release of a suitable drug. The system included input molecules encoding for diagnostic rules, hardware that was comprised of *Fok*I, and software consisting of TMs that were regulated by molecular indicators (Figure 9.15).

The dsDNA input molecule had two main regions: a diagnostic part and a drug administration part. The first part was comprised of several symbols, each sensitive to a certain indicator. The automaton processed the diagnostic part one symbol at a time, determining, in each step, whether or not the corresponding indicator was present. After all symbols were processed, the computing proceeded to the drug administration region. Two input molecules were used, each containing a different drug administration region, capable of releasing either a drug or a drug suppressor. Both consist of dsDNA stem which protects ssDNA loop containing the drug or the drug suppressor. The stem prevented undesired interactions between the drug, drug suppressor, and the target mRNA. After a positive diagnosis, the stem of the drug release region was processed with special Yes-verification TMs and the drug

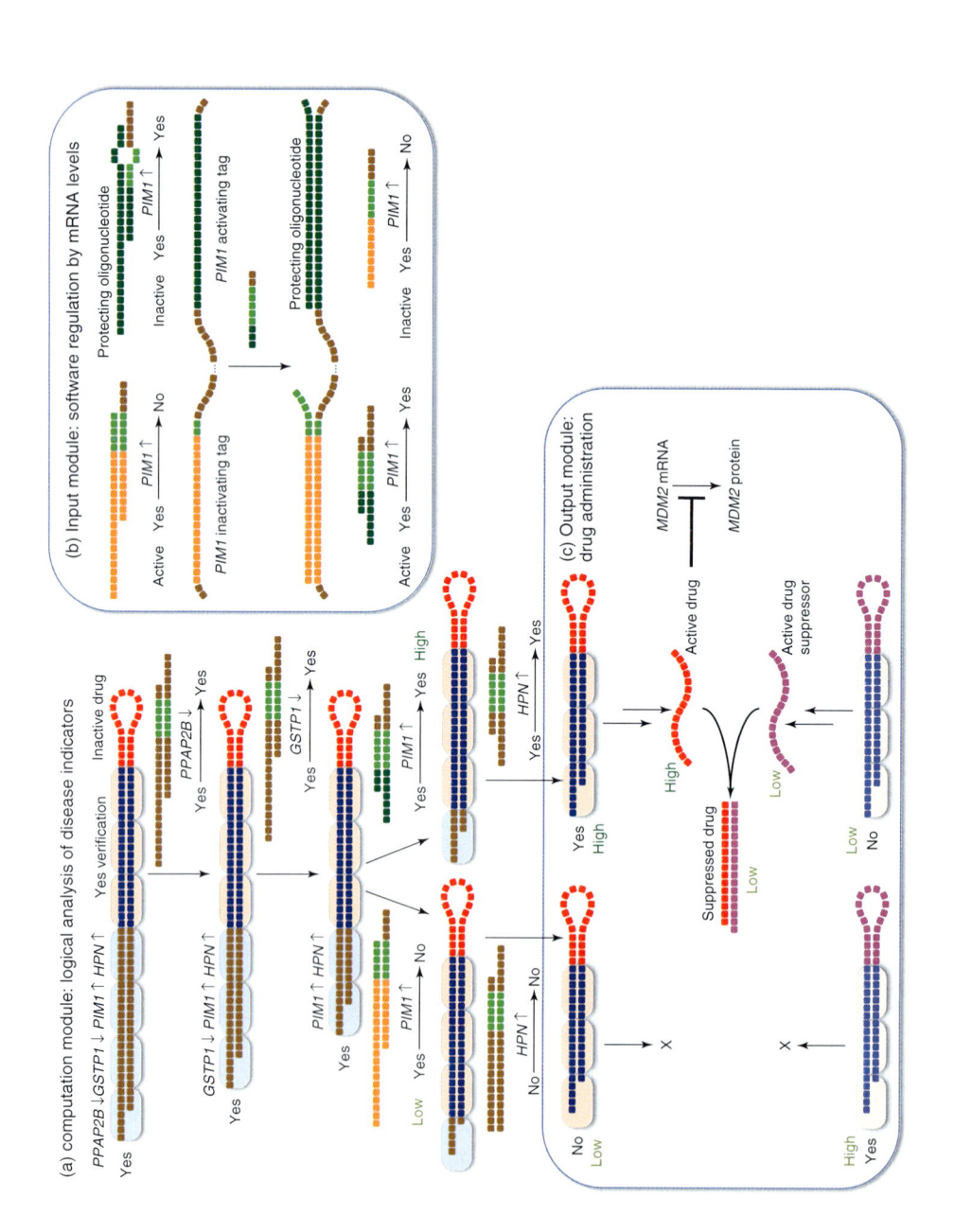

(a) computation module: logical analysis of disease indicators

(b) Input module: software regulation by mRNA levels

(c) Output module: drug administration

is released. At negative diagnosis, this stem remained intact, protecting the drug. After a negative diagnosis, the stem of the drug suppressor release region was processed with special No-verification TMs and released the drug suppressor. At positive diagnosis, this stem remained intact, protecting the drug suppressor.

The molecular automaton was stochastic with two competing transitions, positive and negative for each symbol. This design regulated the probability of each positive transition by the corresponding molecular indicator, so that the presence of the indicator increased the probability of a positive transition, and decreased the probability of its competing negative transition, and vice versa if the indicator was absent.

In order to compute with two input molecules simultaneously, two types of automata were implemented: one that released a drug molecule on positive diagnosis and another that released a drug suppressor molecule on negative diagnosis. The ratio between the drug and drug suppressor molecules released determined the concentration of the active drug.

This work presented a molecular computing device capable of logical analysis of mRNA disease indicators *in vitro* and of controlled administration of biologically active ssDNA molecules. Despite the fact that this work was carried out *in vitro*, it represented a convincing proof of concept concerning the medical applications of BMC.

9.2.6
DNA-Based Automaton with Bacterial Phenotype Output

The next step toward the actual involvement of molecular computing devices with biological systems was the demonstration that the computation output, which was

Figure 9.15 Molecular computing for diagnosis and treatment of prostate cancer. (a) Part of the computing process of the input molecule ending in drug release. The initial input molecule consists of a diagnosis region (highlighted with blue background) and a drug administration region (orange background), which includes an inactive drug loop (red). At each computation step, the most probable transition is shown, except for the processing of the symbol *PIM1*, for which the stochastic choice with two competing transition molecules is shown. (b) Regulation of the two transitions for *PIM1*↑ by subsequences (tags) of overexpressed *PIM1* mRNA, resulting in a relatively high level of the yes-to-yes transition molecule and low level of the yes-to-no transition molecule. Each transition molecule contains regulation (green, orange) and computation (light green, brown) fragments. The inactivation tag of *PIM1* mRNA (orange) displaces the strand of the transition molecule yes-to-no and destroys its functionality. The activation tag of *PIM1* mRNA (green) activates the yes-to-yes molecule. Initially, a protecting oligonucleotide (green) partially hybridizes to the larger strand of the transition molecule and thereby blocks its function. The activation tag displaces the protecting oligonucleotide, allowing annealing of the two strands and rendering an active yes-to-yes transition. (c) Combining of the computation results of both types of input molecules, both with high-Yes and low-No final states, results in high release of drug (red) and low release of drug suppressor (pink), and consequently in the administration of the drug. Arrows pointing up and down represent high and low concentration, respectively.

produced by a molecular finite automaton, could be a visible bacterial phenotype [44]. As was the case with the above-mentioned finite automata (Figures 9.6 and 9.12), the new system also processed linear dsDNA inputs, transforming them into linear output molecules. The difference, however, was that the resultant molecules were transformed into plasmids, thus becoming meaningful genes that could be expressed in *Escherichia coli*, leading to a visible output.

Construction of an extended input molecule was carried out by cloning an insert into a multiple cloning site (MCS) of the vector pUC18. This insert comprised all the previously described components, including several 6-bp symbols, a 6-bp terminator, and a recognition site of the restriction enzyme *Faq*I. In addition to the input string, the inserted computing cassette contained a restriction site for *Mlu*I (ACGCGT), located upstream to the recognition site of *Faq*I (Figure 9.16). Thus, digestion of the plasmid with *Mlu*I, before the computing process, converted the circular plasmid into a linear dsDNA with a GCGC-5' sticky end (the spectator end). This choice dictated that the specific sequence of the terminator would be CGCGCG so that the final obtained sticky end was complementary to the spectator end (Figure 9.17), independent of the final state (5'-CGCG).

The chosen automaton accepted inputs with an even number of b symbols, as described earlier in Figure 9.6, which meant that, if the dsDNA input contained

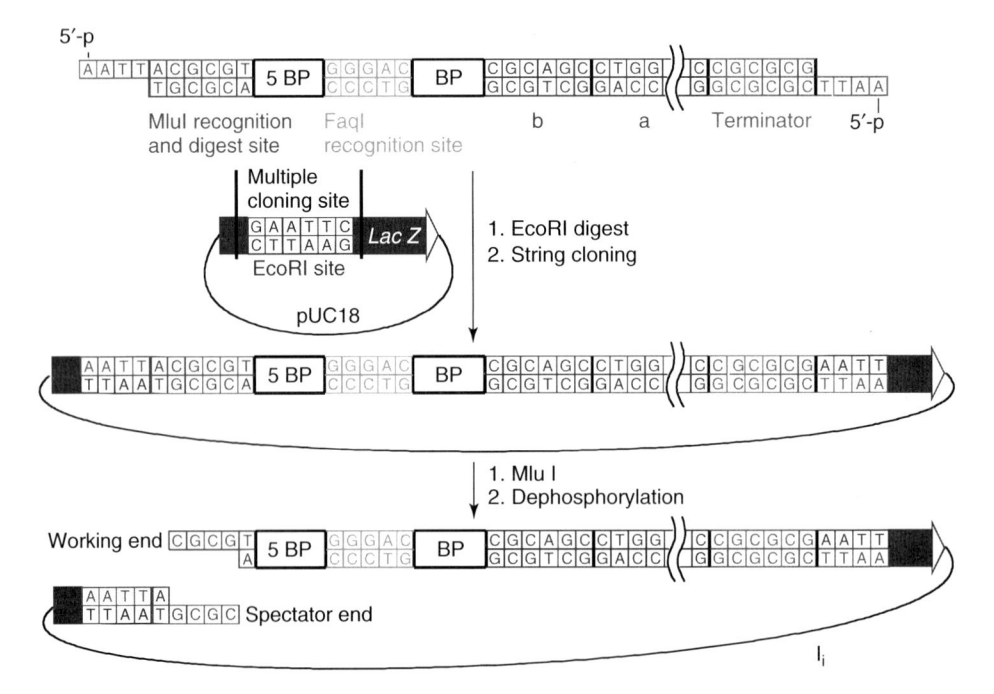

Figure 9.16 Preparation of an input molecule that can generate a bacterial phenotype output. An input insert was first cloned into the multiple cloning site of pUC18, and the resulting plasmid was then converted to a linear dsDNA by digestion with *Mlu*I, thus creating a 5'-CGCG sticky end on the upper string and GCGC-5' on the lower one.

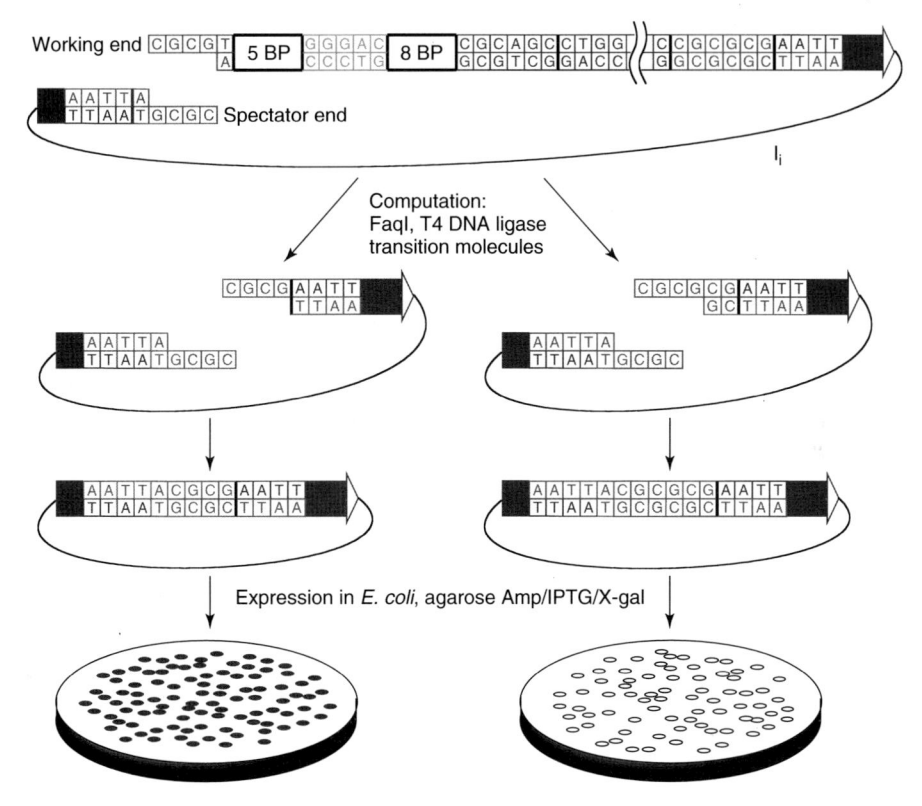

Figure 9.17 Computations that result in bacterial phenotype outputs. Computation results in with an input that contained an even number of b symbols, leading to blue bacterial colonies (appears in gray in this image), meaning S0 output (left), and with an input that contains an odd number of b symbols, leading to white bacterial colonies, meaning S1 output (right).

an even number of b symbols, the computation would lead to a 9-bp insert in the plasmid and formation of blue colonies on the X-gal medium. In contrast, if the input string contained an odd number of b symbols, the computation would result in an 11-bp insert, open reading frame (ORF) shift, and hence in the formation of white colonies on the X-gal medium (Figure 9.18).

9.2.7
Molecular Computing with Plant Cell Phenotype

One step further toward *in vivo* computing in eukaryotic cells was recently demonstrated by the expression of fluorescent proteins in living plant cells [45]. Each of the two possible molecular results of a two-symbol two-state finite automaton led to the creation of a circular plasmid that contained the gene of either the green fluorescent protein (GFP) or the cyan fluorescent protein (CFP). Insertion of these plasmids into living onion cells resulted in either green fluorescent or cyan fluorescent cells as phenotypical output signals.

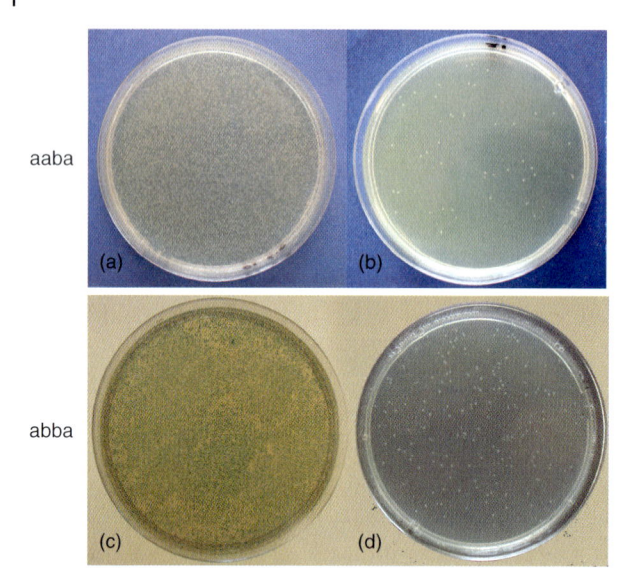

Figure 9.18 (a) Computation with aaba input in the presence of the transition molecules, which resulted in white bacterial colonies (S_1). (b) Control experiment with input aaba without transition molecules. (c) Computation with abba input in the presence of transition molecules, which resulted in blue bacterial colonies (S_0). (d) Control experiment with abba input without transition molecules.

The mathematical model, as well as its implementation using DNA molecules, was based on previously reported finite automata [22, 35, 44]. Two enzymes, endonuclease and a DNA ligase, as well as ATP, represented the hardware, whereas the automaton software was programmed by the choice of several molecules from a library of eight TMs, representing eight transition rules. In the computation process with automaton 1, transition molecules TM_1, TM_2, TM_5, and TM_6 were used, whereas automaton 2 was comprised of TM_3, TM_4, TM_5, and TM_7. Each input molecule was represented by a dsDNA which included the following segments: a recognition site for FokI; a string of 6-bp symbols, either ab or aaa; a 6-bp terminator; and a tail that included a recognition site for the endonuclease *Pst*I. The two detection molecules DM_0 and DM_1 were also dsDNA, each containing a four-base sticky end, complementary to the appropriately restricted terminator; a gene encoding for a fluorescent protein, either eGFP or eCFP; and a recognition site for the endonuclease *Nco*I (Figure 9.19). These DMs were prepared by PCR amplifying the eGFP and eCFP genes.

Mixing together all of the above-described components, including input, hardware, and software, in a single solution resulted in an autonomous cascade of chemical reactions, leading to a specific dsDNA. This dsDNA was digested with *Pst*I and *Nco*I in order to generate unique sticky ends for specific ligation with an appropriate linear vector. This vector was mixed with the DNA content from the computation process in the presence of T4 DNA ligase, giving rise to a circular plasmid (Figure 9.20).

9.2 Biomolecular Finite Automata | 165

Input molecules:

```
TTTTGCCATGCGGGATGCGCTCTGACCTTCGTCGTTGCACAGACAAGTTTACTGCAGCGCATGC
CGGTACGCCCTACGCGAGACTGGAAGCAGCAACGTGTCTGTTCAAATGACGTCGCGTACGTTTT
         FokI        a    b    t          PstI           I₁
```

```
TTTTGCCATGCGGGATGCGCTCTGACCTTCGTCTTCGTCTTCGTACAGACAAGTTTACTGCAGCGCATGC
CGGTACGCCCTACGCGAGACTGGAAGCAGAAGCAGAAGCATGTCTGTTCAAATGACGTCGCGTACGTTTT
         FokI        a    a    a    t          PstI           I₂
```

Transition molecules:

```
TTTTCGAAGGGACCAGGGATGA
    GCTTCCCTGGTCCCTACTGAAG
```
$TM_1 \; (S_0 \xrightarrow{a} S_1)$

```
TTTTCGAAGGGACCAGGGATGAAAAA
    GCTTCCCTGGTCCCTACTTTTTAACG
```
$TM_2 \; (S_1 \xrightarrow{b} S_0)$

```
TTTTCGAAGGGACCAGGGATGAAA
    GCTTCCCTGGTCCCTACTTTGAAG
```
$TM_3 \; (S_0 \xrightarrow{a} S_0)$

```
TTTTCGAAGGGACCAGGGATGA
    GCTTCCCTGGTCCCTACTGCAA
```
$TM_4 \; (S_0 \xrightarrow{b} S_1)$

```
TTTTCGAAGGGACCAGGGATGAAA
    GCTTCCCTGGTCCCTACTTTAGCA
```
$TM_5 \; (S_1 \xrightarrow{a} S_1)$

```
TTTTCGAAGGGACCAGGGATGAAA
    GCTTCCCTGGTCCCTACTTTGCAA
```
$TM_6 \; (S_0 \xrightarrow{b} S_0)$

```
TTTTCGAAGGGACCAGGGATGAAA
    GCTTCCCTGGTCCCTACTTTAACG
```
$TM_7 \; (S_1 \xrightarrow{b} S_1)$

Detection molecules:

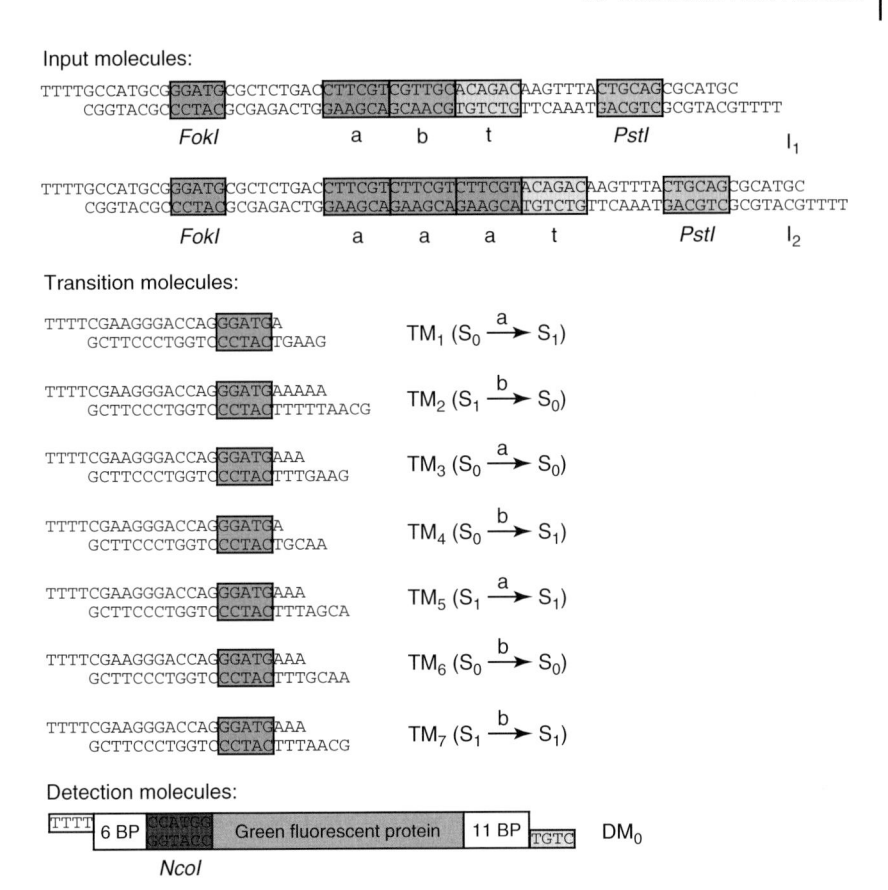

| TTTT | 6 BP | CCATGG / GGTACC | Green fluorescent protein | 11 BP | TGTC | | DM_0 |

NcoI

| TTTT | 6 BP | CCATGG / GGTACC | Cyan fluorescent protein | 11 BP | TCTG | | DM_1 |

NcoI

Figure 9.19 All computation components that lead to functional output in living plant cells. The input molecules contain either two or three symbols and recognition sites for FokI and PstI. The seven transition molecules contain a recognition site for FokI. TM_1, TM_2, TM_5, and TM_6 represent automaton 1 (Aut1), whereas TM_3, TM_4, TM_5, and TM_7 represent automaton 2 (Aut2). The two detection molecules DM_0 and DM_1 contain a fluorescent protein gene and a recognition site for NcoI.

In order to obtain sufficient amounts of plasmid required for insertion into onion cells by particle bombardment, the plasmids were first amplified in *E. coli* in an ampicillin-containing medium. Since only the computation output could form a plasmid by ligation with the linear vector, this procedure provided an opportunity to amplify only the plasmids containing output information, eliminating all other nonrelevant DNA molecules in the mixture.

Computation with input 1 with automaton 1, I_1-A_1, followed by transformation to *E. coli*, yielded several bacterial colonies. Extraction of the plasmid DNA from each of these colonies and subjecting them to DNA sequencing verified the correct

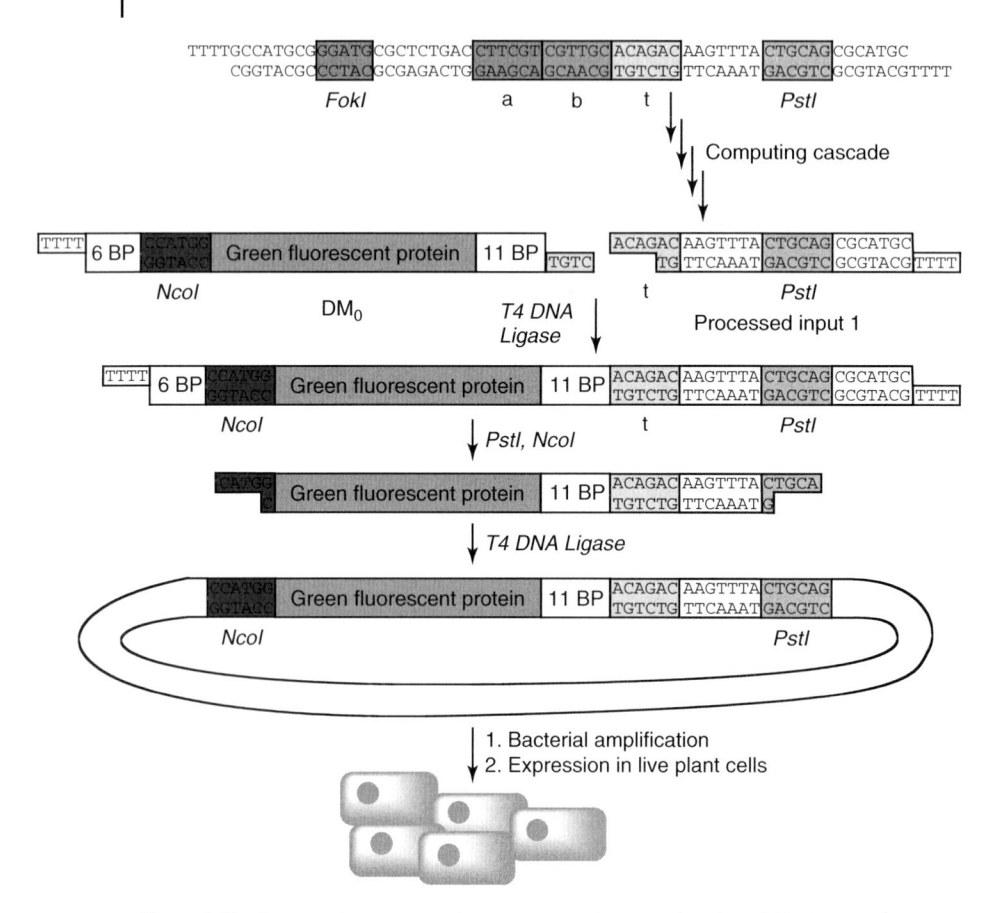

Figure 9.20 Computation process that ends up with green fluorescent onion cells (S$_0$). The processed input molecule undergoes ligation to the proper detection molecule (in this case, S$_0$) following restriction with *Pst*I and *Nco*I to produce two unique sticky ends. The product is then ligated to a linear vector that leads to the expression of a relevant gene in plant cells.

sequence of S_0. Similarly, computation with I_1-A_2 yielded colonies exhibiting the correct plasmid sequence of S_1. Computation with I_2-A_1 yielded two colonies, and both exhibited the correct DNA sequence that represented S_1. The negative control experiments were carried out with mixtures containing all the above components, including either I_1 or I_2, except for the TMs. No bacterial colonies could be found in any of these experiments, indicating the absence of any computation process.

The plasmids extracted from the *E. coli* colonies were coated on tungsten microparticles and used for biolistic delivery into epidermis cells of onion bulbs. In a typical experiment, bombardment with the plasmid containing processed I_1-A_1 resulted in green fluorescent onion cells (Figure 9.21a). Conversely, bombardment with the plasmid containing the processed I_1-A_2 resulted in cyan fluorescent cells (Figure 9.21b).

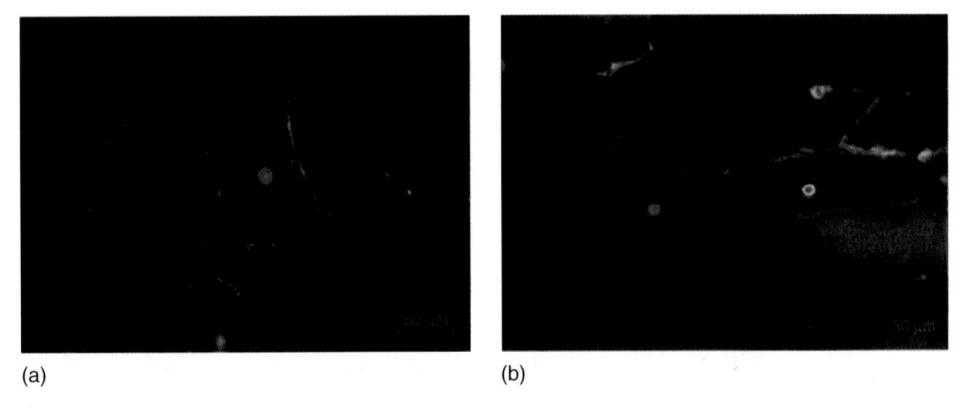

Figure 9.21 Final output in the form of fluorescent onion cells. (a) Bombardment with the plasmid containing processed I_1-A_1 resulted in green fluorescent onion cells 28 h after delivery, representing S_0 output. (b) Same experiment with the plasmid containing the processed I_1-A_2 resulted in cyan fluorescent cells, representing S_1 output.

This study demonstrated the ability of autonomous biomolecular computing devices to interact directly with living organisms without any interface. It also demonstrated that the expression of fluorescent proteins in living plant cells could be utilized as a highly accurate visual output of DNA-based computing devices. A common result of the chemical nature of molecular computing devices is the fact that they produce certain statistical distribution of molecules of which the correct output molecule is the major component. Consequently, the experimenter has to define a threshold value in order to transform the continuous concentration function into a step function that has a strict yes/no value. In this study, the output processing procedure involved the creation of a circular dsDNA, which was an important step of the computation procedure, serving as a quality control gate that transformed a rather noisy output into a clean signal. Out of the mixture of many DNA components produced during the computing process, only those defined as true output molecules could evolve into expressible plasmids.

9.3
Biomolecular Finite Transducer

A transducer is a general computing device that can process data and interconvert different types of information. Unlike the input consuming automata [22, 35, 43, 44], which can only read the input, the transducer is a more advanced and powerful machine because it can read and write information. Explicitly, it gradually transforms the input tape into an output. Moreover, the output can serve as input for subsequent computing by the same or another transducer, enabling computational power equivalent to a UTM [26].

Like the finite-state automaton as a discrete theoretical model of computing machines, the transducer accommodates a finite set of internal states and is guided

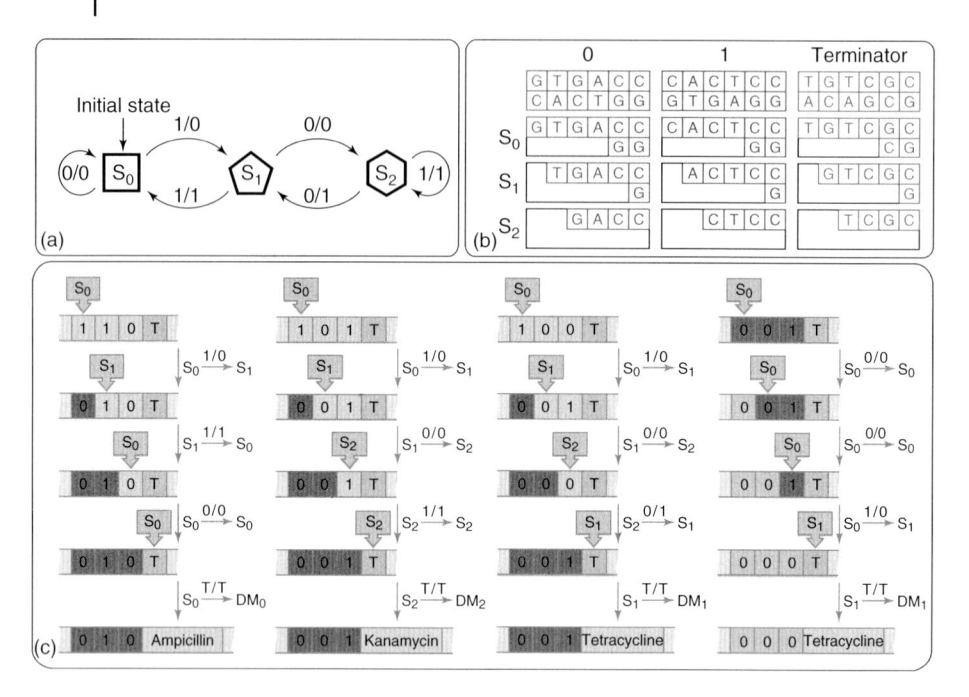

Figure 9.22 Graphical presentation of a transducer that performs long division by 3. (a) The two-symbol three-state transducer with internal states S_0 (the initial state), S_1, and S_2. In each computational step, an arrow from a vertex p to a vertex q (which is labeled a/b) means that, if the device reads an input symbol a while being at state p, it replaces the symbol a by an output symbol b (writes) and switches to the state q. (b) The sequences of each input symbol 1, 0, and terminator and the identity of sticky ends produced on specific restriction (reading), which represents state S_0, S_1, or S_2. (c) A schematic representation of four different division processes. Left to right: 6 ÷ ←3; 5 ÷ ←3; 4 ÷ ←3; and 1 ÷ ←3. The latter process represents the iterative computing performed with the output (001) of 5 ÷ ←3. The computing head with its internal state is shown in light orange above the input string. In the first three operations, the input symbols are shown in light gray and the output symbols in dark gray. In the iterative computing, the input symbols are shown in dark gray and the output symbols in light gray. Every string ended with a terminator region, T.

by a set of transition rules [26]. Formally, the device can be described as a graph with vertices, each representing an internal state, and arrows, representing the transitions between the states. Each arrow (edge) is labeled with a pair of characters that represent the input and output symbols (Figure 9.22a).

A DNA-based molecular transducer that not only computes iteratively but also produces a biologically relevant output was recently realized [47]. The bimolecular device reads and processes a DNA plasmid as input and writes new information by altering the plasmid, displaying the output in the form of a bacterial phenotype and being capable of iterative computing. As a proof of the concepts and their applicability, the machine was programmed to resolve the problem of long division by 3. This task was performed by a transducer with three states, S_0, S_1, and S_2, encoding the three alternative remainders of the division (0, 1, or 2), respectively (Figure 9.22a). A finite-state automaton cannot achieve this computation, which requires multiple tasks, including decoding, encoding, and storage of information.

The operation produced a dual output: a quotient and a remainder. The quotient was represented by the newly written DNA symbols, whereas the remainder, which can be 0, 1, or 2, was exhibited by *E. coli* phenotypes. The iterative power was demonstrated by a recursive application of the transducer on an obtained output.

As was suggested in previous works, each of the input and output symbols was represented by a 6-bp string (Figures 9.22b and 9.23a). The software consisted of a set of TMs in the form of short dsDNA (Figure 9.23b), whereas the hardware comprised restriction enzymes, ligase, and DMs (Figure 9.23c,d). Unlike the previously reported finite automaton [22, 35, 44], which employed linear dsDNA inputs, the transducer input was built in the form of a cyclic plasmid. Successive manipulations of the plasmid were used to read the input symbol on one side

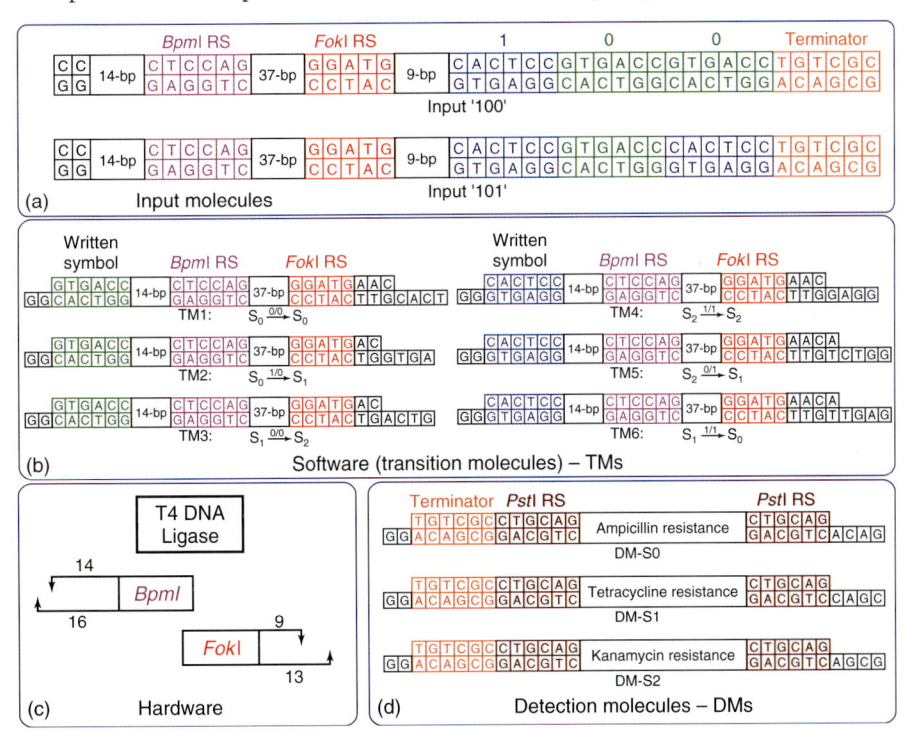

Figure 9.23 Transducer components. (a) Two representative input molecules, 101 and 100, that contain a sequence of 6-pb symbols 0 (green), 1 (blue), and terminator (orange). The recognition sites of the two restriction enzymes performing the computation are also color-coded: *Bpm*I (pink) and *Fok*I (red). (b) Six transition molecules representing the six transition rules, each containing a written symbol, either 0 or 1, and recognition sites of *Bpm*I and *Fok*I. (c) Hardware, which includes the two restriction enzymes and T4-DNA ligase. (d) Three detection molecules, each containing a written terminator (orange), reporter gene, and restriction site of *Pst*I (brown).

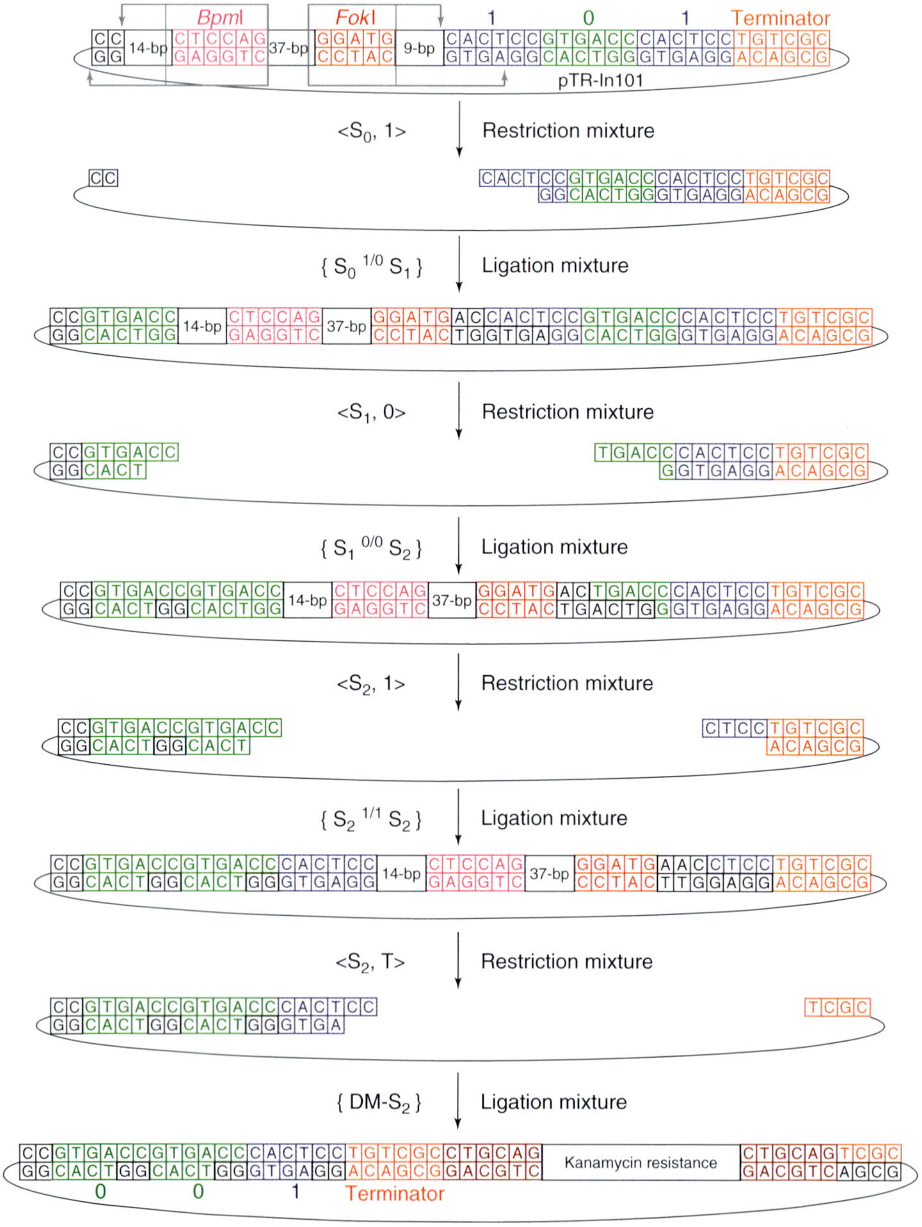

BpmI
FokI
1
0
1
Terminator
14-bp
37-bp
9-bp
pTR-In101
<S0, 1>
Restriction mixture
{ S0 1/0 S1 }
Ligation mixture
<S1, 0>
Restriction mixture
{ S1 0/0 S2 }
Ligation mixture
<S2, 1>
Restriction mixture
{ S2 1/1 S2 }
Ligation mixture
<S2, T>
Restriction mixture
{ DM-S2 }
Ligation mixture
Kanamycin resistance
0
0
1
Terminator

while simultaneously writing the output on the other side (Figure 9.24). This strategy required double restriction and double ligation in each read/write cycle. Computation was carried) out by alternating exposure of the input plasmid to two mixtures: a restriction mixture containing endonucleases and a ligation mixture containing T4 DNA ligase, ATP, TMs, and DMs (Figure 9.24).

Each of the three DMs contained a specific reporter gene, resistant to either ampicillin, kanamycin, or tetracycline, and correlated with one of the three possible terminal states (Figure 9.23d). Each DM could hybridize selectively to a different sticky end produced by a differently cleaved terminator, identifying one of the three states S_0, S_1, or S_2, and hence the remainder. Thus, the output signal was manifested by the specific antibiotic resistance on transformation to bacteria.

Computation with the transducer was demonstrated by long division by 3 of the integers 4, 5, and 6 in their binary format, 100, 101, and 110 (Figure 9.22c). The transducer was implemented by processing the input plasmid via repetitive cycles of restriction and ligation to produce the final state output in the form of a modified cyclic plasmid (Figure 9.24). Division of 5 (binary 101) by 3 resulted in quotient 1 (binary 001) and a remainder of 2. The latter was represented by the final state S_2, which was manifested by *E. coli* resistance to kanamycin (Figures 9.22c, 9.24, and 9.25). Similarly, division of 4 by 3 resulted in the quotient 1 (binary 001), a remainder of 1 and tetracycline resistance, and the division of 6 by 3 resulted in the quotient 2 (binary 010), a remainder of 0 and ampicillin resistance (Figures 9.22c and 9.25).

To demonstrate recursive computing, the output from the primary computing with input 101 was used as a new input for a secondary computing process. The plasmid bearing the output sequence 001 underwent technical adjustments to

Figure 9.24 Long division of 5 (binary 101) by 3. The symbol 0 is shown in green, symbol 1 in blue, terminator in orange, recognition sites of *Bpm*I in pink, and sites of *Fok*I in red. The recognition site of *Bse*RI is located inside the 37-bp spacer. The restriction mixture included *Bpm*I, *Bse*RI, and *Fok*I. The ligation mixture included T4 DNA ligase, ATP, all transition molecules, and all detection molecules. The computation started with downstream cleavage of the input plasmid by *Fok*I with simultaneous upstream cleavage by *Bpm*I. This step, which encoded for the initial state and for reading of the first input symbol, produced a linear plasmid terminated by a four-nucleotide sticky end at one side and a two-nucleotide sticky end on the other. In the following step, the linear plasmid underwent double hybridization and ligation at both sticky ends with TM2 to form a cyclic plasmid. This operation represented the written symbol, the current state, and the next symbol to be read. The latter cyclic plasmid underwent repetitive cycles of double restriction followed by double ligation with TM3 and then with TM4, until restriction occurred at the terminator region, creating a linear plasmid with two sticky ends. At this stage, no TM could match both sticky ends. The only dsDNA that could hybridize and ligate to both ends and form the final cyclic plasmid was one of the three DMs available in solution, DM2 in this case. This division of 5 by 3 resulted in the integer 1 (binary 001) and a remainder of 2 (S_2, Figure 9.1. The integer, which was represented by the sequence of the written symbols, 001, was detected by DNA sequencing of the relevant region on the plasmid. The remainder 2, which was represented by the internal state S_2, was detected by the resistance of *E. coli* expressing the plasmid to kanamycin, see Figure 9.4.

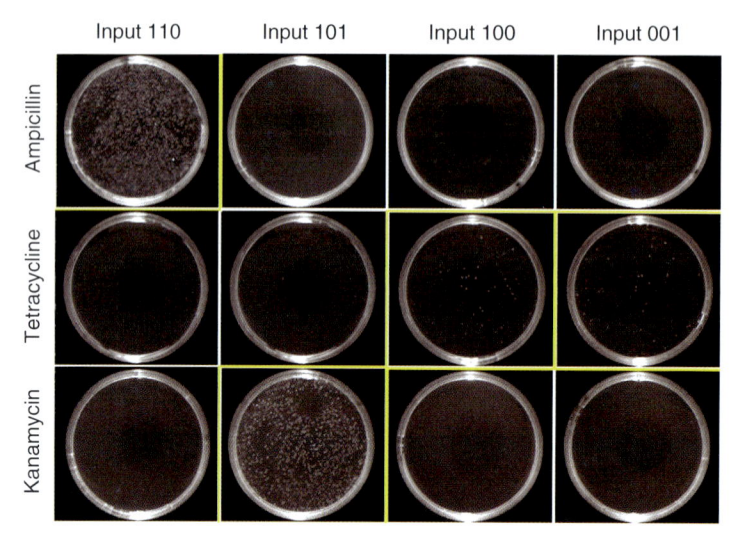

Figure 9.25 Biologically relevant output signals. Computation with inputs 110, 101, and 100 led to bacterial colonies exhibiting resistance to ampicillin (S_0), kanamycin (S_2), and tetracycline (S_1), respectively (marked with yellow frames). Iterative computing with input 001, which was obtained as the output from computing with 101, resulted in tetracycline resistance (S_1).

make it suitable for the consecutive process. First, the antibiotic resistance gene was removed. Second, a restriction cassette, comprising new recognition sites for the restriction enzymes at a location dictating S_0 as the initial state, was cloned into the plasmid. The secondary computing with this modified input resulted in the expected quotient 0 and remainder 1 represented by the final state S_1 (Figures 9.22c and 9.25).

This DNA-based transducer offers exceptional advantages through its ability to read and transform genetic information, its miniaturization to the molecular scale, and its ability to produce an output that interacts directly with living organisms. Although the transducer was employed to solve a specific problem of long division by 3, the general concept and the transducer's capability for iteration and hence composition with other transducers show that it can be applied for any other computational problem. The experimental realization of a powerful molecular computing device capable of algorithmically modifying the genetic code of an organism offers unprecedented opportunities in biomolecular computing.

9.4
Applications in Developmental Biology

The above examples have demonstrated that molecular finite automata can be constructed by employing various biological components and principles. More significantly, these synthetic devices could be inserted into living organisms, thereby interacting with their intrinsic machinery. Alternatively, since every biological system can be described as a computing device, it is conceivable that mathematical models can represent complex biological systems and processes. This is particularly true for decision-making processes in living organisms, which obey distinct rules. The intricate process of myogenesis during embryonic development represents an interesting case of a decision-making course of events [48]. We found that describing this process in terms of the SAT (satisfiability) formalism, particularly in the disjunctive normal form (DNF), is advantageous because it provides an overall view of this phenomenon [49]. This formalism has already been employed in nonbiological disciplines for solving decision-making problems because it can describe the relationship between causes and effects. We applied the DNF methodology and the ternary Łukasiewicz logic to describe the combined signaling effect of four diffusible proteins that leads to myogenesis, which is one of the most fundamental problems in developmental biology.

Creation of an automaton that describes the myogenesis SAT problem has led to a comprehensive overview of this nontrivial phenomenon and also to a hypothesis that was subsequently verified experimentally. The myogenesis example demonstrated the power of applying Łukasiewicz logic in describing and predicting any decision-making problem in general and developmental processes in particular.

We presented myogenesis in SAT formalism with the given concentration set of the morphogens (signaling molecules) in the environment of the forming myogenic tissue. The concentrations of the four morphogens, namely, Wnts, Shh, Noggin, and BMP4 (denoted by the variables w, s, g, and b, respectively (Figure 9.26), represented the input, whereas myogenesis was defined as the output. On the basis of the classical Łukasiewicz logic, the three input values referred to concentration, which was lower than the normal physiological value (state 0), the normal physiological concentration (state 1), or higher than the normal physiological value (state 2). This formalism described appropriately the experimental biological systems in which genes and proteins are either downregulated (0), unchanged (1), or overexpressed (2).

In the more widely known Boolean logic, the digit 1 represents the *true* value and 0 represents the *false* value. In the less used Łukasiewicz logic applied to the myogenesis model, the digit 1 defined the normal physiological concentration of a morphogen, which was sufficient for signaling and therefore represented the *true* value. The digit 0 represented insufficient concentration for signaling, obviously the false value. The digit 2 represented higher concentration than the normal physiological value, obviously defined as true. A system of four variables (proteins w, s, g, and b) and three values (0, 1, and 2), yielded a matrix of $3^4 = 81$ scenarios, which may or may not result in myogenesis.

Twenty out of these 81 scenarios were found to result in myogenesis (*true* value). When expressed in the form of DNF clauses, they satisfied the myogenesis function F (Figure 9.26a). They were abridged to a general formulation, as described in Figure 9.26b. Thus, two clauses were sufficient to describe myogenesis, defining the interplay between all four morphogens in this developmental process. Since BMP4 was known to antagonize myogenesis, its negative effect on myogenesis was

(a)
$F = (w_1 \wedge s_0 \wedge g_2 \wedge b_0) \vee (w_1 \wedge s_0 \wedge g_2 \wedge b_1) \vee (w_1 \wedge s_1 \wedge g_1 \wedge b_0) \vee (w_1 \wedge s_1 \wedge g_1 \wedge b_1) \vee (w_1 \wedge s_1 \wedge g_2 \wedge b_0) \vee (w_1 \wedge s_1 \wedge g_2 \wedge b_1) \vee (w_1 \wedge s_2 \wedge g_1 \wedge b_0) \vee (w_1 \wedge s_2 \wedge g_1 \wedge b_1) \vee (w_1 \wedge s_2 \wedge g_2 \wedge b_0) \vee (w_1 \wedge s_2 \wedge g_2 \wedge b_1) \vee (w_2 \wedge s_0 \wedge g_2 \wedge b_0) \vee (w_2 \wedge s_0 \wedge g_2 \wedge b_1) \vee (w_2 \wedge s_1 \wedge g_1 \wedge b_0) \vee (w_2 \wedge s_1 \wedge g_1 \wedge b_1) \vee (w_2 \wedge s_1 \wedge g_2 \wedge b_0) \vee (w_2 \wedge s_1 \wedge g_2 \wedge b_1) \vee (w_2 \wedge s_2 \wedge g_1 \wedge b_0) \vee (w_2 \wedge s_2 \wedge g_1 \wedge b_1) \vee (w_2 \wedge s_2 \wedge g_2 \wedge b_0) \vee (w_2 \wedge s_2 \wedge g_2 \wedge b_1)$

(b) $F = (w \wedge s \wedge \neg b) \vee (w \wedge g \wedge \neg b)$

Conjunction

$\wedge$	0	1	2
0	0	0	0
1	0	1	1
2	0	1	1

Disjunction

$\vee$	0	1	2
0	0	1	2
1	1	1	2
2	2	2	2

Negation

	0	1	2
$\neg$	2	1	0

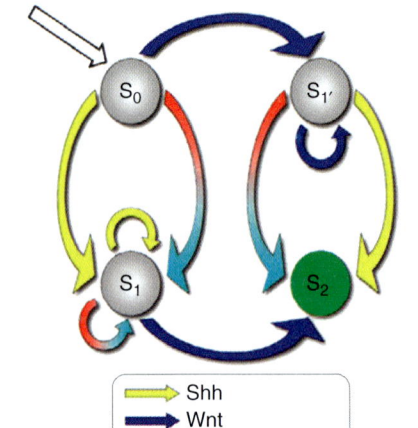

(c)

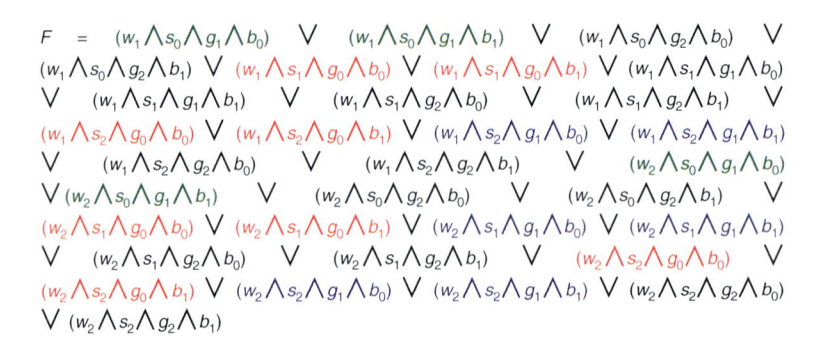

(d)

→ Shh
→ Wnt
→ (Noggin - BMP4)

(e)
$F = (w_1 \wedge s_0 \wedge g_1 \wedge b_0) \vee (w_1 \wedge s_0 \wedge g_1 \wedge b_1) \vee (w_1 \wedge s_0 \wedge g_2 \wedge b_0) \vee (w_1 \wedge s_0 \wedge g_2 \wedge b_1) \vee (w_1 \wedge s_1 \wedge g_0 \wedge b_0) \vee (w_1 \wedge s_1 \wedge g_0 \wedge b_1) \vee (w_1 \wedge s_1 \wedge g_1 \wedge b_0) \vee (w_1 \wedge s_1 \wedge g_1 \wedge b_1) \vee (w_1 \wedge s_1 \wedge g_2 \wedge b_0) \vee (w_1 \wedge s_1 \wedge g_2 \wedge b_1) \vee (w_1 \wedge s_2 \wedge g_0 \wedge b_0) \vee (w_1 \wedge s_2 \wedge g_0 \wedge b_1) \vee (w_1 \wedge s_2 \wedge g_1 \wedge b_0) \vee (w_1 \wedge s_2 \wedge g_1 \wedge b_1) \vee (w_1 \wedge s_2 \wedge g_2 \wedge b_0) \vee (w_1 \wedge s_2 \wedge g_2 \wedge b_1) \vee (w_2 \wedge s_0 \wedge g_1 \wedge b_0) \vee (w_2 \wedge s_0 \wedge g_1 \wedge b_1) \vee (w_2 \wedge s_0 \wedge g_2 \wedge b_0) \vee (w_2 \wedge s_0 \wedge g_2 \wedge b_1) \vee (w_2 \wedge s_1 \wedge g_0 \wedge b_0) \vee (w_2 \wedge s_1 \wedge g_0 \wedge b_1) \vee (w_2 \wedge s_1 \wedge g_1 \wedge b_0) \vee (w_2 \wedge s_1 \wedge g_1 \wedge b_1) \vee (w_2 \wedge s_1 \wedge g_2 \wedge b_0) \vee (w_2 \wedge s_1 \wedge g_2 \wedge b_1) \vee (w_2 \wedge s_2 \wedge g_0 \wedge b_0) \vee (w_2 \wedge s_2 \wedge g_0 \wedge b_1) \vee (w_2 \wedge s_2 \wedge g_1 \wedge b_0) \vee (w_2 \wedge s_2 \wedge g_1 \wedge b_1) \vee (w_2 \wedge s_2 \wedge g_2 \wedge b_0) \vee (w_2 \wedge s_2 \wedge g_2 \wedge b_1)$

related to its concentration. Therefore, a b negation value was used and assigned into the SAT formalism. In contrast, the myogenesis agonists (Wnts, Shh, and Noggin) were assigned their original values.

The universal truth tables described in Figure 9.26c served as a toolbox for defining the function F and assigning the different values to this function (Figure 9.26b). The appropriate combinations of values that satisfied F were expected to result in myogenesis. A three-symbol four-state finite automaton was created to describe this SAT problem (Figure 9.26d). In this device, the morphogens represented the symbols, and the cell types along developmental lineages represented the internal states of the automaton.

This automaton was based on three symbols: (i) Wnts, referring to the local concentration of this morphogen above a certain threshold; (ii) Shh, referring to its local concentration in a similar way; and (iii) Noggin-minus-BMP4, referring to the stoichiometric ratio between these two morphogens. The automaton was described graphically on the basis of the true clauses, each illustrated by symbols, represented by arrows (Figure 9.26d). The states represented by circles described intermediates along the way from a multipotent paraxial cell (S_0, Figure 9.26d) through an unspecified paraxial cell (S_1 or S_1') and, finally, to a determined myogenic cell (S_2).

Although the automaton was created independently of the SAT function F, both logic methods resulted in the same conclusions. Both methods predicted that more clauses than those described in Figure 9.26a could lead to myogenesis (Figure 9.26e). In one such predicted group of clauses (shown in red), $g = 0$ while both w and s are either 1 or 2 and b is either 0 or 1. In the second group (green), w is either 1 or 2, $s = 0$, $g = 1$, and b is either 0 or 1.

Figure 9.26 Myogenesis as a SAT problem presented in DNF using TNS. (a) A summary of all possible scenarios that lead to myogenesis in SAT formalism. w, Wnts level; s, Shh level; g, Noggin level; b, BMP4 level; 0, less than the normal physiological concentration; 1, the normal physiological concentration; and 2, above the normal physiological concentration. (b) Myogenesis SAT problem. Note that in this formulation, values are not assigned to the variables. Any combination of values could be processed by this function using the truth tables in (c). (c) The truth tables of myogenic developmental pathway. (d) A three-symbol four-state finite automaton that describes the SAT problem F. State S_0: multipotent paraxial cell; States S_1/S_1': unspecified paraxial cell; and State S_2: determined paraxial cell (myogenic cell). Uncolored arrow represents the initial state, arrows represent transition rules, and their colors represent the symbols. The green circle represents the accepting state. The order of signals has no importance. Yet, a cell could be exposed to only Wnt signaling, thus accepting state S_1'. Alternatively, a cell could be exposed to only Shh or Noggin minus BMP4, thus accepting state S_1. Note that although we have used four soluble proteins, only three symbols are used because of the close dependence of Noggin and BMP4 signals. Therefore, Noggin minus BMP4 represents the stoichiometric ratio between the two components. If this ratio is negative, then the direction of progression is opposite to the arrow direction. (e) Summary of all possible solutions leading to myogenesis in SAT formalism generated from the function F in 2B. The colored clauses indicate an input that collapses into another. The green clauses represent scenarios that have no biological meaning.

The automaton exhibited two general routes to myogenesis (Figure 9.26d), with the Wnts signaling being essential for proceeding from left to right whereas two alternative signals, either Shh or Noggin, could drive the progression from top to bottom. Similarly, the SAT function F exhibited two clauses (Figure 9.26b), each containing w, b combined with either s or g. Consequently, both the automaton and the SAT formalism indicated that the Shh and Noggin signals were redundant, suggesting an intriguing hypothesis that these two factors were located on the same biochemical pathway.

To examine the hypothesis that Noggin expression required Shh signaling, we carried out three experiments: (i) we inserted a barrier between the midline tissues and the para-somitic mesoderm (PSM); (ii) we ablated the notochord from the PSM anterior area; (iii) we implanted Wnt1-secreting cells lateral to a barrier that was inserted between the midline tissues and the PSM. In all experiments, no Noggin expression could be observed in the experimental side despite the presence of Wnt1 (Figure 9.27f). These results strongly supported our hypothesis that Noggin expression was downstream of Shh signaling. Therefore, it was concluded that there were no situations in which Noggin equals 0 while both Wnts and Shh equal 1 or 2 (red clauses in Figure 9.26e). Moreover, there were no situations in which Noggin was 1 while one of the factors Wnts or Shh equals 2 and the other equals 1 or 2 (blue clauses in Figure 9.26e). Our finding also revealed the unique status of the green scenarios in Figure 9.26e. While mathematically these scenarios must lead to myogenesis, the biological system could not allow Noggin level 1 while Shh level is 0, thus forcing Noggin levels to "collapse" into the level 0. Therefore these unique scenarios did not lead to myogenesis.

We defined this process by the simple but powerful formula: $\text{Wnts} + (-\text{BMP4}) = \text{myogenesis}$, where the Shh levels could control the biological outcome of this formula by regulating Wnts signaling and controlling the expression of Noggin.

9.5
Outlook

The ever-increasing interest in BMC devices has not arisen from the hope that such machines could compete with their electronic counterparts by offering greater computation speed, fidelity, and power or performance in traditional computing tasks. The main advantage of autonomous BMC devices over the electronic computers arises from their ability to interact directly with biological systems and even with living organisms without any interface. The importance of the above-described biologically relevant automata is that they produce a computational output in the form of meaningful, expressible *in vivo* dsDNA. These results demonstrate that an appropriately designed computing machine can produce an output signal in the form of a specific biological function via direct interaction with living organisms. The next steps along this line would be the insertion of a complete computing device into a living cell or a tissue, with the long-term goal of

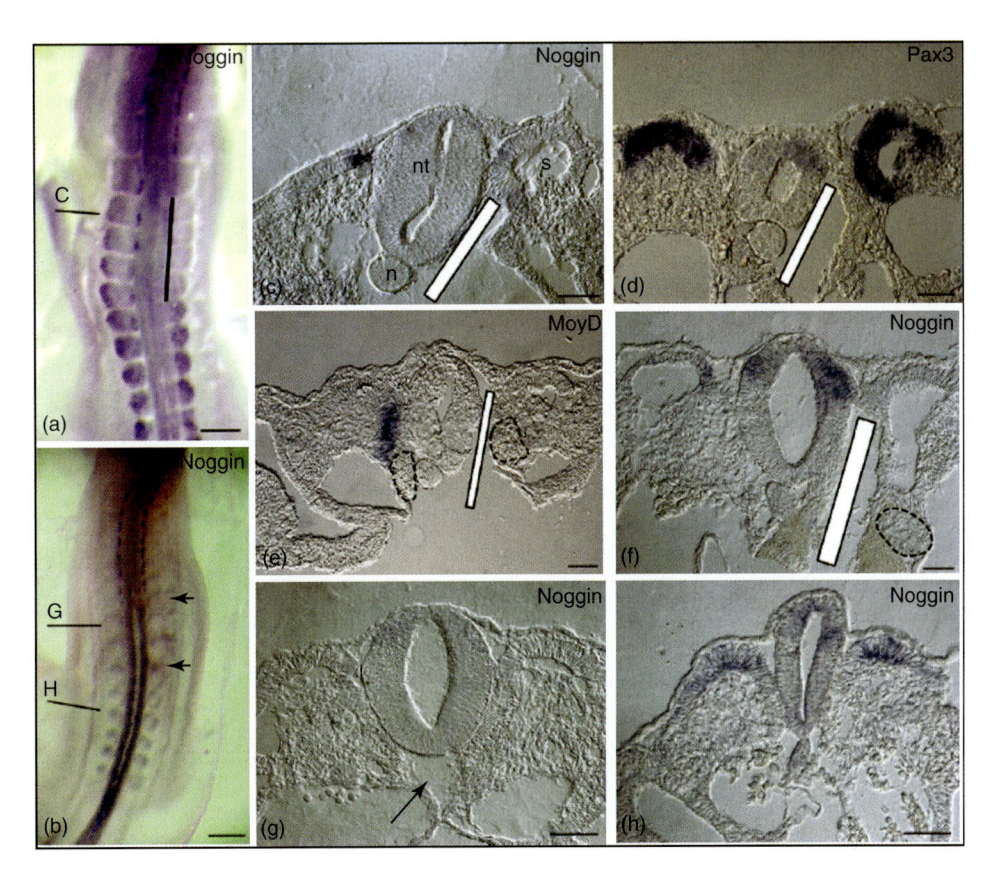

Figure 9.27 Blocking Shh signaling down-regulates Noggin expression in somites. A barrier inserted between the midline tissues and the PSM, or ablation of the notochord from the PSM anterior level, causes a loss of Noggin expression in somites as is evident in whole mount RNA *in situ* hybridization (a,b, respectively). A cross-section through the barrier level reveals no Noggin expression on the operated side (c) while Pax3 expression is expanded (d). When Wnt1-secreting cells were implanted laterally to a barrier that was inserted between the midline tissues and the PSM, no MyoD expression was observed compared to the contralateral side, where implanted Wnt1-secreting cells cause a ventral expansion of this gene (e). Analyzing Noggin in a similar experiment where Wnt1-secreting cells were implanted laterally to a barrier, no Noggin expression was observed (f). Note that in all cases presented in (c–f) the somite retains its epithelial state on the operated side. Cross-sections through regions (G) and (H) in (b) show no Noggin expression in the absence of the notochord (g, arrow) compared to a more posterior region where Noggin expression is normal (h). n, notochord, nt, neural tube, s, somite, scale bar in (a,b) represents 100 μm, and in (c–h) represents 30 μm.

utilizing BMC devices for *in vivo* diagnostics and disease control, and the design of new types of biological regulation.

Although this review focuses mainly on biomolecular finite automata, additional topics are briefly included to highlight the interesting opportunities and to provide the reader with some of the potential applications and extension of the concepts covered. These additional concepts include logic evaluators and logic gates that operate in cells, applications in developmental biology, as well as chemical encoding and processing of alphanumeric information.

References

1. Livstone, S., van Noort, D., and Landweber, L.F. (2003) *Trends Biotechnol.*, **21**, 98–101.
2. Ruben, J.A. and Landweber, L.F. (2000) *Nat. Rev. Mol. Cell Biol.*, **1**, 69–72.
3. Feynman, R. (1961) in *Miniaturization* (ed. D. Gilbert), Reinhold, New York, pp. 282–296.
4. Adleman, L.M. (1994) *Science*, **266**, 1021–1024.
5. Seeman, N.C. (2003) *Chem. Biol.*, **10**, 1151–1159.
6. Reif, J.H. (2002) *Science*, **296**, 478–479.
7. Chen, J. and Wood, D.H. (2000) *Proc. Natl. Acad. Sci. U.S.A.*, **97**, 1328–1330.
8. Ulam, S. (1972) in *Essays on Cellular Automata* (ed. A.E. Burks), University of Illinois Press, Chicago, pp. 219–231.
9. Holland, J.H. (1992) *Adaptation in Natural and Artificial Systems*, MIT Press, Cambridge.
10. Tagore, S., Bhattacharya, S., Islam, M.A., and Islam, M.L. (2010) *J. Proteomics Bioinform.*, **3**, 234–243.
11. Leclerc, E., Kirat, K., and Griscom, L. (2008) *Biomed. Microdevices*, **10**, 169–177.
12. Chang, Y.J., Hu, C.Y., Yin, L.T., Chang, C.H., and Su, H.J. (2008) *J. Biosci. Bioeng.*, **106**, 59–64.
13. Lipton, R.J. (1995) *Science*, **268**, 542–545.
14. Liu, Q., Wang, L., Frutos, A.G., Condon, A.E., Corn, R.M., and Smith, L.M. (2000) *Nature*, **403**, 175–179.
15. Sakamoto, K., Gouzu, H., Komiya, K., Kiga, D., Yokoyama, S., Yokomori, T., and Hagiya, M. (2000) *Science*, **288**, 1223–1226.
16. Faulhammer, D., Cukras, A.R., Lipton, R.J., and Landweber, L.F. (2000) *Proc. Natl. Acad. Sci. U.S.A.*, **97**, 1385–1389.
17. Braich, R.S., Chelyapov, N., Johnson, C., Rothemund, P.W., and Adleman, L. (2002) *Science*, **296**, 499–502.
18. Roweis, S., Winfree, E., Burgoyne, R., Chelyapov, N.V., Goodman, M.F., Rothemund, P.W., and Adleman, L.M. (1998) *J. Comput. Biol.*, **5**, 615–629.
19. Winfree, E., Liu, F., Wenzler, L.A., and Seeman, N.C. (1998) *Nature*, **394**, 539–544.
20. LaBean, T.H., Winfree, E., and Reif, J.H. (1999) in *Descriptive Complexity and Finite Models*, DIMACS series in Discrete Mathematics and Theoretical Computer Science, Vol. **54** (eds E. Winfree and D.K. Gifford), American Mathematical Society, Providence, pp. 123–140.
21. Winfree, E. (1999) *J. Biomol. Struct. Dyn.*, **11**, 263–270.
22. Benenson, Y., Paz-Elizur, T., Adar, R., Keinan, E., Livneh, Z., and Shapiro, E. (2001) *Nature*, **414**, 430–434.
23. Mao, C., LaBean, T.H., Reif, J.H., and Seeman, N.C. (2000) *Nature*, **407**, 493–496.
24. Rose, J.H., Deaton, R.J., Hagiya, M., and Suyama, A. (2002) *Phys. Rev. E. Stat. Nonlinear. Soft Matter Phys.*, **65**, 021910.
25. Komiya, K., Sakamoto, K., Gouzu, H., Yokoyama, S., Arita, M., Nishikawa, A., and Hagiya, M. (2001) in *DNA Computing 6th International Workshop on DNA-Based Computers* (eds A. Condon and G. Rozenberg), Springer-Verlag, Berlin, pp. 19–26.
26. Hopcroft, J.E., Motwani, R., and Ullman, J.D. (2007) *Introduction to*

Automata Theory, Languages, and Computation, Pearson/Addison Wesley.
27. Turing, A. (1936) *Proc. Lond. Math. Soc.*, **42** (2), 230–265.
28. Bennett, C.H. (1973) *IBM J. Res. Dev.*, **17**, 525–532.
29. Bennett, C.H. (1982) *Int. J. Theor. Phys.*, **21**, 905–940.
30. Hjelmfelt, A., Weinberger, E.D., and Ross, J. (1991) *Proc. Natl. Acad. Sci. U.S.A.*, **88**, 10983–10987.
31. Rothemund, P.W.K. (1996) in *DNA Based Computers: Proceedings of a DIMACS Workshop*, vol. **27** (eds R. Lipton and B.B. Eric), American Mathematical Society, pp. 297–302.
32. Radó, T. (1962) *Bell Syst. Tech. J.*, **41**, 877–884.
33. Benenson, Y., Adar, R., Paz-Elizur, T., Livneh, Z., and Shapiro, E. (2003) *Proc. Natl. Acad. Sci. U.S.A.*, **100**, 2191–2196.
34. Adar, R., Benenson, Y., Linshiz, G., Rosner, A., Tishby, N., and Shapiro, E. (2004) *Proc. Nat. Acad. Sci. U.S.A.*, **101**, 9960–9965.
35. Soreni, M., Yogev, S., Kossoy, E., Shoham, Y., and Keinan, E. (2005) *J. Am. Chem. Soc.*, **127**, 3935–3943.
36. Soreni, M. (2004) DNA based advanced parallel biomolecular computing machines. M.Sc. dissertation, Technion – Israel Institute of Technology.
37. Clelland, C.T., Risca, V., and Bancroft, C. (1999) *Nature*, **399**, 533–534.
38. Leier, A., Richter, C., Banzhaf, W., and Rauhe, H. (2000) *BioSystems*, **57**, 13–22.
39. Tanaka, K., Okamoto, A., and Saito, I. (2005) *BioSystems*, **81**, 25–29.
40. Lai, X.J., Lu, M.X., Qin, L., Han, J.S., and Fang, X.W. (2010) *Sci. China Inform. Sci.*, **53**, 506–514.
41. Kim, K.W., Bocharova, V., Halámek, J., Oh, M.K., and Katz, E. (2011) *Biotechnol. Bioeng.*, **108**, 1100–1107.
42. Palacios, M.A., Benito-Peña, E., Manesse, M., Mazzeo, A.D., LaFratta, C.N., Whitesides, G.M., and Walt, D.R. (2011) *Proc. Natl. Acad. Sci. U.S.A.*, doi: 10.1073/pnas.1109554108
43. Shoshani, S., Piran, R., Arava, Y., and Keinan, E. (2012) *Angew. Chem.*, **51**, 2883–2887.
44. Kossoy, E., Lavid, N., Soreni-Harari, M., Shoham, Y., and Keinan, E. (2007) *Chem. Bio. Chem.*, **8**, 1255–1260.
45. Shoshani, S., Wolf, S., and Keinan, E. (2011) *Mol. Biosyst.*, **7**, 1113–1120.
46. Benenson, Y., Gil, B., Ben-Dor, U., Adar, R., and Shapiro, E. (2004) *Nature*, **429**, 423–429.
47. Ratner, T., Piran, R., Jonoska, N., and Keinan, E. in press.
48. Piran, R., Halperin, E., Guttmann-Raviv, N., Keinan, E., and Reshef, R. (2009) *Development*, **136**, 3831–3840.
49. Gopalakrishnan, G.L. (2006) *Computation Engineering Applied Automata Theory and Logic*, Springer, New York.

10
In Vivo Information Processing Using RNA Interference
Yaakov Benenson

10.1
Introduction

10.1.1
Regulatory Pathways as Computations

The impact of synthetic biology on biological sciences has been compared to the effect of synthetic approaches on the hitherto descriptive science of chemistry [1]. While making new genetic material has been possible since the dawn of recombinant DNA technology [2], modern synthetic biology aspires to deliver systems of previously unseen complexity – from whole synthetic genomes [3] to complex pathways [4–6] to artificial cells [7]. In particular, fully or partially engineered biological pathways have been in the spotlight of this research due, on the one hand, to their central role of carrying out useful biological functions and, on the other, to their relative tractability as compared to entire cells or genomes. This latter effort comprises engineering enzymatic and metabolic pathways for bioproduction as well as regulatory and signaling pathways that trigger programmed physiological responses to environmental and internal molecular cues. These pathways have also emerged as the central object of systems biology research that draws the attention of engineers, computer scientists, physicists, and biologists alike.

Regulatory pathways can be conceptualized as reactive systems, or automata, composed of sensors that interact with external stimuli, the processor or a computer that integrates the signals in a meaningful way, and the actuation module that executes the results of signal processing in the cell [8]. While this is a useful abstraction, in nature the pathway structure is less modular; often the same protein combines the role of a sensor with that of a processor and the actuator. Still, it is possible to characterize a pathway by systematically varying its inputs and measuring the response, in a way that will provide enough information about the exact input–output mapping [9] in this pathway. Unfortunately, the actual molecular mechanisms behind these computations are often quite intricate and not easily understood; as such, they may not always be useful guides for constructing pathways that perform new types of computation.

Biomolecular Information Processing: From Logic Systems to Smart Sensors and Actuators, First Edition. Edited by Evgeny Katz.
© 2012 Wiley-VCH Verlag GmbH & Co. KGaA. Published 2012 by Wiley-VCH Verlag GmbH & Co. KGaA.

10.1.2
A Computation Versus a Computer

In the language of computer engineering, naturally evolved pathways are task-dedicated computers that excel in what they do but cannot be easily customized for new objectives. On the evolutionary timescales, regulatory networks have been rewired and reprogrammed [10], and approaches such as directed evolution have been shown to work for synthetic biology in a number of instances [11]. However, relying on evolution alone may not be enough, and these approaches should be combined with an "engineering method" [12], which comprises a set of empirical rules and guidelines that have emerged during hundreds if not thousands of years of experience in engineering nonbiological systems. The method advises the employment of a number of good design features such as modularity, hierarchy, safety mechanisms, and so on. Modern computer engineering has taken these ideas to the extreme, and today computers can be assembled from parts manufactured by hundreds of different companies; at the same time they are capable of running programs written by software experts who may have no idea how a computer is actually built. Modern computers have computation power that is close to universal, and the number of tasks they can perform is endless. This is in stark contrast with how nature organizes its own information processing pathway where the "hardware" and the "software" are mutually optimized and hard-wired. In order to radically increase the power of synthetic biology to offer real-life solutions to longstanding problems, one should strive to borrow as much as possible from traditional engineering in order to engineer biological systems. In the context of biological computing systems, this calls for the capacity to program a molecular computation task to precise specifications with the possibility to perform signal integration of ever-increasing complexity by following a uniform set of design rules and principles, reusing common building blocks for different purposes, operating across a variety of biological "platforms," and so forth. Yet, we should remember that a biological medium is unique and does not lend itself to straightforward application of rules learnt in the world of electromechanics. None of the aforementioned concepts can be implemented to the letter. Therefore, it is crucial to respect the medium in which synthetic biologists operate and learn as much as possible from naturally available systems.

10.1.3
Prior Work on Synthetic Biomolecular Computing Circuits

The bulk of previous work on synthetic regulatory pathways *in vivo* and *in vitro* focused on proof-of-principle systems that demonstrated specific dynamic properties or input–output relations. Impressive results have been obtained in the construction of logic gates [4, 13–15], bistability [6, 16], oscillations [17], molecular memory [18], and state machines [19, 20]. Among these approaches, synthetic transcriptional control based on integration of interchangeable regulatory and protein-coding DNA sequences (i.e., Biobricks [21]) has emerged as a candidate design approach that can implement *in vivo* many of the good engineering practices described above. It has been shown in theory to be fully reprogrammable and scalable [22]. However, practical implementation of these ideas has been progressing slowly because of the idiosyncrasies of promoters, protein–protein and protein–DNA interactions, and so on. Signaling pathways with their post-translational modifications have also been proposed as a basis of engineered circuits in theory [23] and in experiments [24], but, as before, practical implementation of large-scale engineered circuits has not yet been achieved. Regulation mechanisms on the RNA level, notably allosteric riboswitches [25] and ribozymes [26], have been extensively studied as well [15, 27]. As with other approaches, their theoretical potential to form a basis for large-scale computations has been only partially exploited in experiments.

10.2 RNA Interference-Based Logic

10.2.1 General Considerations

The key challenge in designing a scalable approach for a synthetic regulatory pathway is finding the way to build a programmable computing module where multiple inputs are processed in a desired fashion. In the language of mathematics, this normally involves many-to-one mapping, where only one (or two) outcomes are possible for any given combination of inputs. Such architecture is also called *"fan-in"* circuitry in engineering. There are multiple examples of fan-in computations in nature, which involve regulation of a single molecular species by multiple agents in parallel. Many proteins have tens of modification sites, with each set of modifications uniquely defining their activity in different contexts [28]. Gene expression is routinely controlled by a large number of signals, from chromatin structure to distant modulators to transcription factors (TFs), splicing factors, and further along the line, mRNA stability, translation, and protein stability factors [29]. mRNA is itself controlled by numerous other RNA species, most notably microRNA (miR) and long noncoding RNA. In some of these circuits, the outcome is almost binary [30], involving total suppression (or activation) of the protein activity or gene expression; in others, the effect is gradual and yet in additional instances the mapping is many-to-many, for example, with multiple protein modifications activating different action niches of the said protein. When looking at a particular mechanism or a combination of mechanisms to serve as a basis of synthetic efforts, one should ask how difficult it is to engineer a new regulatory link based on prescribed specifications; how difficult it is to integrate a number of such new links within the same target; and how exactly are the links integrated.

We chose RNA interference (RNAi) as such a mechanism because of a combination of highly favorable properties. The molecular details of RNAi are complex and the reader is advised to read a number of excellent reviews on the subject [31]. We build on a few known features of the process in our engineered systems. First, RNAi is essentially a downregulatory link where a small RNA (sRNA) species acts to cause rapid degradation of its target mRNA and/or inhibit protein translation from this mRNA. In mammals, RNAi is elicited by transcribed RNA species called *microRNA* which are encoded as standalone open reading frames or in introns in larger transcripts. A number of complex enzymatic and transport steps are involved in the eventual generation of an active miR species. Those species are single-stranded oligonucleotides 22–24 nt long, and they have partial or full complementarity to their target mRNA, usually in its 3′-untranslated region (3′-UTR) but occasionally in the coding region and 5′-UTR. In the case of partial complementarity, the mRNA is subjected to accelerated exonuclease activity, and, in addition, protein translation from the mRNA is slowed down. In the case of full complementarity, the miR triggers site-specific cleavage of the mRNA, once again leading to rapid mRNA degradation. The fact that the sequence determinant of miR activity (i.e., target sequence) within the mRNA is in most cases just as short as or even shorter than the miR itself, combined with the placement of these determinants in 3′-UTR, already points to the possibility to combine multiple targets in the same transcript without altering its coding region, thereby implementing multisignal control of the gene product.

The endogenous RNAi pathway can be exploited in a number of ways to elicit the downregulatory effect. Special synthetic RNA structures called small interfering RNA (siRNA) as well as synthetic short hairpin RNA (shRNA) can be directly transfected into cells where they knock down gene expression of their targets. These species are commonly used to elucidate gene functions in large-scale screens in multiple organisms. Furthermore, shRNA can be genetically encoded and either transfected to or stably integrated into the host cell, resulting in a constitutive or regulated production of the active species. Finally, endogenous miR-encoding genes can be mimicked with novel active sequences. Remarkably, those genes can be controlled by the same promoters that control protein-coding genes and transcribed by RNA polymerase II. As such, the entire repertoire of mechanisms used to control transcription of protein-coding genes can be harnessed to drive the expression of synthetic mimics of endogenous miRs.

10.2.2
Logic Circuit Blueprint

Among the variety of mechanisms that sRNA uses to regulate gene expression, we chose to focus on the most efficient mode of action in which it elicits site-specific mRNA cleavage: in other words, we only considered fully complementary target sites in the mRNA. We asked ourselves what happens when a number of different targets are integrated into the same transcript. If the sRNAs are considered as inputs and the level of the transcript-encoding protein as the output, then the "computation" of this simple circuit is described by the logic relation of

$$\text{Output} = \text{NOT}(\text{sRNA-A}) \text{ AND NOT}(\text{sRNA-B})$$
$$\text{AND NOT}(\text{sRNA-C}) \text{ AND NOT} \ldots$$

where sRNA-A, B, C ... target the transcript independently and in parallel. In plain language, this means that none of the sRNAs that targets this transcript can be present for the output to be expressed.

We may also imagine that the same output protein is encoded by a separate transcript, targeted by a different set of sRNAs D, E, and F. In this case, at least one intact transcript can generate the output protein (Figure 10.1a). The logic of this extended circuit becomes

Output = (NOT(sRNA-A) AND NOT(sRNA-B)

AND NOT(sRNA-C) AND NOT...)

OR (NOT(sRNA-D) AND NOT(sRNA-E)

AND NOT(sRNA-F) AND NOT...)

Adding more transcripts will extend the number of OR operators of the computation, leading to the general form

Output = (NOT(A) AND NOT(B) AND...) OR (NOT(·)

AND NOT(·)...) OR(·)...

Even in these simple circuits one sees the potential of programmability and scalability. We can implement AND, OR, and NOT logic operators, and we can alter the logic formula by changing the number of sRNA targets in each transcript and by adding more transcripts. However, there are insufficiently flexible for two reasons: First, the input signals are sRNA species and no others. Second, all the inputs are inverted by the NOT operator. For fully programmable logic networks, the NOT operators need to be used selectively. To address both issues, we proposed that the sRNA inputs themselves be controlled by additional signals in one of the two common ways – activating or inhibitory. A signal that activates a particular sRNA will then become an input to the circuit and, similar to the sRNA itself, be inverted by the NOT operator. A signal that inhibits an sRNA will have to be present or active for this sRNA to be absent; such signal will enter the computation without the NOT inversion. To summarize, let us assume that signals A and C inhibit sRNA-A and sRNA-C, respectively, while signal B activates sRNA-C. Let us also suppose that signals D and F inactivate sRNAs D and F, while signal E activates sRNA-E. Overall, the computation of this new circuit will be described by the equation

Output = (A AND NOT(B) AND C) OR (D AND NOT(E) AND F) (Figure 10.1b)

In fact, the approach is even more powerful because the logic computation one might wish to perform does not have to be confined to the specific arrangement of AND, OR, and NOT operators. It turns out that the specific way of writing logical expression using NOT operators on individual inputs, combining them into groups connected by AND operators and finally combining those using OR, is a universal way to represent any logic expression. It is known as the disjunctive normal form (DNF). Therefore, one might consider a circuit performing an XOR (exclusive OR)

(a)

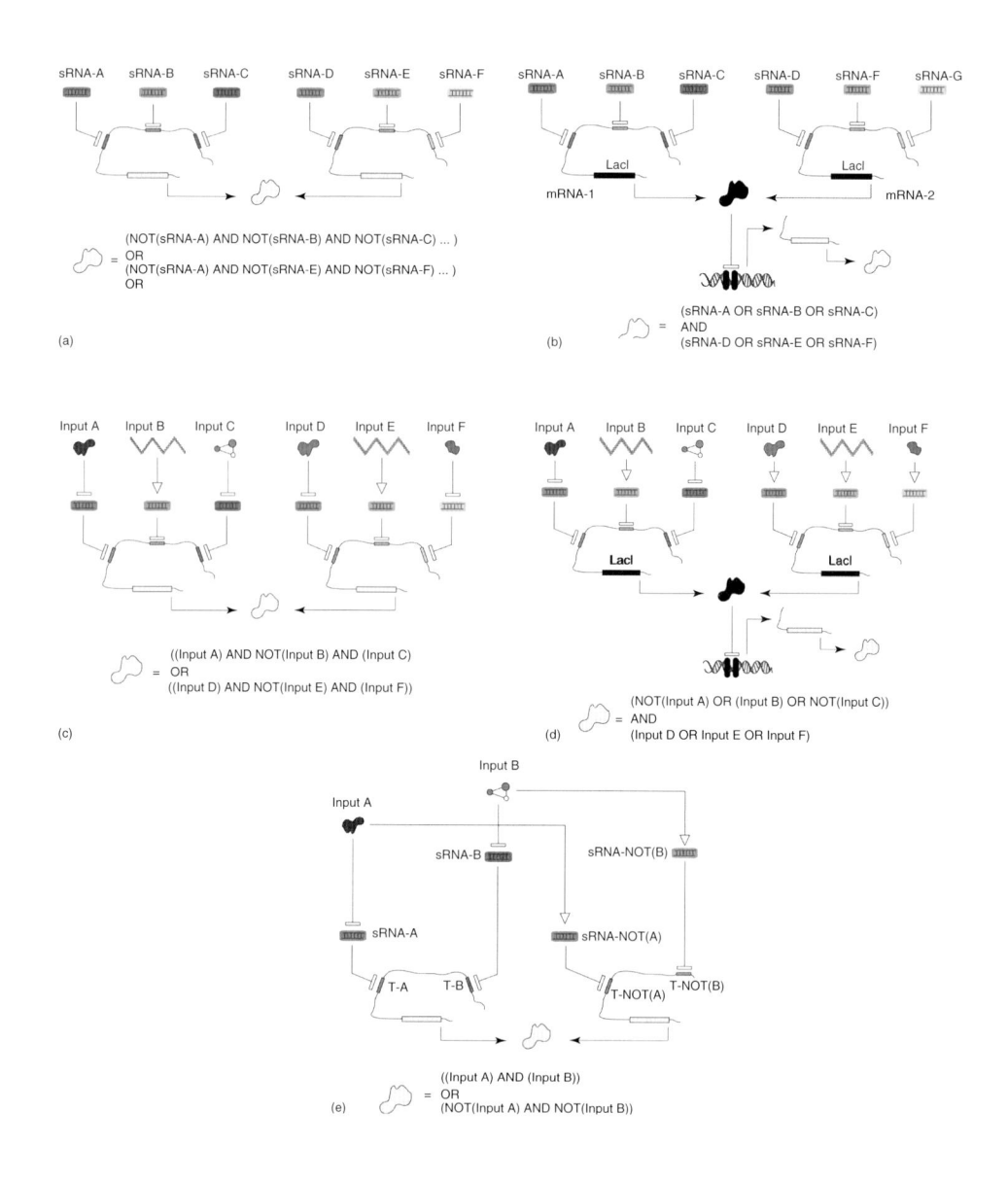

$$\text{\large{🧬}} = \begin{array}{l} \text{(NOT(sRNA-A) AND NOT(sRNA-B) AND NOT(sRNA-C) ...)} \\ \text{OR} \\ \text{(NOT(sRNA-A) AND NOT(sRNA-E) AND NOT(sRNA-F) ...)} \\ \text{OR} \end{array}$$

(b)

$$\text{\large{🧬}} = \begin{array}{l} \text{(sRNA-A OR sRNA-B OR sRNA-C)} \\ \text{AND} \\ \text{(sRNA-D OR sRNA-E OR sRNA-F)} \end{array}$$

(c)

$$\text{\large{🧬}} = \begin{array}{l} \text{((Input A) AND NOT(Input B) AND (Input C)} \\ \text{OR} \\ \text{((Input D) AND NOT(Input E) AND (Input F))} \end{array}$$

(d)

$$\text{\large{🧬}} = \begin{array}{l} \text{(NOT(Input A) OR (Input B) OR NOT(Input C))} \\ \text{AND} \\ \text{(Input D OR Input E OR Input F)} \end{array}$$

(e)

$$\text{\large{🧬}} = \begin{array}{l} \text{((Input A) AND (Input B))} \\ \text{OR} \\ \text{(NOT(Input A) AND NOT(Input B))} \end{array}$$

logic, such as A XOR B. The normal form representation of this expression is

(A AND NOT(B)) OR (NOT(A) AND B)

It turns out that DNF is not the only universal form of logic expressions that uses AND, OR, and NOT operators. A complementary form called the conjunctive normal form (CNF) inverts the location of AND and OR operators, and its general form is

(A OR NOT(B) OR..) AND (... OR ... OR ...) AND (... OR ... OR ...)...

We attempted to design RNAi-based circuits for this kind of expressions. To do so, we had to integrate some transcriptional regulation in the circuits. Let us consider a repressible promoter that is regulated by a repressor protein. We can further suppose that the repressor protein is translated from a number of mRNA transcripts and that the dose is calibrated such that the repressor protein produced from at least one of the transcripts will fully repress the output. The basic logic of this circuit is

Output = NOT(mRNA-1) AND NOT(mRNA-2) AND...

Let us suppose that each of these mRNAs is controlled by a number of sRNAs, for example, sRNA-A, B, and C downregulate mRNA-1; C, D, and E downregulate mRNA-2; and so on. From the previous construction, the following relation holds

Figure 10.1 Basic blueprint of normal form logic circuits. (a) The core computational architecture of a disjunctive normal form (DNF) logic functions. A number of mRNA molecules code for the same protein (gray blob). Each mRNA is targeted by a distinct set of small RNAs via specific targets fused in their 3′-UTR. Each inhibitory integration is indicated by a blunted arrow. The logic relation between the presence of sRNA and the output protein is given. (b) An interface between the DNF computational core and the external inputs, whereby each sRNA is either activated or inhibited by its cognate signal. Inclusion of these sensory links allows computing any logic expression in a disjunctive normal form. (c) Core computational architecture of a conjunctive normal form (CNF) logic functions. A number of mRNA molecules code for the same repressor protein (black blob). Each repressor mRNA is targeted by a distinct set of small RNAs via specific targets fused in their 3′-UTR. The repressor affects the promoter that drives the expression of the output (gray blob). Implicit in the figure is the assumption that the amount of repressor produced from either one of the repressor-coding mRNAs is enough to efficiently repress the promoter. The logic relation between the presence of sRNA and the output protein is given. (d) Interface between the CNF computational core and the external inputs, whereby each sRNA is either activated or inhibited by its cognate signal. Inclusion of these sensory links allows computing any logic expression in a conjunctive normal form. (e) Construction of a circuit computing a given logic formula in a DNF form. Each group of AND operators is associated with an output-encoding mRNA molecule. Each input, either direct or negated, is associated with an sRNA target and consequently with an sRNA. Targets and sRNAs that represent an input and its negation are chemically and physically distinct. Finally, appropriate regulatory links are introduced between the inputs and the sRNAs, such that an sRNA representing an input is inhibited by the input, and an sRNA representing input negation is activated by it. Each input can have at most two emerging regulatory links.

between the levels of sRNAs and the mRNA:

mRNA-1 = NOT(sRNA-A) AND NOT(sRNA-B) AND NOT(sRNA-C)

mRNA-2 = NOT(sRNA-D) AND NOT(sRNA-E) AND NOT(sRNA-F)

Substituting the latter equations into the former, one obtains

Output = NOT(NOT(sRNA-A) AND NOT(sRNA-B)
AND NOT(sRNA-C)) AND
NOT(NOT(sRNA-D) AND NOT(sRNA-E) AND NOT(sRNA-F))

According to logic rules known as *De Morgan laws*, this is equivalent to

Output = (sRNA-A OR sRNA-B OR sRNA-C)
AND (sRNA-D OR sRNA-E OR sRNA-F) (Figure 10.1c)

While this resembles the CNF, there are no NOT operators here. In order to introduce them, we make the same step as with the DNF circuit and require that each sRNA be linked to an additional input signal. A signal that activates an sRNA will enter the equation directly; a signal that inhibits an sRNA will be inverted. Therefore if A, B, and C are, respectively, inhibitory, activating, and inhibitory while D, E, and F are all activating, the computation will become

Output = (NOT(A) OR B OR NOT(C)) AND (D OR E OR F) (Figure 10.1d)

We can, or course, work in the opposite direction and first describe a specific logic computation we want the circuit to perform. As mentioned above, this computation need not be in a normal form. For example, it could be Output = A ⇔ B, meaning that the output will be produced when the two inputs are the same (either both On or both Off). In the DNF, this equation becomes

Output = (A AND B) OR (NOT(A) AND NOT(B))

In this formula, the same variable (input) appears more than once. Yet, we can construct the circuit using our standard approach. The key here is to use one sRNA to represent the On state of the input, and another to represent the Off state. Each input A and B will need to inhibit an sRNA representing the On state and activate an sRNA representing the Off state. Overall, the circuit will require four sRNAs with two regulatory links generated by each input (Figure 10.1e).

This completes the blueprint construction of the logic computations using RNAi as the computational module and sensory links as the interface with the external environment.

10.2.3
Experimental Confirmation of the Computational Core

Our initial experimental work focused on quantifying the performance of the computational "core" of the circuit [32]. We asked ourselves whether multiple sRNA inputs could indeed be integrated in a small number of transcripts in a digital fashion. For example, it was unclear whether multiple binding sites could be combined at will in the 3'-UTR. It was also not clear whether highly efficient RNAi could be achieved, given that commonly the reporter knock-down values are around 70–80%.

We decided to use siRNA as the sRNA species in these experiments because of their perceived high efficiency. We looked for a set of siRNAs that would not cross-react with each other's targets and elicit efficient knockdown in the presence of the intended target sequences. We based our set on sequences that had already been reported to work against non-mammalian genes, such as fluorescent proteins and firefly luciferase. We also used empirical rules for siRNA efficiency determination in order to improve these sequences. Overall, we compiled the structures and, in parallel, designed integrator mRNA molecules with their targets in their 3'-UTR. In order to test the digital nature of the circuits, we used all possible combinations of siRNA inputs (32 combinations for 5 inputs and 16 for 4 inputs) and checked whether we were obtaining only very low (zeros) or very high (ones) output levels. However, when we did this experiment, the observed output values took the entire range between 0 and 1; in other words, the system behaved as an analog, rather then digital, circuit. We had to revisit our naïve assumptions about the way multiple siRNAs act on mRNA targets. It turned out to be the mRNA secondary structure at and around the target region. When the target sites were redesigned with the secondary structure in mind, the performance improved considerably and became very close to the anticipated digital separation (Figure 10.2a). Yet, among the 32 outputs there were a few outliers, which, while not violating the overall trend, were tending toward the middle of the scale (Figure 10.2b). We were unable to provide a satisfactory explanation to these outliers, especially since they appeared only when the entire system was assembled but not during preliminary tests.

We also tested the CNF configuration with LacI constructs targeted by the same siRNAs. However, this system proved much more resistant to scaling up. We showed a satisfactory digital response with up to two inputs. We hypothesized that the residual LacI levels were still able to exert substantial repression of the output, and the anticipated elimination of transcriptional repression had not taken place.

10.3
Building the Sensory Module

The next step in constructing the full-fledged circuits would be the implementation of the appropriate sensory/regulatory links between the diverse biological inputs and the sRNA species. One approach was to directly affect an sRNA via specific interactions, a strategy that had been used successfully with small molecule inputs [33]. Another was to control them indirectly, for example, by using miR expression as an intervention point.

D1: (A AND B AND C) OR (D AND E)

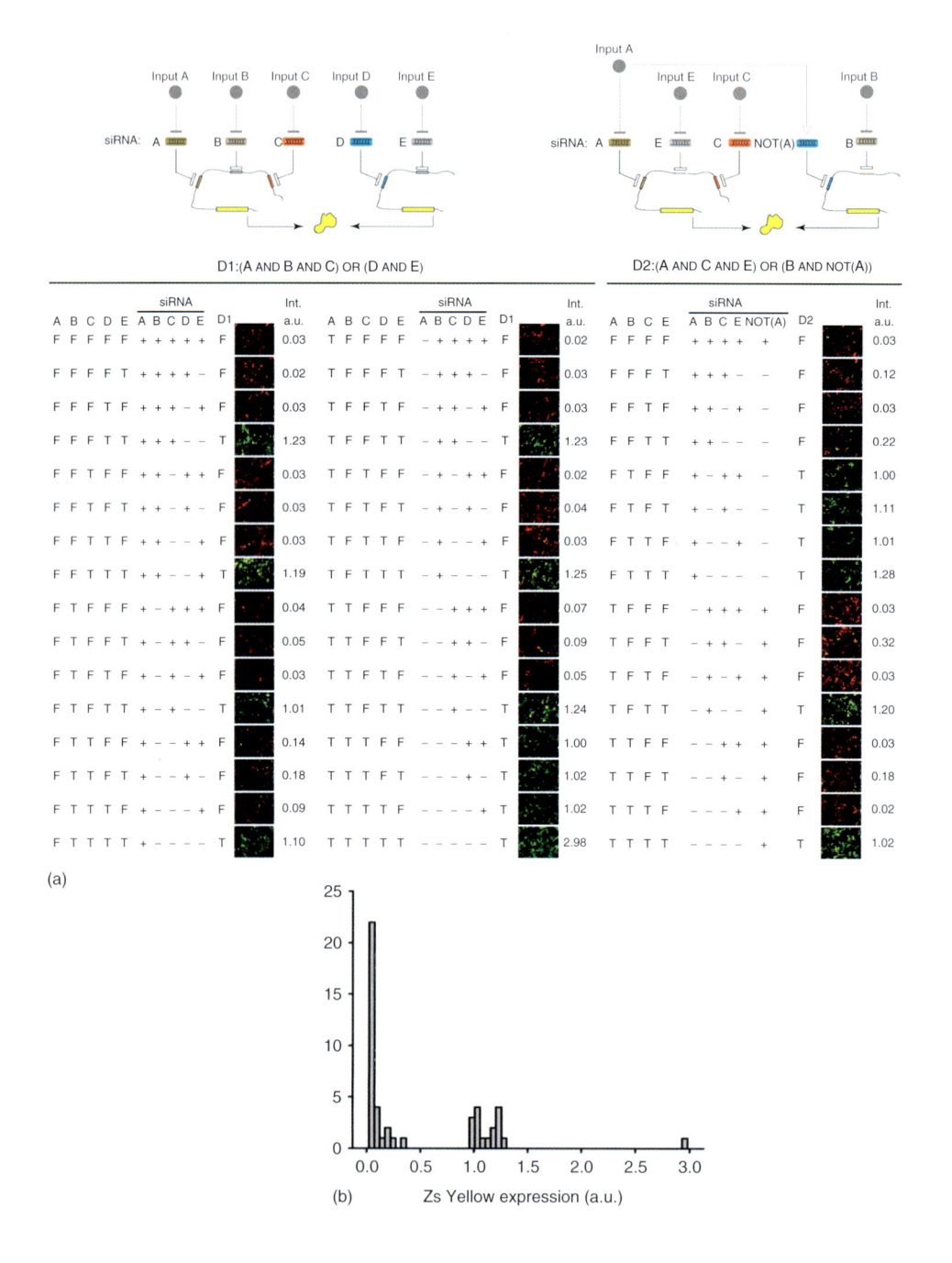

D2: (A AND C AND E) OR (B AND NOT(A))

(a)

A	B	C	D	E	A	B	C	D	E	D1	Int. a.u.	A	B	C	D	E	A	B	C	D	E	D1	Int. a.u.	A	B	C	E	A	B	C	E	NOT(A)	D2	Int. a.u.
							siRNA												siRNA											siRNA				
F	F	F	F	F	+	+	+	+	+	F	0.03	T	F	F	F	F	−	+	+	+	+	F	0.02	F	F	F	F	+	+	+	+	+	F	0.03
F	F	F	F	T	+	+	+	+	−	F	0.02	T	F	F	F	T	−	+	+	+	−	F	0.03	F	F	F	T	+	+	+	−	−	F	0.12
F	F	F	T	F	+	+	+	−	+	F	0.03	T	F	F	T	F	−	+	+	−	+	F	0.03	F	F	T	F	+	+	−	+	−	F	0.03
F	F	F	T	T	+	+	+	−	−	T	1.23	T	F	F	T	T	−	+	+	−	−	T	1.23	F	F	T	T	+	+	−	−	−	F	0.22
F	F	T	F	F	+	+	−	+	+	F	0.03	T	F	T	F	F	−	+	−	+	+	F	0.02	F	T	F	F	+	−	+	+	−	T	1.00
F	F	T	F	T	+	+	−	+	−	F	0.03	T	F	T	F	T	−	+	−	+	−	F	0.04	F	T	F	T	+	−	+	−	−	T	1.11
F	F	T	T	F	+	+	−	−	+	F	0.03	T	F	T	T	F	−	+	−	−	+	F	0.03	F	T	T	F	+	−	−	+	−	T	1.01
F	F	T	T	T	+	+	−	−	+	T	1.19	T	F	T	T	T	−	+	−	−	−	T	1.25	F	T	T	T	+	−	−	−	−	T	1.28
F	T	F	F	F	+	−	+	+	+	F	0.04	T	T	F	F	F	−	−	+	+	+	F	0.07	T	F	F	F	−	+	+	+	+	F	0.03
F	T	F	F	T	+	−	+	+	−	F	0.05	T	T	F	F	T	−	−	+	+	−	F	0.09	T	F	F	T	−	+	+	−	+	F	0.32
F	T	F	T	F	+	−	+	−	+	F	0.03	T	T	F	T	F	−	−	+	−	+	F	0.05	T	F	T	F	−	+	−	+	+	F	0.03
F	T	F	T	T	+	−	+	−	−	T	1.01	T	T	F	T	T	−	−	+	−	−	T	1.24	T	F	T	T	−	+	−	−	+	T	1.20
F	T	T	F	F	+	−	−	+	+	F	0.14	T	T	T	F	F	−	−	−	+	+	T	1.00	T	T	F	F	−	−	+	+	+	F	0.03
F	T	T	F	T	+	−	−	+	−	F	0.18	T	T	T	F	T	−	−	−	+	−	T	1.02	T	T	F	T	−	−	+	−	+	F	0.18
F	T	T	T	F	+	−	−	−	+	F	0.09	T	T	T	T	F	−	−	−	−	+	T	1.02	T	T	T	F	−	−	−	+	+	F	0.02
F	T	T	T	T	+	−	−	−	−	T	1.10	T	T	T	T	T	−	−	−	−	−	T	2.98	T	T	T	T	−	−	−	−	+	T	1.02

(b) Zs Yellow expression (a.u.)

10.3.1
Direct Control of siRNA by mRNA Inputs

We were interested in using mRNA molecules as inputs to the circuits because mRNA levels are important indicators of both normal and abnormal (disease) cell states. Previously, mRNAs were successfully used as inputs to a DNA-based *in vitro* molecular computer [19, 34]. We built on those results to implement a regulatory link between mRNA and siRNA [35].

We based the mechanism on the process of base recognition and strand migration between the mRNA sequence and the RNA oligonucleotides comprising the sRNA sensor. We implemented an activating interaction, whereby a high mRNA concentration triggered high levels of an active siRNA. The "sensor" was a combination of two RNA molecules, one single-stranded and another partially double-stranded. The partially double-stranded component contained an antisense strand of an siRNA molecule hybridized to another RNA oligomer functioning as a protecting strand. The single-stranded component was the sense strand of the siRNA. The protecting strand was designed to be complementary to a specific "trigger" sequence motif in the mRNA molecule, such that, in the presence of high enough mRNA concentration, the protecting strand would migrate to that mRNA, leaving the single-stranded antisense strand free to hybridize with the single-stranded sense strand to form a functional siRNA. That siRNA would then be capable of downregulating target gene expression via the RNAi pathway (Figure 10.3a,b).

While originally planned for *in vivo* implementation, we took the first experimental steps in the cell-free extract of *Drosophila* embryos. This extract, or lysate, is a very popular biochemical system for RNAi studies, and it works much better than mammalian cell extracts [36]. The lysate offered the advantage of performing well-controlled experiments where parameters such as concentration dependence, kinetics, and so on, could be delineated before testing the system in a much less predictable setting of living cells.

The system proved to be very sensitive to the exact nucleotide sequence of its components. We had to screen multiple structural variants to achieve low

---

Figure 10.2 Experimental testing of RNAi-based logic computational core. (a) Two expressions in DNF form are evaluated for all possible variable assignments as indicated in the figure. The complete circuits whose logic core is being evaluated are shown above the images. The hypothetical inputs (gray circles) are assumed to control the siRNA mediators in certain ways (gray arrows). Output-encoding genes are introduced in all the experiments. Depending on the truth-value assignments of the hypothetical inputs A–E (columns on the left; T, True or On; F, False or Off) and the direction of the hypothetical regulatory link, we determine the states of the siRNAs (+/On or −/Off) that are actually added or withheld in each particular measurement. Red pseudocolor shows DsRed protein fluorescence used as a transfection marker. Green pseudocolor indicates fluorescence of the ZsYellow protein, used as the output. (b) The quantitative results corresponding to the images obtained using flow cytometry show bimodal output distribution.

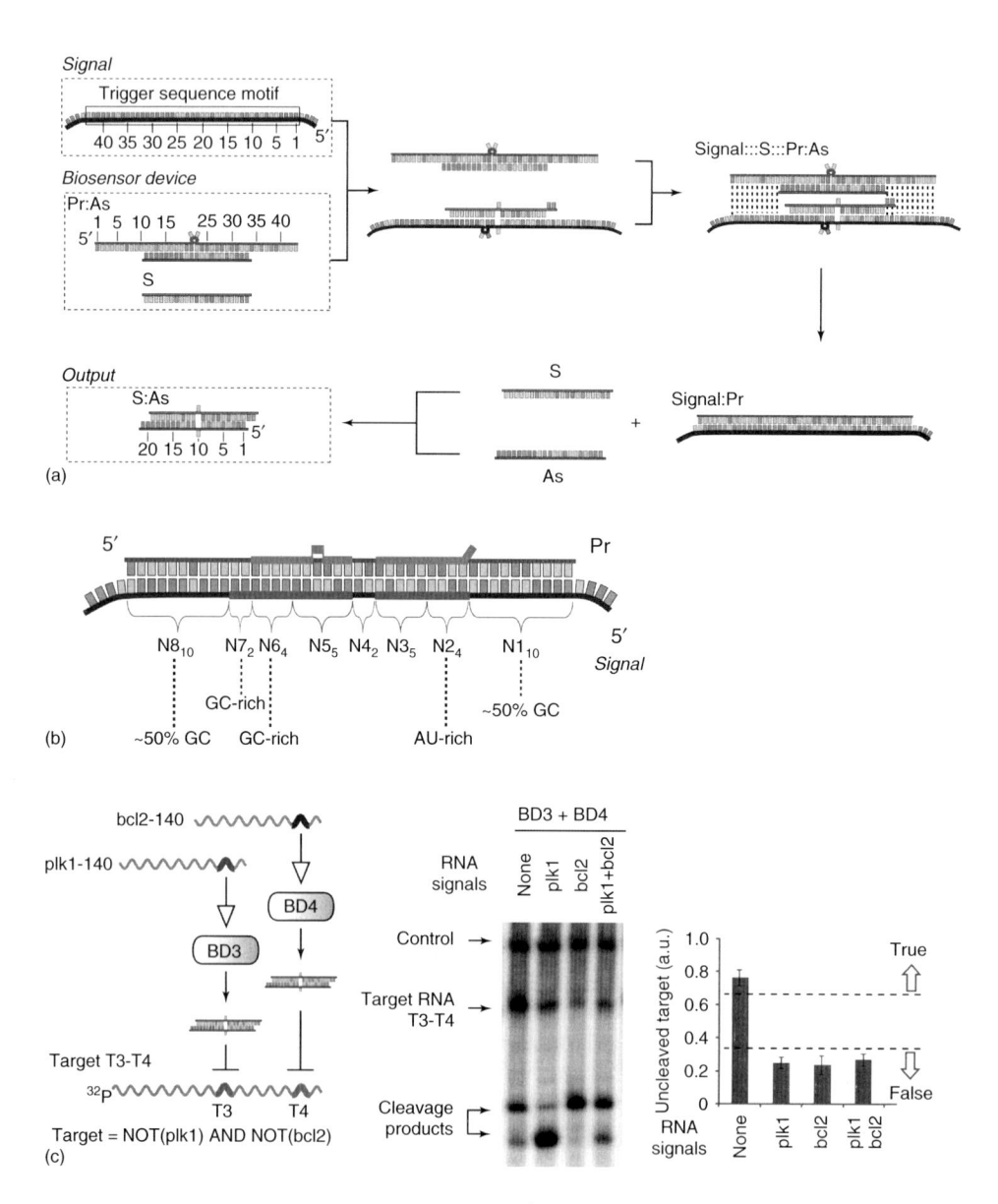

192 *10 In Vivo Information Processing Using RNA Interference*

background in the Off state and rapid and efficient transition to the On state in the presence of the cognate signal. In the first step, we investigated the switching process in a simple buffer solution by running native acrylamide gels that resolved various molecular species in the mixture, such as the single-stranded sense oligomer, the hybrid between the protecting and the antisense oligomer, and the active siRNA duplex. We also measured the rates using a continuous assay based on fluorescent labels. In our initial attempts, we observed a high rate on nonspecific strand exchange (in the absence of input) and a low rate of specific switching. We proceeded by systematic study of the different sensor components and found an optimal length of single-stranded overhangs that nucleate to the mRNA. Other factors affecting the rates of background and correct switching were the AT contents of the different parts of the sensor, inclusion of bulges in the duplexes, and even chemical modifications of the backbone. Eventually, we were able to construct a number of sensors with a very high ratio of specific to nonspecific switching that responded rapidly and to sense very low concentrations of the signal.

We then went on to combine the sensors with the computational module in the lysate, which presented a special challenge because the switching had to be followed by RNAi directed toward a target RNA. The main difficulty presented by the lysate system was rapid RNA degradation. While in mammalian cells it

Figure 10.3 Design and operation of the biosensor devices for mRNA signals. (a) Mechanism of action. The mRNA signal contains a "trigger-sequence motif" 43 nt long. The biosensor device consists of a "protecting" (Pr) strand preannealed to an "antisense" (As) strand, along with a single "sense" (S) strand. The Pr strand is a chemically modified ribooligonucleotide fully complementary to the trigger-sequence motif. The As strand is complementary to the nucleotides 11, ..., 20 and 23, ..., 33 of the Pr strand, generating a 2-nt "bulge" in the Pr:As duplex. (This bulge serves to accelerate the sensing process and can also be 3 nt long.) The Pr:As duplex has two 10-nt single-stranded overhangs, which serve as nucleation sequences during its interaction with the trigger-sequence motif. The S strand RNA contains, starting from its 3′-terminus, two unpaired nucleotides, a segment complementary to the nucleotides 1, ..., 9 of the As strand, one unpaired nucleotide, and another segment complementary to the nucleotides 11, ..., 19 of the As strand. During the interaction between the signal and the biosensor, the Pr strand migrates over to the trigger-sequence motif, and the released As strand hybridizes with the single-stranded S strand to form a canonical siRNA duplex S:As with a single-nucleotide mismatch that is not detrimental to RNAi efficiency but serves to modulate the energy balance of the process. (b) Structural details of the trigger motif and the biosensor device. Dark gray strand is the input and light gray strand is the fully complementary Pr. The requirements from various subsequences of the trigger motif are indicated. The bases of the trigger motif that serve to form an As strand are highlighted in green, and the bases of the Pr strand that form the S strand are shown in thick light gray. (c) The "AND" gate between bcl2 and plk1 signals. Left panel: the schematics of the circuit and the logic formula it computes. Middle panel: denaturing gel electrophoresis analysis of the circuit's function with different signal combinations. Right panel: quantified amounts of uncleaved target RNA normalized to the control radiolabeled RNA in the sample (mean ± SD). High and low amounts represent the circuit's True and False logic outputs, respectively. The True/False separation is a qualitative illustration of the gate's performance under observed sample-to-sample variability.

is possible to increase siRNA stability by chemical modifications, the lysate does not tolerate them well. We were able, however, to introduce modifications on the protecting strand, resulting in a stably bound antisense oligomer. This still left the single-stranded sense oligomer unprotected, and its lifetime was measured to be less than 5 min. We compensated this rapid degradation by adding fresh aliquots of the sense oligomer every few minutes to replenish its concentration. In combination with fast switching kinetics, this was enough to generate active siRNA in the lysate and incorporate it in the RNAi machinery. We then combined sensors for two different mRNAs in two-input logic gates and observed correctly computed outputs (Figure 10.3c).

This study represented one of the first examples of a molecular computing network operating in cell-free extracts, representing the middle ground between simple buffer solutions and live cells. We learned valuable lessons: for example, RNA-based systems are more difficult to construct and rationalize than comparable DNA-based systems and cell-free extracts present unique challenges. Nevertheless, cell-free extracts remain an important testing ground for synthetic biomolecular constructs.

10.3.2
Complex Transcriptional Regulation Using RNAi-Based Circuits

In a different study, we wanted to apply our approach to the longstanding challenge in synthetic biology and gene regulation in general – implementation of arbitrarily specified transcription regulatory programs. Previous efforts had encountered the following dilemma: On the one hand, many natural regulatory circuits implement highly complex regulatory functions with multiple inputs. On the other, these complex regulatory circuits are difficult to re-engineer. In other words, while it is possible to find in nature or engineer a promoter or promoter cascade that implements a complex logic relation between *some* TFs and a target gene, it is difficult to do so for *any* arbitrarily chosen set of TFs [4, 14, 37]. Therefore we applied RNAi-based approaches to this challenge [38]. We took advantage of a well-known fact that miR molecules, which are endogenous species that take part in the RNAi pathway, are expressed in cells in the same way as protein-coding genes. Their transcription is performed by RNA polymerase II, and the promoter structure is identical to those of protein-coding genes. Moreover, many miRs are expressed from introns and can therefore be co-transcribed with protein-coding mRNA. In another recent development, it was shown that it is possible to engineer mimics of endogenous miR genes by following a few simple design rules [39]. Furthermore, there has been substantial progress in elucidation of DNA sequences that interact specifically with TFs. To summarize these developments, it has become increasingly plausible to engineer promoters that are selectively regulated (activated or repressed) by individual endogenous TFs via integration of specific binding sites, and use these promoters to drive the expression of engineered, nonendogenous miR molecules. In such constructs, we implement a one-to-one link between the activity of a given TF and the intracellular level of an engineered miR molecule. This represents one sensory link in our large-scale logic circuits. Combining multiple regulated constructs in a logic circuit would lead to logic integration of multiple TFs. In this setup, the requirement to build a complex promoter is replaced by a requirement to build a number of simple promoters, which is a much more feasible task. Moreover, there is no *a priori* requirement that the TFs must fulfill apart from the ability to selectively control their response elements. Therefore, the task of tailored transcriptional circuits becomes much more feasible (Figure 10.4a).

To test the approach proposed above, we chose a more modest task of building complex circuits with well-characterized TFs, which are often used in transgene expression and gene regulation studies. The advantage of using simple factors was the availability of well-characterized simple promoters, which they regulate. Another advantage was the possibility to artificially create different combinations of factors in cells and thoroughly test the system performance. We then designed and built a number of synthetic miR genes. In order to facilitate circuit testing, we fused the miR molecules as introns in the coding frame of fluorescent reporters. In this way, expression of an miR is accompanied by the expression of a fluorescent protein. This allowed us to track miR expression in single cells and verify that the logic circuit worked properly (Figure 10.4b).

At the onset of the project, we naively overlooked one system component that was supposed to be relatively unimportant – the promoter that drives the output. Our design calls for a constitutive baseline expression, since all the regulation is directed via RNAi and 3′-UTR of the output mRNA. However, in our experiments we observed that the promoter is of critical importance. We started with a commonly used cytomegalovirus (CMV) promoter and quickly discovered that downregulation of the output by a miR was inefficient. The problems got aggravated when we increased circuit complexity and added two and three TF inputs at the same time. We were unsure how to proceed until, almost by accident, we found that a different promoter, EF1a, leads to a much better repression. While we were not sure what caused the difference, replacing CMV with EF1a in the circuits dramatically improved their performance (namely, quantitative ratios between the On and the Off signals). At the same time, performance issues did not disappear and they became more and more apparent as we moved toward more complex circuit. Yet we could successfully design regulatory programs with up to three TF inputs.

10.4
Outlook

Our work so far with RNAi-based computing circuits has clearly shown the advantages of the method, especially when it comes to scalability and programmability of specific computations. Much more work remains to be done before the method can be used to solve real-life problems. The experiments described here used artificially provided inputs, a far cry from the original intention to use endogenous molecular signals to trigger molecular computation. Such incorporation of endogenous inputs, for example, TFs, is the next great challenge. Another challenge pertains to circuit analysis in single cells. So far we have limited ourselves to transient DNA

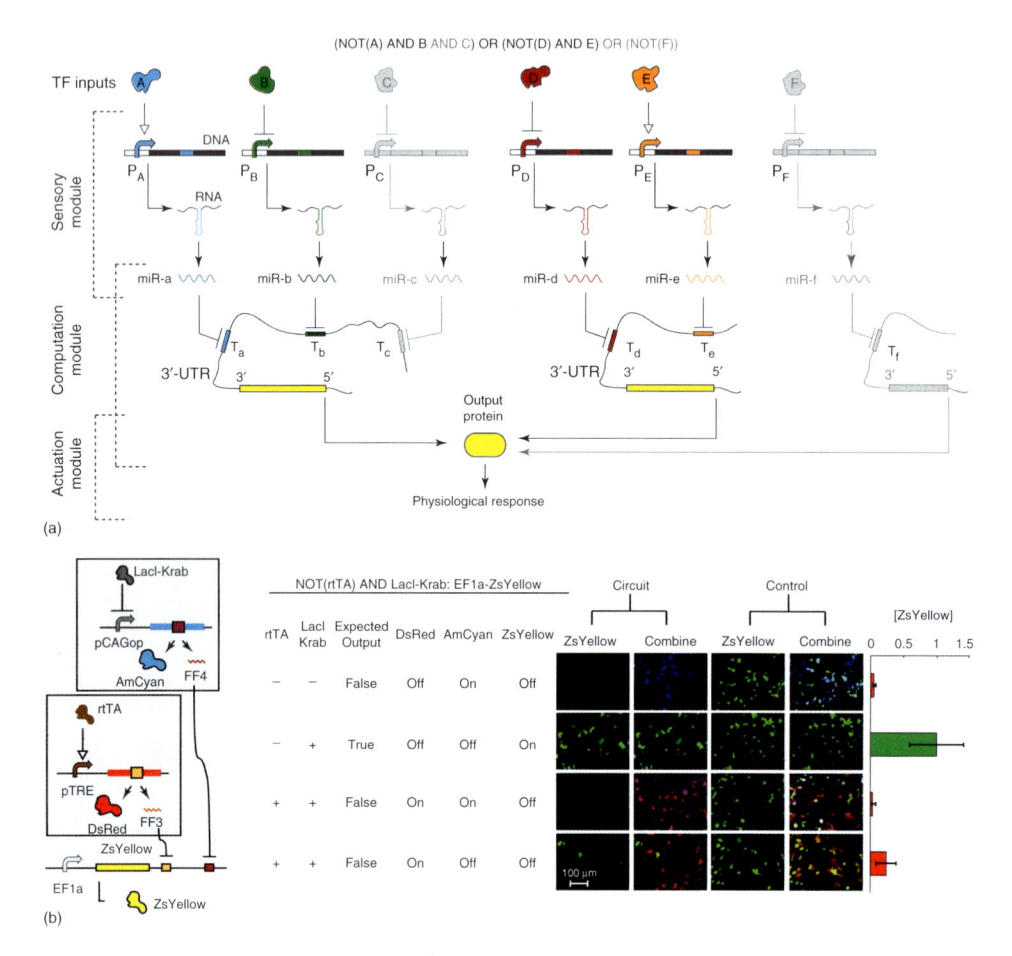

Figure 10.4 Transcription regulatory programs with RNAi logic circuits. (a) Blueprint of a logic circuit with multiple TF inputs and a fluorescent ZsYellow protein output. Sensory, computation, and actuation modules of the system are indicated. TF inputs A–F, promoters P_A–P_F, miRs miR-a to miR-f and output-encoding mRNA transcripts containing miR targets T_a–T_f are indicated. Arrows with arrowheads denote activation, and blunt arrows represent repression. Elements in gray denote potential opportunities for circuit scale-up. (b) Experimental implementation of a regulatory program "ZsYellow = NOT(rtTA) AND LacI-Krab." Only the results for EF1a-driven output are shown. rtTA is a transcriptional activator, and LacI-Krab is a transcriptional repressor. From left to right are circuit schematics, anticipated output for different input combinations, representative microscopy snapshots, and quantitative performance of the circuit. TFs, their promoters, the synthetic microRNAs, coexpressed fluorescent proteins, output-driving promoter, and the output protein are all indicated in the scheme.

delivery and population-based measurements. While those are inevitable at the early stages of circuit testing, eventually the circuits need to be analyzed in single cells and their delivery has to be uniform from cell to cell as well as highly efficient. We expect that detrimental effects of biological noise will become evident at this point, and one could face the challenge of curbing those effects or, alternatively, exploiting them. It is important to understand how synthetic circuits evolve, and whether directed evolution could augment rational design. All these questions are only starting to be addressed by the community, but their prominence will grow as synthetic systems transition from the proof-of-concept stage to bioengineering and biomedical applications.

References

1. Benner, S.A. and Sismour, A.M. (2005) *Nat. Rev. Genet.*, **6**, 533–543.
2. (a) Khorana, H.G., Agarwal, K.L., Buchi, H., Caruther, M.H., Gupta, N.K., Kleppe, K., Kumar, A., Ohtsuka, E., Rajbhandary, U.L., Vandesan, J.H., Sgaramel, V., Terao, T., Weber, H., and Yamada, T. (1972) *J. Mol. Biol.*, **72**, 209–217; (b) Jackson, D.A., Berg, P., and Symons, R.H. (1972) *Proc. Natl. Acad. Sci. U.S.A.*, **69**, 2904–2909.
3. Gibson, D.G., Benders, G.A., Andrews-Pfannkoch, C., Denisova, E.A., Baden-Tillson, H., Zaveri, J., Stockwell, T.B., Brownley, A., Thomas, D.W., Algire, M.A., Merryman, C., Young, L., Noskov, V.N., Glass, J.I., Venter, J.C., Hutchison, C.A., and Smith, H.O. (2008) *Science*, **319**, 1215–1220.
4. Guet, C.C., Elowitz, M.B., Hsing, W.H., and Leibler, S. (2002) *Science*, **296**, 1466–1470.
5. Martin, V.J.J., Pitera, D.J., Withers, S.T., Newman, J.D., and Keasling, J.D. (2003) *Nat. Biotechnol.*, **21**, 796–802.
6. Gardner, T.S., Cantor, C.R., and Collins, J.J. (2000) *Nature*, **403**, 339–342.
7. Chen, I.A., Roberts, R.W., and Szostak, J.W. (2004) *Science*, **305**, 1474–1476.
8. (a) Benenson, Y. (2009) *Mol. Biosyst.*, **5**, 675–685; (b) Wiener, N. (1948) *Cybernetics*, MIT Press, New York; (c) Regev, A. and Shapiro, E. (2002) *Nature*, **419**, 343–343.
9. Kaplan, S., Bren, A., Zaslaver, A., Dekel, E., and Alon, U. (2008) *Mol. Cell*, **29**, 786–792.
10. Ihmels, J., Bergmann, S., Gerami-Nejad, M., Yanai, I., McClellan, M., Berman, J., and Barkai, N. (2005) *Science*, **309**, 938–940.
11. (a) Yokobayashi, Y., Weiss, R., and Arnold, F.H. (2002) *Proc. Natl. Acad. Sci. U.S.A.*, **99**, 16587–16591; (b) Wang, H.H., Isaacs, F.J., Carr, P.A., Sun, Z.Z., Xu, G., Forest, C.R., and Church, G.M. (2009) *Nature*, **460**, 894–U133.
12. Koen, B.V. (1984) *Eng. Educ.*, **75**, 150–155.
13. (a) Seelig, G., Soloveichik, D., Zhang, D.Y., and Winfree, E. (2006) *Science*, **314**, 1585–1588; (b) Niazov, T., Baron, R., Katz, E., Lioubashevski, O., and Willner, I. (2006) *Proc. Natl. Acad. Sci. U.S.A.*, **103**, 17160–17163; (c) Balzani, V., Credi, A., and Venturi, M. (2003) *Chemphyschem*, **4**, 49–59; (d) Weiss, R. and Basu, S. (2002) NSC-1: The First Workshop of Non-SiliconComputing, Boston, MA, 2002; (e) Ashkenasy, G. and Ghadiri, M.R. (2004) *J. Am. Chem. Soc.*, **126**, 11140–11141; (f) Privman, V., Strack, G., Solenov, D., Pita, M., and Katz, E. (2008) *J. Phys. Chem. B*, **112**, 11777–11784; (g) Stojanovic, M.N. and Stefanovic, D. (2003) *Nat. Biotechnol.*, **21**, 1069–1074.
14. (a) Weiss, R., Homsy, G.E., and Knight, T.F. (1999) in *Evolution as Computation: DIMACS Workshop* (eds L.F. Landweber and E. Winfree), Springer, pp. 275–295; (b) Kramer, B.P., Fischer, C., and Fussenegger, M. (2004) *Biotechnol. Bioeng.*, **87**, 478–484.

15. Win, M.N. and Smolke, C.D. (2008) *Science*, **322**, 456–460.
16. Kramer, B.P., Viretta, A.U., El Baba, M.D., Aubel, D., Weber, W., and Fussenegger, M. (2004) *Nat. Biotechnol.*, **22**, 867–870.
17. (a) Elowitz, M.B. and Leibler, S. (2000) *Nature*, **403**, 335–338; (b) Stricker, J., Cookson, S., Bennett, M.R., Mather, W.H., Tsimring, L.S., and Hasty, J. (2008) *Nature*, **456**, 516–U539; (c) Tigges, M., Marquez-Lago, T.T., Stelling, J., and Fussenegger, M. (2009) *Nature*, **457**, 309–312; (d) Danino, T., Mondragon-Palomino, O., Tsimring, L., and Hasty, J. (2010) *Nature*, **463**, 326–330.
18. Friedland, A.E., Lu, T.K., Wang, X., Shi, D., Church, G., and Collins, J.J. (2009) *Science*, **324**, 1199–1202.
19. Benenson, Y., Paz-Elizur, T., Adar, R., Keinan, E., Livneh, Z., and Shapiro, E. (2001) *Nature*, **414**, 430–434.
20. Soreni, M., Yogev, S., Kossoy, E., Shoham, Y., and Keinan, E. (2005) *J. Am. Chem. Soc.*, **127**, 3935–3943.
21. Canton, B., Labno, A., and Endy, D. (2008) *Nat. Biotechnol.*, **26**, 787–793.
22. (a) Ben-Hur, A. and Siegelmann, H.T. (2004) *Chaos*, **14**, 145–151; (b) Marchisio, M.A. and Stelling, J. (2008) *Bioinformatics*, **24**, 1903–1910; (c) Knight, T.F. and Sussman, G.J. (1998) in *Unconventional Models of Computation* (eds C.S. Calude, J. Casti, and M.J. Dinneen), Springer-Verlag Singapore Pte Ltd, Singapore, pp. 257–272.
23. (a) Hjelmfelt, A., Weinberger, E.D., and Ross, J. (1991) *Proc. Natl. Acad. Sci. U.S.A.*, **88**, 10983–10987; (b) Dueber, J.E., Yeh, B.J., Chak, K., and Lim, W.A. (2003) *Science*, **301**, 1904–1908.
24. Bashor, C.J., Helman, N.C., Yan, S.D., and Lim, W.A. (2008) *Science*, **319**, 1539–1543.
25. Tucker, B.J. and Breaker, R.R. (2005) *Curr. Opin. Struct. Biol.*, **15**, 342–348.
26. Tang, J. and Breaker, R.R. (1997) *Chem. Biol.*, **4**, 453–459.
27. (a) Sudarsan, N., Hammond, M.C., Block, K.F., Welz, R., Barrick, J.E., Roth, A., and Breaker, R.R. (2006) *Science*, **314**, 300–304; (b) Benenson, Y. (2009) *Curr. Opin. Biotechnol.*, **20**, 471–478.
28. Appella, E. and Anderson, C.W. (2001) *Eur. J. Biochem.*, **268**, 2764–2772.
29. Maniatis, T. and Reed, R. (2002) *Nature*, **416**, 499–506.
30. Gregor, T., Tank, D.W., Wieschaus, E.F., and Bialek, W. (2007) *Cell*, **130**, 153–164.
31. (a) Siomi, H. and Siomi, M.C. (2009) *Nature*, **457**, 396–404; (b) Brodersen, P. and Voinnet, O. (2009) *Nat. Rev. Mol. Cell Biol.*, **10**, 141–148.
32. Rinaudo, K., Bleris, L., Maddamsetti, R., Subramanian, S., Weiss, R., and Benenson, Y. (2007) *Nat. Biotechnol.*, **25**, 795–801.
33. (a) Beisel, C.L., Bayer, T.S., Hoff, K.G., and Smolke, C.D. (2008) *Mol. Syst. Biol.*, **4**: 224; (b) Benenson, Y. (2008) *Mol. Syst. Biol.*, **4**: 227; (c) An, C.I., Trinh, V.B., and Yokobayashi, Y. (2006) *RNA-a Publ. RNA Soc.*, **12**, 710–716.
34. (a) Shapiro, E. and Benenson, Y. (2006) *Sci. Am.*, **294**, 44–51; (b) Benenson, Y., Adar, R., Paz-Elizur, T., Livneh, Z., and Shapiro, E. (2003) *Proc. Natl. Acad. Sci. U.S.A.*, **100**, 2191–2196; (c) Benenson, Y., Gil, B., Ben-Dor, U., Adar, R., and Shapiro, E. (2004) *Nature*, **429**, 423–429; (d) Adar, R., Benenson, Y., Linshiz, G., Rosner, A., Tishby, N., and Shapiro, E. (2004) *Proc. Natl. Acad. Sci. U.S.A.*, **101**, 9960–9965.
35. Xie, Z., Liu, S.J., Bleris, L., and Benenson, Y. (2010) *Nucleic Acids Res.*, **38**, 2692–2701.
36. Elbashir, S.M., Martinez, J., Patkaniowska, A., Lendeckel, W., and Tuschl, T. (2001) *EMBO J.*, **20**, 6877–6888.
37. (a) Anderson, J.C., Voigt, C.A., and Arkin, A.P. (2007) *Mol. Syst. Biol.*, **3**: 145; (b) Cox, R.S., Surette, M.G., and Elowitz, M.B. (2007) *Mol. Syst. Biol.*, **3**, 11; (c) Ellis, T., Wang, X., and Collins, J.J. (2009) *Nat. Biotechnol.*, **27**, 465–471.
38. Leisner, M., Bleris, L., Lohmueller, J., Xie, Z., and Benenson, Y. (2010) *Nat. Nanotechnol.*, **5**, 666–670.
39. Stegmeier, F., Hu, G., Rickles, R.J., Hannon, G.J., and Elledge, S.J. (2005) *Proc. Natl. Acad. Sci. U.S.A.*, **102**, 13212–13217.

11
Biomolecular Computing Systems

Harish Chandran, Sudhanshu Garg, Nikhil Gopalkrishnan, and John H. Reif

11.1
Introduction

The field of biomolecular computation started 3.5 billion years ago on earth when the first life forms evolved. What distinguished these simple organisms from a collection of inanimate molecules was the ability to process information, and it is this ability of chemistry-based computation that powers life in its various forms even today. Our attempts at exploiting this rich molecular tool set for computation have just begun. This chapter recounts some of the amazing experiments, starting from 1994, which demonstrated how one could process information and perform computation at the molecular scale using DNA.

11.1.1
Organization of the Chapter

We begin (Section 11.2) by discussing our molecule of choice DNA: why it is ideally suited for nanoscale assembly and computation, its structure, and key reactions. Section 11.3 describes the initial work in DNA computing, including Adleman's experiment of solving a small-scale Hamiltonian path problem via DNA computing and the extensions of that work to solve small-scale instances of other hard search problems. Section 11.4 discusses algorithmic assembly via DNA tiling lattices as well as algorithmic tiling theory based on various abstract models of self-assembly and error correction schemes for tilings. Section 11.5 describes experimental advances in purely hybridization-based computation, including hairpin systems, seesaw gates, as well as speedups using localized reactions. Section 11.6 discusses experimental advances in enzyme-based DNA computing. Section 11.7 discusses DNA reaction networks, including various amplifiers, switches, and oscillators. Section 11.8 concludes the chapter with a discussion of future challenges.

11.2
DNA as a Tool for Molecular Programming

DNA systems are relatively easy to design, fairly predictable in their geometric structures, and chemically stable, and have been experimentally implemented in a growing number of laboratories around the world. Abstractions for programming DNA systems and software tools that aid in the design, verification, and simulation of such systems have further expanded the horizons of what are feasible using DNA. Most DNA systems are autonomous: they can execute steps with no external mediation after starting, they are programmable, and the tasks executed can be modified without entirely redesigning the system. In contrast, lipids and carbohydrates are not programmable, the function and structure of proteins are hard to design, and RNA is unstable.

11.2.1
DNA Structure

Single-stranded DNA (ssDNA) is a polymer made from repeating units called *nucleotides*. The nucleotide repeats contain both the segment of the backbone of the molecule, which holds the chain together, and a base. ssDNA has asymmetries along its backbone which gives it a directionality. The asymmetric ends of ssDNA are called the 5′ *and* 3′ *ends*. The four bases found in DNA are adenine (abbreviated A), cytosine (C), guanine (G), and thymine (T) (Figure 11.1).

These bases form the alphabet of DNA; the specific sequence of an ssDNA comprises its information content. In living organisms, DNA does not usually exist as a single molecule, but instead as double-stranded DNA (dsDNA): a pair of oppositely directed ssDNA held together via *complementary base binding*, with A hydrogen-bonded preferentially to T, and C hydrogen-bonded preferentially to G. These two long strands entwine like vines, forming a double helix. As hydrogen bonds are not covalent, they can be broken and rejoined relatively easily, for example, by heating. The two types of base pairs form different numbers of hydrogen bonds, AT forming two hydrogen bonds and GC forming three hydrogen bonds. The association strength of hybridization depends on the sequence of complementary bases, the stability increasing with the length of the DNA sequence, and the GC content. This association strength can be approximated by software packages.

11.2.2
Review of DNA Reactions

We review a few key reactions that allow DNA to execute molecular programs. Toehold-mediated strand displacement is the displacement of a single strand of DNA from a double helix by an incoming strand with a longer complementary region to the template strand. The incoming strand has a toehold, an empty single-stranded region on the template strand complementary to a subsequence of the incoming strand, to which it binds initially. It eventually displaces the outgoing

11.2 DNA as a Tool for Molecular Programming

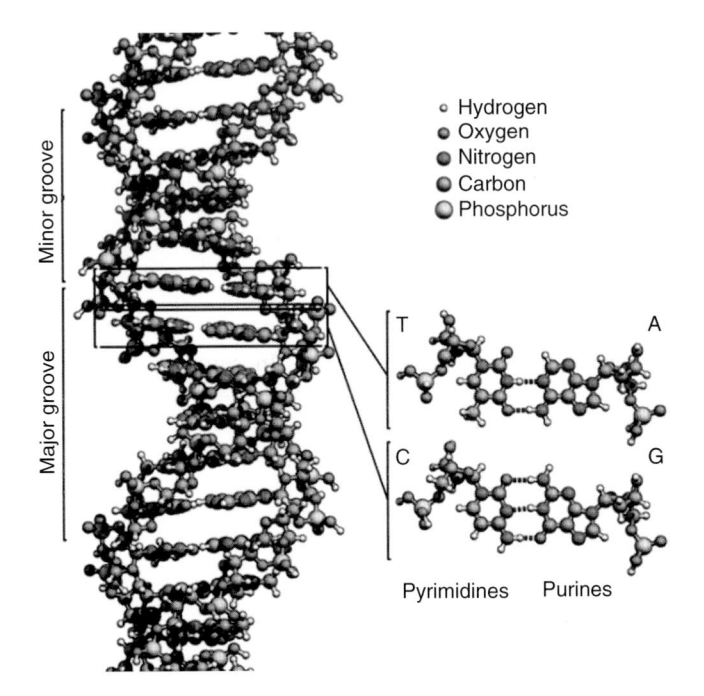

Figure 11.1 Structure of a DNA double helix. (Image by Richard Wheeler, reproduced under the Creative Commons Attribution-Share Alike 3.0 Unported license.)

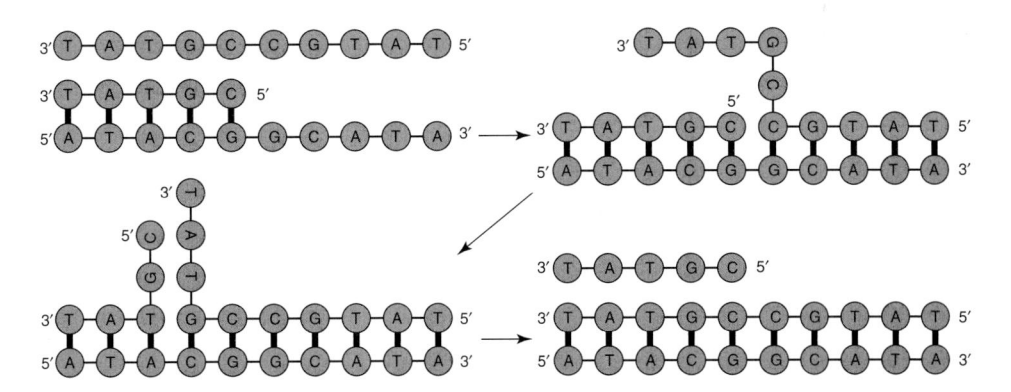

Figure 11.2 Toehold-mediated strand displacement reaction.

strand via a kinetic process modeled as a one-dimensional random walk. Strand displacement is a key process in many of the DNA protocols for running DNA autonomous devices. Toehold exchange is a similar strand displacement reaction mediated by a toehold, with the exception that both the incoming and outgoing strands have distinct, short toeholds on the template strand. Thus, either strand can initiate strand displacement. See [1] for details about the kinetics of the toehold exchange process (Figure 11.2).

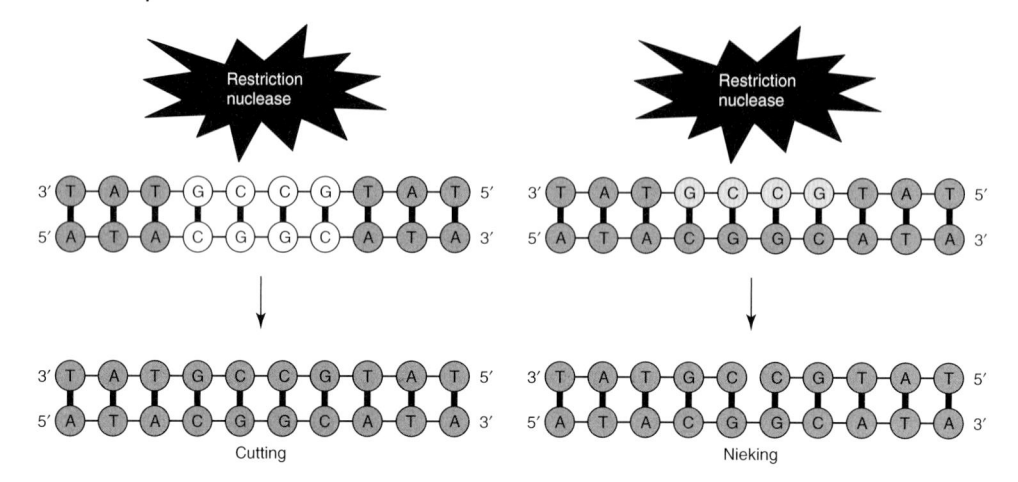

Figure 11.3 Example of restriction enzyme cuts of a single-stranded DNA sequence. The subsequence recognized by the nuclease is unshaded.

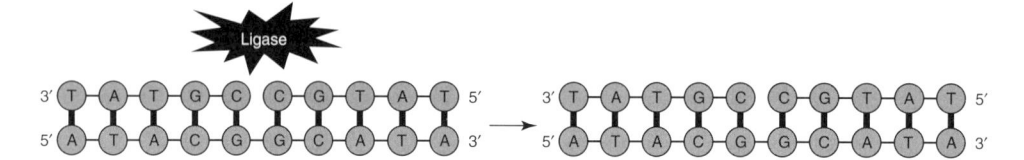

Figure 11.4 Ligase repairing a single-stranded nick.

DNA restriction (Figure 11.3) is the cleaving of phosphodiester bonds by a class of enzymes called *nucleases*. *Endonucleases* cleave the phosphodiester bond within a polynucleotide chain between the nucleotide subunits at specific locations determined by short (4–8 bases) sequences, whereas *exonucleases* cleave the phosphodiester bond at the end of a polynucleotide chain. Some nucleases have both these abilities. Some restriction enzymes cut both the strands of a DNA double helix, whereas others cut only one of the strands (called *nicking*).

DNA ligation (Figure 11.4) is repair of the phosphodiester bond between nucleotides by the class of enzymes known as *ligases*. *Deoxyribozymes* (DNAzymes) are DNA strands that possess enzymatic activity – they can, for example, cleave specific target RNA strands. Typically, they are discovered by *in vivo* evolution search. They have had some use in DNA computations [2]. *DNA polymerases* (Figure 11.5) are a class of enzymes that catalyze the polymerization of nucleoside triphosphates into a DNA strand. The polymerase extends a *primer* strand attached to a DNA template. The newly synthesized sequence is complementary to the template strand.

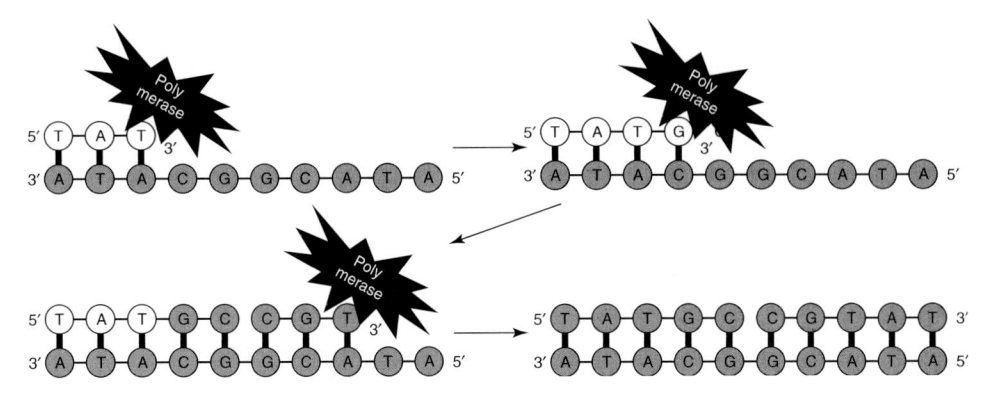

Figure 11.5 DNA synthesis by a polymerase enzyme.

11.3
Birth of DNA Computing: Adleman's Experiment and Extensions

11.3.1
NP-Complete Problems

Anyone who has attempted to solve a Sudoku puzzle would have realized that it is fairly easy to check whether a purported solution is correct, while finding a correct solution might not be. The class of combinatorial decision problems that can be solved by efficient[1] algorithms is formally captured in the computational complexity class P (polynomial time). The computational complexity class nondeterministic polynomial time (NP) captures a larger class of decision problems, those whose solution can be efficiently (in polynomial time) verified. Note that the class P is a subset of the class NP. It is widely believed that P does not equal NP: that is, there are problems, like Sudoku, whose solutions can be efficiently verified but cannot be solved by an efficient algorithm. *NP-complete problems* [3] are the hardest problems in the class NP, in the sense that, if there exists an efficient algorithm that can solve some NP-complete problem, then there exist efficient algorithms to solve any problem in NP and thus P equals NP. Examples of NP-complete problems include Boolean formula satisfiability, Hamiltonian path problem, and Sudoku. Computer scientists were first attracted to DNA computing as a means of solving NP-complete problems, not via efficient algorithms, but rather with the hope of efficiently executing inefficient algorithms in a hugely parallel manner. This hope was belied, but paved the way for DNA molecular computing.

---

1) Efficient is generally taken to mean that the algorithm's running time is some polynomial function of the size of the problem.

11.3.2
Hamiltonian Path Problem via DNA Computing

Richard Feynman [4] in 1959 envisioned the use of molecules to perform computation but stopped short of proposing concrete methods for doing molecular-scale computation. Thirty-five years later, Adleman [5] provided a dramatic first experimental demonstration of molecular-scale computation via DNA molecules, which gave birth to the field of DNA computing. His experiment solved a small seven-vertex instance of the Hamiltonian path problem. Given a graph, the problem asks if there exists a path that visits each vertex exactly once. Adleman carefully designed one set of DNA sequences to encode the vertices of the graph and another set of bridge DNA sequences to encode edges between these vertices. These synthetic DNA sequences were annealed together in a test tube and formed nanostructures encoding all possible paths in the graph. Note that the huge number of copies of each sequence provides the necessary parallelism to explore this state space. A series of biochemical manipulations were used to isolate DNA nanostructures that encoded any existing Hamiltonian paths. These nanostructures were analyzed to read out the actual path information. The computation was highly energy- and space-efficient (ignoring the energy to isolate the Hamiltonian path), leading to much excitement about its potential uses. However, scientists soon realized that molecular DNA computers could not compete with conventional general-purpose silicon hardware because of issues of error rates, speed, difficulties in reading the output, and so on, which limited the scalability of such methods (Section 11.3.4).

11.3.3
Other Models of DNA Computing

Inspired by Adleman's success, many models were proposed to perform computation with DNA strands. Adleman *et al.* [6] developed a class of methods, known as *sticker-based methods*, for solving Boolean formula satisfiability problems. The sticker-based methods created a combinatorial library of DNA sequences that encoded possible Boolean variable assignments. Then a series of hybridization reactions were employed to select out those DNA sequences whose encoding satisfied all the clauses of the Boolean formula.

Lipton [7] developed a method, termed the *test tube model*, that provided a general-purpose instruction set for DNA computing. The test tube model included various biochemical operations, such as merging test tubes and selecting out from a test tube DNA sequences with specified subsequences. The test tube model was theoretically capable of solving NP search problems in polynomial time.

11.3.4
Shortcomings and Nonscalability of Schemes Using DNA Computation to Solve NP-Complete Problems

Using these approaches and a number of related methods, DNA computation was used to solve a number of relatively small instances of NP problems. These methods require a number of DNA molecules that grow as an exponential function of the size of the problem instance and hence are not scalable to large instances in practice (see [8] for a discussion of these scaling limitations). In addition, these approaches are tailored to specific problems, are use-once systems, and do not have the power of general-purpose programmable computers. Hence the idea of using DNA computation to solve large instances of NP problems was eventually discarded, but the insights gained seeded the field of DNA molecular programming, in particular the computational power of self-assembly. Winfree [9] developed the tile assembly model (TAM) as a Turing universal model of DNA tile-based molecular computation, which captures algorithmic growth processes, as discussed in Section 11.4.

11.4
Computation Using DNA Tiles

Rapid advancements in experimental DNA self-assembly in conjunction with Winfree's TAM (Section 11.4.1) gave rise to a new computational paradigm, namely, computing using self-assembly [54]. DNA nanostructures can be programmed to form tiles that self-assemble in the test tube to form large lattices as shown by Winfree *et al.* [10] with the DX tile and LaBean *et al.* [11] with the TX tile. The DX and TX tiles have pads that specify their interaction with other tiles. The pads are ssDNA sequences that attach via hybridization to pads with complementary sequences. Mao *et al.* [12] performed a laboratory demonstration of computation via tile assembly using TX tiles. Yan *et al.* [13] performed parallel XOR computation in a test tube using Winfree's DX tile. Other simple computations have also been demonstrated. Yan *et al.* [14] demonstrated the 4×4 tile which had pads in two linearly independent directions in contrast to the earlier DX and TX tiles. This tile assembled into two-dimensional (2D) lattices with square holes. These achievements led to a new field, namely, algorithmic tiling theory.

11.4.1
TAM: an Abstract Model of Self-Assembly

Winfree [9] defined a model of algorithmic self-assembly, the TAM. The model studies unit-sized square tiles that interact via specific glues along their edges. The type of glues along a tile's four edges decides its tile type. A set of tile types specifies a tile assembly program. Growth starts from a specified *seed* tile and continues via a series of single tile additions. The number of copies of each tile type is assumed to be limitless. A tile can be added to an empty position if its interactions with its neighboring tiles are sufficiently strong. All possible maximal assembled structures that are assembled from this process are the output of the tile assembly program. Winfree showed that for every possible Turing machine there exists a tile assembly program that uniquely assembles the complete history of the machine's execution, thus laying a theoretical foundation for a form of DNA-based computation, in particular, molecular computation via assembly of DNA lattices with tiles in the form of DNA nanostructures.

11.4.2
Algorithmic Assembly via DNA Tiling Lattices

Mao et al. [12] demonstrated one of the first examples of using DNA tiles to compute a function. They implemented a cumulative XOR of 4 bits using the TX molecule. Their work is discussed more extensively in Section 11.7. Perhaps the simplest nontrivial tiling 2D construction is the Sierpinksi triangle, a fractal structure. In TAM, a set of tile types performing binary XOR can be used to assemble the Sierpinksi triangle. Rothemund [15] designed plastic tiles (about 1 cm in size) that assemble on a fluid layer via surface tension. The assembly was designed to mimic the Sierpinski triangle pattern. The glue interactions of the tiles were specified by hydrophilic and hydrophobic patches along the tile edges. The system took a long time to assemble (60 h) and was beset by errors. In addition, the size of the tiles made it infeasible to get millions or billions of tiles to assemble.

11.4.2.1 Source of Errors

Implementing an XOR computation requires a tile to add itself to an assembly only if two neighboring glues match. There are competing tile types that have one matching glue and these sometimes add spuriously to the assembly. These incorrect attachments may be unstable at first but may later be stabilized by further "correct" attachments. Such errors are called *cooperative binding errors*. Another major source of errors is spurious nucleation: all assemblies are assumed to arise from a seed assembly. However, there are sets of tiles not containing the seed assembly but form stable assemblies. These assemblies often grow to give partial or even incomplete assemblies. These types of errors are not limited to XOR computation and have to be overcome for most nontrivial algorithmic tiling constructions.

Winfree [17] introduced the kinetic tile assembly model (kTAM), a reversible TAM that models tile dissociation, to study the effect of binding strength and tile concentration on the rate of errors in the assembly. The kTAM predicts that cooperative binding errors may be reduced by growing the assemblies slowly at slightly supersaturated tile concentrations, while nucleation errors may be reduced by providing a wide nucleating assembly. Schulman et al. [18] attempted to grow a Sierpinski tiling from DNA tiles by assembling the lattices at close to the melting temperature of cooperative binding and also to seed the assembly using tiles. This strategy was only partially successful since the self-assembly of border tiles proved problematic. Rothemund et al. [16] instead used a long ssDNA to serve as a scaffold for the assembly of a row of input tiles (Figure 11.6). This same strategy was used to implement tilings that perform binary counting and copying by Barish et al. [19]. This nucleating strategy proved more successful; however, a new kind of error was revealed: facet nucleation errors. Spurious growth occurred at places along the crystal facet where no tiles were designed to stably attach. Fujibayashi et al.

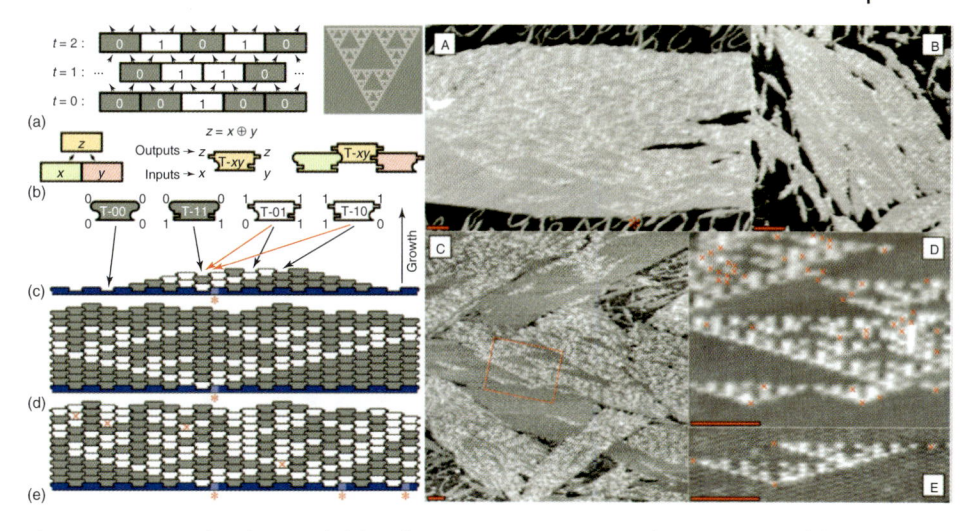

Figure 11.6 Sierpinksi tiling seeded by a long ssDNA [16]. (a) Logical programming of the Sierpinkski triangle. (b) Tiles that achieve this programming. (c), (d) and (e) Self-assembly process. On the right: Atomic force microscopic images of Sierpinksi tiling.

[20] solved this by using border tiles along the facet that did not have sticky ends on their outside, thus preventing any facet nucleation errors. The combination of all these strategies was used by Barish et al. [21] to perform binary copying and counting at high yields and low error rates.

11.4.3
Algorithmic Error Correction Schemes for Tilings

As discussed in Section 11.4.2, errors in tiling were mitigated using techniques that optimized physical conditions for decreasing the probability of erroneous tile attachments. An alternate approach, introduced in Winfree and Bekbolatov [22], was to design tile sets that were robust to assembly errors by exploiting the mechanism of cooperative bindings. They replaced each tile type by a $k \times k$ block of tile types such that, if an erroneous tile was incorporated in the assembly, there could not be further growth without additional tile mismatches (see Figure 11.7 for an example). Thus, assemblies with incorrect tiles grew much slower and allowed more time for the erroneous attachments to dissociate before the error was locked into place. This *proofreading* could reduce error rates by a square factor over tile sets that did not implement proofreading. However, there is an inherent scale blowup associated with this technique. Reif et al. [23] eliminated this scale blowup by giving a compact method to perform proofreading. While these schemes reduce cooperative binding errors, they do not protect against facet nucleation errors and do not scale well with increased k. Chen and Goel [24] introduced the snaked-proofreading technique that guards against both cooperative binding and facet nucleation errors and proved that error rates drop exponentially as a function of k. Both conventional and

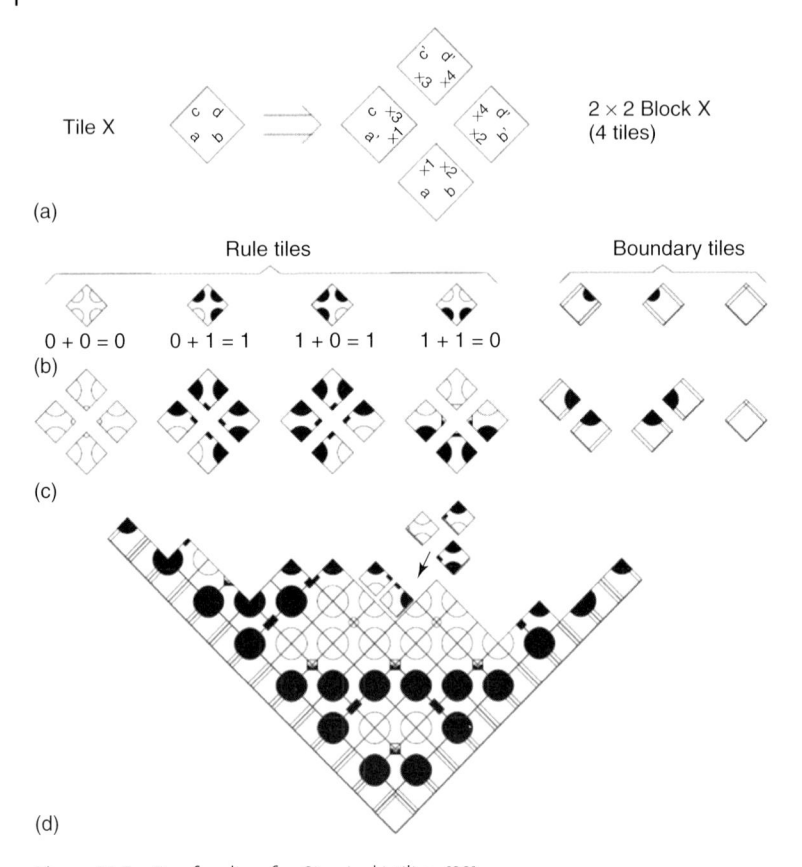

Figure 11.7 Proofreading for Sierpinski tiling [22].

snaked-proofreading systems were experimentally tested out by Chen et al. [25], and a 2 × 2 snaked-proofreading system was shown to reduce facet nucleation errors fourfold as compared to the conventional proofreading technique.

Snaked proofreading reintroduced the problem of scale blowup, but Soloveichik and Winfree [26] fixed this by describing a compact method to implement it. The drawback is that, for certain patterns, this results in an exponential blowup in the descriptional complexity of the pattern (as measured by the number of tile types that assemble it) (Figure 11.8).

Given a tile set, what are the relative concentrations of tile types that minimize errors? Chen and Kao [27] proved that, for rectilinear patterns, carefully setting the concentrations of tile types allows one to achieve minimum errors with the fastest possible assembly time. Instead of working with static tiles, one may imagine dynamic tiles that change state based on some timed signals. Fujibayashi et al. [20] and Majumder et al. [28] suggested techniques to implement such dynamic tiles and how they might be used to reduce errors in tile assembly.

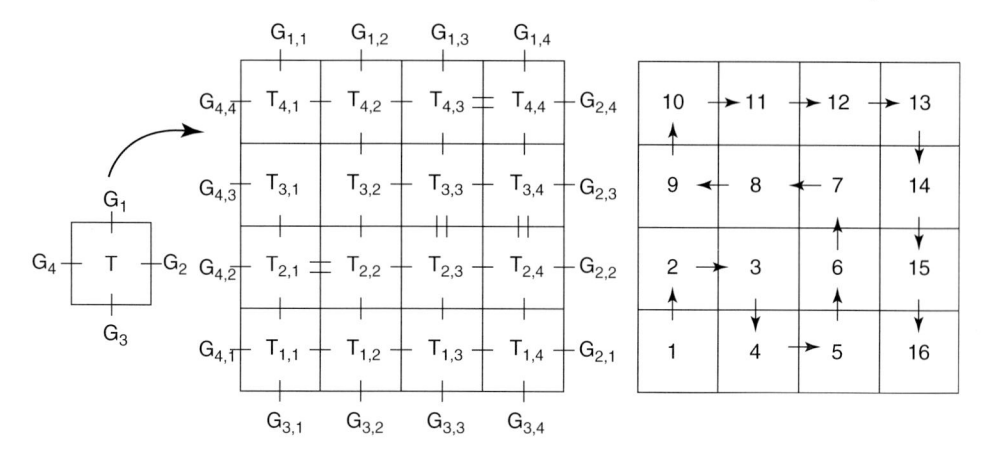

Figure 11.8 Snaked proofreading [24].

11.5
Experimental Advances in Purely Hybridization-Based Computation

Hybridization is the simplest DNA reaction; yet it is powerful enough for performing a wide array of computations. In this section, we will study a few systems built purely on DNA hybridization networks. A DNA nanostructure is formed via the self-assembly of multiple interacting DNA strands. The actual sequence of incorporation of strands into the nanostructure is not always clear. Control over this pathway, in a process termed *"directed self-assembly"*, was demonstrated by Yin et al. [29]. Their basic unit of construction was the hairpin motif.

Recall that the single-stranded bulge-loop of hairpins undergoes hybridization extremely slowly and thus provides a method for hiding information (making a subsequence unreactive can be thought of as hiding). This information is revealed (activating a subsequence and making it reactive can be thought of as revealing) when the hairpin is opened up via a toehold-mediated strand displacement. The newly revealed subsequence can now reveal other subsequences. Thus one can achieve precise control on the order in which strands get activated and information gets revealed. Yin et al. [29] provided an abstract symbolic representational language for programming pathways using the hairpin motifs. Using this language, they programmed and implemented a variety of dynamic functions: catalytic formation of branched junctions, autocatalytic duplex formation via a cross-catalytic circuit, nucleated dendritic growth, and autonomous locomotion of a bipedal walker. Figure 11.9 shows the directed assembly of the catalytic branched junctions.

The notion of hiding and revealing information allows control of the reaction pathway and this idea can be extended to control the computational pathway of a program implemented by DNA molecules. Qian and Winfree [30] built complexes, which they termed *"seesaw"* gates, that allowed them to hide and reveal information along a computational pathway. Figure 11.10 shows how to achieve the AND and OR logic via seesaw reactions.

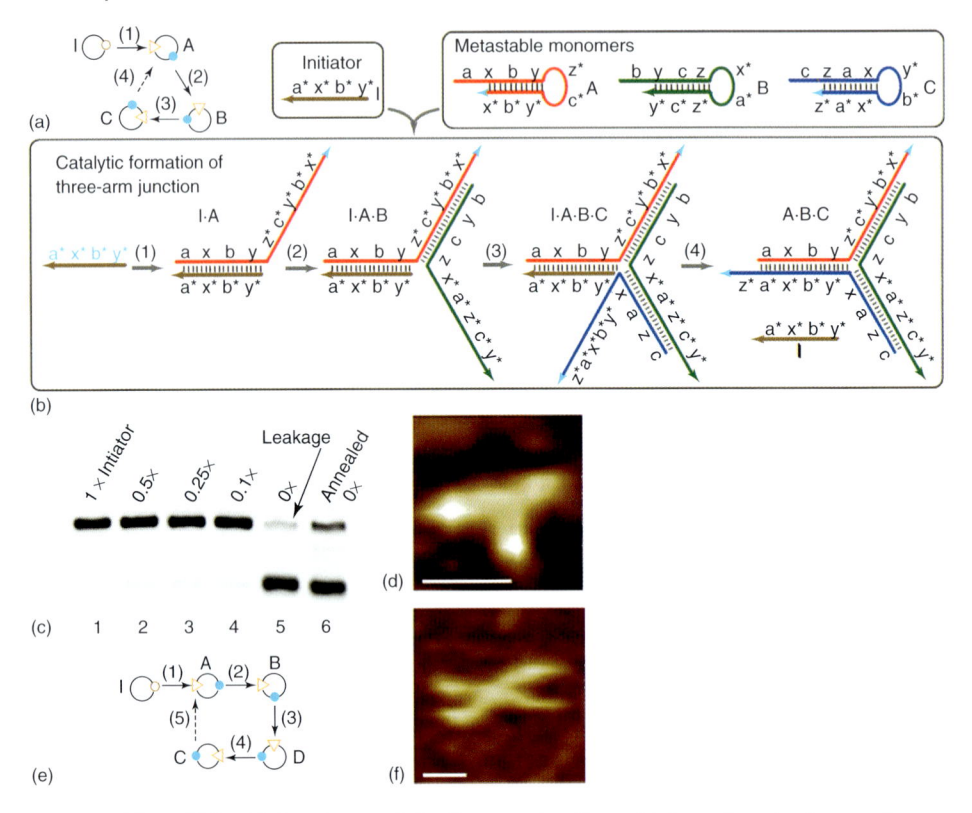

Figure 11.9 (a) Reaction graph for three-arm junctions. (b) Secondary structure mechanism. (c) Agarose gel electrophoresis demonstrating catalytic self-assembly for the three-arm system. (d) AFM image of a three-arm junction. Scale bar: 10 nm. (e) Reaction graph and (f) AFM image for a four-arm junction. Scale bar: 10 nm [29].

The process is represented in an abstract form in Figure 11.10a. Signals x_1 and x_2 are inputs to gate 2 and produce identical outputs that feed into the threshold gate 5. The threshold level th determines the logic implemented by this circuit; setting $th = 0.6$ makes the circuit compute the OR of x_1 and x_2, while setting $th = 1.2$ makes the circuit compute the AND of x_1 and x_2. Threshold gates absorb their input signal up to the level specified. Any signal beyond the threshold level gets catalytically amplified to the logical high. Any signal below this threshold is absorbed by the threshold gate and this provides for digital signal restoration in the circuit. Figure 11.10b shows the actual domain-level strand design of the system. The single strands in the system are the signal and fuel strands (that drive the catalytic amplification), while the double-stranded complexes are the various gates. Notice that the gates have two small blue domains T′. These are the toeholds of the gate complexes. In addition, notice how one of the toeholds is revealed while the other is hidden. Seesaw gates essentially work by hiding and revealing these toeholds. Given a circuit consisting of a large number of gates, a set

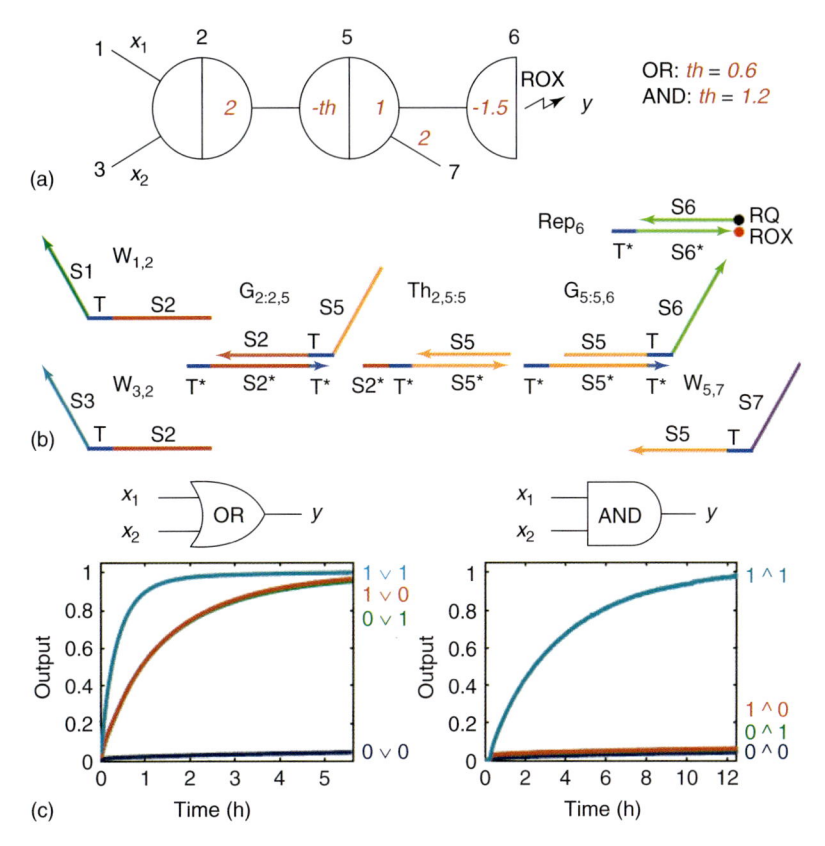

Figure 11.10 Implementing AND and OR logic via seesaw gates. (a): An abstract view of OR and AND seesaw gates. Gate 2 is an integrator, gate 5 is a thresholded catalytic amplifier and gate 6 acts as a downstream drain and releases a fluorescent reporter molecule. (b): The corresponding strand diagram for the gates. (c): Time traces indicating the operation of the OR and AND gates [31].

of input strands will sequentially reveal and hide a specific subset of the toeholds in a specific order, mimicking the process of traversing a computing tree. Qian and Winfree [31] showed how to use the AND and OR gate modules described above to modularly construct large circuits and demonstrated the operation of 4 bit square-root circuits comprised of 130 DNA strands. Furthermore, they showed how to modify the thresholding logic to implement a series of linear threshold gates that can act as a neural network [32]. Their approach allowed them to implement a Hopfield associative memory with four fully connected artificial neurons that, after training *in silico*, remembered four ssDNA patterns and recalled the most similar one when presented with an incomplete pattern.

The scale and complexity of these demonstrations of computing via seesaw gates were truly remarkable. Unfortunately, further scaling of these circuits seems problematic. First, seesaw circuits take a long time (6–10 h) to perform moderately complex (4 bit square-root) computations. This is primarily due to the slow diffusion-based hybridization reactions that power the system. Diffusion-based bimolecular chemistry is usually speeded up by increasing concentrations of the reactants. Unfortunately, increasing the concentration of the various species in the circuit increases the rate of spurious reactions, yielding higher leak rates. Even at low concentrations, leaks become a significant issue for deep circuits. To avoid spurious interaction, Qian and Winfree [31] carefully designed the sequences of various strands in the system. Ultimately, the scale, and hence the complexity, of seesaw circuits is limited by the number of noninteracting DNA strands one can design. Chandran *et al.* [33] proposed a solution to these scalability issues by organizing circuits on an addressable 2D substrate. This allowed them to achieve fast non-diffusion-based unimolecular kinetics, to control leaks via careful placement of gates on the surface to ensure that spurious interaction were minimized, and to reuse DNA sequences in spatially separate locations.

11.6
Experimental Advances in Enzyme-Based DNA Computing

Enzymes are powerful, naturally evolved protein catalysts that are widely employed by life to perform complex tasks. So it was only natural to adapt some of the enzymes to perform useful computational tasks. One of the simplest computational tasks is to copy a given string, and, unsurprisingly, one of the earliest biomolecular protocols that were developed was to make many copies of a given string. Polymerase chain reaction (PCR) developed in 1983 by Kary Mullis [34] is a protocol for rapidly creating multiple copies of a given DNA sequence (referred to as *template*) via the use of the enzyme polymerase. Polymerase synthesizes a new DNA strand complementary to the DNA template strand by adding free nucleotides in solution that are complementary to the template in the 5′ to 3′ direction. This addition is done to the 3′ end of a primer strand that is hybridized to the template. The newly synthesized strand is then heat-denatured from the template and now both these strands can serve as templates for the creation of their respective complementary strands. Note that in this procedure both the strand and its reverse complements are produced.

The basic PCR technique was adapted for finite-state computations by Sakamoto *et al.* [35] known as *whiplash PCR* (Figure 11.11). The technique uses a strand of DNA that essentially encodes a finite-state machine; the strand is comprised of a sequence of "rule" subsequences each encoding a state transition rule. They are each separated by a stopper sequence that stops the action of DNA polymerase. On each step of the computation, the 3′ end of the DNA strand has a terminal subsequence encoding the state of the computation. A computation step is executed when this 3′ end hybridizes to a portion of a "rule" subsequence, and the action of DNA polymerase extends the 3′ end to a further subsequence encoding a new state. The complex is now thermally denatured and primed to execute the next step of computation.

11.6 Experimental Advances in Enzyme-Based DNA Computing

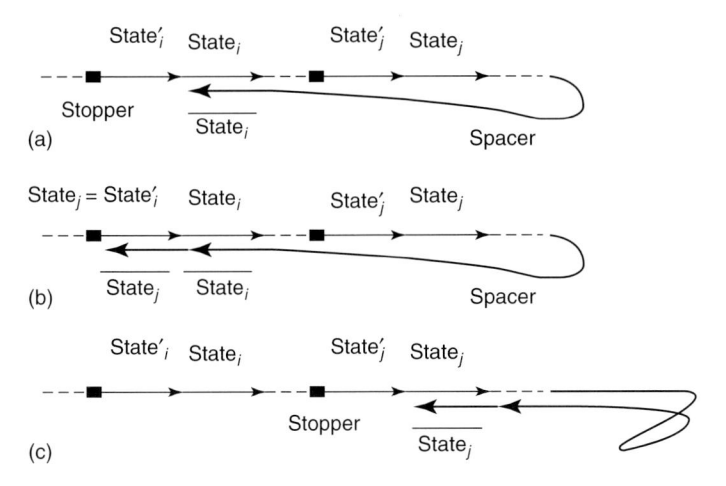

(a)
State$'_i$ State$_i$ State$'_j$ State$_j$
Stopper
State$_i$
Spacer

(b)
State$_j$ = State$'_i$ State$_i$ State$'_j$ State$_j$
State$_j$ State$_i$
Spacer

(c)
State$'_i$ State$_i$ State$'_j$ State$_j$
Stopper
State$_j$

Figure 11.11 Whiplash PCR state transitions. (a) The current state is annealed onto the transition table by forming a hairpin structure. (b) The current state is then extended by polymerase and the next state is copied from the transition table. (c) After denaturation, the new current state is annealed to another part of the transition table to enable the next transition.

Note that whiplash PCR executes a local molecular computation, whereas most methods for autonomous molecular computation (such as those based on the self-assembly of tiles) are global molecular computations as they require multiple distinct molecules that interact to execute each step of the computation.

Neither the original PCR protocol nor the whiplash PCR executes autonomously: they require thermal cycling for each step of their protocols. To overcome this, Walker et al. [36] developed isothermal methods for PCR known as "strand displacement amplification" (SDA) using a strand-displacing DNA polymerase. In this method, strands displaced by the DNA polymerase are used for the further stages of the amplification reaction. Reif and Majumder [37] developed an isothermal, autonomously executing version of whiplash PCR that makes use of a strand-displacing polymerase.

The first experimental demonstrations of computation via DNA tile assembly were done in 2000 by Mao et al. [12] with the help of the ligase enzyme. Ligases repair single-stranded discontinuities in dsDNA molecules. Mao et al. [12] demonstrated a two-layer linear assembly of TX tiles that executed bit-wise cumulative XOR computation. This computation takes as input n bits and computes n output bits, where the ith output bit is the XOR of the first i input bits. This computation frequently occurs when one determines the output bits of a full-carry binary adder circuit found in most microprocessors.

These experiments provided initial answers to some of the most basic questions of how autonomous molecular computation might be done: How can one provide data input to a molecular computation using DNA tiles? In this experiment, the input sequence of n bits was defined using a specific series of "input" tiles, with the input bits (1's and 0's) encoded by distinct short subsequences. Two different tile types (depending on whether the input bit was 0 or 1, these had specific sticky ends and also specific subsequences at which restriction enzymes can cut the DNA backbone) were assembled according to specific sticky-end associations, forming the blue input layer illustrated in Figure 11.12.

Figure 11.12a shows a unit TX tile and Figure 11.12b shows the sets of input and output tiles with geometric shapes conveying sticky-end complementary matching. The tiles of Figure 11.12b execute binary computations depending on their pads, as indicated by the table in Figure 11.12b. The (blue) input layer and (green) corner condition tiles were designed to assemble first (see example computational assemblies in Figure 11.12c,d). The (red) output layer then assembles specifically starting from the bottom left using the inputs from the blue layer. The tiles were designed such that an output reporter strand ran through all the n tiles of the assembly by bridges across the adjoining pads in input, corner, and output tiles. This reporter strand was pasted together from the short ssDNA sequences within the tiles using ligase. When the solution was heated, this output strand was isolated and identified. The output data was read by experimentally determining the sequence of cut sites (see below). In principle, the output could be used for subsequent computations.

The next question of concern is: How can one execute a step of computation using DNA tiles? To execute steps of computation, the TX tiles were designed to have pads at one end that encoded the cumulative XOR value. In addition, since the reporter strand segments ran through each such tile, the appropriate input bit was also provided within its structure. These two values implied that the opposing pad on the other side of the tile would be the XOR of these two bits.

A final question is: How can one determine and/or display the output values of a DNA tiling computation? The output in this case was read by determining which of the two possible cut sites (endonuclease cleavage sites) was present at each position in the tile assembly. This was executed by first isolating the reporter strand, and then digesting separate aliquots with each endonuclease separately and the two together. Finally, these samples were examined by gel electrophoresis and the output values were displayed as banding patterns on the gel. Another method for output is the use of atomic force microscopy (AFM) observable patterning. The patterning was made by designing the tiles computing a bit 1 to have a stem loop protruding from the top of the tile. This molecular patterning was clearly observable under appropriate AFM imaging conditions.

An alternative method for autonomous execution of a sequence of finite-state transitions was subsequently developed by Shapiro and Benenson [38, 53]. Their technique essentially operated in the reverse of the assembly methods described above, and can be thought of as disassembly. They began with a linear dsDNA nanostructure whose sequence encoded the inputs, and then they executed a series of steps that digested the DNA nanostructure from one end (Figure 11.13).

On each step, a sticky end at one end of the nanostructure encoded the current state, and the finite transition was determined by hybridization of the current sticky end with a small "rule" nanostructure encoding the finite-state transition rule. Then a restriction enzyme, which recognized the sequence encoding the current input as

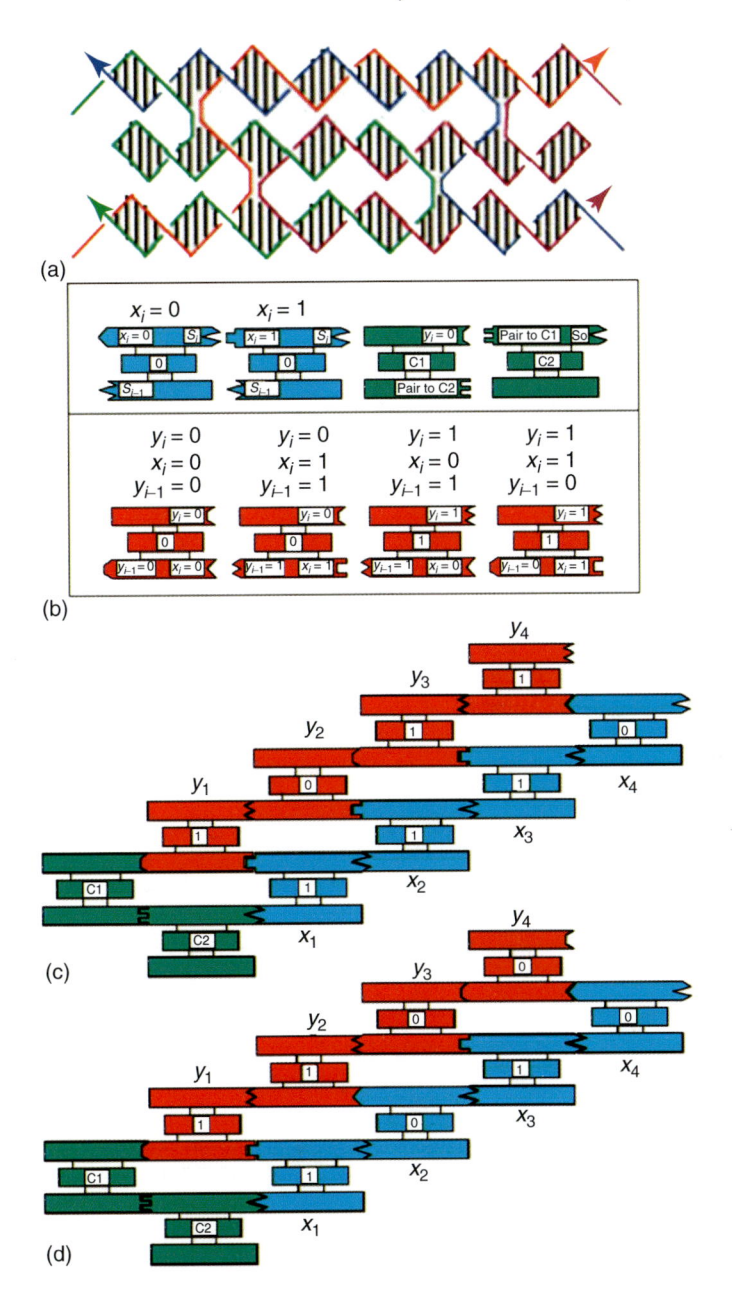

Figure 11.12 Sequential Boolean computation via a linear DNA tiling assembly. (a) TX tile used in assembly. (b) Set of TX tiles providing logical programing for computation. (c,d) Example of resulting computational tilings. (Reprinted by permission from Macmillan Publishers Ltd: Nature Mao, C. et al., Logical computation using algorithmic self-assembly of DNA triple-crossover molecules, 407, 49–496. © 2000.)

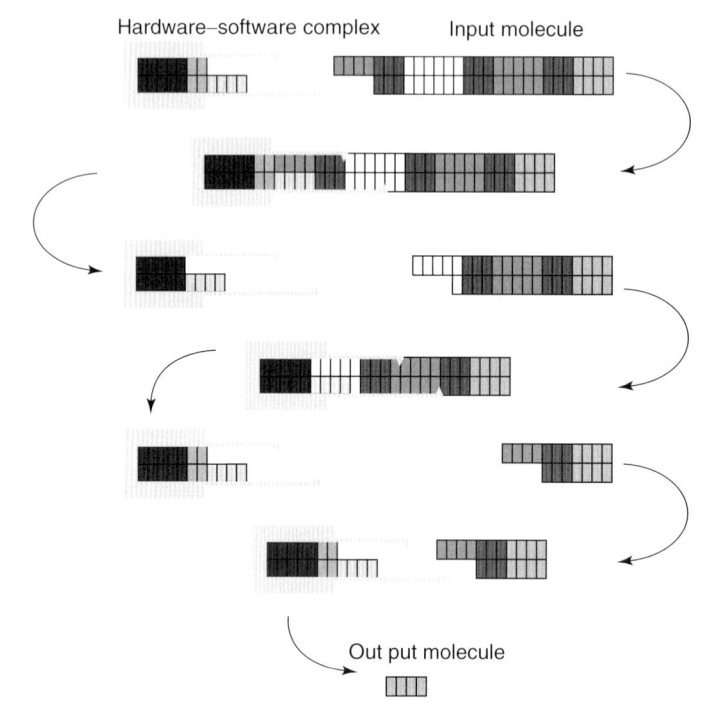

Figure 11.13 Autonomous finite-state computations via disassembly of a double-stranded DNA nanostructure.

well as the current state, cut the appended end of the linear DNA nanostructure to expose a new sticky end encoding the next state. The hardware–software complex for this molecular device is composed of dsDNA with an ssDNA overhang (shown at top left ready to bind with the input molecule) and a protein restriction enzyme (shown as gray pinchers). This ingenious design is an excellent demonstration that there is often more than one way to do any task at the molecular scale. Adar et al. [39] demonstrated in a test tube a potential application of such a finite-state computing device to medical diagnosis and therapeutics.

Exonucleases are enzymes that cleave nucleotides one at a time from the end of a polynucleotide chain. Techniques using a solid support and exonucleases for biomolecular computing have also been explored. Liu et al. [39] demonstrated a technique that involves the immobilization and manipulation of combinatorial mixtures of DNA stands on a support. A set of DNA molecules encoding all possible candidate solutions to the combinatorial problem of interest was synthesized and attached to the surface. Successive rounds of hybridization operations and exonuclease digestion were employed to identify and eliminate those strands that were not solutions to the problem. Upon completion of all the rounds, the solution to the problem was identified using PCR to amplify the remaining molecules. This method was used to solve an NP-complete problem, namely a four-variable 3SAT problem.

Deoxyribozymes (or DNAzymes) are DNA molecules that possess catalytic enzymatic activity. A particular class of DNAzymes acts as RNA nicking enzymes and possesses site-specific restriction activity. Using this class of DNAzymes, Stojanovic and Stefanovic [2] demonstrated an automaton that played a perfect game of tic-tac-toe against a human opponent. The automaton was implemented as a Boolean network of three types of DNAzymes gates: YES, NOT, and AND gates.

11.7
Biochemical DNA Reaction Networks

Controlling biochemical processes at the cellular and subcellular levels is a key endeavor of modern biology. These processes are at the very heart of life itself, and understanding and manipulating them would revolutionize our understanding of what makes life tick. These biochemical processes are often best understood as chemical reaction networks (CRNs). In this section, we show that DNA has the potential to implement arbitrary CRNs, and thus mimic arbitrary biochemical processes. It might need the help of RNA and protein enzymes, but most of the heavy lifting is done by the ingenious use of DNA strand displacement reactions. The path to programming *in vivo* biochemical processes might very well be through implementing interesting *in vitro* DNA systems such as amplifiers, switches, and oscillators.

Soloveichik et al. [40] asked, given a fully specified CRN does there exist a DNA hybridization-based system that can mimic, *in vitro*, the behavior of this network. Note that, if we can implement CRNs as DNA hybridization systems, we can use CRNs as a programming language and thus exploit the large preexisting body of work that tells us how to encode dynamic behavior as CRNs. Soloveichik *et al.* [40] gave a general scheme to do this: modulo the time and concentration of species. They used their scheme to design DNA hybridization-based oscillators (Figure 11.14), 2 bit pulse counters, chaotic systems, and integral counters. The correctness of their schemes assumed certain idealizations of how DNA hybridization systems work, which in practice do not exactly hold. How well their systems perform in practice will be determined by how gracefully actual DNA hybridization systems deviate from their ideal.

Kim *et al.* [41] implemented an *in vitro* bistable DNA switch regulated by RNA signals that repress transcription. The transcriptional system was fueled by RNA polymerase to create signals and ribonuclease to degrade RNA signals. The key advantage of this switch is that it is designed to be modular and composable. Thus, in principle it can execute arbitrary Boolean computation and also implement neural networks [42]. In fact, this same switch was used by Kim and Winfree [43] to construct three kinds of oscillators: a two-switch negative feedback oscillator, an amplified version of a negative feedback oscillator, and a three-switch ring oscillator. Franco *et al.* [44] have recently shown (Figure 11.15) that the two-switch negative feedback oscillator can drive a load. They used it to open and close a DNA tweezer and also released a functional RNA molecule.

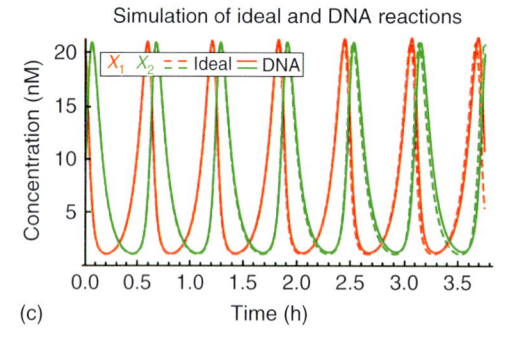

Figure 11.14 Lotka–Volterra chemical oscillator CRN using DNA [40].

11.8
Conclusion: Challenges in DNA-Based Biomolecular Computation

We have discussed many methods for DNA-based biomolecular computation, including some very impressive experimental demonstrations, but there are many challenges still remaining.

11.8.1
Scalability of Biomolecular Computations

Scalability of biomolecular computations are currently limited by a number of highly interconnected issues. We briefly discuss them below.

- The rate of errors: We have discussed approaches for reducing various types of errors in biomolecular computations including leakages of logical gates in the case of hybridization reaction circuits and also errors in the case of algorithmic self-assembly of DNA tiles. While some of these methods have been implemented experimentally, most of them remain as experimental challenges. These error-correction methods need to be improved to make them experimentally implementable and effect further improvements in the complexity of biomolecular computations.

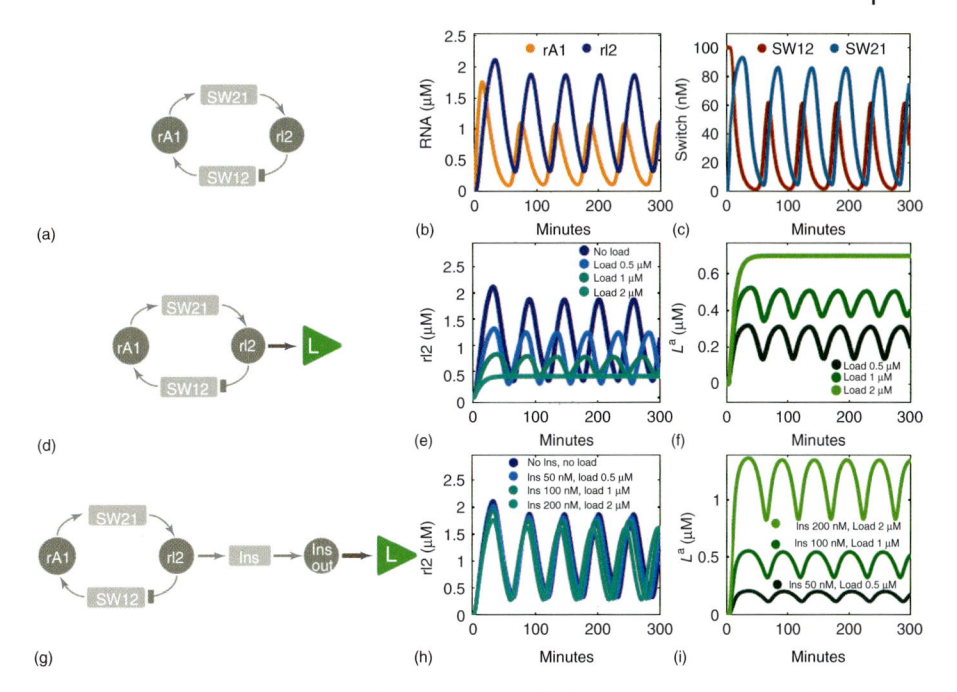

Figure 11.15 Two-switch negative feedback oscillator driving a load. Caption adapted from [44] (a): Diagram for the simple model for the oscillator. (b): Time traces for the oscillator species rA1 and rI2. (c): Time traces for the oscillator species SW12 and SW21. (d): Oscillator scheme with consumptive load coupled to rI2. (e, f): Time traces for the oscillator and load for consumptive coupling on rI2. (g): Oscillator scheme with consumptive insulating circuit and consumptive load. (h, i): Time traces for the oscillator and load when the insulating genelet is used to amplify rI2.

- The speed of biomolecular computation: The speed of many biomolecular computations is limited principally by diffusion delay times – the time for distinct molecules to find each other so as to initiate a reaction. These delays are substantially increased by the number of molecules that need to find each other to initiate a reaction, and hence bimolecular reactions are generally preferred to reactions involving more than two molecules at a time. The diffusion delay can be decreased by increasing the concentration of the reactants, but this can increase the rate of errors as well. One promising approach to improve the rate of biomolecular computation that we have discussed is to make the reactions local, for example, tethering the reacting molecules so they stay in the same relative vicinity.
- The modularity of biomolecular computations: Currently, biomolecular computations are only weakly modular, and there is considerable cross-talk between distinct reactions in a test tube. The modularity may be increased by the use of techniques such as artificial liposomes (45, 46) or DNA nanostructure boxes [47], which segregate the DNA material and their reactions. These may need to be designed to merge and split in a dynamic and/or programmable

fashion. Already, there are theoretical models of such methods termed "*membrane-based computation*" [48].

11.8.2
Ease of Design and Programmability of Biomolecular Computations

Design and simulation software has been very critical for building complex electronic devices such as computer processor chips. Design and simulation software have already been developed for various specific types of biomolecular computations, such as hybridization reaction networks [31], DNA tiling systems [10], and DNA origami [49]. Such software will be increasingly critical for future larger scale biomolecular computations, where the scale exceeds the ability of humans to predict their behavior and/or optimize their performance.

However, these software systems are limited to restricted subclasses of DNA nanostructures and their time and space complexity grows very rapidly with the size of the system. The software systems need to be extended to larger classes of DNA nanostructures. In addition, improved methods need to be developed for decreasing time and space complexity, for example, by decomposing the systems into weakly interacting, small systems. In addition, we expect the programmability of biomolecular computations will also be improved with increased use of robotic manipulation of reagents (see [50]) and microfluidics technologies [51].

11.8.3
In Vivo Biomolecular Computations

Some of the most potential promising applications of biomolecular computations are *in vivo* applications. These may include the idea of a DNA doctor [38] that makes use of a biomolecular device within a cell that monitors the cells and diagnoses diseases determined by conditions such as underexpression or overexpression of some of the cell's messenger RNA and responds by the release of nucleic acids that mitigate the disease. However, only a very few biomolecular computations [52] have been demonstrated to have *in vivo* viability. It is an open challenge to develop robust techniques for executing biomolecular computations within a living cell. One recent promising technique is the use of liposomes within the cell to protect synthetic DNA devices from digestion by the cell.

11.8.4
Conclusions

In spite of the many considerable challenges enumerated in this section, there has been very impressive scalability of experimental demonstrations of biomolecular computations, riveling the rate of improvements in VLSI technologies and computer architectures in the 1970s. The community of scientists working on biomolecular computations has also to likewise considerably increase, both in numbers and diversity of their disciplines, to overcome the many challenges that will need interdisciplinary approaches.

Acknowledgment

The authors wish to thank the support of NSF under grants **NSF CCF-1141847**, **CCF-0829797**, and **CCF-0829798**.

References

1. Zhang, D.Y. and Winfree, E. (2009) Control of DNA strand displacement kinetics using toehold exchange. *J. Am. Chem. Soc.*, **131** (48), 17303–17314.
2. Stojanovic, M.N. and Stefanovic, D. (2003) A deoxyribozyme-based molecular automaton. *Nat. Biotechnol.*, doi: 10.1038/nbt862
3. Garey, M.R. and Johnson, D.S. (1979) *Computers and Intractability: A Guide to the Theory of NP-Completeness*, W.H. Freeman and Company.
4. Feynman, R. (1959) There's Plenty of Room at the Bottom.
5. Adleman, L. (1994) Molecular computation of solutions to combinatorial problems. *Science*, **266**, 1021–1024.
6. Roweis, S., Winfree, E., Burgoyne, R., Chelyapov, N.V., Goodman, M.F., Rothemund, P.W.K., and Adleman, L.M. (1996) A sticker based model for DNA computation, *Proceedings of the Second Annual Meeting on DNA Based Computers*, American Mathematical Society, 1–29.
7. Lipton, R. (1995) DNA solution of hard computational problems. *Science*, **268**, 542–545.
8. Reif, J.H. (2002) DNA computation – perspectives: successes and challenges. *Science*, **296**, 478–479.
9. Winfree, E. (1998b) Algorithmic self-assembly of DNA. PhD Thesis. Caltech.
10. Winfree, E., Liu, F., Wenzler, L.A., and Seeman, N.C. (1998) Design and self-assembly of two-dimensional DNA crystals. *Nature*, **394** (6693), 529–544.
11. LaBean, T., Yan, H., Kopatsch, J., Liu, F., Winfree, E., Reif, J., and Seeman, N. (2000) Construction, analysis, ligation, and self-assembly of DNA triple crossover complexes. *J. Am. Chem. Soc.*, **122** (9), 1848–1860.
12. Mao, C., Labean, T., Reif, J., and Seeman, N. (2000) Logical computation using algorithmic self-assembly of DNA triple-crossover molecules. *Nature*, **407**, 493–496.
13. Yan, H., Feng, L., LaBean, T., and Reif, J. (2003a). Parallel molecular computation of pair-wise XOR using DNA string tile. *J. Am. Chem. Soc.*, **12**.
14. Yan, H., Feng, L., LaBean, T., and Reif, J. (2003) Parallel molecular computations of pairwise exclusive-or (XOR) using DNA string tile self-assembly, *Nature*, **125** (47), 14246–14247.
15. Rothemund, P.W.K. (2000) Using lateral capillary forces to compute by self-assembly, *Proc. Natl. Acak. Sci.*, **97** (3), 984–989, doi: 10.1073/pnas.97.3.984
16. Rothemund, P.W.K., Papadakis, N., and Winfree, E. (2004) Algorithmic self-assembly of DNA sierpinski triangles, *PLoS Biol.*, **2** (12), 424–436.
17. Winfree, E. (1998a) Simulations of Computing by Self-Assembly. Technical report. California Institute of Technology, CaltechCSTR:1998.22Persistent http://resolver.caltech.edu/CaltechCSTR: 1998.22.
18. Schulman, R., Lee, S., Papadakis, N., and Winfree, E. (2004) in *DNA Computing*, Vol. **9** (eds J Chen and J Reif), Springer-Verlag, Berlin, pp. 108–125.
19. Barish, R.D., Rothemund, P.W.K., and Winfree, E. (2005) *Nano Lett.*, **5**, 2586–2259.

20. Fujibayashi, K., Zhang, D.Y., Winfree, E., and Murata, S. (2009) Error suppression mechanisms for DNA tile self-assembly and their simulation. *Nat. Comput.*, **8** (3), 589–612.
21. Barish, R.D. et al. (2009) An information-bearing seed for nucleating algorithmic self-assembly, *Proc. Natl. Acak. Sci.*, **106** (15), 6054–6059.
22. Winfree, E. and Bekbolatov, R. (2003) Proofreading tile sets: error correction for algorithmic self-assembly. *DNA Comput.*, LNCS, **2943**, 126–144.
23. Reif, J., Sahu, S., and Yin, P. (2004) Compact error-resilient computational DNA tiling assemblies. *DNA Comput.*, **3384**, 293–307.
24. Chen, H.L. and Goel, A. (2005) Error free self-assembly using error prone tiles, *DNA Computing*, **3384**, 702–707.
25. Chen, H.L., Schulman, R., Goel, A., and Winfree, E. (2007) Reducing facet nucleation during algorithmic self-assembly, *Nano Letters* **7**, 2913–2919.
26. Soloveichik, D. and Winfree, E.. (2006) Complexity of compact proofreading for self-assembled patterns. *Proceedings of DNA Computing 11*, Springer-Verlag Berlin Heidelberg, LNCS **3892**, 305–324.
27. Chen., H.L. and Kao., M.Y. (2011) *DNA Computing and Molecular Programming*, Lecture Notes in Computer Science, Vol. **6518**, Springer, Berlin, Heidelberg, p. 13, ISBN: 978-3-642-18304-1.
28. Majumder, U. et al. LaBean, T.H., and Reif, J. (2008) *Activatable Tiles for Compact, Robust Programmable Assembly and other Applications, DNA 13*, Springer-Verlag, LNCS 4848, 15–25, Newyork.
29. Yin, P., Choi, H.M., Calvert, C.R., and Pierce, N.A. (2008) Programming biomolecular self-assembly pathways, *Nature*, **451** (7176), 318–322.
30. Qian, L. and Winfree, E. (2011) A simple DNA gate motif for synthesizing large-scale circuits, *J. R. Soci. Interface*, http://dx.doi.org/10.1098/rsif.2010.0729, doi: 10.1098/rsif.2010.0729
31. Qian, L. and Winfree, E. (2011) Scaling up digital circuit computation with DNA strand displacement cascades, *Science*, **332** (6034), 1196–1201.
32. Qian, L., Winfree, E., and Bruck, J. (2011) Neural network computation with DNA strand displacement cascades. *Nature*, **475**, 368–372.
33. Chandran, H., Gopalkrishnan, N., Phillips, A., and Reif, J. (2011) Localized hybridization circuits, in *DNA17*, Springer-Verlag Berlin, Heidelberg, 64–83.
34. Mullis, K., Faloona, F., Scharf, S., Saiki, R., Horn, G., and Erlich, H. (1986) Specific enzymatic amplification of DNA in vitro: the polymerase chain reaction. *Cold Spring Harb. Symp. Quant. Biol.*, **51**, 263.
35. Sakamoto, K., Kiga, D., Momiya, K., Gouzu, H., Yokoyama, S., Ikeda, S., Sugiyama, H., and Hagiya, M. (1998) State transitions with molecules. Proceedings of the 4th DIMACS Meeting on DNA Based Computers, held at the University of Pennsylvania, June 16–19, 1998.
36. Terrance Walker, G., Fraiser, M.S., Schram, J.L., Little, M.C., Nadeau, J.G., and Malinowski, D.P. (1992) Strand displacement amplification---an isothermal, in vitro DNA amplification technique. *Nucleic Acids Res.*, **20** (7), 1691–1696.
37. Reif, J.H. and Majumder, U. (2009) Isothermal Reactivating Whiplash PCR for Locally Programmable Molecular Computation, *DNA Computing, Lecture Notes in Computer Science* Vol. 5347 (eds A. Goel, F. Simmel, P. Sosík, Springer Berlin, Heidelberg, pp. 41–56, ISBN: 978-3-642-03075-8, http://dx.doi.org/10.1007/978-3-642-03076-5_5.
38. Shapiro, E. and Benenson, Y. (2006) Bringing DNA computers to life. *Sci. Am.*, **294**, 44–51.
39. Liu, Q., Wang, L., Frutos, A.G., Condon, A.E., Corn, R.M., and Smith, L.M. (2000) DNA computing on surfaces. *Nature*, **403**, 175–179.
40. Soloveichik, D., Seelig, G., and Winfree, E. (2010) DNA as a universal substrate for chemical kinetics.
41. Kim, J., White, K.S., and Winfree, E. (2006) Construction of an in vitro bistable circuit from synthetic transcriptional switches, 2 (Art. no: 68), *Mol. Syst. Biol.*, doi: 10.1038/msb4100099

42. Kim, J., Hopfield, J.J., and Winfree, E. (2004) Neural network computation by in vitro transcriptional circuits. *Adv. Neural Inf. Process. Syst. (NIPS)*, **17**, 681–688.
43. Kim, J. and Winfree, E. (2011) Synthetic in vitro transcriptional oscillators.
44. Păun, G. (1998) Computing with Membranes. *J. Comput. Syst. Sci.*, **61**, 108–143.
45. Hamada, T., Sugimoto, R., Vestergaard, M.C., Nagasaki, T., and Takagi, M. (2010) Membrane disk and sphere: controllable mesoscopic structures for the capture and release of a targeted object. *J. Am. Chem. Soc.*, **132** (30), 10528–10532.
46. Thompson, M.P., Chien, M.P., Ku, T.H., Rush, A.M., and Giannneschi, N.C. (2010) Smart lipids for programmable nanomaterials. *Nano Lett.*, **10** (7), 2690–2693.
47. Andersen, E.S., Dong, M., Nielsen, M.M., Jahn, K., Subramani, R., Mamdouh, W., Golas, M.M., Sander, B., Stark, H., Oliveira, C.L.P., Pedersen, J.S., Birkedal, V., Besenbacher, F., Gothelf, K.V., and Kjems, J. (2009) Self-assembly of a nanoscale DNA box with a controllable lid. *Nature*, **459**, 73–76.
48. Păun, G. (1998) Computing with Membranes. TUCS Report 208, Turku Center for Computer Science.
49. Rothemund, P.W.K. (2006) Folding DNA to create nanoscale shapes and patterns. *Nature*, **440**, 297–302.
50. Shapiro, E. and Robolab (2009) A Robot Programming Language.
51. Kirby, B.J. (2010) *Micro- and Nanoscale Fluid Mechanics: Transport in Microfluidic Devices*, Cambridge University Press.
52. Modi, S., Swetha, M.G., Goswami, D., Gupta, G.D., Mayor, S., and Krishnan, Y. (2009) A DNA nanomachine maps spatiotemporal pH changes in living cells. *Nat. Nanotechnol.*, **4**, 325–330.
53. Benenson, Y., Gil, B., Ben-Dor, U., Adar, R., and Shapiro, E. (2004) An autonomous molecular computer for logical control of gene expression. *Nature*, **429** (6990), 423–429.
54. Winfree, E., Yang, X., and Seeman, N. (1996) Universal computation via self-assembly of DNA: some theory and experiments. *DNA Based Computers II*, DIMACS, **44**, 191–213.

12
Enumeration Approach to the Analysis of Interacting Nucleic Acid Strands

Satoshi Kobayashi and Takaya Kawakami

12.1
Introduction

The importance of the analysis by computer algorithms of interacting nucleic acid strands is rapidly increasing. In DNA computing community, the analysis of interacting nucleic acid strands is important for the design of DNA sequences that will assemble into a desired nano-scale structure [1, 2]. In bioinformatics community, since the secondary structure plays an important role in the biological functioning of many RNAs, the prediction of interacting RNA structures is an important research topic [3].

One of the most effective methods for such prediction is to use dynamic programming (DP) to obtain a minimum free-energy (mfe) structure [4, 5]. The DP method has been extensively applied also to the calculation of equilibrium structure ensembles of RNA secondary structures [6, 7]. These algorithms, however, can deal with an mfe structure or the thermodynamic equilibrium of a single nucleic acid strand, but not with the secondary structures of interacting strands. Algorithms for finding the mfe structure among a combinatorial set of nucleic acid sequences were given by Andronescue *et al.* [1, 8]. Kobayashi gave an algorithm for finding the mfe structure from among all pseudoknot-free secondary structures of a regular set of nucleic acid strands [10, 11]. Efficient algorithms have been presented for finding the mfe structures from among secondary structures of a finite combinatorial set of strands [8, 11].

However, finding the mfe structure is not sufficient for the design and analysis of nucleic acid strand interactions. It is important to know concentrations of various interacting structures at equilibrium. For this purpose, an efficient method was proposed by Dirks *et al.* [12]. They nicely extended Mackaskill's partition function computation algorithm for a single RNA molecule with a pseudoknot-free structure [6] to the case of multiple strands. Furthermore, they succeeded in converting the total free-energy minimization problem (FEMP) of an interacting strand system to a convex programming problem, after computing the partition functions of all strand complexes. However, their method still has a combinatorial explosion problem, since the number of all strand complexes is exponential with respect to the number of interacting strands.

Kobayashi proposed a general method, called the *symmetric enumeration method* (SEM), for efficiently computing the equilibrium state of a complex chemical reaction system. By applying the SEM, we can efficiently compute the equilibrium of a *hybridization reaction system* (HRS), in which molecules interact in various ways to produce an exponential number of assemblies [13, 14]. This framework gives us a general method to reduce the number of variables of the total FEMP of such a complex reaction system.

In this chapter, we will apply SEM to nucleic acid strand interaction system. In Reference [12], the authors deal with pseudoknot-free interacting secondary structures. In this chapter, we will restrict the class of interacting secondary structures to the class of secondary structures that do not contain hairpin loops and branches (or, multiple loops). Although we should place this kind of restriction on the class of secondary structures, we *theoretically* show that, in this restricted case, the number of variables needed to solve the equilibria can be upper-bounded by polynomial with respect to the cardinality K of the strand set S put into the test tube, the maximum length L of the strands in S, and the maximum number N of strands that assemble together to produce an interacting secondary structure. More precisely, we will show that given a set S of nucleic acid strands, the initial concentration of each strand of S, the maximum number N of interacting strands, and the secondary structure T of interacting strands, we can compute the concentration of T at equilibrium with the the class of secondary structures restricted to the above subclass, by (i) first solving a convex programming problem, defined as FEMP in Section 12.4.5, with the number of variables bounded by some polynomial with respect to N, K, and L; (ii) and then calculating the concentration of T based on the obtained minimizer of the FEMP. The calculation of the second step requires $O(MKL^2)$ number of arithmetic operations, where M is the number of base pairs in T.

Thus, the proposed method opens up a new direction of the research on nucleic acid strand interaction analysis.

In Section 12.2, we give some definitions and notations related to multiset theory, which are necessary to formally define the problem of computing chemical equilibrium. Section 12.3 explains what chemical equilibrium is and what a HRS means. On the basis of these definitions and notations, we review the SEM proposed by Kobayashi [13, 14] in Section 12.4, in which we also give a typical example of the enumeration scheme for the folding reaction system of an RNA molecule. Finally, we apply SEM to nucleic acid strand interactions in Section 12.5.

12.2
Definitions and Notations for Set and Multiset

In order to formally introduce a mathematical definition of chemical reaction and its equilibrium, we need some basic definitions and notations related to *set* and *multiset*.

For a set X, we denote the cardinality of X by $|X|$. For sets X_1 and X_2, we define $X_1 - X_2 = \{x \in X_1 \mid x \notin X_2\}$. For a set N of numbers, we denote the subsets of N consisting of all nonnegative and positive numbers in N by N_+ and N_{++}, respectively. By $\mathbf{R}$ and $\mathbf{Z}$, we denote the set of real numbers and integers, respectively.

Let X be some underlying set of elements. A *multiset* S over X is a pair $(X, \mathbf{I}_S)$ where $\mathbf{I}_S$ is a mapping from X to $\mathbf{Z}_+$, and is called a *multiplicity function* of S. For an element $x \in X$, $\mathbf{I}_S(x)$ denotes the number of occurrences of x in the multiset S. Let $S = (X, \mathbf{I}_S)$ be a multiset over X. We say that $x \in X$ is in S or x is an element of S, written $x \in S$, if $\mathbf{I}_S(x) > 0$. For multisets S_1 and S_2, its union and difference are defined as follows: $\mathbf{I}_{S_1 \cup S_2}(x) = \mathbf{I}_{S_1}(x) + \mathbf{I}_{S_2}(x)$, and $\mathbf{I}_{S_1 - S_2}(x) = \mathbf{I}_{S_1}(x) - \mathbf{I}_{S_2}(x)$ if $\mathbf{I}_{S_1}(x) - \mathbf{I}_{S_2}(x) \geq 0$ holds, and $\mathbf{I}_{S_1 - S_2}(x) = 0$ otherwise. The symmetric difference $S_1 \triangle S_2$ of multisets S_1 and S_2 is defined as $S_1 \triangle S_2 = (S_1 - S_2) \cup (S_2 - S_1)$.

For instance, consider the multisets $S_1 = (X, \mathbf{I}_{S_1})$ and $S_2 = (X, \mathbf{I}_{S_2})$, where

$$X = \{a, b, c, d\}$$
$$\mathbf{I}_{S_1}(a) = 1, \quad \mathbf{I}_{S_1}(b) = 2, \quad \mathbf{I}_{S_1}(c) = 0, \quad \mathbf{I}_{S_1}(d) = 3$$
$$\mathbf{I}_{S_2}(a) = 0, \quad \mathbf{I}_{S_2}(b) = 2, \quad \mathbf{I}_{S_2}(c) = 3, \quad \mathbf{I}_{S_2}(d) = 1$$

It is sometimes convenient to represent multisets by enumerating elements, where we allow multiple occurrences of elements. For instance, S_1 and S_2 can be represented by $S_1 = \{a, b, b, d, d, d\}$ and $S_2 = \{b, b, c, c, c, d\}$, respectively. Then, we have $S_1 \cup S_2 = \{a, b, b, b, b, c, c, c, d, d, d, d\}$, $S_1 - S_2 = \{a, d, d\}$, $S_2 - S_1 = \{c, c, c\}$, $S_1 \triangle S_2 = \{a, c, c, c, d\}$.

For a multiset $S = (X, \mathbf{I}_S)$ and any function f from X to $\mathbf{R}$, we define

$$\sum_{x \in S} f(x) \stackrel{\text{def}}{=} \sum_{x \in X} \mathbf{I}_S(x) \cdot f(x)$$

$$\prod_{x \in S} f(x) \stackrel{\text{def}}{=} \prod_{x \in X} f(x)^{\mathbf{I}_S(x)}$$

For instance, for the above example multisets S_1 and S_2, we have

$$\sum_{x \in S_1} f(x) = f(a) + 2f(b) + 3f(d) = f(a) + f(b) + f(b) + f(d) + f(d) + f(d)$$

$$\prod_{x \in S_2} f(x) = f(b)^2 \times f(c)^3 \times f(d) = f(b) \times f(b) \times f(c) \times f(c) \times f(c) \times f(d)$$

Thus, the sum and product over a multiset count the elements redundantly for their multiple occurrences.

12.3
Chemical Equilibrium and Hybridization Reaction System

We start from an *informal* explanation of the chemical equilibrium by taking a simple example reaction system. Consider two molecules α and β, and a compound, or assembly, $\alpha\beta$ consisting of molecules α and β. We then consider a reaction rule $\alpha + \beta \rightleftharpoons \alpha\beta$. The equilibrium state of this reaction system is determined by free energies $E(\alpha)$, $E(\beta)$, and $E(\alpha\beta)$ of α, β, and $\alpha\beta$, respectively. More precisely, the concentrations $[\alpha]$, $[\beta]$, and $[\alpha\beta]$ at the equilibrium state should satisfy in this chapter, $E(X)$ is a dimensionless quantity. For instance, $E(X)$ is the free energy per mol of X divided by the physical quantity RT, where R is gas constant and T is the absolute temperature of the reaction system. the following:

$$\frac{[\alpha\beta]}{[\alpha][\beta]} = e^{-(E(\alpha\beta)-(E(\alpha)+E(\beta)))}$$

which is called an *equilibrium equation*.

Then, the problem is to find $[\alpha]$, $[\beta]$, and $[\alpha\beta]$ satisfying

$$e^{E(\alpha\beta)} \times [\alpha\beta] = e^{E(\alpha)+E(\beta)} \times [\alpha][\beta], \quad [\alpha] + [\alpha\beta] = [\alpha]_0, \quad [\beta] + [\alpha\beta] = [\beta]_0$$

where $[\alpha]_0$ and $[\beta]_0$ are initial concentrations of α and β, respectively. This solution is called an *equilibrium state* of the reaction system.

We will formulate these concepts in a more general setting and define a *HRS*.

Let M be a set of *molecules* and A be a set of *assemblies of molecules* consisting of some elements in M. For $x \in M$ and $X \in A$, we denote the number of molecules x contained in the assembly X by $\#_x(X)$. A *reaction rule* over A is a pair of X_1 and X_2 of multisets consisting of elements of A satisfying the following equation:

$$\sum_{X \in X_1} \#_x(X) = \sum_{X \in X_2} \#_x(X), \qquad (\forall x \in M) \tag{12.1}$$

where the sum over a multiset counts elements redundantly for their multiple occurrences. This equality constraint (12.6) corresponds to the law of conservation of each molecule. A reaction rule (X_1, X_2) is usually denoted by $X_1 \rightleftharpoons X_2$. In case of $X_1 = \{X_1, ..., X_{n_1}\}$ and $X_2 = \{Y_1, ..., Y_{n_2}\}$, where multiple occurrences of a same assembly are allowed, we often write

$$X_1 + \cdots + X_{n_1} \rightleftharpoons Y_1 + \cdots + Y_{n_2}$$

A *distribution* of a set U is a function from U to $\mathbf{R}_+$. Usually, we use notations $[\,]$, $[\,]_1, [\,]_2, ...$, for representing distributions. For example, for a distribution $[\,]$ of A and an assembly $X \in A$, $[X]$ represents concentration of the assembly X. A *distribution of M* is especially called an *initial distribution*. If we have a set M of molecules with its initial distribution $[\,]_0$, then any distribution $[\,]$ of A should satisfy the following equation:

$$\sum_{X \in A} \#_x(X) \cdot [X] = [x]_0. \qquad (\forall x \in M) \tag{12.2}$$

This equality constraint corresponds to the law of conservation of each molecule.

In summary, a HRS is defined by $P = (M, A, \{\#_x \mid x \in M\}, R, E, [\,]_0)$, where M is a nonempty set of molecules, A is a nonempty set of assemblies consisting of molecules in M, $\#_x$ is a function from A to $\mathbf{Z}_+$ such that $\#_x(X)$ indicates the number of molecules x contained in the assembly X, R is a set of reaction rules satisfying the Eq. (12.1), E is a free-energy function from A to $\mathbf{R}$, and $[\,]_0$ is an initial distribution of M. In the case where M, A, and R are finite, we say that P is a *finite HRS*.

In this general setting, for a reaction rule $X_1 \rightleftharpoons X_2$, its equilibrium equation is given by

$$e^{\sum_{X \in X_1} E(X)} \times \Pi_{X \in X_1}[X] = e^{\sum_{X \in X_2} E(X)} \times \Pi_{X \in X_2}[X] \quad (12.3)$$

where the product over a multiset multiplies the value $[X]$ of elements X redundantly for their multiple occurrences.

The problem of interest is to find an equilibrium state of P, that is, a distribution $[\,]$ of A, satisfying equilibrium equations (12.3) of all $r \in R$ and conservation laws (12.2) of all molecules $x \in M$. Such a distribution $[\,]$ is called an *equilibrium state* of P.

An HRS is said to be *consistent* if there is a distribution of A satisfying Eq. (12.2) for all $x \in M$. It is said to be *inconsistent* if it is not consistent. In actuality, we can construct an inconsistent HRS. However, we can ignore such a possibility in most of real applications in the sense discussed below.

In most of real applications, assemblies of molecules are usually constructed by reactions starting from molecules in M. Therefore, the elements of M take part in the reactions defined by the rule set R. The concepts of *normal* HRSs introduced below formulates this idea.

First of all, we assume that $M \subseteq A$ holds. We will define recursively the notion of assemblies *normally generated* from M as follows:

1) Every $x \in M$ is normally generated from M.
2) $X \in A$ is normally generated from M if there exists a reaction rule $X_1 \rightleftharpoons X_2$ such that X is in $X_1 \Delta X_2$ and every element of $X_1 \Delta X_2$ except for X is already normally generated from M, where $X_1 \Delta X_2$ denotes the symmetric difference of multisets X_1 and X_2.

We say that an HRS is *normal* if $M \subseteq A$ holds and every assembly in A is normally generated from M. Then, we have the following proposition:

Proposition 1 *A normal, finite HRS is consistent.* ∎

An initial distribution $[\,]_0$ of M is said to be *normal* if $[x]_0 > 0$ holds for every $x \in M$.

In the sequel of this chapter, we will assume the following condition **(A0)**: **(A0)** Any HRS in this chapter is a normal, finite HRS with a normal initial distribution.

The result in Section 12.4 assumes the above condition **(A0)** [14]. So, in order to apply it to nucleic acid strand interaction system, the system should satisfy the condition **(A0)**. It is clear that the nucleic acid strand interaction system considered in this chapter satisfies **(A0)**, since it is a system such that strands (corresponding to *molecules*) are put into the test tube and they interact together to produce various interacting structures (corresponding to *assemblies*) step by step from nucleic acid strands (i.e., molecules). Therefore, we can apply the theory in Section 12.4 to the interaction system.

12.4
Symmetric Enumeration Method

An important idea for tackling the combinatorial explosion problem is the *locality* of HRSs. Locality intuitively corresponds to the physical property that the free energy of an assembly X of molecules can be computed as the sum of free energies of all local substructures of X. For instance, the free energy of a single RNA molecule at the secondary structure level can be calculated as the sum of free energies of all local substructures such as hairpin loops, bulge loops, internal loops, multiple loops, etc. In this section, a theoretical formulation, which was proposed in the author's previous work [14], will be introduced for defining the locality of HRSs using graph theory.

12.4.1
Enumeration Graph

In the original paper [13, 14] on SEM, hypergraph theory was utilized to formulate the locality of HRSs. In this chapter, however, we need only an ordinary graph to enumerate target secondary structures. So, we introduce a specialized formulation of SEM using only ordinary graphs, without using hyperarcs.

A *directed graph* G is a pair (V, Eg), where V is a finite set of *vertices*, and Eg is a finite set of *arcs* associated with two functions $t : Eg \to V$ and $h : Eg \to V$. An arc e is interpreted as an arrow from a *tail* $t(e)$ to a *head* $h(e)$. A directed graph is simply called a *graph* in this chapter.

In this definition, we allow multiarcs: that is, there can be more than one distinct arcs with the same head and the same tail. However, for two vertices v_1 and v_2, in case there exists only one arc e such that $t(e) = v_1$ and $h(e) = v_2$, we simply write $e = (v_1, v_2)$ or $e = v_1 \to v_2$. For a vertex v, an arc e such that $v = t(e)$ ($v = h(e)$) is called an *outgoing* (*entering*) arc of v. For a vertex v, by v_{out} (v_{in}), we denote the set of outgoing (entering) arcs of v. For a set W of vertices, we define $W_{\text{out}} = \cup_{v \in W} v_{\text{out}}$ and $W_{\text{in}} = \cup_{v \in W} v_{\text{in}}$. By V_0 and V_f, we denote the set of vertices $v \in V$ such that $v_{\text{in}} = \emptyset$ and $v_{\text{out}} = \emptyset$, respectively. Elements of V_0 and V_f are called *initial vertices* and *final vertices*, respectively.

A *path* from s to u in G is a sequence $s = v_1, e_1, v_2, e_2, \ldots, e_q, v_{q+1} = u$ of vertices v_i ($i = 1, \ldots, q+1$) and arcs e_i ($i = 1, \ldots, q$) such that $v_i = t(e_i)$ and $v_{i+1} = h(e_i)$ for $i = 1, \ldots, q$. If $s = u$ holds, the path is called a *cycle*. We say G is *acyclic* if it contains no cycles. A path is said to be *trimmed* if it contains at least one arc: that is, $q \geq 1$.

A graph $G' = (V', Eg')$ is called a *subgraph* of $G = (V, Eg)$ if $V' \subseteq V$ and $Eg' \subseteq Eg$ hold. Note that a path of a graph G can be regarded as a subgraph of G. For a subgraph $G' = (V', Eg')$ of G, and a vertex v (an arc e), we write $v \in G'$ ($e \in G'$) if $v \in V'$ ($e \in Eg'$).

For a graph $G = (V, Eg)$, we denote by $PT(G)$ the set of all trimmed paths from some root $r \in V_0$ to some sink s with $s \in V_f$.

12.4.2
Path Mappings

In this subsection, we will define for a given graph G, a *path mapping*, which maps a path to another one. In particular, a set Φ of path mappings such that each element of Φ is an isomorphism or an anti-isomorphism plays an important role in the theory of enumeration method reported in Ref. [14].

Let G be an acyclic graph $G = (V, Eg)$. Let ϕ be an injective and surjective mapping from $V \cup Eg$ to $V \cup Eg$ such that $\phi(V) = V$ and $\phi(Eg) = Eg$ hold. For a subgraph $G' = (V', Eg')$ of G, we denote by $\phi(G')$ a graph $(\phi(V'), \phi(Eg'))$. For a set S of subgraphs of G, we define a set $\phi(S) = \{\phi(G') \mid G' \in S\}$. If ϕ satisfies $\phi(PT(G)) = PT(G)$, ϕ is called a *path mapping*. Let Φ be a set of path mappings of G. Φ is called a *path mapping group* if Φ constitutes a group under composition.

For instance, consider a graph G_e illustrated in Figure 12.1, and let V_e and Eg_e be the set of vertices and arcs of the graph. We define for symbols in $\{L, R\}$ and $\{l, r\}$, the bar notations as $\overline{R} = L$, $\overline{L} = R$, $\overline{r} = l$, and $\overline{l} = r$.

Consider an identity mapping ϕ_0 and mappings ϕ_1 and ϕ_2 from $V_e \cup Eg_e$ to $V_e \cup Eg_e$ defined by

$$\phi_1(X(i,j)) = \overline{X}(i,j), \quad \text{(for } X(i,j) \in V_e\text{)}$$
$$\phi_1(x(i,j)) = \overline{x}(i,j), \quad \text{(for } x(i,j) \in Eg_e\text{)}$$
$$\phi_2(X(i,j)) = \begin{cases} \overline{X}(i,j) & \text{if } j = 2 \\ X(i,j) & \text{otherwise} \end{cases} \quad \text{(for } X(i,j) \in V_e\text{)}$$
$$\phi_2(x(i,j)) = x(i,-j). \quad \text{(for } x(i,j) \in Eg_e\text{)}$$

The mappings ϕ_1 and ϕ_2 are path mappings. The sets $\Phi_1 = \{\phi_0, \phi_1\}$, $\Phi_2 = \{\phi_0, \phi_2\}$, and $\Phi_3 = \{\phi_0, \phi_1, \phi_2, \phi_1 \circ \phi_2\}$ are path mapping groups.

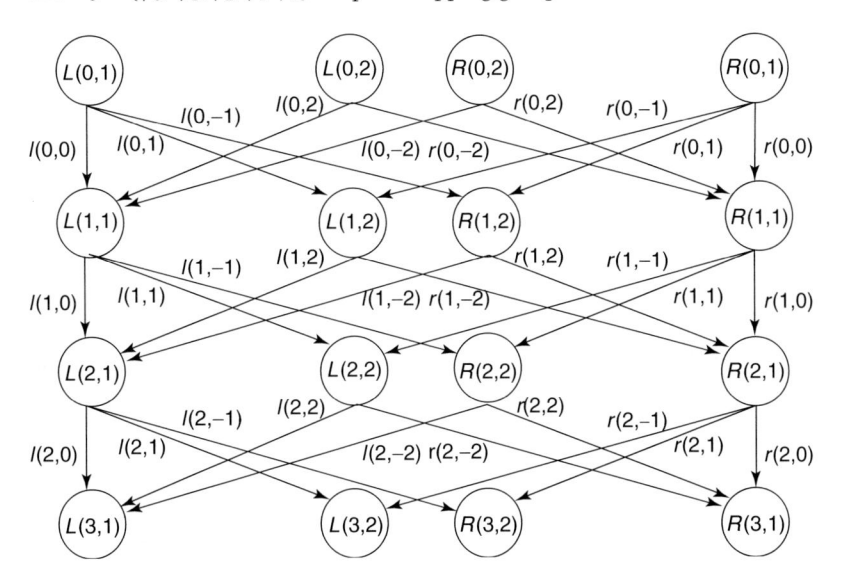

Figure 12.1 Graph G_e.

We say that a path mapping ϕ is an *isomorphism* if for every $e \in Eg$, $\phi(t(e)) = t(\phi(e))$ and $\phi(h(e)) = h(\phi(e))$ hold. Note that the path mappings $\phi_0, \phi_1, \phi_1 \circ \phi_2$ given in the above example are all isomorphisms. We say that a path mapping ϕ is an *anti-isomorphism* if for every $e \in Eg$, $\phi(t(e)) = h(\phi(e))$ and $\phi(h(e)) = t(\phi(e))$ hold.

12.4.3
Enumeration Scheme

Let G be an acyclic graph $G = (V, Eg)$. A *weight function* w of G is a function from Eg to $\mathbf{R}$. A function f from $PT(G)$ to $\mathbf{R}$ is said to be *locally definable* if there exists a weight function $\bar{f}$ from Eg to $\mathbf{R}$ such that, for every $\gamma \in PT(G)$,

$$f(\gamma) = \sum_{e \in Eg \text{ s.t. } e \in \gamma} \bar{f}(e)$$

holds.

Now, we will define the key concept of *locality* by using the above definitions and notations. Let $P = (M, A, \{\#_x \mid x \in M\}, R, E, [\,]_0)$ be an HRS and consider an acyclic graph $G = (V, Eg)$. Let ψ be a surjective function from $PT(G)$ to A. For $X \in A$, we define $\psi^{-1}(X) = \{\gamma \in PT(G) \mid \psi(\gamma) = X\}$. For $X \in A$ and $\gamma \in PT(G)$, we define $PT(X) = \psi^{-1}(X)$, $r_X = |\psi^{-1}(X)|$, and $r_\gamma = |\psi^{-1}(\psi(\gamma))|$, where r_X and r_γ are called a *rank* of X and a *rank* of γ, respectively. Given a triple $S = (P, G, \psi)$, a *modified free-energy function* E_r from A to $\mathbf{R}$ is defined as

$$E_r(X) = E(X) + \log r_X$$

This function E_r is a free-energy function with compensation logarithmic factor related to the rank. This modified free-energy function plays an important role in establishing the theory of equilibrium computation. Note that $E_r = E$ holds if ψ is a one-to-one mapping.

A triple $S = (P, G, \psi)$ is called an *enumeration scheme* for P if $E_r \circ \psi$ and $\#_x \circ \psi$ are locally definable for every $x \in M$. For an enumeration scheme $S = (P, G, \psi)$, we use the symbol ϵ to denote a weight function that locally defines $E_r \circ \psi$, and, for every $x \in M$, we use the symbol σ_x to denote a weight function that locally defines $\#_x \circ \psi$. Then, we define a set $W(G)$ of weight functions associated with G as $W(G) = \{\epsilon\} \cup \{\sigma_x \mid x \in M\}$. An enumeration scheme $S = (P, G, \psi)$ is said to be *simple* if ψ is a one-to-one mapping.

For a set Φ of path mappings and a subgraph G' of G, we denote the set $\{\phi(G') \mid \phi \in \Phi\}$ by $\Phi(G')$. An enumeration scheme $S = (P, G, \psi)$ for P is said to be *self-isomorphic* if there exists a path mapping group Φ such that

1) every element of Φ is an isomorphism, and
2) $\Phi(\gamma) = \psi^{-1}(\psi(\gamma))$ holds for every $\gamma \in PT(G)$.

An enumeration scheme $S = (P, G, \psi)$ for P is said to be *semi-self-isomorphic* if there exists a path mapping group Φ such that

1) every element of Φ is an isomorphism or an anti-isomorphism, and
2) $\Phi(\gamma) = \psi^{-1}(\psi(\gamma))$ holds for every $\gamma \in PT(G)$.

The following proposition is clear from the definitions.

Proposition 2 *The following statements hold true:*
1. *An enumeration scheme is semi-self-isomorphic if it is self-isomorphic.*
2. *An enumeration scheme is self-isomorphic if it is simple.* ∎

12.4.4
An Example of Enumeration Scheme – Folding of an RNA Molecule

We will give an example of a *simple* enumeration scheme for a reaction system of an RNA molecule folding, since this is a typical good example for understanding how we can enumerate structures by using a graph. Furthermore, this example will help in understanding the enumeration scheme that will be defined in Section 12.5.

Let $X = x_1, x_2, \cdots, x_n$ be an RNA sequence, each letter $x_i 0$ being an element of $\Sigma = \{A, C, G, U\}$, ordered from the 5' to the 3' direction. It is known that every pair of bases in $WC = \{(A, U), (U, A), (C, G), (G, C), (G, U), (U, G)\}$ may form a hydrogen bond, resulting in a stable structure, called a *secondary structure*. A secondary structure of X is a finite set T of pairs (i, j) of integers with $1 \leq i < j \leq n$, $i + 3 < j$, and $(x_i, x_j) \in WC$ such that, for any $bp_1 = (i_1, j_1)$ and $bp_2 = (i_2, j_2)$ in S, either $i_1 = i_2$ or $j_1 = j_2$ implies $bp_1 = bp_2$. The condition $i + 3 < j$ reflects the fact that a hairpin loop should contain at least three unpaired bases.

A secondary structure T is said to be *pseudoknotted* if there exist base pairs (i, j) and (k, l) in S such that $i < k < j < l$ (Figure 12.2a). A secondary structure T is said to be *pseudoknot-free* if it is not pseudoknotted. The secondary structures in Figure 12.2b, c, and d are pseudoknot-free. Although there are some experimental reports on the structural roles of pseudoknotted structures in biological functions, the computational analysis of secondary structures including them is time consuming

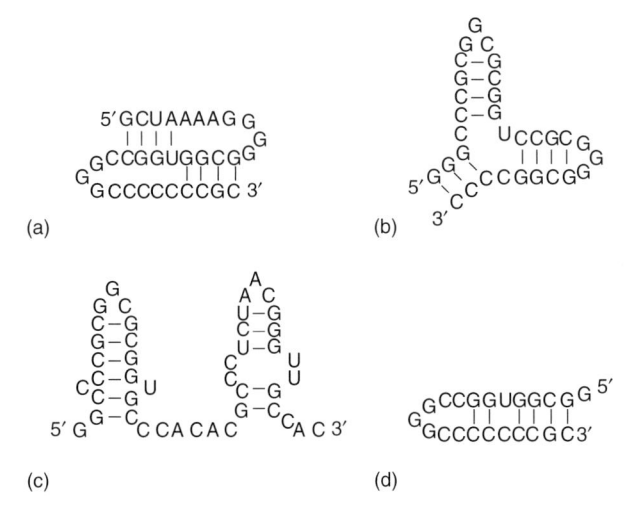

Figure 12.2 Secondary structures.

[15] and thus it is often the case that we focus on pseudoknot-free structures as in Refs [6, 12].

In this example, we will focus on a much smaller class of secondary structures than the class of pseudoknot-free secondary structures. For two base pairs (i, j) and (p, q), we say (i, j) *surrounds* (p, q), written as $(p, q) < (i, j)$, if $i < p < q < j$ holds. A secondary structure T is said to be *linear* if base pairs of T can be linearly ordered with respect to the relation $<$. In this example, we will focus on the class of linear secondary structures. The secondary structures in Figure 12.2a–c are not linear, but the secondary structure in Figure 12.2d is linear.

We now give an example of graphs by which we can enumerate all linear secondary structures of an RNA sequence.

Let $X = x_1, \cdots, x_n$ be an RNA sequence. Then, we prepare a set of vertices corresponding to base pairs that may form in the sequence X. Moreover, we use two additional special vertices: an initial vertex s and a final vertex f. The construction of edge set is as follows: We draw an edge from a base pair (i, j) to a base pair (k, l) if and only if $(k, l) < (i, j)$ holds. Furthermore, for every base pair bp, we place an edge from s to bp and an edge from bp to f. Formally, we can define a graph $G = (V, Eg)$ for the sequence X:

$$BP = \{(i, j) \mid 1 \leq i < j \leq n, i + 3 < j, (x_i, x_j) \in WC\}$$
$$V = \{s, f\} \cup BP$$
$$Eg = \{(s, bp) \mid bp \in BP\} \cup \{(bp, f) \mid bp \in BP\} \cup$$
$$\{((i, j), (k, l)) \mid (i, j), (k, l) \in BP, (k, l) < (i, j)\}$$

A path in $PT(G)$ for G defined above naturally corresponds to a linear secondary structure consisting of base pairs contained in it. An example of graphs for enumerating secondary structures of the sequence $X =$ GGAAACUU is given in Figure 12.3.

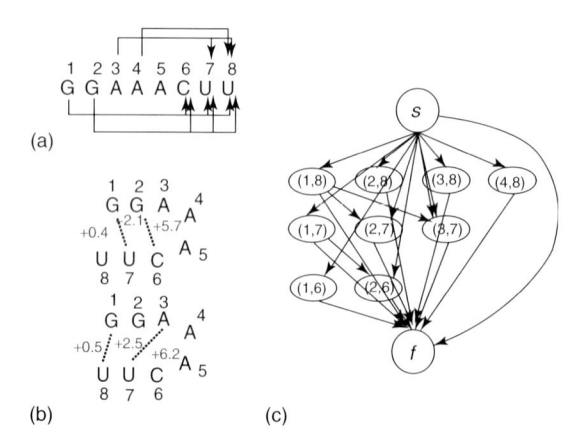

Figure 12.3 Example of enumeration graphs. (a) Base pairs, (b) structures, and (c) enumeration graph.

Figure 12.3a illustrates all possible base pairs of the sequence X. Figure 12.3c shows an enumeration graph for the sequence X. A path $s \to (1,7) \to (2,6) \to f$ corresponds to the upper secondary structure in Figure 12.3b. A path $s \to (1,8) \to (3,7) \to f$ corresponds to the lower secondary structure in Figure 12.3b. In this way, we can enumerate all linear secondary structures of X. The mapping from a path to its corresponding secondary structure is denoted by ψ.

As is clear from the above example, an edge between the base pairs (i,j) and (k,l) in the graph corresponds to a local loop structure (either of stacked base pairs, a bulge, or an internal loop) surrounded by (i,j) and (k,l). An edge between s (f) and a base pair (i,j) corresponds to a free end loop outside (a hairpin loop close by) the base pair (i,j). Thus, the weight $\epsilon(e)$ of an edge e is defined as the free energy of the corresponding local secondary structure of e. For instance, the free-energy values of local secondary structures are given as real values in Figure 12.3c. Thus, the weights of edges $s \to (1,7), (1,7) \to (2,6), (2,6) \to f, s \to (1,8)$, and $(1,8) \to (3,7)$, $(3,7) \to f$ are given by $+0.4, -2.1, +5.7, +0.5, +2.5$, and $+6.2$, respectively.

It is left to define the weight function σ_x for each $x \in M$. Note that M is a singleton since this example deals with a monomolecular reaction. Let $M = \{x\}$. Then, since each path $\gamma \in PT(G)$ contains exactly one molecule x, the weight σ_x can be defined as follows: $\sigma_x((v_1, v_2)) = 1$ if $v_1 = s$; $\sigma_x((v_1, v_2)) = 0$ if otherwise.

Note that it is not necessary to define a path mapping group Φ, since ψ is one-to-one, and thus, the above enumeration scheme is *simple*.

12.4.5
Convex Programming Problem for Computing Equilibrium

Let $S = (P, G, \psi)$ be a semi-self-isomorphic enumeration scheme, where $G = (V, Eg)$. We will prepare a variable w_e on $\mathbf{R}_+$ for each $e \in Eg$. Furthermore, for convenience, we will define for each $v \in V - V_0 - V_f$, $w_v = \sum_{e \in v_{out}} w_e$. Note that w_v is not a new variable, but an *expression* representing the sum of variables w'_es such that $e \in v_{out}$.

Consider the following minimization problem:

Free-Energy Minimization Problem (FEMP)
minimize:

$$FE_3(P, (w_e \mid e \in Eg)) \stackrel{def}{\equiv} \sum_{e \in Eg} \epsilon(e) \cdot w_e + \sum_{e \in Eg} w_e(\log w_e - 1)$$
$$- \sum_{v \in V - V_0 - V_f} w_v(\log w_v - 1)$$

subject to:

$$\sum_{e \in Eg} \sigma_x(e) \cdot w_e = [x]_0, \qquad (\forall x \in M)$$

$$\sum_{e \in v_{in}} w_e = \sum_{e \in v_{out}} w_e, \qquad (\forall v \in V - V_0 - V_f\})$$

$$w_e = w_{\phi(e)}, \qquad (\forall e \in Eg \; \forall \phi \in \Phi)$$

$$w_e \geq 0. \qquad (\forall e \in Eg)$$

For each $e \in Eg$, let $\tilde{w}_e$ be the value of w_e of a minimizer of FEMP. By $\tilde{w}$, we denote the vector $(\tilde{w}_e \mid e \in Eg)$. We say $\gamma \in PT(G)$ is *positive* with respect to $\tilde{w}$ if $\tilde{w}_e > 0$ holds for every $e \in \gamma$.

Define a distribution $[\,]_{+,G}(\tilde{w})$ of $PT(G)$ as follows:

$$[\gamma]_{+,G}(\tilde{w}) = \begin{cases} \dfrac{\prod_{e \in Eg \text{ s.t. } e \in \gamma} \tilde{w}_e}{\prod_{v \in V - V_0 - V_f \text{ s.t. } v \in \gamma} \tilde{w}_v} & \text{if } \gamma \text{ is positive} \\ & \text{with respect to } \tilde{w} \\ 0 & \text{otherwise} \end{cases} \quad (12.4)$$

where the product Π of empty set of reals is defined as 1.

Furthermore, define a distribution $[\,]_{*,G}(\tilde{w})$ of A based on the distribution $[\,]_{+,G}(\tilde{w})$ of $PT(G)$ as follows:

$$[X]_{*,G}(\tilde{w}) = \sum_{\gamma \in PT(X)} [\gamma]_{+,G}(\tilde{w}) \quad (12.5)$$

Then, we have the following theorems.

Theorem 1 *[14] Assume that there exists a semi-self-isomorphic enumeration scheme $S = (P, G, \psi)$ for an HRS P. Let $\tilde{w} = (\tilde{w}_e \mid e \in Eg)$ be a minimizer of the FEMP corresponding to S. Then, the distribution $[\,]_{*,G}(\tilde{w})$ is an equilibrium state of P.*

Theorem 2 *[14] The objective function of FEMP is convex over $\mathbf{R}_{++}^m$, where $m = |Eg|$. In particular, the Hessian matrix of the objective function of FEMP at any point in $\mathbf{R}_{++}^m$ is positive definite.*

12.5
Applying SEM to Nucleic Acid Strands Interaction

In this section, we will propose an enumeration scheme for nucleic acid strands interaction at the secondary structure level, where the interacting strand secondary structures are restricted to some class $L_N(S)$ of secondary structures defined in Section 12.5.1. The definitions of enumeration scheme will be defined in Sections 12.5.3, 12.5.4, and 12.10.

By demonstrating the construction of the enumeration scheme, we will show that, given a set S of nucleic acid strands, initial concentration of each strand of S, the maximum number N of strands contained in an interacting secondary structure, and a secondary structure T of interacting strands, we can compute the concentration of T at equilibrium with the class of secondary structures restricted to $L_N(S)$ by the following two steps:

1) Solve a convex programming problem, defined as FEMP in Section 12.4.5, with the number of variables bounded by some polynomial with respect to N, $|S|$, and maximum length L of strands in S.

2) Calculate the concentration of T based on the expression (12.5) and the obtained minimizer of FEMP.

The calculation of the second step requires $O(M \cdot |S| \cdot L^2)$ number of arithmetic operations, where M is the number of base pairs in T.

12.5.1
Target Secondary Structures

Let S be a set of nucleic acid strands. In this chapter, for a positive integer N, we consider a subclass $L_N(S)$ of secondary structures of interacting nucleic acid strands. The class $L_N(S)$ is the set of all secondary structures T consisting of at most N copies of strands in S such that T does not contain branches (multiple loops) and hairpin loops.

An example of secondary structures that do not contain branches and hairpin loops is given in Figure 12.4. An example of secondary structures that contain a branch and a hairpin loop, that is, nontarget structure of this chapter, is given in Figure 12.5.

12.5.2
Introducing Basic Notations

Let S be a set of strands interacting in a test tube or in a cell. We will consider only secondary structures in $L_N(S)$ consisting of at most n copies of strands of S, where we allow multiple occurrences of a same strand in S.

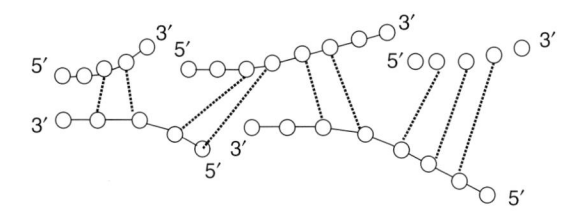

Figure 12.4 Target structure.

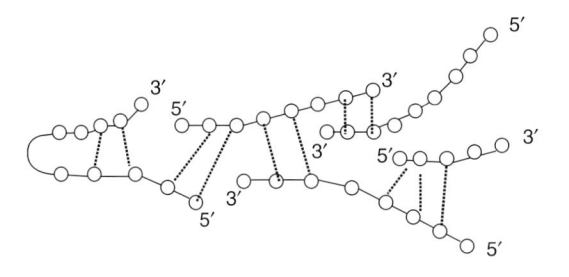

Figure 12.5 Nontarget structure.

A graph to enumerate such secondary structures is given by $G = (V, Eg)$, where V is a set of vertices and Eg is a set of directed arcs, and there are two kinds of associated weight functions ϵ and σ_x ($x \in S$), which locally define $E_r \circ \psi$ and $\#_x \circ \psi$ ($x \in S$), respectively.

Before moving on to the definition of G, we will introduce some basic notations that are necessary for the rest of this chapter.

For $\alpha \in S$, by l_α we denote the length of α. For i with $1 \leq i \leq l_\alpha$, by $\alpha[i]$, we denote the ith letter (in $\Sigma_{DNA} = \{A, C, G, T\}$) of α from the 5' end. On the other hand, by α_i, we mean the ith position of base sequence α from the 5' end. Intuitively, $\alpha[i]$ represents the ith base *letter* of the base sequence α, but α_i represents the ith base *itself* contained in α.

First, we consider a sequence $\alpha_1, \alpha_2, \cdots, \alpha_m$ ($\alpha_i \in S$, $i = 1, ..., m$) of m nucleic acid strands ordered from the 5' end to the 3' end (Figure 12.6). By $m\_p\_\alpha_p\_i$, we denote the ith base itself of the pth strand $\alpha_p \in S$ from the 5' end in the sequence.

For two bases $m\_p\_\alpha\_i$ and $m\_p'\_\beta\_i'$, we write $m\_p\_\alpha\_i < m\_p'\_\beta\_i'$ if and only if the following condition holds:

$$m\_p\_\alpha\_i < m\_p'\_\beta\_i' \equiv \begin{cases} p < p' \\ \text{or} \\ p = p' \wedge \alpha = \beta \wedge i < i' \end{cases}$$

This relation $m\_p\_\alpha\_i < m\_p'\_\beta\_i'$ means that the base $m\_p\_\alpha\_i$ is located in the 5' end direction from the base $m\_p'\_\beta\_i'$.

Next we consider an interaction of two sequences of nucleic acid strands, where we have a sequence $\alpha_1, \alpha_2, \cdots, \alpha_s$ of s upper strands and a sequence $\beta_1, \beta_2, \cdots, \beta_t$ of t lower strands (Figure 12.7). By $(s\_p\_\alpha_p\_i, t\_q\_\beta_q\_j)$, we denote the base pair between the base $s\_p\_\alpha_p\_i$ and the base $t\_q\_\beta_q\_j$ (indicated by $*$ in Figure 12.7).

Let $WC = \{(A, T), (T, A), (G, C), (C, G)\}$. Using the above notations, we define a set BP of base pairs that can exist in a secondary structure of interacting n nucleic

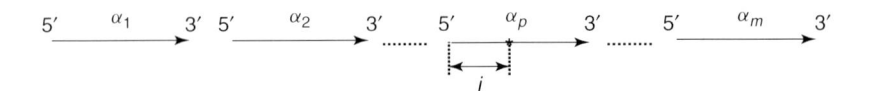

Figure 12.6 Strand sequence $\alpha_1, \alpha_2, \cdots, \alpha_m$.

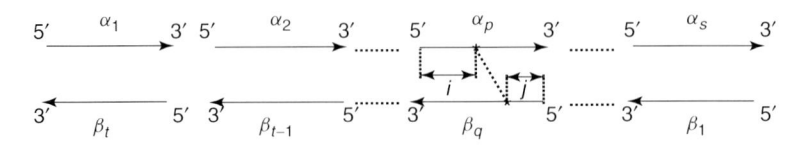

Figure 12.7 Base pair representation of nucleic acid strands interaction.

acid strands as follows:

$$BP = \{(s\_p\_\alpha\_i, t\_q\_\beta\_j) \mid 1 \le s < n, 1 \le t \le n-s, 1 \le p \le s,$$
$$1 \le q \le t, \alpha, \beta \in S, 1 \le i \le l_\alpha, 1 \le j \le l_\beta, (\alpha[i], \beta[j]) \in WC\}.$$

For two base pairs $(s\_p\_\alpha\_i, t\_q\_\beta\_j)$ and $(s\_p'\_\alpha'\_i', t\_q'\_\beta'\_j')$, we define the relation $\sim$ as follows:

$$(s\_p\_\alpha\_i, t\_q\_\beta\_j) \sim (s\_p'\_\alpha'\_i', t\_q'\_\beta'\_j') \stackrel{def}{=}$$
$$\begin{cases} p = p' \wedge \alpha = \alpha' \wedge 0 \le q - q' \le 1 \\ \quad\quad\quad\quad or \\ q = q' \wedge \beta = \beta' \wedge 0 \le p' - p \le 1. \end{cases}$$

The relation $(s\_p\_\alpha\_i, t\_q\_\beta\_j) \sim (s\_p'\_\alpha'\_i', t\_q'\_\beta'\_j')$ means that the base pairs $(s\_p\_\alpha\_i, t\_q\_\beta\_j)$ and $(s\_p'\_\alpha'\_i', t\_q'\_\beta'\_j')$ can be connected and exist in the same interacting structure.

Then, we can represent the relation that the base pairs $(s\_p\_\alpha\_i, t\_q\_\beta\_j)$ and $(s\_p'\_\alpha'\_i', t\_q'\_\beta'\_j')$ can be located adjacently in an interacting structure by the following expressions:

$$((s\_p\_\alpha\_i, t\_q\_\beta\_j) \sim (s\_p'\_\alpha'\_i', t\_q'\_\beta'\_j'))$$
$$\wedge ((s\_p\_\alpha\_i < s\_p'\_\alpha'\_i') \wedge (t\_q'\_\beta'\_j' < t\_q\_\beta\_j)).$$

12.5.3
Definition of Enumeration Graph Structure

By these definitions and notations, we will define a graph to enumerate all secondary structures in $L_N(S)$ of interacting nucleic acid strands as follows:

$$G = (V, Eg)$$
$$V = BP \cup \{I, F\}$$
$$Eg = \{((s\_p\_\alpha\_i, t\_q\_\beta\_j), (s\_p'\_\alpha'\_i', t\_q'\_\beta'\_j'))$$
$$\quad \mid ((s\_p\_\alpha\_i, t\_q\_\beta\_j), (s\_p'\_\alpha'\_i', t\_q'\_\beta'\_j')) \in BP)$$
$$\quad \wedge ((s\_p\_\alpha\_i, t\_q\_\beta\_j) \sim (s\_p'\_\alpha'\_i', t\_q'\_\beta'\_j'))$$
$$\quad \wedge ((s\_p\_\alpha\_i < s\_p'\_\alpha'\_i') \wedge (t\_q'\_\beta'\_j' < t\_q\_\beta\_j))\}$$
$$\cup \{(I, bp) \mid bp = (s\_1\_\alpha\_i, t\_t\_\beta\_j) \in BP\}$$
$$\cup \{(bp, F) \mid bp = (s\_s\_\alpha\_i, t\_1\_\beta\_j) \in BP\}$$
$$\cup \{e_\alpha \mid \alpha \in S\}$$

where $t(e_\alpha) = I$ and $h(e_\alpha) = F$ for any $\alpha \in S$.

In the above graph, the set of vertices contain the set BP of all possible base pairs, and two vertices in BP are connected by an arc if and only if those can be located adjacently in an interacting structure. Furthermore, an initial vertex I and a final vertex F are added to the vertex set, and all the base pairs that can be contained in the first free end structure are connected from I by arcs, and all the

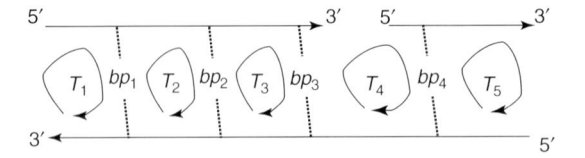

Figure 12.8 Correspondence between the arc e_i and the local structure T_i.

base pairs that can be contained in a last free end structure are connected to F by arcs. Furthermore, multiarcs e_α ($\alpha \in S$) exist between I and F.

A mapping ψ from $PT(G)$ to the set $\mathcal{A}$ of all secondary structures of interacting n strands is defined as follows:

First, for any $\alpha \in S$, the path $I \xrightarrow{e_\alpha} F$ represents a strand α itself with no base pairs.

Next, let us consider a path $\gamma : I \xrightarrow{e_1} bp_1 \xrightarrow{e_2} bp_2 \xrightarrow{e_3} \cdots \xrightarrow{e_k} bp_k \xrightarrow{e_{k+1}} F$ on the graph G. See Figure 12.8 for an example of the case $k = 4$. For each $i = 1, 2, 3, 4, 5$, the arc e_i corresponds to the local structure T_i in Figure 12.8. More precisely, for each $i = 2, 3, 4$, the arc $e_i = (bp_{i-1}, bp_i)$ corresponds to the local structure T_i surrounded by the base pairs bp_{i-1} and bp_i, and the arc e_1 (e_5) corresponds to a free end structure T_1 (T_5).

In this way, ψ maps every given path $\gamma : I \to bp_1 \to bp_2 \to \cdots \to bp_k \to F$ to the secondary structure consisting of a free end structure containing the base pair bp_1, local structures surrounded by the base pairs bp_i and bp_{i+1} ($i = 1, \ldots, k-1$), and a free end structure containing the base pair bp_k.

It should be noted that the graph G defined above might contain some vertices (or base pairs) that cannot be reached from the initial vertex I or cannot reach to the final vertex F. For instance, a vertex of the form

$$v = (s\_p\_\alpha\_l_\alpha, t\_q\_\beta\_1)$$

does not have any arc going out from v, since the base pair uses the last and the first base of upper and lower strands, respectively. These *nonuseful* vertices should be removed from the graph G. Such reachability test can be done by using a standard DP method. For instance, we can directly transform Dijkstra's algorithm for computing the shortest path problem to the algorithm for removing nonuseful vertices from G.

The mapping ψ is possibly a many-to-one mapping. For instance, consider the linear secondary structures in Figure 12.9. Both the structures illustrated in it correspond to the same secondary structure, since the 180° rotation of the upper structure is equivalent to the lower structure. It should be noted that the paths γ_1 and γ_2 corresponding to the upper and the lower structures, respectively, are different from each other. So, in general, two different paths can be mapped by ψ to a same linear structure. An exceptional case is a linear structure that is symmetric (Figure 12.10). For such a symmetric structure $X \in \mathcal{A}$, there exists only one path $\gamma \in PT(G)$ such that $\psi(\gamma) = X$.

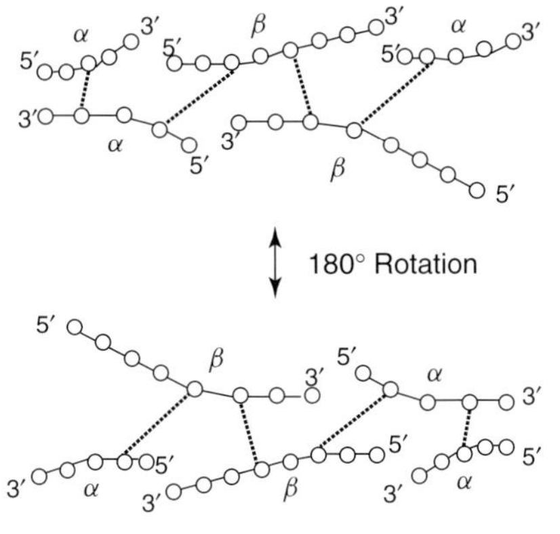

Figure 12.9 Asymmetric structure.

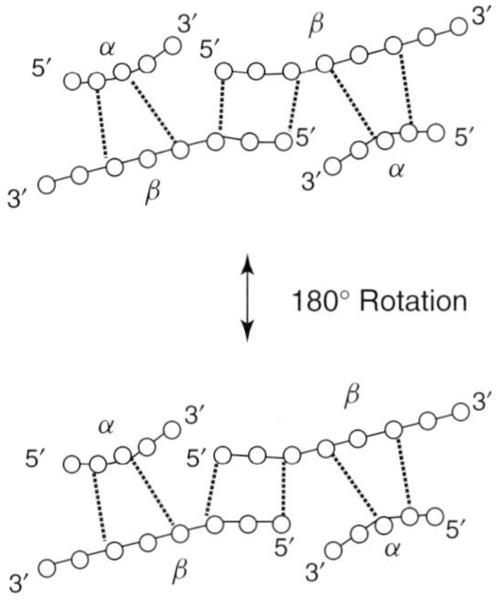

Figure 12.10 Symmetric structure.

12.5.4
Associated Weight Functions

The weight functions ϵ and $\sigma_x (x \in \mathcal{M})$ to locally define E_r and $\#_x$ should satisfy the following expressions:

$$E_r(\psi(\gamma)) = \sum_{e \in \text{Egs.t.} e \in \gamma} \epsilon(e), \qquad (12.6)$$

$$\#_x(\psi(\gamma)) = \sum_{e \in \text{Egs.t.} e \in \gamma} \sigma_x(e). \qquad (12.7)$$

The weight function ϵ satisfying the expression (12.6) can be defined as follows: Recall that the left-hand side $E_r(\psi(\gamma))$ of the expression is defined as $E(\psi(\gamma)) + \log r_{\psi(\gamma)}$. If $\psi(\gamma)$ is symmetric, then $r_{\psi(\gamma)} = 1$ holds, and otherwise, $r_{\psi(\gamma)} = 2$ holds, since, as already mentioned, there are two paths corresponding to $\psi(\gamma)$. On the other hand, statistical physics says that the free energy of an asymmetric structure should be the sum of the free energies of its local structures, and that the free energy of a symmetric structure should be the sum of the free energies of its local structures plus log 2. Thus, in both cases (symmetric and asymmetric structures), $E_r(\psi(\gamma))$ is the sum of the free energies of local structures of $\psi(\gamma)$ plus log 2.

Therefore, the weight function ϵ can be defined so that $\epsilon(e)$ might be the free energy of the local structure corresponding to $e \in Eg$ if $t(e) \neq I$ or $e = e_\alpha$ for some $\alpha \in S$, and $\epsilon(e)$ might be the free energy of the local structure corresponding to $e \in Eg$ plus log 2 if $t(e) = I$ and $e \neq e_\alpha$ for any $\alpha \in S$.

The definition of the weight function $\sigma_x(e)$ can be defined as follows: $\sigma_x(e) = 1$ if the local structure corresponding to $e \in Eg$ contains the 5' end of a nucleic acid strand $x \in S$, $\sigma_x(e) = 0$ otherwise.

12.5.5
Symmetric Properties

Consider a path mapping ϕ as defined below:

$$\phi(I) = F$$
$$\phi(F) = I$$
$$\phi((s\_p\_\alpha\_i, t\_q\_\beta\_j)) = (t\_q\_\beta\_j, s\_p\_\alpha\_i)$$
$$\phi((I, (s\_1\_\alpha\_i, t\_t\_\beta\_j))) = ((t\_t\_\beta\_j, s\_1\_\alpha\_i), F)$$
$$\phi((s\_s\_\alpha\_i, t\_1\_\beta\_j), F) = (I, (t\_1\_\beta\_j, s\_s\_\alpha\_i))$$
$$\phi(((s\_p\_\alpha\_i, t\_q\_\beta\_j), (s\_p'\_\alpha'\_i', t\_q'\_\beta'\_j'))) = ((t\_q'\_\beta'\_j', s\_p'\_\alpha'\_i'), (t\_q\_\beta\_j, s\_p\_\alpha\_i)).$$

Then, it is clear by the definition that ϕ is an anti-isomorphism, and that ϕ maps a given path γ to a path γ' such that $\psi(\gamma) = \psi(\gamma')$. Therefore, the set $\Phi = \{\phi_0, \phi\}$ of path mappings, where ϕ_0 is an identity mapping, satisfies $\Phi(\gamma) = \psi^{-1}(\psi(\gamma))$ for any $\gamma \in PT(G)$. Furthermore, it is clear that Φ constitutes a group under composition. Thus, we have the following theorem.

Theorem 3 *The triple (P, G, ψ) is a semi-self-isomorphic enumeration scheme for the nucleic acid strands interaction system considered in this chapter.*

12.5.6
Complexity Issues

Let S be a strand set; define $K = |S|$ and $L = max\{l_\alpha \mid \alpha \in S\}$. Let N be the maximum number of strands contained in a structure.

Then, the number of vertices of the enumeration graph defined above is upper-bounded by $O(N^4 K^2 L^2)$. The number of arcs of the enumeration graph is upper-bounded by $O(N^4 K^3 L^4)$, since for an arc

$$((s\_p\_\alpha\_i, t\_q\_\beta\_j), (s\_p'\_\alpha'\_i', t\_q'\_\beta'\_j')) \in Eg$$

the relation

$$(s\_p\_\alpha\_i, t\_q\_\beta\_j) \sim (s\_p'\_\alpha'\_i', t\_q'\_\beta'\_j')$$

requires either

$$p = p' \wedge \alpha = \alpha' \wedge 0 \leq q - q' \leq 1$$

or

$$q = q' \wedge \beta = \beta' \wedge 0 \leq p' - p \leq 1$$

When we apply Dijkstra's algorithm for removing nonuseful vertices and arcs, the time complexity for this removal process is $O(|V|\log|V| + |Eg|) = O(N^4 K^3 L^4 + N^4 K^2 L^2 \log N)$.

After obtaining a minimizer of FEMP, we will apply the expressions (12.4) and (12.5) for a given secondary structure T. Let M be the number of base pairs in T. Note that the paths in $\psi^{-1}(T)$ have vertices each corresponding to a base pair in T. Thus, the calculation of numerator of (12.4) requires only $O(M)$ number of arithmetic operations.

On the other hand, the calculation of the denominator of (12.4) requires computation of the sum w_v's. Consider a vertex

$$v = (s\_p\_\alpha\_i, t\_q\_\beta\_j)$$

and any arc

$$e = ((s\_p\_\alpha\_i, t\_q\_\beta\_j), (s\_p'\_\alpha'\_i', t\_q'\_\beta'\_j')) \in Eg$$

going out from v. Because of the relation

$$(s\_p\_\alpha\_i, t\_q\_\beta\_j) \sim (s\_p'\_\alpha'\_i', t\_q'\_\beta'\_j')$$

the number of possible e's is bounded by $O(KL^2)$. Therefore, the calculation of the denominator of (12.4) requires $O(MKL^2)$ number of arithmetic operations.

In summary, the calculations of (12.4) and (12.5) require $O(MKL^2)$ number of arithmetic operations.

12.6 Conclusions

In this chapter, we have applied the SEM to nucleic acid strand interaction systems with the restriction of the class of interacting secondary structures to the class of secondary structures that do not contain hairpin loops and branches. We have *theoretically* shown that, in this restricted case, the number of variables needed to solve the equilibria can be upper-bounded by a polynomial with respect to the number N of interacting strands, the maximum length L of the strands, and the number K of the elements in the strand set.

In the proposed enumeration scheme, we exclude the existence of hairpin loops from the secondary structures. So, it is interesting to extend the proposed method to contain hairpin loops. It does not seem difficult to carry out such an extension by adding some more vertices and arcs to the enumeration scheme that was presented in Section 12.5. Recall that SEM utilizes hypergraph theory instead of graph theory in its original form Refs [13, 14]. Therefore, it can deal with tree structures in a natural way. Thus, one of the interesting future researches can be an extension of the class of secondary structures to pseudoknot-free ones. Since some class of pseudoknotted structures can be represented by tree grammars [15], it is also natural to ask how we can enumerate them using a hypergraph in an enumeration scheme.

References

1. Condon, A.E. (2003) Problems on RNA secondary structure prediction and design. Proceedings of ICALP'2003, LNCS 2719, Eindhoven, The Netherlands, 22–32.
2. Brenneman, A. and Condon, A.E. (2002) Strand design for bio-molecular computation. *Theor. Comput. Sci.*, **287**, 39–58.
3. Andronescu, M., Zhang, Z., and Condon, A. (2005) Secondary structure prediction of interacting RNA molecules. *J. Mol. Biol.*, **345**, 987–1001.
4. Nussinov, R. and Jacobson, A. (1980) Fast algorithm for predicting the secondary structure of single stranded RNA. *Proc. Natl. Acad. Sci. U.S.A.*, **77**, 6309–6313.
5. Zuker, M. and Steigler, P. (1981) Optimal computer folding of large RNA sequences using thermodynamics and auxiliary information. *Nucleic Acids Res.*, **9**, 133–148.
6. McCaskill, J. (1990) The equilibrium partition function and base pair binding probabilities for RNA secondary structure. *Biopolymers*, **29**, 1105–1119.
7. Dirks, R. and Pierce, N. (2004) An algorithm for computing nucleic acid base-pairing probabilities including pseudoknots. *J. Comput. Chem.*, **25**, 1295–1304.
8. Andronescu, M., Dees, D., Slaybaugh, L., Zhao, Y., Condon, A., Cohen, B., and Skiena, S. (2002) Algorithms for testing that sets of dna words concatenate without secondary structure. Proceedings of the 8th International Meeting on DNA Based Computers, LNCS 2568, 182–195.
9. Kobayashi, S., Yokomori, T., and Sakakibara, Y. (2004) An algorithm for testing structure freeness of biomolecular sequences. *Aspects of Molecular Computing – Essays dedicated to Tom Head on the occasion of his 70th birthday*, Springer-Verlag, LNCS 2950, pp. 266–277.
10. Kobayashi, S. (2004) Testing structure freeness of regular sets of biomolecular sequences. Preliminary Proceedings of 10th International Meeting on DNA Based Computers, Milan, Italy, 395–404.
11. Kijima, A. and Kobayashi, S. (2005) Efficient algorithms for testing structure freeness of finite set of biomolecular sequences. Preliminary Proceedings of 11th International Meeting on DNA Based Computers, London, Canada, 278–288.
12. Dirks, R., Bois, J., Schaeffer, J., Winfree, E., and Pierce, N. (2007) Thermodynamic analysis of interacting nucleic acid strands. *SIAM Rev.*, **49**, 65–88.
13. Kobayashi, S. (2007) A new approach to computing equilibrium state of combinatorial hybridization reaction systems. Proceedings of Workshop on Computing and Communications from Biological Systems: Theory and Applications, CD-ROM, paper 2376. Budapest, Hungary.
14. Kobayashi, S. (2008) Symmetric enumeration method: a new approach to computing equilibria, CS 08-01, Department of Computer Science, University of Electro-Communications.
15. Uemura, Y., Hasegawa, A., Kobayashi, S., and Yokomori, T. (1999) Tree adjoining grammars for RNA structure prediction. *Theor. Comput. Sci.*, **210**, 277–303.

13
Restriction Enzymes in Language Generation and Plasmid Computing
Tom Head

13.1
Introduction

In the early 1980s, it appeared to me that the most profound new developments in science would be coming from molecular biology. Could formal language theorists get into the act based on the string aspect of the informational macromolecules? I spent 1983–1984 intensely studying the beautiful book by Benjamin Lewin, *Genes* [1], hoping to find a way to link formal language theory with molecular biology. Could the action of enzymes be modeled as rules that act in a generative capacity on sets of DNA molecules? The action of restriction enzymes accompanied by a ligase on double-stranded DNA molecules was described in that book with such specificity that it suggested that a description in formal language terms might be appropriate. The abstract concept of a *splicing system* was developed in [2] based only on this textbook description. In the early 1990s, Gheorghe Păun and numerous of his coworkers saw how to augment splicing systems with many different control structures and proved that many of these augmented systems have the same generating power as a Turing machine. They went on to develop *abstract* programmable universal computers using the splicing concept [3, 4]. To my knowledge, none of these has yet been realized as a "wet" biochemical computing system. During these same 1990s, Yokomori and coworkers [5] in Japan and Siromoney and coworkers [6] in India also began to investigate splicing systems.

When Leonard Adleman's 1994 paper [7, 8] appeared reporting a "wet" solution to an instance of the directed Hamiltonian path problem, all of us who had thought about splicing systems were jerked to attention, wondering how we might carry out "wet" algorithms, possibly using some of the abstract splicing concepts that had been developed. My interest shifted almost exclusively to actual wet laboratory computing but I did not see how to solve algorithmic problems in a wet laboratory using splicing. I needed a new, simple but flexible, computing idea which would use the resources that I had already thought about: double-stranded DNA, restriction enzymes, and a ligase. This new idea was first summarized in [9] and later called *aqueous computing*. It was initially realized in computations carried

Biomolecular Information Processing: From Logic Systems to Smart Sensors and Actuators,
First Edition. Edited by Evgeny Katz.
© 2012 Wiley-VCH Verlag GmbH & Co. KGaA. Published 2012 by Wiley-VCH Verlag GmbH & Co. KGaA.

out with several collaborators in the laboratories of Herman Spaink in Leiden and Susannah Gal in Binghamton.

Splicing theory in the original restricted sense of [2] is exposited in [10] and splicing in this and its various extended senses is exposited in the fundamental volume [4] by Păun *et al*. An exposition of aqueous computing is available in [11]. Rosalba Zizza [12] curates the valuable web page: Scholarpedia – splicing systems. Since up-to-date detailed references on splicing systems in the original sense [10] and on aqueous computing [11] are available, I have chosen to write here in an informal style describing the contexts in which these two concepts arose.

13.2
Wet Splicing Systems

We call a test tube of water (or an appropriate buffer) that contains a set of restriction enzymes, a ligase enzyme, and an appropriate energy source (perhaps ATP) a 'wet' *splicing scheme*. We think of such a scheme as a biochemical operator that is ready to act on whatever collection of DNA molecules one might wish to add. Suppose now that we add multiple copies (perhaps a picomole) of each of several DNA molecules to a wet splicing scheme. We idealize by assuming that there is no bound on the number of each of the DNA molecules we have added. We call the contents of this tube a 'wet' *splicing system*. The *splicing language* generated by this system is the collection of all molecular varieties of *well formed fully double-stranded* DNA molecules *having no 'sticky ends'* that can *potentially* arise under the action of the specified enzymes. We provide two simple examples:

Example 1. (A wet splicing language consisting of 10 molecular varieties). For our set of restriction enzymes we choose two: *Bam*HI and *Bgl*II.

$\quad\quad$ *Bam*HI cuts at the site 5'...G|GATCC...3' giving 5'...G and GATCC...3'
$\quad\quad\quad\quad\quad\quad\quad\quad\quad\quad\quad$ 3'...CCTAG|G...5' $\quad\quad$ 3'...CCTAG $\quad$ G...5'

with 'sticky ends' 5'GATC.

{Three periods $\cdots$ indicate the possible occurrence of additional nucleotides.}

$\quad\quad$ *Bgl*II cuts at site $\quad$ 5'...A|GATCT...3' $\quad$ giving $\quad$ 5'...A and GATCT...3'
$\quad\quad\quad\quad\quad\quad\quad\quad\quad\quad\quad$ 3'...TCTAG|A..5' $\quad\quad\quad\quad\quad$ 3'...TCTAG $\quad$ A..5'

with the *same* 'sticky ends', namely 5'GATC.

Suppose that we now add two DNA molecules (perhaps a picomole of each) having the forms:

$\quad\quad\quad\quad$ 5'...u...GGATCC...v...3' $\quad$ and $\quad$ 5'...x...AGATCT...y...3'
$\quad\quad\quad\quad$ 3'...u'..CCTAGG...v'..5' $\quad\quad\quad\quad\quad\quad$ 3'...x'..TCTAGA...y'..5'

where $u/u', v/v', x/x'$, and y/y' represent unspecified pairs of complementary substrings.

Suppose further that the exhibited sites for *Bam*HI and *Bgl*II are the only ones that occur in the given molecules. Recall that the language generated by the splicing system consists of all those fully double stranded DNA molecules with no sticky ends that can *potentially* arise. Since the 'sticky ends' produced by our two enzymes are identical, the language generated by this splicing system consists of the following ten molecular varieties:

The two *original* molecules. The four 'special' molecules:

$$5'...u...GGATCT...y...3'$$
$$3'...u'..CCTAGA...y'..5'$$

$$5'...u...GGATCT...x'...3'$$
$$3'...u'..CCTAGA...x...5'$$

$$5'...x...AGATCC...v...3'$$
$$3'...x'..TCTAGG...v'..5'$$

$$5'...x...AGATCC...u'..3'$$
$$3'...x'...TCTAGG...u...5'$$

and the four (*sometimes overlooked*) molecules

$$5'...u...GGATCC...u'..3'$$
$$3'...u'..CCTAGG...u...5'$$

$$5'...v'..AGATCT...v...3'$$
$$3'...v...TCTAGA..v'..5'$$

$$5'...x...AGATCT...x'..3'$$
$$3'...x'...TCTAGA..x...5'$$

$$5'...y'..GGATCC...y...3'$$
$$3'...y...ACTAGG...y'..5'.$$

Of these 10 molecules, 6 have sites at which either *Bam*HI or *Bgl*II will again cut. Notice, however, that none of the four 'special' molecules has a site for either *Bam*HI or *Bgl*II. Suppose that in the system we have just discussed we have provided exactly the same molarity of each of the two original DNA molecules. Now temporarily give up the idealization that there is always an unbounded supply of each molecule, but keep the idealization that the enzymes operate exactly as prescribed. Note that one can predict that the percentage of DNA molecules that have no sites will go to 100. See [13] for an experimental demonstration and a discussion of the concept of an *adult* language. See [14] for a further related wet demonstration and a discussion of the two related concepts of an adult language and a *limit* language. See [15] for further adventures with limit behavior.

Example 2. (An infinite wet splicing language). For our set of restriction enzymes, we choose the same two enzymes: *Bam*HI and *Bgl*II. But this time, we add the single DNA molecule (in multiple copies):

$$5'...x...GGATCC...y...AGATCT...z...3'$$
$$3'...x'..CCTAGG...y'..TCTAGA...z'..5'$$

where x/x', y/y', and z/z' represent unspecified pairs of complementary substrings. Assume that the two displayed sites are the only sites for *Bam*HI and *Bgl*II that occur in this molecule. Multiple copies of this molecule can splice together to give arbitrarily long molecules:

$$5'...x...GGATCC...y...A|GATCT...z...3' \quad \text{and}$$
$$3'...x'..CCTAGG...y'..TCTAG|A...z'..5'$$

5′...x...G|GATCC...y...AGATCT...z...3′   give
3′...x′..CCTAG|G...y′..TCTAGA...z′..5′

5′...x...GGATCC...y...AGATCC...y...AGATCT...z...3′   and
3′...x′..CCTAGG...y′..TCTAGG...y′..TCTAGA...z′..5′

5′...x...GGATCT...z...3′
3′...x′..CCTAGA...z′..5′.

Continuing in this way, we see that the splicing language contains, for every non-negative integer n,

5′...x...GGATCC...y...[AGATCC...y...]^{n}AGATCT...z...3′
3′...x′..CCTAGG...y′..[TCTAGG...y′..]^n TCTAGA...z′..5′

However, these are not the only molecules in the language, as many more arise in the manner in which the four 'sometimes overlooked' molecules arose in Example 1. Can you determine which molecules in this language have no site for either *Bam*HI or *Bgl*II?

We clarify the significance of *an additional defining condition* that we have avoided until now. Some restriction enzymes cut leaving sticky ends with no overhang at all. Examples:

*Alu*I cuts DNA at the site   AG|CT   and A*fe*I cuts at the site   AGC|GCT
                              TC|GA   TCG|CGA

each leaving blunt ends. Ligase connects blunt ends as well as those having matching sticky overhangs. Note that any two blunt ends are compatible. But ligation is only possible if the phosphate remains at each of the 5′ locations on the sticky ends. Restriction enzymes (should) leave the required phosphate in place. Finally we state one last requirement on wet splicing systems: The *initial molecules of each wet splicing system are assumed not to have phosphates at their 5′ ends*. This assumption is made so that a splicing system will not ligate the initial molecules – (which would make each splicing language automatically a subsemigroup under concatenation). We want ligation only of sticky ends that have been created by the restriction enzymes of the system.

Circular DNA molecules. Circular DNA molecules occur in nature especially as genomes of bacteria. Moreover, DNA molecules can arise in wet splicing systems: The sticky ends of

5′GATCC...y...[AGATCC...y...]^{n}A
  G...y′..[TCTAGG...y′..]^{n}TCTAG5′

arise when *Bam*HI and *Bgl*II cut at the remaining sites of a molecule previously displayed above. When n is large enough to allow sufficient flexibility of such a fragment, a circular molecule can arise, as the matching overhangs at each end anneal and are ligated. This was pointed out in [2] and elaborated on in [16]. In this introductory summary of splicing theory, we treat only linear (non-circular) strings since these are the most often considered objects in classical formal language theory. The preponderance of the literature on dry splicing is concerned with linear (non-circular) strings. Consequently, in this introductory article we take the liberty of ignoring circular molecules and strings. Please see [17] and the bibliography it contains regarding dry splicing of circular strings. Complete generality would allow the initial set of DNA molecules of a wet splicing system to include both linear and circular molecules in which case we would consider the language generated to include both linear and circular molecules (as mentioned in [16]).

13.3
Dry Splicing Systems

Rather than detail the steps by which the notion of a wet splicing system evolved into the concept of a dry splicing system, we choose to present the current concept of a dry splicing system and then explain how a wet splicing system may be regarded as a biomolecular realization of a dry splicing system.

A (dry) *splicing scheme* (A, R) uses a finite set, A, as an *alphabet* of indivisible *letters* from which *strings* of letters are formed. The set of all strings of letters of finite length is denoted as A^*. We include in A^* a string of length zero. Each subset of A^* is called a *language*. A (splicing) *rule* is an ordered quadruple of strings (p, q, u, v), where each of p, q, u, v is in A^*. A splicing scheme uses a *finite* set, R, of such rules.

A rule $r = (p, q, u, v)$ allows the creation of a new string from any pair of strings having appropriate factored forms: Given strings having factored forms $wpqx$ and $yuvz$, where w, x, y, z are in A^*, r allows the formation of the (spliced or recombinant) string $r(wpqx, yuvz) = wpvz$. Let S be a finite subset of A^*. Let $R'S$ be the set of all strings s in A^* for which there is an r in R, and strings having factorizations $wpqx$ and $yuvz$ in S, for which $s = r(wpqx, yuvz)$. Let $R^0 S = S$. For each nonnegative integer n, define $R^{n+1}S$ to be the union of $R^n S$ with $R'(R^n S)$. Let R^*S be the union of all the $R^n S$, n being a nonnegative integer. This makes R^*S the iterative closure of S under the action of all the rules in R.

A (dry) *splicing system* (A, I, R) consists of a splicing scheme (A, R) and a *finite* initial subset, I, of strings in A^*. R^*I is said to be the *splicing language* generated by the splicing system (A, I, R).

{Note: The rule (p, q, u, v) is denoted in several different ways in the splicing literature. Examples that are still in use are: $p\#q\$u\#v$, $(p, q; u, v)$, and $(p, q)(u, v)$.}

Example 3. (For ease of reading we insert the symbol | in appropriate strings.) First a simple abstract example: Let $A = \{a, b\}$, $I = \{aab\}$, $r = (aa, b, a, ab)$, and $R = \{r\}$. Then $r(aab, aab) = r(aa|b, a|ab) = aa|ab = aaab$, $R'I = \{aaab\}$, and $R^1I = \{aab, aaab\}$. Since $r(aaab, aab) = r(aaa|b, a|ab) = aaaab$ and also $r(aaab, aaab) = r(aaa|b, aa|ab) = aaaab$ we have $R^2I = \{aab, aaab, aaaab\}$. Apparently (or by an induction argument), the splicing language, R^*I, is $\{a^n b | n \geq 2\}$.

Example 4. (For ease of reading, we insert blank spaces into long strings.) An abstract example that has a suggestive relationship with Example 2: $A = \{a, c, g, t\}$, I is the singleton set $\{a^9 \text{ ggatcc } a^9 \text{ agatct } a^9\}$, $r = (a, gatct, g, gatcc)$ and $R = \{r\}$. Then

$r(a^9 \text{ ggatcc } a^9 \text{ a}|\text{gatct } a^9, a^9 \text{ g}|\text{gatcc } a^9 \text{ agatct } a^9) = a^9 \text{ ggatcc } a^9 \text{ agatcc } a^9 \text{ agatct } a^9$. Thus $R'I = \{a^9 \text{ ggatcc } a^9 \text{ agatcc } a^9 \text{ agatct } a^9\}$ and $R^1 I = I$ union $R'I = \{a^9 \text{ ggatcc } a^9 \text{ agatct } a^9, a^9 \text{ ggatcc } a^9, \text{ agatcc } a^9, \text{ agatct } a^9\}$. Since $r(a^9 \text{ ggatcc } a^9 \text{ a}|\text{gatct } a^9, a^9 \text{ g}|\text{gatcc } a^9 \text{ agatcc } a^9 \text{ agatct } a^9) = a^9 \text{ ggatcc } a^9 \text{ agatcc } a^9 \text{ agatct } a^9$ and $r(a^9 \text{ ggatcc } a^9 \text{ agatcc } a^9 \text{ a}|\text{gatct } a^9, a^9 \text{ g}|\text{gatcc } a^9 \text{ agatcc } a^9 \text{ agatct } a^9) = a^9 \text{ ggatcc } a^9 \text{ agatcc } a^9 \text{ agatcc } a^9 \text{ agatct } a^9$, $R^2 S = \{a^9 \text{ ggatcc } a^9 \text{ [agatcc } a^9]^n \text{ agatct } a^9 : 0 \leq n \leq 3\}$. Hopefully, it is now clear that the splicing language $R^* I$ generated by this (A, I, R) is $\{a^9 \text{ ggatcc } a^9 \text{ [agatcc } a^9]^n \text{ agatct } a^9 \mid 0 \leq n\}$.

Not all languages can be generated by splicing systems. The language having alphabet $\{a,b\}$ and consisting of all strings that contain *exactly two* occurrences of b cannot be so generated. Neither can the language having the singleton alphabet $\{a\}$ and consisting of all strings of even length [18].

We now wish to create a dry model of the wet splicing system in Example 1. A basic decision had to be made in [2] concerning the modeling of wet splicing by dry splicing: Should we view double-stranded DNA molecules as *two mated strings* each over the alphabet {A,C,G,T} or should we view such molecules as *single strings* over the compound alphabet {A/T, C/G, G/C, T/A}? Since the classical results of formal language theory are stated for strings, rather than paired strings, I made the latter choice. This allows splicing languages to be mixed with other languages in a seamless way. For example, the first deep theorem on splicing languages asserts that they are all regular languages. {For reading this article it is enough to know that the class of regular languages is a well-understood class. It is the bottom (simplest) level of Chomsky's classical four-level hierarchy. However, if a reference is desired, see [19].} The meaning of this statement is immediately clear to all formal language theorists. What would this result become if stated in terms of double strings [20]? This decision is also the source of our focusing attention on "fully double-stranded" DNA. A DNA molecule with sticky *overhanging* ends is not a string over the compound alphabet. If we wished to allow DNA molecules that are not fully double-stranded, we could expand the alphabet by adjoining the eight extra symbols {A/-, C/-, G/-, T/-, -/A, -/C, -/G, -/T}. I preferred to regard all molecules with sticky ends as intermediate temporary molecules. Since we start with fully double-stranded molecules, for every molecule having a sticky end there is always a molecule with a complementary sticky end. Thus, ideally, a final period with ligase but no restriction enzymes present could (in theory) complete all molecules to fully double-stranded form. For ease in reading (and writing!), we rename A/T, C/G, G/C, T/A as: *a, c, g, t*, respectively. Caution: as hydrogen bonded nucleotides, A/T and T/A (hence *a* and *t*) are the same molecule – but in context they are not: *ccacc* is not *cctcc*. Now that upper case *A* is no longer being used as a letter in an alphabet, we may use it as the notation of the alphabet of our splicing system: Thus we now have $A = \{a,c,g,t\}$. Strings "live on a line" but molecules live in "three-space." This difference requires that we exert special care in setting up an (A, I, R) model of a wet splicing system. It is helpful to define a function ' from *A* onto *A*: let ' be the function from $A = \{a,c,g,t\}$ onto A defined by $a' = t, c' = g, g' = c$, and $t' = a$. Define the operation of ' on *strings* so that ' *reverses* the order, as illustrated: $(ac)' = c'a'$,

$(acg)' = g'c'a'$, $(cctgagacaa)' = a'a'c'a'g'a'g't'c'c' = ttgtctcagg$. The significance of $'$ is seen when one realizes that, when, for example, the string $s = cgagt$ represents a DNA molecule, then the string $s' = t'g'a'g'c' = actcg$ represents the SAME molecule: By rotating the molecule represented by s through $180°$ in the plane of the paper you obtain the molecule represented by s'. Thus each DNA molecule has *two distinct representations* (except those having dyadic symmetry which have only one). When we formulate a dry splicing system (A,I,R) as a model of a wet splicing system, we must be careful to list in I both forms (s and s') of each initial DNA molecule that does not have dyadic symmetry. Failure to list both representations of an initial molecule can lead to the formal model "overlooking" molecules such as those four "sometimes overlooked" molecules in Example 1 in Section 13.2.

We say that a splicing scheme (A,R) is *reflexive* if for each rule $r = (p,q,u,v)$ in R the rules (p,q,p,q) and (u,v,u,v) are also in R. We say that (A,R) is *symmetric* if for each rule $r = (p,q,u,v)$ in R the rule (u,v,p,q) is also in R. A dry splicing system that models a wet splicing system is inevitably both reflexive and symmetric. This becomes clear when one considers why a rule $r = (p,q,u,v)$ has been included in R: (p,q,u,v) has been included because there are enzymes E and F that cut DNA at subsegments pq and uv, respectively, leaving compatible sticky ends. Then F and E cut uv and pq leaving compatible sticky ends. Thus (A,R) must be symmetric. The enzyme E that cuts at the subsegment pq leaves the same sticky ends on every occurrence of pq and produces exactly the same sticky ends; consequently (p,q,p,q) belongs in R. Likewise (u,v,u,v) belongs in R and R is reflexive. Researchers studying *abstract splicing* are not obligated to assume reflexivity and symmetry. The original definition of a splicing system [2] contained reflexivity and symmetry as implicit features since the goal in [2] was to model wet splicing. I thank Gh. Păun for liberating splicing theory and simplifying its exposition by creating the definition used here which is much more convenient for theoreticians wishing to write concise and clear proofs. If one wants to refer only to dry splicing systems that are potential models of wet splicing systems one need only add the reflexivity and symmetry hypotheses and close the initial set under $'$.

The language having alphabet $\{a,b\}$ and consisting of all strings having *at most two* occurrences of b is a splicing language, but it cannot be generated by any splicing language that is either reflexive or symmetric [21, 22].

Example 5. (Modeling a wet system by a dry system). For the wet splicing system, we will take *Bam*HI and *Bgl*II as our set of enzymes and, as our initial set, the single molecule (where we again insert spaces for ease of reading):

AAAAAAAAA GGATCC AAAAAAAAA AGATCT AAAAAAAAA
T TT TT TT TT CCTAGG T TT TT TT TT TCTAGA T TT TT TT TT.

We use an alphabet $A = \{a,c,g,t\}$ where a, c, g, and t correspond to (model) A/T, C/G, G/C, and T/A, respectively. The initial molecule will be modeled by $s = a^9$ ggatcc a^9 agatct a^9. However, as we have seen, $s' = t^9$ agatct t^9 ggatcc t^9 also models this same molecule. Thus we use $t = \{s, s'\}$. That *Bam*HI and *Bgl*II cut leaving compatible sticky ends gives us the rule $r = (a, gatct, g, gatcc)$. But for R we need also (g, gatcc, a, gatct), recognizing symmetry, and both (a, gatct, a, gatct) and (g, gatcc, g, gatcc), recognizing reflexivity. Thus for R we have $R = \{(a, gatct, g, gatcc),$ $(g, gatcc, a, gatct), (a, gatct, a, gatct), (g, gatcc, g, gatcc)\}$. This dry splicing system generates precisely the set of strings that represent molecules produced by the wet splicing system.

The generative activity of any wet splicing system can be modeled by a dry splicing system in much the same manner in which we constructed the dry splicing system from the wet system in Example 5. This constitutes our explanation of how the formal language theoretic (dry) splicing formalism was created by abstracting the action of restriction enzymes and a ligase as described in Lewin's book [1]. An extensive literature on splicing theory grew quickly [4] and additional results continue to appear on a regular basis [17, 23]. Numerous doctoral theses were devoted either in whole or in part to splicing theory. In the earliest days of the study of membrane systems [24, 25], the processes taking place within individual membranes were often chosen to be splicing activities. Much of the excitement about splicing systems has now been transferred to membrane systems which have been under intense study since Gh. Păun introduced the concept in [26].

13.4
Splicing Theory: Its Original Motivation and Its Extensive Unforeseen Developments

The motive for creating splicing theory was to provide a paradigm for the representation of the generative (creative) activity of enzyme systems as they modify biologically significant polymers. My first thought was to consider homologous crossover and my second was to consider the behavior of transposons. The description of neither of these seemed to be sufficiently incisive at the time to allow a precise formalism. The cut and paste activity of restriction enzymes and a ligase appeared to be the place to start. A first *test question* to determine whether a formal theory of splicing systems says something of interest about wet splicing is: Can one decide whether a given DNA molecule can be constructed from a given finite set of DNA molecules using a given finite set of restriction enzymes and a ligase? The satisfying answer, "Yes," is a consequence of the first major theorem of splicing theory treated by several authors beginning with [20]. A rigorous demonstration and proof is given in the fundamental work of Dennis Pixton [4, 10, 27, 28]: *The language generated by each splicing system is a (constructively) regular language.* Formal language theorists recognize that this means that there is a constructible input/output machine, called a *finite automaton*, that, when reading a string as input, will specify immediately on completing the reading whether the string is or is not in the language. This provides the encouraging "Yes" above. Moreover, it follows that *each splicing language can be expressed in a closed form as a (Kleene) regular expression* [19].

The next natural question might be: Which regular languages are generated by dry splicing systems? *An algorithm for deciding whether a given regular language is a reflexive splicing language has been given by* Pixton and Goode [10, 22].

Can other schemes of enzymatic activities be formalized in such a way that algorithmic processes of formal language theory can be applied to say something of interest? This does not seem to have been a major concern of formal language theorists. The great thrust in splicing theory has been concerned with the formal analysis of the (dry) splicing systems and the languages they define. Strong new results concerning (finite) splicing systems as defined here continue to appear [17, 23]. In the future, (finite) splicing languages might blend into abstract algebra through the concept of the syntactic monoid (semigroup) of a language [22]. By replacing the finiteness requirements on I and R, many new theorems have been developed [4]. Quite early, additional structures and hypotheses were added that extended the languages generated from a subclass of the regular languages as far as the class of all recursively enumerable languages. With these adaptations and controls, formal models for universal computing were soon given [4]. I did not anticipate that such results would be generated from the splicing concept and I have been quite pleased and dazzled by this extensive literature. I give special thanks to my colleague Ron Gatterdam [18, 29] for his interest, ideas, and encouraging conversations as the splicing concept was being developed during 1984–1985.

13.5
Computing with Plasmids

Adleman's solution in 1994 [7] of an instance of the directed Hamiltonian path problem used the self-assembly of many *single strands* of DNA into double-stranded form. This gave birth to DNA computing. An irritating feature of single-stranded DNA is that the strands can form partial matchings having unmatched "bumps." Moreover a strand may "double over" and create a partial matching with itself. This has led to an extensive literature reporting sets of single strands that provide the least troublesome mismatches among themselves [30]. I took an alternate approach to DNA computing by *avoiding* single-stranded DNA with its propensity for mismatchings. My approach was to use only one molecular variety of *double-stranded* DNA (in many copies) and to "write on" the molecules in massive parallelism with enzymes. In this way each molecule becomes a scratch pad.

Satoshi Kobayashi gave me the good news that one can easily recircularize previously cut circular molecules without worrying that the cut circles will link into longer linear molecules. This made it attractive to use circular DNA molecules (plasmids) as the scratch pads. In conversations with Peter Kaplan, I was encouraged to carry out an alternate treatment of the solution of the instance of the maximal clique problem reported by Kaplan and coworkers in [31]. While visiting Gregorz Rozenberg at Leiden University, Herman Spaink agreed to have our first aqueous computation carried out in his laboratory at the Leiden Institute of Molecular Plant Science. We treated the same instance of the maximal clique problem (dually, the maximum independent set problem) that was treated in [31]. The result of this effort has appeared as [32]. On returning to Binghamton, Susannah Gal agreed to solve an instance of the Boolean satisfiability problem (SAT) in her laboratory.

This computation was designed by Gal in collaboration with our visitor, Masayuki Yamamura, from Tokyo Institute of Technology. This result appeared in [33, 34]. Christiaan Henkel, again in Herman Spaink's laboratory, solved an instance of the minimal dominating set problem in the aqueous style and developed a major new reading procedure using the generation of peptides followed by application of mass spectroscopy [35, 36].

All three of these computations treated only *small* problem instances that provided "proof of concept" examples. Here I will discuss even *tinier* instances of the first two algorithmic problems in the belief that they will allow you to quickly pick up the essence of the aqueous computing concept. If your interest continues, please see [32–34], and [35, 36] which include all the chemical details in solving slightly larger instances of these problems. These are not the only ways one can write on DNA. Yamamura has illustrated in [37] a process of writing by allowing short sequences of PNA to interfere with the hydrogen bonding of bases in double-stranded DNA. My wish to write with a tunable laser on (modified) DNA has been abandoned in favor of writing by xeroxing onto transparencies [38].

Since fluid memory, as used in aqueous computing, may be valuable in other contexts, I have treated it separately in Section 13.6 which is preliminary to our detailed examples of aqueous computing given in Section 13.7.

13.6
Fluid Memory

Suppose that one has a test tube T containing water (or an appropriate buffer) in which are dissolved molecules in the structure of which information has been encoded. Suppose that each molecular variety in the memory is present in many copies (perhaps a femtomole or more of each variety). We can make two copies of the entire memory by simply pouring half the contents of tube T into a second tube T' and half into a third tube T''. That T' and T'' contain the same set of molecular varieties is of extremely high probability because of diffusion and the huge number of molecules of each variety in T. If we wish to merge two memories held in tubes T' and T'', we need only pour T' and T'' into T. *Thus replication and merging are one-step operations.* The plasmid computing we will describe makes use of these fascinating features of fluid memory. Note that such memory uses no addresses. We will be computing by adding enzymes to test tubes containing DNA molecules each of which is interpreted as a carrier of information. When we add an enzyme to a tube, it will act in an ultraparallel manner on all the DNA molecules that are susceptible to its action. Our computations will begin with a tube T containing only one single variety of DNA molecules. This is our tube of "scratch pads." We will pour one half of tube T into each of tubes T' and T'' and we will then introduce distinct enzymes into the two tubes. We will then pour tube T' and T'' back into T. T may now contain two distinct molecular varieties. This process of subdividing, treating separately, and reuniting can be repeated several times resulting in T finally containing as many as 2^n distinct molecular varieties where n is the number of times the process has been carried out. We have called our method of computing "aqueous computing" to recognize that the most fundamental feature of our method is *the use of these two remarkable features of fluid memory*. An exposition in Japanese of fluid memory has been provided by Yamamura et al. [39].

An aqueous computation must be planned so that one is confident that the final tube T will have a content that provides the solution of the problem considered. Moreover, one must have planned a procedure for reading this content. In the next section, we sketch aqueous solutions of instances of three standard algorithmic problems. Four distinct reading schemes are given. Biochemists will know, and nonchemists should be warned, that some standard laboratory procedures which must be carried out to realize the steps in our computations are being ignored here. {A single example: after the action of an enzyme on the DNA in a tube T is complete, the enzyme and the DNA in T must be separated and only the appropriate DNA retained.}

13.7
Examples of Aqueous Computations

Example 6. A SAT problem [40]. Let p and q be Boolean (logical) variables which can take only the values True (T) and False (F). Let p' and q' be negations of p and q. Does there exist a truth setting for the variables p and q for which the each of the three propositions $p \vee q$, $p' \vee q$, $p' \vee q'$ evaluates to T?

Of course, this example is so tiny that one sees immediately that there exists such a truth setting and that $p = F$ and $q = T$ is the unique such setting. However, our purpose here is to encode this problem in DNA and solve it biochemically. We do this in such a manner that the reader will see how to solve similar SAT problems with n variables and m propositions and in essentially $n + m$ steps. Choose a standard cloning plasmid C. {A cloning plasmid is a circular DNA molecule of, perhaps 2000 bp. It incorporates code for important standard uses. It can replicate when inserted in a bacterium. It contains code that generates at least one antibiotic resistance factor. It contains a "multiple cloning" site which is a collection of, perhaps 20, distinct unique sites at which specified restriction enzymes can cut.} Choose four enzyme pairs (P, M) (P', M') (Q, N) (Q', N') where P, P', Q, Q' are restriction enzymes whose sites occur in the multiple cloning site of the plasmid C, and M, M', N, N' are the corresponding methylase enzymes having the same sites as P, P', Q, Q', respectively. These pairs are chosen so that, when the methylase of a pair is applied to the cloning plasmid C, its corresponding restriction enzyme will no longer cut C. Let the relevant sites at which P, P', Q, Q' act be denoted as s, s', t, t', respectively. {If, for example, P is chosen to be *Bam*HI, then we would have $s = $ ggatcc.} Let U be a test tube containing many copies (perhaps a picomole) of the plasmid C. We will regard C as an information-containing molecule. The information is contained at the four sites, s, s', t, t' at which the restriction enzymes can act. We will be modifying the molecules C during our computation, but we will continue to call the molecules C even after they have been the altered. A site will be considered to represent True (T) if its associated restriction enzyme will cut at the site and to represent False (F) otherwise. In its initial stage, C is read as a representation of TTTT, that is, $p = T$, $p' = T$, $q = T$, $q' = T$.

Our first task is to eliminate the two logical contradictions, $p = p' = T$ and $q = q' = T$, while producing all the versions of C that represent only logically consistent truth settings. This can be done in a number of steps equal to the number of variables, two in the present case, as follows: U initially contains only plasmids that read TTTT. Pour half of U into each of two test tubes U' and U''. Add methylases M to tube U' and M' to U''. {U' now contains only plasmids C that read FTTT and U'' contains only plasmids C that read TFTT.} Pour U' and U'' into tube U. {U now contains two plasmids reading FTTT and TFTT.} Pour U into tubes U' and U''. Add methylase N to U' and N' to U''. {U' now contains plasmids reading FTFT and TFFT and U'' contains FTTF and TFTF.} Pour U' and U'' into tube U. {U now contains FTFT, TFFT, FTTF, and TFTF.} Having treated each logical variable, we now have in U the complete set of all logically consistent truth settings for our logical variables. Notice that the number of these uniform compound steps is the number of variables.

Our second task is to eliminate from U all those plasmids that fail to satisfy the given propositions. We treat the propositions one by one starting with $p \vee q$: We must cut each plasmid for which *both $p' = T$ and $q' = T$*. To do this, pour half of U into tube U' and half into U''. Add enzyme P' to U' and enzyme Q' to U''. {U' now contains only plasmids C that read TFFT and TFTF and U'' contains only FTTF and TFTF.} Pour U' and U'' into U. {U now contains TFFT, TFTF, and FTTF.} We have ensured that $p \vee q$ evaluates to T for all plasmids in U. Treat the next proposition, $p' \vee q$: We must cut each plasmid for which *both $p = T$ and $q' = T$*. Pour half of U into each of U' and U''. Add enzyme P to U' and Q' to U''. {U' now contains only FTTF but U'' contains both TFTF and FTTF.} Pour U' and U'' into U. {U now contains FTTF and TFTF.} We have now ensured that both $p \vee q$ and $p' \vee q$ evaluate to T for all plasmids in U. Treat the final proposition $p' \vee q'$: We must cut each plasmid for which *both $p = T$ and $q = T$*. Pour half of U into each of U' and U''. Add P to U' and Q to U''. {U' now contains only FTTF and U'' contains no plasmids at all.} Pour U' and U'' into U. U now contains all plasmids (if any) that satisfy each of the propositions. {U contains only FTTF.} A test for the presence of plasmids provides the answer to the yes/no question asked. Thus "plasmids present" –> Yes; "no plasmids present" –> No. {In the present example: Yes.} Notice that the number of these uniform steps is the number of propositions. Thus this algorithm is linear in the length of the problem description. {Remark: Our propositions contained only two literals. With three literals in a clause, one simply pours U into tubes U', U'', and, say, V.}

Although the classical SAT problem asks only the yes/no question, when there is a solution it is natural to ask for a truth setting for which all the given propositions evaluate to True. In general, there may be several truth settings that satisfy all the given propositions. Here there is only one solution, and when there is only one solution, the following Insert, which uses entirely standard microbiology techniques, can be skipped. {Insert: Absorb the plasmids into competent bacteria. Plate the bacteria on a gel that contains the antibiotic for which the plasmid has the resistance factor. Choose one of the resulting clones and collect the plasmids from that clone. All the plasmids from a single clone will be of the same variety (or so biologists have insisted to me).} Now we have a single plasmid variety C to work with. Let C be in solution in a tube U. Choose four tubes $U(p)$, $U(p-)$, $U(q)$, $U(q-)$, and pour a fourth of U into each of the four tubes. Cloning plasmids have an auxiliary site for a restriction enzyme that is well away from the multiple cloning site. Let us call this auxiliary enzyme K. Add K to each of the four tubes cutting and *linearizing* all the plasmids. Now add P to $U(p)$, P' to $U(p-)$, Q to $U(q)$, and Q' to $U(q')$. What will the result be? Each of the sites s, s', t, t' on our plasmid C will be cut in two in its appropriate tube if the site represents T and will not be cut if it represents F. Thus we complete the production of an example truth setting by placing the contents of our four tubes in separate columns of a gel and doing a gel separation. In columns associated with sites representing T, there will be two bands on the gel. Columns associated with sites representing F will provide only a single band. Thus in the case of our present problem, we will see in our four columns: Under $U(p)$ – one band telling us p is False. Under (p') – two bands us that p' if True. Under $U(q)$ – two bands telling us that q is True. Under $U(q')$ – one band telling us that q' is False. Certainly we could have merely used $U(p)$ and $U(q)$ – as the other information is redundant. For those of us who actually carry out such experiments, it is delightfully reassuring to have confirming redundancy! There are so many things that can go wrong in laboratory work.

The use of methylase enzymes in computing as illustrated here has been explored by Gal and coworkers in [41]. In fact, our first complete solution of an instance of a SAT problem done in Gal's laboratory did not use methylation. The function of each methylase enzyme in the procedure above was simply to "lock" its site against the possibility of its being cut by its companion restriction enzyme. Methylation was simply our tool for altering a site reading T so that forever after the site reads F. There are other ways of "locking" a site against cutting by its restriction enzyme. In our first treatment of an instance of the SAT problem [33, 34], we locked sites by altering the DNA sequence slightly at the site. Here is a specific example: Suppose in (P,M) above that P is *Bam*HI and M is its corresponding methylase. The site at which they act is $s =$ ggatcc. Here is the replacement of M used in [33, 34]: We used three steps to lock s against P. First, we cut the plasmid C with P linearizing C and leaving 5' overhangs 5'-GATC at each end. Second, we added DNA polymerase which added the 4 bp 3'-CTAG under the 5' overhangs to fill out each end of the molecule to produce the fully double-stranded form having the double-stranded form *gatc* making each end blunt. Third, we added ligase to recircularize the linear molecules. But notice that where the original site was *ggatcc*, the same region of C has now been lengthened by 4 bp and reads *ggatcgatcc*. Thus there is no longer a site for *Bam*HI. This three-step procedure has locked C against the action of *Bam*HI fully as effectively as applying the methylase M.

Example 7. A maximal independent subset problem [40]. Let G be an undirected graph having vertex set $V = \{u,v,w\}$ and undirected edges (u,v) and (v,w). Determine the cardinality of a largest independent subset of V.

Of course, this example is so tiny that one sees immediately that the required cardinal is 2 and in fact that the subset $\{u,w\}$ is the only independent subset of that cardinal. However, our purpose here is to encode this problem in DNA and solve it biochemically. Choose four restriction enzymes E, E', E'', K that cut at distinct sites s, s', s'', and k, respectively. Choose a standard cloning plasmid and replace its multiple cloning site by the double-stranded DNA sequence $ka^9 sc^{30} sa^9 s' g^{30} s' a^9 s'' t^{30} s'' a^9 k$. This inserted subsequence will be the information-containing region of the plasmid and we will denote it as INFO. The plasmid itself will be denoted by P. Let U be a test tube containing P in solution. We will be modifying INFO by "writing on" it using E, E', or E'', but we will continue to call it INFO even after altering it. We will regard INFO, in each of its forms, as representing a subset of V. Writing on INFO will involve removing one of the subsegments c^{30}, g^{30}, or t^{30}. The presence of c^{30}, g^{30}, or t^{30} in any state of INFO will mean that u, v, or w, respectively, is contained in the subset represented by INFO. Thus initially INFO denotes the entire set $\{u,v,w\}$. It is convenient to represent subsets of V using their characteristic functions. Thus INFO in its current state is now represented as 111. How can we represent 011? To tube U add enzyme E. This cuts P into a long linear segment (perhaps 2000 bp) terminating at each end with matching halves of the site s, and a tiny segment consisting of c^{30} with matching halves of s at each end. We recircularize by adding ligase. The ligase recircularizes the long segment into a molecule now having as INFO $ka^9 sa^9 s' g^{30} s' a^9 s'' t^{30} s'' a^9 k$ which we read as 011, that is, as $\{v,w\}$. The $c^{30}s$ fragment is too small to be of further concern. I trust that it is now clear how to express each of the subsets of V by "writing zeros" in the appropriate places in INFO by using the appropriate enzymes. So how do we solve the problem?

Let U be a test tube containing many copies (perhaps a picomole) of our special plasmid P. Each molecule in P represents 111. We take up the first edge $\{u,v\}$ and realize that no subset that contains both u and v is independent. Any independent subset must either fail to contain u or fail to contain v. Thus we must eliminate each molecule from U that contains *both u and v*. We pour one half of U into a tube U' and the other half into tube U''. We introduce enzyme E into U' and E' into U''. We now introduce a ligase into both U' and U'' recircularizing all cut molecules in both U' and U''. $\{U'$ now contains 011 and U'' contains 101.$\}$ We pour U' and U'' into U. $\{U$ now contains 011 and 101$\}$. We take up the next edge $\{v,w\}$ and treat it similarly. We pour half of U into each of U' and U''. We add E' to U' and E'' to U''. We add ligase to U' and to U''. $\{U'$ now contains 001 and 101. U'' now contains 010 and 100.$\}$ Pour U' and U'' into U. $\{U$ now contains 001, 101, 010, and 100$\}$. Since we have now treated all the edges of the graph, we know that U contains only independent subsets and must contain all independent subsets of maximum cardinal. The solution to our problem is in U, but we must separate it out now from the nonsolutions.

Add the enzyme K to U which will cut all the plasmids into a comparatively short segment containing the information content of each plasmid and an *extremely* long segment (perhaps 2000 bp) consisting of the remainder of the plasmid. Make an appropriate gel separation of the contents of U. The *extremely* long segments will remain out of our way very near the negative end of the gel. The comparatively short information-containing segments will spread out according to their lengths. The length in base pairs of the longest of these information-containing segments can be read from a calibration column on the gel. From the length, we then calculate how many of the three subsegments c^{30}, g^{30}, and t^{30} remain in the segments of greatest length. This number will be the cardinal of the largest independent subset. {In this example, there will be two relevant bands in the gel. The band of longer segments will contain only 101 and the other band will contain 001, 010, and 100.} The longer segments will contain two of the 30-bp intervals telling us that the maximum cardinal that we seek is 2.

If one wishes to exhibit an example of an independent set of maximum cardinals, then one can excise from the gel the band containing the subsegments of maximum length. In the fortunate case in which there is only a single maximum independent subset (as is the case of our example), we may sequence the molecule to determine the specific set represented. Otherwise, we can cut the solution band from the gel and reinstall these linear molecules back into their original plasmid form (of perhaps 2000 bp). Pass these plasmids through bacteria and plate as in the discussion in Example 6. Choose one clone from the plate and retrieve the plasmids. Cut out the INFO segments using K and sequence the resulting unique molecular variety.

Example 8. A minimum dominating subset problem [40]. Let G be an undirected graph having vertex set $V = \{u,v,w,x,y,z\}$ and undirected edges (u,w), (v,w), (w,x), (x,y), and (x,z). Determine the cardinality of the smallest dominating subset of V.

This is precisely the instance of the minimal dominating subset that was treated by Christiaan Henkel in [35, 36]. Recall the subset D of V is a dominating subset if, for every p in V, either p is in D or there is a d in D for which (p,d) is an edge of G. One can easily see that {w,x} is the smallest dominating subset of G and therefore the required cardinal is 2. We will only comment on the encoding of this problem in DNA and the intermediate steps in solving the problem. Our purpose here is to discuss the novel methods given by Henkel for reading the solution of this problem and potentially many others.

The solution of this problem was begun by modifying a cloning plasmid by replacing the multiple cloning site by a segment built from six inserts, one for each vertex. We will call this segment INFO. Each of the six inserts was bounded by a pair of restriction enzyme sites just as in Example 7. By choosing the inserts to have (roughly) the same length, the present problem was solved in a manner rather similar to the way Example 7 was solved with the solution read from a gel separation. The solution was given by the longest of the INFO segments. However, this was not the major aspect of Henkel's work. The code for the six inserts was chosen with great care: Each of these inserts held the genetic code for 3–5 amino acids! Moreover, the plasmid incorporated an initiation site for RNA polymerase just upstream of INFO. RNA transcriptions of the INFO segments were made by adding RNA polymerase. With ribosomes present, polypeptides were then translated from the RNA transcripts. The heaviest of these polypeptides encoded the solution. The heaviest was determined using mass spectroscopy, which also allowed the specific amino acids present to be determined. Thus mass spectroscopy gave not only the cardinal number 2 but also allowed the determination that $\{w,x\}$ was the minimal dominating set.

Finally, the amino acids encoded by the six inserts were also chosen so that they could be *recognized and distinguished by antibodies*. Thus Henkel's work illustrated three quite different ways to read solutions in aqueous computing: gel separation, antibody recognition, and mass spectroscopy.

13.8
Final Comments about Computing with Biomolecules

The possibility of "scaling up" any method of wet biomolecular computing to compete successfully with standard computers seems unlikely as was explained by Hartmanis [42] already in 1995. If only small computations are possible, then one must look for places where small computations can be important and where standard computers cannot go, such as into living organisms. We can only hope that the ideas developed in biomolecular computing will have value in other contexts. There are three such contexts that I regard as hopeful: (i) computing inside (or among) living cells, (ii) extending the understanding of biology by viewing naturally occurring biological processes as computations, and (iii) advancing nanotechnology.

(i) One goal here is to develop a cell that could live in the human digestive tract and, on encountering a toxin, could compute an appropriate output (probably a protein) to defeat the source of the toxin. This explains our special interest in Henkel's computation [35, 36] using protein output. See Ratner and Keinan [43] and their bibliography for a plausible beginning of the sort of computing that may be required.

(ii) Landweber and Kari began in [44] the application of computation theory to the understanding of a remarkably complex biomolecular behavior that takes place in ciliates. This has developed into a substantial body of research stimulating to both computer scientists and ciliate biologists [45]. A recent valuable introduction to this work has been given by Daley and Domaratzki [46] who include splicing systems as a tool in their exposition.

(iii) The progression from biomolecular computing to the building of structures on the nanoscale using biomolecules (mainly DNA) is perhaps the most significant development from DNA computing. Recall that Adleman's 1994 solution of a computational problem using DNA [7] involved such a construction. With Adleman's guidance, Winfree and Rothemund immediately became leading contributors to the resulting new *bionanotechnology*. Nadrian C. Seeman and his students had been carrying out stunning biomolecular constructions for several years [47] before DNA computing was initiated. He has joined the DNA computing community as a leading member.

There are now international societies with annual meetings and published proceedings that have grown from the sequence of "Workshops on DNA-Based Computers" that began in 1995 in Princeton. This first "Workshop" was given as a setting for Adleman to introduce his 1994 result to a wider community of enthusiasts. The "International Society for Nanoscale Science, Computation, and Engineering (ISNSCE)," founded in 2004 by N.C. Seeman, encompasses two major organizations: "Foundations of Nanoscience (FNANO)" and the renamed version of the original "Workshop" now called *"DNA Computing and Molecular Programming."* Yet another annual international conference began in 2010: "Nanomedicine." All developments in *bionanotechnology* are within the purview of each of these conferences.

References

1. Lewin, B. (1983) *Genes*, John Wiley & Sons, Inc., New York (but many newer editions have appeared!).
2. Head, T. (1987) Formal language theory and DNA: an analysis of the generative capacity of specific recombinant behaviors. *Bull. Math. Biol.*, **49** (49), 737–759.
3. Păun, G., Rozenberg, G., and Salomaa, A. (1996) Computing by splicing. *Theor. Comput. Sci.*, **168**, 321–336.
4. Păun, G., Rozenberg, G., and Salomaa, A. (1998) *DNA Computing – New Computing Paradigms*, Springer, Berlin.
5. Yokomori, T. and Kobayashi, S. (1995) DNA evolutionary linguistics and RNA structure modeling: a computational approach. IEEE Conference on Intelligence in Neural and Biological Systems, Herndon, Washington, pp. 38–45.
6. Siromoney, R., Subramanian, K.G., and Dare, V.R. (1992) in *Parallel Image Analysis*, LCNS, Vol. **654**, Springer-Verlag, pp. 260–273.
7. Adleman, L. (1994) Molecular computations of solutions of combinatorial problems. *Science*, **266**, 54–61.
8. Adleman, L. (1998) Computing with DNA. *Sci. Am.*, **297**, 54–61.
9. Head, T. (2000) in *Pattern Formation in Biology, Vision and Dynamics* (eds A. Carbone, M. Gromov, and P. Prusinkiewicz), World Scientific, Singapore and London, pp. 325–335.
10. Head, T. and Pixton, D. (2006) in *Recent Advances in Formal Languages and Applications*, Studies in Computational Intelligence, Vol. **25** (eds Z. Esik, C. Martin-Vide, and V. Mitrana), Springer, pp. 119–147.
11. Head, T. and Gal, S. (2006) in *Nanotechnology: Science and Computation* (eds J. Chen, N. Jonoska, and G. Rozenberg), Springer, pp. 321–331.
12. Zizza, R. Curator of: Scholarpedia – Splicing Systems, Current.
13. Laun, E. and Reddy, K.J. (1997) in *Proceedings of the 3rd DIMACS Workshop on DNA Based Computers*, University of Pennsylvania, Philadelphia, pp. 115–126.
14. Heng, F.W. Sarmin, N.H. Wahab, M.F.A. Rashid, N.A.A. (2007) Modeling of splicing system in DNA. *International Conference on Mathematical Sciences* 2007, Bangi-Putrajaya, Malaysia pp. 712–718.
15. Goode E. and Pixton, D. (2004) in *Aspects of Molecular Computing*, LCNS, Vol. **2950** (eds N. Jonoska, G. Păun, and G. Rozenberg), Springer, pp. 189–201.
16. Head, T. (1992) in *Lindenmayer Systems – Impacts on Theoretical Computer Science, Computer Graphics, and Developmental Biology* (eds G. Rozenberg and A. Salomaa), Springer-Verlag, Berlin, pp. 371–383. (Also in (1992) *Nanobiology*, **1**, 335–342).

17. Bonizzoni, P., De Felice, C., and Zizza, R. (2010) A characterization of (regular) circular languages generated by monotone complete splicing systems. *Theor. Comput. Sci.*, **48**, 4149–4161.
18. Gatterdam, R. (1989) Splicing systems and regularity. *Int. J. Comput. Math.*, **31**, 63–67.
19. Kozen, D.C. (1997) *Automata and Computability*, Springer-Verlag, New York.
20. Culik, K. II and Harju, T. (1991) Splicing semigroups of dominoes and DNA. *Discrete Appl. Math.*, **31**, 261–277.
21. Goode, E. (1999) *Constants and Splicing Systems*. PhD thesis. Binghamton University.
22. Goode, E. and Pixton, D. (2006) Recognizing splicing languages: syntactic monoids and simultaneous pumping. *Discrete Appl. Math.*, **155**, 989–1006.
23. Bonizzoni, P. and Jonoska, N. (2011) in *Developments in Language Theory*, LCNS, Vol. **6795** (eds G. Mauri and A. Leporati), Springer, pp. 82–92.
24. Păun, G. (2002) *Membrane Computing – An Introduction*, Springer, Berlin.
25. Frisco, P. (2009) *Computing with Cells – Advances in Membrane Computing*, Oxford University Press.
26. Păun, G. (2000) Computing with membranes. *J. Comput. Syst. Sci.*, **61**, 108–143.
27. Pixton, D. (1996) Regularity and splicing systems. *Discrete Appl. Math.*, **69**, 101–124.
28. Pixton, D. (2000) Splicing in abstract families of languages. *Theor. Comput. Sci.*, **234**, 135–166.
29. Denninghoff, K.L. and Gatterdam, R.W. (1989) On the undecidability of splicing systems. *Int. J. Comput. Math.*, **27**, 133–145.
30. Arita, M. (2004) in *Aspects of Molecular Computing*, LCNS, Vol. **2950** (eds N. Jonoska, G. Păun, and G. Rozenberg), Springer, pp. 23–35.
31. Ouyang, Q., Kaplan, P.D., Liu, S., and Libchaber, A. (1997) DNA solution of the maximal clique problem. *Science*, **278**, 446–449.
32. Head, T., Rozenberg, G., Bladergroen, R., Breek, C.K.D., Lomerese, P.H.M., and Spaink, H. (2000) Computing with DNA by operating on plasmids. *Bio Syst.*, **57**, 87–93.
33. Head, T., Chen, X., Nichols, M.J., Yamamura, M., and Gal, S. (2002) in *DNA Computing – 7th International Workshop on DNA-Based Computers*, LNCS, Vol. **2340** (eds N. Jonoska and N.C. Seeman), Springer, pp. 191–202.
34. Head, T., Chen, X., Yamamura, M., and Gal, S. (2002) Aqueous computing: a survey with an invitation to participate. *J. Comput. Sci. Tech.*, **17**, 672–681.
35. Henkel, C. (2005) *Experimental DNA computing*. Doctoral Thesis. University of Leiden, The Netherlands.
36. Henkel, C., Bladergroen, R., Balog, C., Deelder, A., Head, T., Rozenberg, G., and Spaink, H. (2005) Protein output for DNA computing. *Nat. Comput.*, **4**, 1–10.
37. Yamamura, M., Hiroto, Y., and Matoba, T. (2002) in *DNA Computing – 7th Workshop, June 2001*, LCNS, Vol. **2340** (eds N. Jonoska and N.C. Seeman), Springer, Berlin, pp. 213–222.
38. Head, T. (2011) Computing with light: toward parallel Boolean algebra. *Int. J. Found. Comput. Sci.*, **22**, (2011), 1635–1637.
39. Yamamura, M., Head, T., and Gal, S. (2000) in *Genetic Algorithms*, vol. **4** (ed. H. Kitano), Sangyo-Tasho Pub.Co.Ltd pp. 49–73. (in Japanese).
40. Garey, M.R. and Johnson, D.S. (1979) *Computers and Intractability – A Guide to the Theory of NP-Completeness*, W.H. Freeman, San Francisco.
41. Gal, S., Monteith, N., Shkalim, S., Huang, H., and Head, T. (2007) in *Current Developments in Mathematical Biology*, Series in Knots and Everything, Vol. **38** (eds K. Mahdavi, R. Culshaw, and J. Boucher), World Scientific, pp. 1–14.
42. Hartmanis, J. (1995) On the weight of a computation. *Bull. Eur. Assoc. Theor. Comput. Sci.*, **55**, 136–138.
43. Ratner, T. and Keinan, E. (2009) in *Algorithmic Bioprocesses* (eds A. Condon, D. Harel, L.N. Kok, A. Salomaa, and E. Winfree), Springer-Verlag, Berlin, pp. 505–516.
44. Landweber, L.F. and Kari, L. (1999) The evolution of cellular computing: nature's

solution to a computational problem. *BioSystems*, **52**, 3–13.
45. Ehrenfeucht, A., Harju, T., Petre, I., Prescott, D., and Rozenberg, G. (2004) *Computation in Living Cells: Gene Assembly in Ciliates*, Springer, Berlin.
46. Daley, M. and Domaratzki, M. (2009) in *Algorithmic Bioprocesses* (eds A. Condon, D. Harel, L.N. Kok, A. Salomaa, and E. Winfree), Springer-Verlag, Berlin, pp. 117–137.
47. Chen, J. and Seeman, N.C. (1991) The synthesis from DNA of a molecule with the connectivity of a cube. *Nature*, **350**, 631–633.

14
Development of Bacteria-Based Cellular Computing Circuits for Sensing and Control in Biological Systems

Michaela A. TerAvest, Zhongjian Li, and Largus T. Angenent

14.1
Introduction

The term *"biocomputing"* encompasses many different types of unconventional computing, including DNA, RNA, and enzyme computing and others. One of the newest subareas of biocomputing is cellular computing, or the use of whole, live cells for computation. Cellular computing is currently in development in both eukaryotic and bacterial systems. Here we will discuss the current state and potential of cellular computing in bacterial systems. We will focus on the current implementations of bacterial cellular computing, which we will refer to as *"cellular computing circuits"* (i.e., bacterial strains engineered to perform basic computing functions, such as logic operations). Cellular computing circuits have been created to perform several basic computing functions through the use of engineered genetic circuits (i.e., the engineered gene sequence that confers computational functionality on the cell).

While cellular computing circuits can be noisier and more difficult to control than enzymatic or DNA-based computing systems, they also have several advantages. Whole, live cells are more resilient to environmental damage than pure enzymes or DNA because of their ability to maintain intracellular conditions and repair themselves. Cellular computing circuits also have the advantage that they reproduce on their own, and a bacterial culture is much less expensive to grow than enzyme purification or DNA synthesis. Another benefit of cellular computing circuits is that they are capable of self-optimizing if given the proper selective pressure – something that DNA and enzymes cannot do because they do not reproduce or mutate. This has been demonstrated by directed evolution of cellular computing circuits, which can turn a nonfunctional genetic circuit into a functional one using bacterial mutation and reproduction.

Directed evolution has been used successfully to optimize a genetic circuit in a study by Yokobayashi *et al.* [1]. Previously, Weiss [2] constructed a nonfunctional A IMPLY B gate (Table 14.1) using a modified lac repression system in *Escherichia coli*. In the initial study, the cellular computing was optimized by rational design using computer models to predict favorable mutations and testing each one in the laboratory. Yokobayashi *et al.* [1] used a simpler approach to functionalize

Table 14.1 Cellular computing circuit logic gate operations.

Component	Function	References
AND gate	True when A and B are true	[3–7]
NAND gate	True except when A and B are true	[4]
OR gate	True when A, B, or both are true	[4]
NOR gate	True when neither A nor B is true	[4]
XOR gate	True when A or B but not both are true	[4]
A gate	True when A is true (filter)	[4]
NOT A gate	True when A is not true (inverter)	[4]
A IMPLY B gate	True when A is true and B is false	[4]
A NIMPLY B gate	True except when A is true and B is false	[4]
TRUE gate	Always true	[4]
EQUAL gate	True when A and B are equal	[4]

the original nonfunctional strain. Instead of using computer models, the authors simply introduced random mutations into the strain and screened for mutants with the desired logic output. Several such mutants were discovered, and sequencing revealed more than one pathway to functionalize the genetic circuit. This is clearly a powerful advantage of cellular computing circuits and will likely become an important part of designing these systems in the future.

Recombination can also be used for self-optimization of cellular computing circuits, and this has been demonstrated by the success of a cellular computing circuit created by Baumgardner *et al.* [8]. This strain contained the genes required for fluorescent protein synthesis, but these genes were mixed up so that they could not function until specific recombination events had occurred. These recombination events were induced by flagging the mixed up DNA region with recombination sites. On growing this strain, the authors found colonies that had successfully produced fluorescence as the output signal by recombination during growth. This system was conceptualized as a Hamiltonian-path solver (i.e., it solves implementations of the traveling salesman problem) and the authors suggested uses for cellular computing circuits in solving problems similar to those for silicon-based computers. However, owing to exponentially increasing processing speeds in silicon computing, it is not likely that biocomputing will ever be competitive in processing speed or versatility. For this reason, we will focus on existing cellular computing circuits and their possible application in biosensing and in controlling biological systems.

Even for simpler computing functions, such as biosensing, a detailed understanding and manipulation of biological systems is necessary. Whole cells are much more complex than enzyme or DNA sequences and in engineering new functions there are often unintended consequences. This is why researchers have mainly used the relatively simple system of *E. coli* with fluorescent protein in cellular computing circuits. Introducing exogenous pathways that do not normally occur in *E. coli* greatly decreases the chance that the genetic circuit will interfere with normal cellular processes, or vice versa. In addition, fluorescent protein is an easily detectable output that does not interfere with cellular processes. However, our work has demonstrated that working within this narrow system is not necessary for creation of functional cellular computing circuits. We will use our recent work as a case study to explain how native bacterial regulation and function can be used in cellular computing circuits when genetic engineering is combined with rational system design to reduce complexity.

14.2
Cellular Computing Circuits

14.2.1
Genetic Toolbox

14.2.1.1 Engineered Gene Regulation
Genetic modification of bacteria is an important tool in the development of cellular computing circuits. A common modification approach has been the use

of inducible promoters (i.e., promoters that can be turned on and off by a specific chemical or environmental signal) to control expression of an output behavior. A wide variety of these promoters exist in nature and may respond to environmental signals, such as pH, temperature, sugar, pollutants, and light [9]. Extracting useful inducible promoters from wild-type bacteria and recombining them in novel ways has resulted in many regulatory networks for controlling cell behaviors. The different types of genetic circuits that have been created, thus far, have been comprehensively reviewed by Voigt [9]. The resulting engineered genetic circuits can control different functions in the bacterium, including the production of chemicals or behaviors, such as cell death. However, cellular computing circuits have not reached application yet, and those developed have mostly used fluorescent protein for the output signal as a proof of concept. Fluorescent protein has, indeed, been a useful output signal because it is simple to detect and does not interfere with normal cellular processes.

There are a few exceptions to this, and a striking example is a set of *E. coli* strains designed by Anderson *et al.* [10] to detect and invade cancer cells. These strains detect one input (arabinose, homoserine lactone (HSL), or oxygen) and link that output with invasin expression, which allows the cells to invade cancer cell lines. Another example is a population control strain created by You *et al.* [11]. This strain could detect population density and induce production of a killing protein to reduce population when the density became too high. These designs are clearly application-driven, but have limited utility because they are capable of detecting only one input, and are not yet optimized for real-world applications.

Increased complexity and functionality can be gained by combining different promoters in creative ways (i.e., to control not only cell functions, but also each other). This opens the possibility for complex Boolean logic function to simultaneously process multiple inputs. For example, an *E. coli* strain that could count three inputs was created by combining three inducible promoters to control expression of fluorescent protein [12]. This strain counts pulses of three different inducers (arabinose, tetracycline, and IPTG) and responds by producing fluorescent protein only after receiving these three signals in the correct order. The three promoters were combined using three single invertase memory modules in series. This allowed fluorescent protein production only after DNA inversion steps, which were controlled by the inducible promoters, and, because of their arrangement in series, the signals work only in the correct order. Combination of the inducible promoters with the single invertase memory modules demonstrates that combining different types of regulation can be useful for creation of complex genetic circuits. Using the same approach, the authors also created a strain that responded to three pulses of a single inducer (arabinose). This is in interesting implementation because it could be used to monitor pulses of a single substance, but the caveat is that the pulse length must be strictly controlled to avoid fluorescent protein expression after one long pulse. These counters were a successful proof of concept for counting function in bacterial cells; however, this system is subject to the stochastic nature of biology and the different behavior of each individual bacterial cell. Discrepancies between model predictions and the counters' performance show that there are still issues in obtaining a strain that performs as expected, and in unison. Bacterial communication (e.g., quorum sensing) has, thus far, been useful in partially overcoming this obstacle.

14.2.1.2 Quorum Sensing

One of the most widespread forms of bacterial communication is quorum sensing. Quorum-sensing communication can be used for synchronization of one population or communication among populations in cellular computing circuits, and both these uses have already been demonstrated. Synchronization of a population increases the quality of the output signals by causing all the individual cells to act in unison, resulting in a higher signal maximum and lower signal background. Communication among different populations can be used to link cellular computing circuits together to create more complex functions [3, 4]. Quorum-sensing systems are applicable in synchronization and communication because these pathways control extracellular concentrations of bacterial communication chemicals, which can be detected by all cells in close vicinity. HSLs are used for quorum-sensing communication in gram-negative organisms and some of these HSLs are species specific while others are recognized among different bacterial species. The Lux system of *Vibrio fischeri* is the model gram-negative quorum-sensing pathway, and it controls luciferase expression. Lux-type pathways are found in many other bacterial species and consist of a LuxI (autoinducer) homolog and a LuxR (receptor) homolog [13]. LuxI produces an autoinducer HSL (in a positive feedback loop), which is transported freely across the cell membrane, resulting in equal HSL concentrations inside and outside the cell. When the concentration of the HSL in the medium (and therefore in the cell) is above a certain threshold, the HSL binds to LuxR and the LuxR–HSL complex induces expression of genes that are under control of this system, including *LuxI*. All cells that are exposed to the same solution will experience the same HSL concentration, and therefore, they will have synchronous expression of any genes under quorum-sensing control. Using different combinations of these systems (Lux-type systems have been found in at least 25 different bacteria [13]) will facilitate synchronization and communication for improvement of cellular computing circuits.

14.2.2 Implementations

14.2.2.1 Oscillators

Danino *et al.* [14] engineered an *E. coli* strain to act as synchronized oscillators with a fluorescence output signal using the concept of synchronization through quorum sensing. The genetic circuit was constructed using the Lux pathway, an HSL-degrading protein, and a fluorescent protein. These genes were arranged so that HSL and fluorescent protein production were directly related to HSL concentration. This would have created a simple positive feedback loop without the addition of the HSL-degrading protein, which is induced by HSL after a short delay. This delay caused HSL and the fluorescent protein to accumulate before the HSL-degrading protein was expressed, resulting in fluorescence oscillations. The cells were grown in microfluidic devices with small trapping chambers where groups of cells could be monitored. In smaller trapping chambers, the group oscillated as a whole, while diffusion limitations in larger chambers caused waves of fluorescence to propagate through the group (see the original publication for videos). To change the frequency and amplitude of the oscillations, the authors had only to alter the flow rate in the microfluidic device. This oscillation function could someday be applicable in biomedical or industrial implementations because it could potentially be adapted for periodic drug or chemical delivery.

An earlier implementation of a cellular oscillator was performed by Elowitz and Leibler [15], and comparing this work to the oscillator created by Danino *et al.* [14] underscores the importance of quorum sensing to proper function. In this study, oscillation was based on regulation within each cell and the response was not averaged by the quorum-sensing system. This resulted in high cell-to-cell variability, making it very difficult to read the results of the oscillator from a bacterial population. The oscillation could only be measured by tracking the fluorescence of single cells microscopically. While this cellular computing circuit could potentially find use in some single-cell research applications, the quorum-sensing-regulated oscillator by Danino *et al.* is clearly a more functional oscillator on a population scale and is more likely to see further development in biotechnology applications.

14.2.2.2 Switches

An important part of computing function is switchable elements (i.e., circuits with bimodal stability). Gardner *et al.* [16] have produced cellular computing circuits that act as toggle switches. The authors created *E. coli* strains containing a simple genetic circuit that switches between high and low fluorescent protein expression based on inputs of IPTG and high temperature using inducible promoters. The switch functioned properly and could be reset. Switching from low to high condition took 6 h, while switching from high to low took only 35 min. This was a result of the different properties of the IPTG- and temperature-sensitive promoters. Although this switch is conceptualized as a memory component, the authors also suggest applications in gene therapy.

14.2.2.3 AND Logic Gates

Although there may be exciting applications for single input cellular computing circuits, an important research goal in this area is the development of strains that simultaneously process two or more inputs. Multi-input cellular computing circuits will be important for detection of several biomarkers simultaneously, which is often required to make a conclusive decision (e.g., disease state, security threat, and environmental contamination). A basic example of a multi-input strain is cellular AND logic gates. The AND Boolean function delivers a true output signal only when both input signals are true (Table 14.1).

AND logic gates in bacteria have previously been demonstrated. For example, Ramalingam *et al.* [6] constructed plasmid-encoded AND logic gates in *E. coli* with IPTG and tetracycline as the inputs and fluorescent protein as the output. The

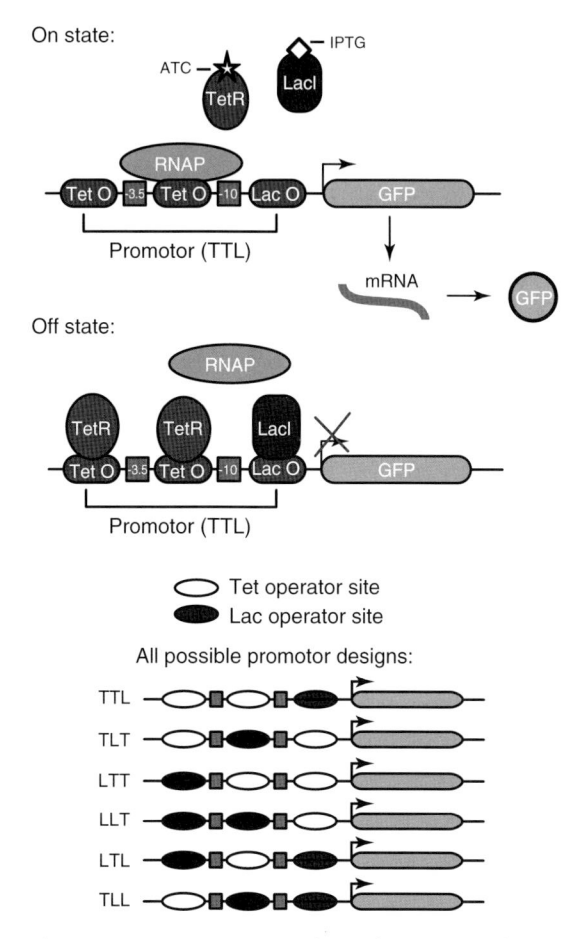

Figure 14.1 Genetic circuit and AND logic gate mechanism. (From [6] with permission.)

authors used six different combinations of two inducible promoters to construct their genetic circuits (Figure 14.1).

Comparison of the performance of the different genetic circuits showed that fine details of the genetic engineering approach have an impact on the characteristics of the strain as a whole. While none of the logic gates was 100% accurate, some logic gates performed better than others, indicating that alteration of the circuit arrangement can improve performance of cellular computing circuits.

14.2.2.4 Edge Detector

For more complex signal processing in one strain, it is sometimes necessary to incorporate more than one logic function. Tabor *et al.* [3] constructed an edge detection program in *E. coli* and this required the combination of three different logic gates. To create this program, the authors combined an AND logic gate with two NOT logic gates (Figure 14.2).

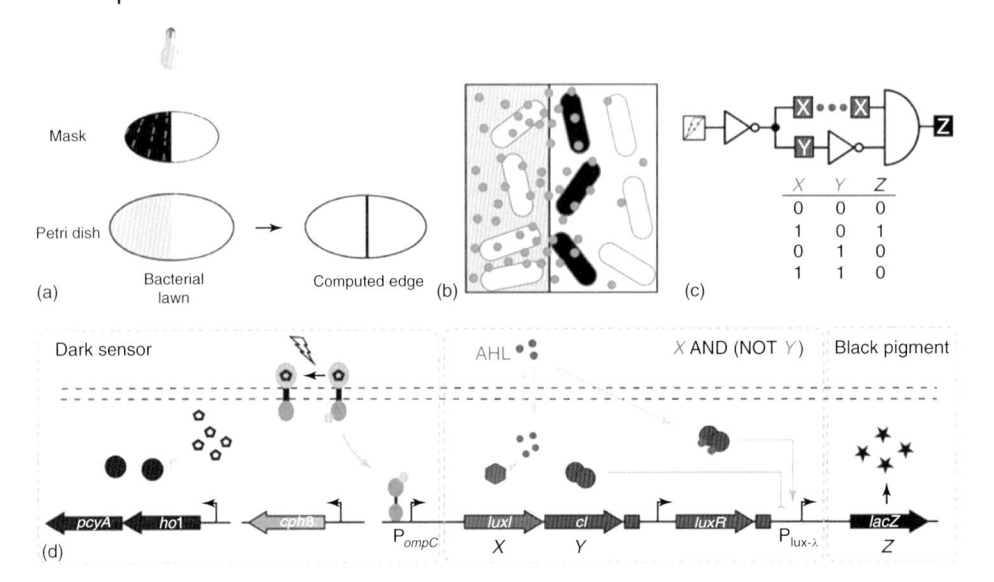

Figure 14.2 Genetic circuit and mechanism of genetically encoded edge detector showing overall input and output (a), diffusion of signal chemicals (b), logical function (c), and overview of the genetic circuit (d). (From [4] with permission.)

The three logic gates used carefully chosen inputs and outputs to create a complete genetic circuit with the desired output characteristics. This circuit caused transformed *E. coli* cells to produce HSL only in the dark, and pigment only in the presence of HSL and light. Therefore, in populations of this strain HSL and light will only coincide at light/dark edges, and pigment will be produced only at these edges. Black and white edge detection is a very specific function, but this approach has already been extended to engineer a cellular computing circuit that can also distinguish between red and green light [17].

14.2.2.5 Complex Logic Functions with Multiple Strains

When a function becomes very complex, it may become difficult to encode the entire processing pathway into a genetic circuit small enough and simple enough to be introduced into one strain. In this case, different parts of the genetic circuit can be introduced into different strains, provided that the strains can communicate and interact with each other. This concept has been demonstrated by Tamsir et al. [4], who engineered different genetic circuits into eight unique *E. coli* strains. Each individual strain carried only one simple logical function, but more complex processing was achieved by combining up to four of the individual strains through the Las and Rhl quorum-sensing pathways, which are other important quorum-sensing systems besides the Lux type [13]. Combinations were achieved by placing colonies of different strains near each other on agar plates so that HSLs could diffuse between them. Through these combinations, the authors were able to construct more complex logic operations, such as XOR, EQUAL, NAND, and A IMPLY B, some of which have not yet been demonstrated in single-strain cellular computing circuits.

The library of inputs contained two different HSL molecules (3-oxododecanoyl homoserine lactone (3-oxo-C12-HSL) and *N*-butyryl homoserine lactone (C4-HSL)), tetracycline, and arabinose, while the output library contained the same HSL molecules and fluorescent protein. This system can be conceptualized as a biocomputing platform with a library of interchangeable parts, although the specificity of connections between the strains will be an obstacle in future design [4].

An alternative use for cellular computing circuits that communicate and interact with each other (rather than creation of complex logic function) is control of environmental and industrial microbial communities. For example, cellular computing circuits could be introduced to industrial fermenters or environmental remediation sites to monitor and control activity of other bacterial populations. Brenner *et al.* [18] created a first step in functional bacterial communication by designing two *E. coli* strains that could sense each other and produce fluorescent protein as a signal. Each strain required one of two HSL molecules (3-oxo-C12-HSL and C4-HSL) to be produced by the other strain to generate the fluorescent protein. This is the first demonstration of a cellular computing circuit that uses other bacteria as an input signal. This study is an important first step toward cellular computing circuits that can sense, decide, and act, based on the cues of a microbial community.

A similar example is a pulse-generating cellular computing circuit designed by Basu *et al.* [19]. Rather than two co-dependent strains, this system consisted of a sender and a receiver strain. The sender strain produced a constant output of an HSL signal, while the receiver strain responded to the HSL by producing a pulse of fluorescent protein. The specifics of the genetic circuit used also caused the receiver strain to be sensitive to the rate of signal increase. Because of this property and the separation of this function into two strains, the authors were able to use patterning of the two strains on solid medium to generate specific patterns of output fluorescence. Essentially, the receiver strain is capable of detecting the sender strain within a certain proximity, suggesting that this cellular computing circuit could be used to detect distance-sensitive variables in biotechnological applications. Indeed, the authors later published another study in which the receiver cells were engineered to detect specific concentration ranges of HSL, and the two strains were used to create complex fluorescence patterns [20].

14.2.3
Transition to *In Silico* Rational Design

A common element of most current cellular computing circuits is that the basic structure of the genetic circuit has been designed manually. This is a time-consuming endeavor, requiring both great depth and great breadth of knowledge of biochemical regulatory mechanisms. However, this process can be aided by computational design of genetic circuits. Rational redesign of existing circuits is also an important part of this process, whereby algorithms design modifications of existing genetic machinery to make it more favorable for implementation in cellular computing circuits [21]. Preliminary studies also show that computational models can do more than just model the behavior of a defined circuit, but also design some genetic circuits, if given the right building blocks. Batt *et al.* [22] have shown that genetic circuits can be improved by characterizing and compiling genetic circuits, which can be combined and tested *in silico*.

A recent study by Purcell *et al.* [23] demonstrates *in silico* design of genetic circuits for cellular computing functions. The authors used mathematical computer modeling to design a genetic circuit with multiple functions: frequency multiplication, oscillation, and switching. The different functionalities could be used with different input concentrations and temporal profiles. The frequency multiplier is a particularly interesting function because it has not yet been reported in a cellular computing circuit. Unfortunately, the authors did not base the computer model on existing genetic components, but rather idealized components. Thus, implementation of this abstract genetic circuit will not be simple, since real genetic components will not behave like the idealized components used in the model. Future efforts toward *in silico* design will need to incorporate not only design parameters of the desired system but also properties of available genetic circuit components.

In silico design opens the door toward much faster strain design, testing, and creation. A pipeline for future creation of cellular computing circuits through computer-aided design is already under way, which will allow faster tailoring of cellular computing circuits toward real-world applications. In addition, increased complexity will likely be achieved through use of sophisticated design algorithms. Computer-aided design will be necessary for development of this field because of the extreme complexity of living cells. In contrast, enzyme computing has already reached a higher level of complexity without *in silico* design owing to the simplicity and specificity of single enzymes.

14.2.4
Transition from Enzyme Computing to Bacteria-Based Biocomputing

Although enzyme computing has already reached greater complexity than bacterial computing, there are several reasons to develop bacterial computing. To overcome the disadvantages of enzyme-based logic gates (e.g., sensitivity to environmental factors and inability to reproduce), transition from enzyme-based Boolean logic gates to bacteria-based logic gates is desirable. Although enzyme-based Boolean logic gates are more selective and specific, bacteria-based logic gates are self-renewing and more robust. These advantages make bacteria-based logic gates more suitable for certain applications, especially when the computation speed is not critical. In addition, the ability to produce more complex output signals than enzyme-based Boolean logic gates allows bacteria-based logic gates to perform more complex tasks than enzymatic cascades. Therefore, we translated an enzyme-based Boolean logic gate with a biofuel cell (a type of bioelectrochemical system (BES)) that is able to produce a direct electric current digital signal into a bacteria-based logic gate by replacing the electrochemically active enzyme with an electrochemically active bacterium [24]. Since direct electric current is an easily measurable and codified digital signal, the translation steps for converting analog signal to digital signal (e.g., fluorescence measurement) can be omitted. This will simplify the biocomputing process, decrease error propagation, and simplify the bacteria–machine communication interface.

Our system was built with a three-electrode potentiostatically controlled BES with *Pseudomonas aeruginosa* PA 14 as the core part of the logic gate. In our previous work, quorum sensing had proven to be an effective mechanism for controlling electric current production in BES [25]. Here, we used a *P. aeruginosa* PA14 $\Delta lasI/\Delta rhlI$ double mutant with both quorum-sensing chemicals (3-oxo-C12-HSL and C4-HSL) as input signals for our system because we used a mutant strain without the self-secretion autoinducers LasI and RhlI, but with functional LasR and RhlR receptors [5]. This mutant cannot produce the signal but can still detect it and trigger the quorum-sensing cascade to induce phenazine production. Phenazines are electron mediators for electron transfer between *P. aeruginosa* PA14 $\Delta lasI$ and electrodes [25–27], resulting in an electric current signal. Thus, the phenazine production could be upregulated only by adding these two quorum-sensing signals. Indeed, our result showed that the electric current increased dramatically only when both input signals were present in the system. By setting a proper threshold, the logic gate gave an output signal 1 only when both input signals were true (presence of the input signal is defined as true and absence of the input signal is defined as false), which makes the logic gate an AND logic gate. This result illustrates that, although replacing the enzyme with bacteria increases the system complexity, highly specific input signals can reduce the uncertainty in signal processing and deliver expected and clear output signals.

The AND logic gate was built in two BES configurations: microbial fuel cell (MFC) and microbial three-electrode cell (M3C). In MFC mode, a natural potential difference is utilized for electric current production and no external power supply is required, which makes this AND logic gate a self-powered biosensor [5, 28]. To obtain a clear output signal, at least 115 h were needed to allow the output signal definition to increase to the desired level. The very slow processing speed makes this biosensor suitable only for the areas where the computing speed is not critical. Another problem that should be noted is that in MFC mode, the output signal can be affected by both anode and cathode performance. Thus, both the anode and cathode noise propagate through the entire signal processing and can cause a decrease in the output signal definition (Figure 14.3).

We circumvented these problems and attained a more robust output signal by using a M3C in which the working electrode potential is controlled at a stable level by a potentiostat, resulting in a better defined electrochemical environment than MFC. As expected, the output signal definition was improved in the M3C mode (Figure 14.4).

The direct digital output signal is one of the major improvements in this system compared to other bacterial logic gates with analogous output signals (e.g., fluorescent protein, optical density). These analogous signals require additional equipment to detect, and thus might decrease the signal definition. The electrochemically active bacteria are core parts to create these logic gates. Using the native regulatory systems of electrochemically active bacteria is a good strategy for

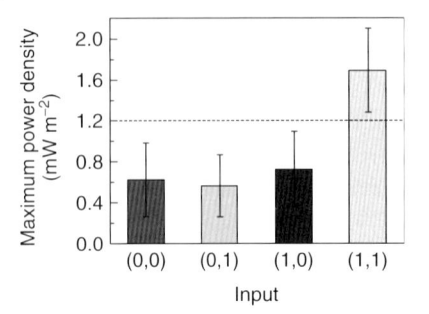

Figure 14.3 Bar diagram showing maximum MFC power density for all input combinations. (From [5] with permission.)

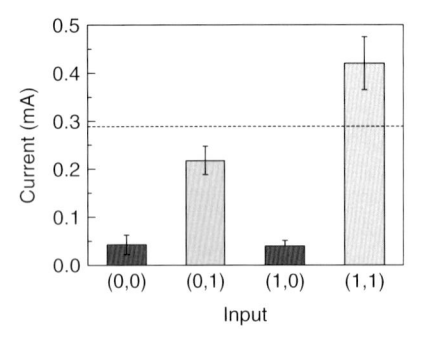

Figure 14.4 Bar diagram showing maximum M3C current production for all input combinations. (From [5] with permission.)

building bacterial logic gates because the electron transfer pathways are not fully understood and cannot easily be transferred to other organisms at this time. Similar to *P. aeruginosa* PA14 used here, another electrode respiring bacterium, *Shewanella oneidensis*, shuttles electrons between cells and electrodes via riboflavin, and is an alternative option [29]. We propose to use different types of electrochemically active bacteria to develop logic gates for various applications

14.3
Conclusion

Cellular computing circuits are in an early stage of development, although increased speed of development is likely with emerging computer-aided design. Based on the current implementations discussed here, we expect development of cellular computing circuits for application in the near future. Cellular computing circuits are likely to be used for biosensing and biocontrol in environments, such as the human gut, industrial fermenters, and in the natural environment. The development of cellular computing circuits with a direct electrical output signal will be particularly useful in biosensing applications, however, that do not require immediate response times. We predict that combining this concept with cutting edge strain design will lead to great improvement in biosensing and biocontrol through cellular computing circuits.

Acknowledgment

This work was supported through an NSF Career Grant # 0939882 to LTA.

References

1. Yokobayashi, Y., Weiss, R., and Arnold, F.H. (2002) *Proc. Natl. Acad. Sci.*, **99**, 16587–16591.
2. Weiss, R. (2001) *Cellular Computation and Communications Using Engineered Genetic Regulatory Elements*. PhD thesis. Massachusetts Institute of Technology, Cambridge, MA.
3. Tabor, J.J., Salis, H.M., Simpson, Z.B., Chevalier, A.A., Levskaya, A. *et al.* (2009) *Cell*, **137**, 1272–1281.
4. Tamsir, A., Tabor, J.J., and Voigt, C.A. (2011) *Nature*, **469**, 212–215.
5. Li, Z.J., Rosenbaum, M.A., Venkataraman, A., Tam, T.K., Katz, E., and Angenent, L.T. (2011) *Chem. Commun.*, **47**, 3060–3062.
6. Ramalingam, K.I., Tomshine, J.R., Maynard, J.A., and Kaznessis, Y.N. (2009) *Biochem. Eng. J.*, **47**, 38–47.
7. Anderson, J.C., Voigt, C.A., and Arkin, A.P. (2007) *Mol. Syst. Biol.*, **3**, 133.
8. Baumgardner, J., Acker, K., Adefuye, O., Crowley, S.T., Deloache, W. *et al.* (2009) *J. Biol. Eng.*, **3**, 11.
9. Voigt, C.A. (2006) *Curr. Opin. Biotechnol.*, **17**, 548–557.
10. Anderson, J.C., Clarke, E.J., Arkin, A.P., and Voigt, C.A. (2006) *J. Mol. Biol.*, **355**, 619–627.
11. You, L., Cox, R.S., Weiss, R., and Arnold, F.H. (2004) *Nature*, **428**, 868–871.
12. Friedland, A.E., Lu, T.K., Wang, X., Shi, D., Church, G., and Collins, J.J. (2009) *Science*, **324**, 1199–1202.
13. Miller, M.B. and Bassler, B.L. (2001) *Annu. Rev. Microbiol.*, **55**, 165–199.
14. Danino, T., Mondragon-Palomino, O., Tsimring, L., and Hasty, J. (2010) *Nature*, **463**, 326–330.
15. Elowitz, M.B. and Leibler, S. (2000) *Nature*, **403**, 335–338.
16. Gardner, T.S., Cantor, C.R., and Collins, J.J. (2000) *Nature*, **403**, 339–342.
17. Tabor, J.J., Levskaya, A., and Voigt, C.A. (2011) *J. Mol. Biol.*, **405**, 315–324.
18. Brenner, K., Karig, D.K., Weiss, R., and Arnold, F.H. (2007) *Proc. Natl. Acad. Sci.*, **104**, 17300–17304.
19. Basu, S., Mehreja, R., Thiberge, S., Chen, M.-T., and Weiss, R. (2004) *Proc. Natl. Acad. Sci.*, **101**, 6355–6360.
20. Basu, S., Gerchman, Y., Collins, C.H., Arnold, F.H., and Weiss, R. (2005) *Nature*, **434**, 1130–1134.
21. Purnick, P.E.M. and Weiss, R. (2009) *Nat. Rev. Mol. Cell Biol.*, **10**, 410–422.
22. Batt, G., Yordanov, B., Weiss, R., and Belta, C. (2007) *Bioinformatics*, **23**, 2415–2422.
23. Purcell, O., di Bernardo, M., Grierson, C.S., and Savery, N.J. (2011) *PLoS ONE*, **6**, e16140.
24. Willner, I. and Katz, E. (2003) *J. Am. Chem. Soc.*, **125**, 6803–6813.
25. Venkataraman, A., Rosenbaum, M., Arends, J.B.A., Halitschke, R., and Angenent, L.T. (2010) *Electrochem. Commun.*, **12**, 459–462.
26. Rabaey, K., Boon, N., Siciliano, S.D., Verhaege, M., and Verstraete, W. (2004) *Appl. Environ. Microbiol.*, **70**, 5373–5382.
27. Venkataraman, A., Rosenbaum, M.A., Perkins, S.D., Werner, J.J., and Angenent, L.T. (2011) *Energy Environ. Sci.*, **4**, 4550–4559.
28. Fornero, J.J., Rosenbaum, M., and Angenent, L.T. (2010) *Electroanalysis*, **22**, 832–843.
29. Lies, D.P., Hernandez, M.E., Kappler, A., Mielke, R.E., Gralnick, J.A., and Newman, D.K. (2005) *Appl. Environ. Microbiol.*, **71**, 4414–4426.

15
The Logic of Decision Making in Environmental Bacteria

Rafael Silva-Rocha, Javier Tamames, and Víctor de Lorenzo

15.1
Introduction

Under natural conditions, environmental microorganisms are continuously exposed to changes that compromise their survival unless they respond and then adapt to shifting physicochemical and nutritional scenarios [1]. Soil bacteria constitute a remarkable example of environmental adaptation. Despite their deceiving simplicity compared to eukaryotic cells, many types of bacteria are able to colonize a large number of niches and to deal with continuously variable conditions [2]. The reasons for this extraordinary flexibility reside not only in the catalog of regulatory and structural genes encoded in their genomes but also, more decisively, in the way regulatory networks sense external conditions and adjust cell physiology to ever-changing circumstances. Such a sensorial ability is reflected in the repertoire of transcriptional factors (TFs) available in the genomes of archetypical soil microbes (e.g., Pseudomonads) for controlling expression of both metabolic and stress–response functions. In this respect, the genomic complement of generalist species that thrive in natural environments has a much larger share of regulatory genes encoded than counterparts that inhabit stable niches (e.g., endosymbionts [2–4]). Needless to remark here that TFs do not act in isolation but are hierarchically connected [5], allowing the cell to integrate different stimuli and build proper responses for prevailing under new settings. The overall flow of signal propagation through regulatory networks is sketched in Figure 15.1. Note that for the rest of the Chapter we refer to such networks as biological devices composed of connected nodes in each of which given inputs are converted into distinct outputs, which then become the inputs of other downstream nodes [6]. In this respect (and for the sake of illustration of this concept), the material nature of such inputs/outputs is not important, provided that they can be computed at the corresponding nodes [7]. As discussed below, this simplification is the key for merging transcriptional and metabolic events with the same formalisms.

Biomolecular Information Processing: From Logic Systems to Smart Sensors and Actuators, First Edition. Edited by Evgeny Katz.
© 2012 Wiley-VCH Verlag GmbH & Co. KGaA. Published 2012 by Wiley-VCH Verlag GmbH & Co. KGaA.

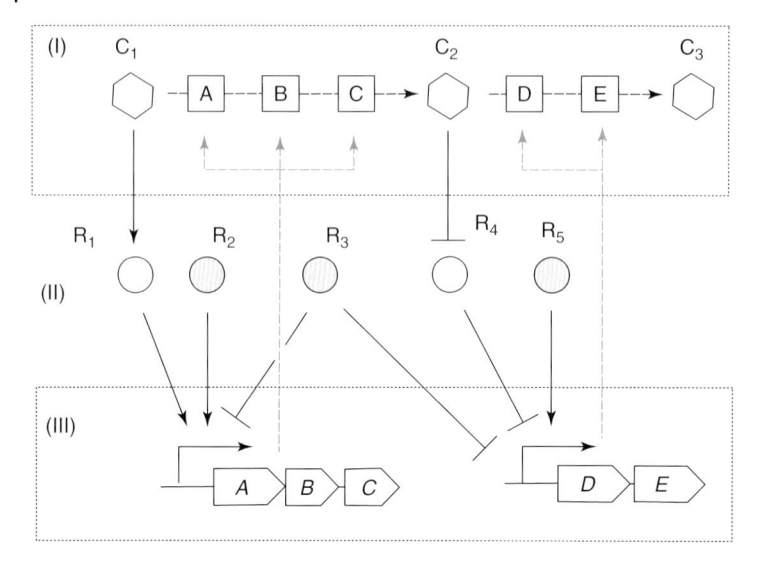

Figure 15.1 Flow of signals through typical regulatory networks. The figure represents one network controlling a metabolic pathway. (I) In the metabolic part of the system, compound c1 is converted to c3 by the action of several enzymes through the intermediate species c2. Both c1 and c2 are signal molecules which trigger the expression of the cognate pathways. (II) At the transcriptional factor level, signals c1 and c2 are sensed by specific TFs (R1 and R4, respectively) which themselves can integrate inputs from global regulators (R2, R3, and R5) in order to trigger the expression of the two operons sketched. (III) Finally, at the gene expression level, the binding/unbinding of the different TFs determines the production of enzymes (A–E) which, in turn, may feedback the first level (metabolism), closing the loop of signal propagation.

Although the main function of regulatory networks is signal integration, the specific architectures of their constituents endow the corresponding systems with distinct dynamic properties that make a difference in the final response. The structure of network motifs (e.g., feedback and feed forward loops (FFLs), multi-input modules, and many others [8]) not only enables them to compute fixed inputs into equally fixed outputs but also determines important properties such as response time, shape of the response generated and pulsing, or monotonic outcome, independently of specific parameters [9, 10]. Intricate integration phenomena encompassing extracellular compounds, intracellular metabolic sensors, and signal propagation by small molecules often appear in regulatory networks that control biodegradative pathways for recalcitrant compounds [11]. This makes sense, as bacteria that inhabit polluted sites have to make decisions between different nutrients-to-be on the basis of many endogenous and exogenous factors (i.e., compound availability, physiological state of the cells, the flux of carbon through the metabolism, final electron acceptors, physicochemical circumstances, etc.) [12].

The most characterized cell-wide regulatory network is that of the central metabolism of carbon in the model organism *Escherichia coli* [13]. In this case, the hierarchy of global and specific TFs that control expression of the genes encoding large metabolic blocks originates network topologies that are optimal for processing the signals mentioned earlier and bring about the most advantageous physiological result [13, 14]. The bottom line is, in any case, that easy-to-degrade carbon sources are consumed first over compounds more difficult to metabolize [15]. In contrast, nutrient choices in environmental bacteria are not, for example, between a palatable glucose and a less edible glycerol, but between carbon and nitrogen compounds with unusual molecular structures that often act themselves as chemical stressors [16]. In these cases, it cannot come as a surprise that the corresponding regulatory architectures become more intricate. The set of TFs that control biodegradative pathways has not only to recognize a xenobiotic or recalcitrant compound as a nutrient-to-be but also to ensure that the trade-off between metabolic gain and stress endurance is not detrimental to the general cell physiology [11, 17].

The complete understanding of the properties of extant regulatory networks opens new perspectives for the engineering of cell factories [18]. This is because identification of general design principles should allow the correct integration of new regulatory circuits into preexisting systems [6], one of the key objectives of contemporary synthetic biology [19]. A growing approach to address the functions and properties of regulatory networks involves the formulation of models that translate the molecular interactions known for a given system into a set of equations that afford a simulation of the entire lot of physical and functional interplays [20]. During the last few years, a number of methods have been implemented and validated for building and simulating such models. In the sections that follow, some of these methods are briefly discussed in regard to their potential application for engineering new metabolic pathways (whether for biodegradation or biosynthesis) of interest for bioremediation and biocatalysis. The power of Boolean models to this end is examined in more detail using as an example the system for biodegradation of *m*-xylene borne by the TOL plasmid of the soil bacterium *Pseudomonas putida* mt-2. In this context, we argue that adoption of simple logic-gate formalisms for describing the regulatory and metabolic actions embodied in the TOL regulatory network [21] permits conversion of a complex but still *soft* model into a rigorous logic circuit that can be accurately described by means of piecewise-linear (PL) equations.

15.2
Building Models for Biological Networks

The initial step for analyzing cellular networks involves the elaboration of a relational diagram comprising all components of the system and as much data as possible on the interactions among them, whether physical or functional. As many different types of biological networks exist and the components and interactions can be of very different nature, so are the experimental techniques aimed at uncovering them. In typical *metabolic networks*, the nodes are the substrates and products on which the enzymes act on, while the edges represent enzymatic activities themselves. Merely descriptive models with nodes (components) and edges (connection between nodes) can then be enriched with stoichiometric coefficients, kinetic constants, and thermodynamic information (e.g., reversibility). The list of enzymatic activities can be grossly derived from genomic annotations based on homology search [22–24]. Ideally, such lists can then be refined by manual curation, for instance by looking for missing activities in the corresponding metabolic map [25]. A large number of additional problems can appear at every step of genome-based metabolic reconstructions [26, 27]; see [28] for a compendium. Experimental kinetic and thermodynamic data have to be determined either by direct biochemical analyses or indirectly by means of parameterization algorithms that convert *omics* data into apparent kinetic constants [29, 30]. The volume and quality of such *wet* information is highly variable depending on the organism, and in many cases it is not available at all. Besides, this information often pertains *in vitro* data, which often reflect poorly true intracellular behavior.

How to make then reasonable metabolic models in view of the frequent dearth of data? To overcome this difficulty, genome-scale metabolic reconstruction and simulations are often performed through constraint-based approaches (like flux-balance analysis), which do not need such information [22, 28]. Instead, by describing the entire set of metabolic and transport reactions in an organism under the assumption of steady state and by balancing mass and charge, they attempt to find an optimal set of flux distributions across the metabolic network solution space, thereby producing an essentially static view of the metabolism under given conditions and constraints. Data from different experimental *omics* approaches, such as metabolomics, fluxomics, transcriptomics, and proteomics, provide additional information with which to constrain the models and thus to reduce the solution space and increase accuracy [28, 31]. These models, while lacking the detail of kinetic representations, provide a valuable framework with which to explore the metabolic space and capabilities of the organisms involved, to generate testable hypotheses of the relationship between genotype and phenotype, and to test the effect of external and internal perturbations (such as nutrients or other microbial species).

In contrast to metabolic counterparts, regulatory networks capture information on the influence that particular genes (i.e., those encoding TFs) exert on the expression or activity of others. As a consequence, the nodes in this case are composed of genes (often assimilated to their encoded proteins), while the edges express regulatory connections, for example, either activation or inhibition. Regulatory diagrams of this type facilitate the visualization of coordinated regulatory effects, which appear as sets of TFs acting on the same target [32]. Unlike the metabolic scenarios discussed above, the reconstruction of the regulatory network of a given organism on the mere basis of its genomic sequence is a very challenging task. While metabolism (specially the core of central enzymatic reactions) is relatively conserved in many different organisms, similar regulatory outcomes can originate from altogether unlike TFs and regulatory modules [33]. Beyond a few model organisms, the volume of information on regulatory interactions is very scarce, making it very difficult to translate the knowledge obtained for one bacterium into others, even in the case of close species. Bona fide orthologs of typical *E. coli*'s TFs such as the catabolite regulatory protein (CRP) or the integration host factor (IHF) have been found to govern entirely different sets of functions in *P. putida* through an evolutionary exaptation process [34]. In reality, the targets and functions of orthologous TFs are often not the same in different species [35], thereby making it necessary to examine regulatory interactions experimentally on a case-by-case basis.

Several procedures are available for identifying target genes for each of the known TFs of an organism. Binding sites can be determined by a ChIP-chip assay, where the DNA-bound TF is made out by means of a cognate antibody and the corresponding DNA sequence determined by microarray hybridization or direct sequencing [36]. Although this procedure has been used for mapping TF targets in several organisms, especially yeasts [37], one bottleneck is the necessity of antibodies against the TFs at stake, which often flaws the whole procedure. Microarray expression profiling of the wild-type strain versus mutants lacking (or overexpressing) different TFs has been also used to infer possible regulatory interactions [38, 39]. The main difficulty in this case is to discriminate between the direct effects of TF binding to primary targets and the indirect results due to downstream expression of other TFs or intracellular signals. While information from expression profiles is useful to identify conserved *cis*-acting regulatory motifs, it is also necessary to verify the hypothetical regulatory network with alternative methods. Alas, given the difficulty of the corresponding experiments, very little biochemical data on regulatory nodes is available from transcription experiments *in vitro*. Fortunately, the popularization of high-throughput methods for measuring promoter activity with optical reporter genes [6] helps us to remediate the shortage of such *hard* regulatory data. Determination of regulatory transfer functions in specific nodes of the network can in this way be easily deciphered *in vivo*. As was the case with metabolic networks above, the corresponding output (typically fluorescence of the green fluorescent protein, GFP) can be translated into kinetic parameters with suitable algorithms [40].

15.3
Formulation and Simulation of Regulatory Networks

A wide variety of mathematical approaches have been developed in recent years for simulating genetic regulatory systems of the type just discussed. These models attempt to both describe the behavior of relational objects and to make predictions on them on the basis of models obtained as described above. These simulations, which attempt to determine the dynamics and outcomes of gene regulation, can be classified according to several criteria [41]. They are labeled *static* if they capture just a snapshot of the active system, or *dynamic* if they instead display the progression of the network along a time course as a succession of discrete or continuous states. An important divide between the various models and their corresponding simulations is the choice between deterministic and stochastic systems [42, 43]. In the first approach, the state of expression/concentration/activity of given genes and the regulatory interactions between them at a particular time point unambiguously determine the states of the same genes at the next temporal state. In such systems, there is only one path leading from a certain gene-expression state to another. Alternatively, stochastic systems contemplate that a certain state of expression in a regulatory node or module can bring about more than one subsequent condition, the occurrence of each possible situation being expressed as a probability. Theoretical considerations and experimental results support the notion that genetic networks in single cells are inherently stochastic while populations tend to behave in a deterministic manner [43]. Some of the most popular methods for formalization and simulation of networks are briefly addressed below.

15.3.1
Stochastic Versus Deterministic Models

On one side of the range of modeling possibilities we find *stochastic master equations* [42, 44]. They are instrumental in describing regulatory scenarios in which the net number of molecules at play is small (e.g., in single cells) and the corresponding systems are subject to a considerable level of noise and uncertainty in their scale of operation [45]. Unlike deterministic equations discussed below, stochastic master equations express the probability that the system contains n_i molecules of each of the molecular species i at a given time point. These equations thus express the probabilities for the system to be in a particular state at a given time interval. Although equations of this type surely describe more accurately the dynamics of a regulatory system at single cell level, they are more difficult to solve by analytical means than their deterministic counterparts (see below). Therefore, their actual implementation may become problematic and will be feasible only if the system is small and the sought-after granularity (i.e., the temporal scale and definition of the results) is limited [41]. Besides, experimental validation of stochastic models is still cumbersome and often not at all possible.

Deterministic approaches, for instance through *ordinary differential equations* (ODEs) offer a workable alternative to simulate regulatory systems. Given that concentrations of the system's components cannot be negative, the levels/activities of each of the molecular species (genes, proteins, RNAs, etc.) that participate in the network are represented by continuous time variables. Typically, production of a component is expressed as a function of the presence of other components, which is articulated by a rate equation of the form

$$\frac{dx_i}{dt} = f_i x$$

where $x = [x_1 \ldots x_n]$ is a vector representing the concentrations of molecular species and f_i is a (most often) nonlinear function expressing the dependence of x_i on other components. f_i thus embodies different regulatory interactions and allows their modeling and follow-up over time. In this way, the concentration of each of the molecular species at any given moment can be determined by solving the corresponding set of differential equations. Adequate mathematical and computational tools are available for such a task [46, 47], and simulations based on ODEs are used to examine many regulatory systems [48, 49]. Unfortunately, the ODE approach is not devoid shortcomings, as equations and models have to be fed with kinetic parameters (e.g., the Hill coefficients of sigmoid functions or Michaelis–Menten coefficients of hyperbolic functions) that affect rate equations. As the complexity of a network grows nonlinearly with the number of components, having complete sets of such parameters is time consuming and a difficult fastidious endeavor. Furthermore, the error associated to experimental determination propagates and, for large systems, such as genome-scale models, this becomes impossible and meaningless with the technologies currently available.

15.3.2
Graphical Models

One way of overcoming such a shortage of kinetic information involves the application of graphical models, such as those involving Bayesian probabilities, to describe the state of the components of the biological object under study [50]. The angle here is that the starting probability of a given state is deduced from the data on the upstream and the downstream conditions. This formalism provides a standard set of procedures and formulae to perform such calculations, which results in a probabilistic model that represents a set of random variables and their conditional dependences. Such variables (the nodes of the network) include observable qualities/quantities, latent properties, or even unknown parameters. Each node is associated with a probability function that takes as input a particular set of values for the node's parent variables and gives the probability of the variable represented in that point. Edges then reflect conditional dependencies between the components. Under this scheme, genetic networks can be constructed such that the expression level each of the genes is expressed as a conditional probability of the presence of its direct regulators. Although Bayesian networks are static and usually deterministic, their statistical basis makes it possible to include stochastic effects [41]. The appeal of Bayesian networks is that they can be formalized even when *wet* biological information on the system is scarce. Furthermore, this formalism has the bonus that probabilities can be inferred with algorithms applicable to experimental data, especially from gene expression information [50]. A separate type of graphical models that can be merged very productively with experimental data is based on Boolean logic and binary gates, as explained in more detailed below.

15.4
Boolean Analysis of Regulatory Networks

Despite the various descriptive languages and modeling options just enumerated, the lack of benchmarking for measuring transfer functions in various laboratories [6] reduces the actual data that can be used to feed simulations. Typically, one group will employ *lacZ* as reporter to measure promoter activity, another will prefer GFP, and still others may rely on DNA array technology or RT-PCR. The lack on anything comparable to a conversion table between measurement units *in vivo* very often leads to soft consensus descriptions of regulatory circuits that for the most part employ arrows (→) or hammers (T) for expressing either positive or negative interactions between system components. Assigning numbers to the arrows [40, 51] is in fact one of the current challenges of contemporary systems biology.

15.4.1
Translating Biological Networks into Logic Circuits

A simple alternative to display and simulate genetic circuits when little or no information on transfer functions between nodes is accessible involves the adoption of Boolean concepts. Binary logic is in this case sufficiently instrumental for describing the states of the components of the biological system under scrutiny. In Boolean networks, the status of any specified gene is characterized by only two possible values (true or false: 1 or 0) which reflect whether that node of the circuit is active or not. Regulatory interactions can then be accurately entered in the network as logic gates that execute Boolean functions such as AND, OR, NOR, and so on [52], thereby describing expression of any gene as a result of the presence or absence of other genes and small molecules that act as regulators [53]. This type of descriptive language allows the layout of dynamic and deterministic models in which known inputs are computed into just discrete outputs [54]. In this respect, the sole architecture and hierarchy of network components endows the system with intrinsic signal computation capacities at the nodes of the circuit and fixes a signal propagation itinerary through the entire setup. These features are ultimately shaped by connectivity and the sign of the interplay between the interacting components of the network. For instance, in an FFL module [5] in which two TFs activate directly and indirectly the same target gene, the final shape of the response curve is very different depending on whether both regulators are equally efficient in the activation of the third gene or they have to cooperate for generating the output [9]. These alternative FFL scenarios can be easily translated into different classes of Boolean operators. Specifically, the first instance is equivalent to an OR gate acting on both TFs (i.e., the presence of just one TF is sufficient to bring about the final effect), while the second corresponds to an AND operator (both TFs are necessary to activate the target gene). In this way, logic operators (gates) describe rigorously the sign of the relationship between interacting components of the regulatory system and fix the outcome resulting from these interactions. Values of 0 or 1 can be assigned to both the inputs and outputs of the circuit, which becomes a signal computation device reminiscent of those made with transistors [52]. This is not yet a quantitative description of the system, but it allows penetrating its inner logic and move further than the typical arrows/hammers depiction of regulatory networks.

The logic gates abstraction applied to genetic circuits is not only useful for the analysis of existing systems [55] but also opens a range of possibilities for the rational design and implementation of synthetic counterparts [56]. One could envisage that a collection of well-characterized molecular devices with a known logic would be suitable for constructing artificial cellular networks with predictable functions, in a similar way that an engineer builds a digital circuit in a computer. Alas, most logic gates found in extant regulatory networks are often biased to just a few classes (i.e., AND, OR, and NOT), and more complex types of *natural* computation are infrequent [52, 53]. This state of affairs has recently motivated a considerable effort to design and validate novel logic gates by means of modifying existing regulatory elements. In a remarkable seminal case [57], the logic of the *lac* promoter of *E. coli* was swapped from an AND gate to an OR gate by introducing a few point mutations within its *cis*-regulatory region. In a more recent example [58], the SOS system of *Escherichia* was rewired in a manner that made the output of both the *recA* and *lexA* promoters to faithfully follow the pattern of a binary composite OR-NOT gate (ORN) in which the inputs are DNA damage (e.g., nalidixic acid addition) and IPTG acts as an exogenous signal. Unlike other nonnatural gates whose implementation requires changes in genes and promoters of the genome of the host cells, such an ORN node was brought about by the sole addition of wild-type bacteria with a plasmid encoding a module for LacIQ-dependent expression of *lexA*. It is thus possible to artificially interface autonomous cell networks with a predetermined logic by means of Boolean gates built with regulatory elements already functioning in the recipient organism. But one can also leave behind such a rational design and adopt combinatorial procedures to explore the entire *logic space* of given TFs. For instance, after shuffling *cis*-regulatory elements of the *gal* circuit, the authors of [59] were able to generate 11 out of the possible 16 logic gates responsive to two inputs. The bonus of promoter shuffling is that the experimental setup scans the entire combinatorial landscape of *cis*-regulatory sequences, thereby accessing solutions to simple computation *traps* which may not be easy to figure out by rational design.

15.4.2
Integration of Regulatory and Metabolic Logic in the Same Boolean Circuit

Two extreme abstractions have to be adopted for conversion of any regulatory circuit into a logic network. First, the components can hold only either of the state values (0 and 1). Second, the material nature of the same components is entirely disregarded as long as they do their job in computing set inputs into determined outputs. Although not always explicit, the near-exclusive components of networks of this type consist of TFs and inducing signals (whether exogenous molecules or physicochemical conditions). In real cells, however, regulatory devices operate on the background of an active metabolism. Not only is such biochemical activity controlled by dedicated genetic circuits, but also the enzymes and substrates/products can physically or functionally interact with TFs, creating regulatory interplays between the transcriptome and the metabolome [13]. Interestingly, the organization of an enzymatic network can also be formalized as a whole of logic gates (i.e., biochemical computing) in which both the inputs and the outputs consist of enzymes and metabolites rather than TFs and inducers [60–62]. Since the activity states of the components of such enzymatic systems are equally abstracted to binary values 1 and 0, it is then perfectly feasible to merge regulatory and biochemical networks in the same logic circuit and examine its structure as a unique biological object [63]. An example is described in more detail below regarding the logic architecture of the entire regulatory/enzymatic network for environmental *m*-xylene biodegradation. Note that dual inputs that can be processed by a node/gate in a merged network of this type may include (i) two TFs, (ii) one TF and one metabolite, and (iii) one enzymatic activity and one substrate (e.g., a nutrient or a metabolite). By the same token, outputs might consist of proteins (whether TFs or enzymes) and small molecules (reaction products, intermediate metabolites, and signaling chemicals, e.g., autoinducers). This type of abstractions allows us to uncover emergent properties of biological networks which are not noticeable if the regulatory and the enzymatic connections of the same system are addressed separately, let alone if the properties of each of the components are examined out of their context.

15.4.3
From Digital Networks to Workable Models

While elaboration of a logic map of a natural regulatory network of the type just described might be the first step in the global analysis of a whole system, the final objective of any modeling is the formulation of equations that represent all key interactions between components. Boolean approaches can be enriched with methods for incorporating stochastic effects [64], but digital genetic/metabolic networks and their cognate simulations ultimately ignore kinetic parameters. As a result, all components of the simulated system update their state in a synchronous manner, which is evidently far from biological reality. Strategies such as PL approximations [65] have been developed to remediate this caveat. The underlying concept in this case is that the switch-like behavior of gene regulation affords to grossly match the nonlinear function of ODEs (see above) to step functions that reflect downstream gene expression at a given concentration of the corresponding upstream regulator:

$$f_i(x_j) = k_i * s^+(x_j, \theta_j)$$

This equation indicates that the synthesis of product *i* is a function of the presence of the effector *j*. In this equation, s^+ is a step function, which is a Boolean operator that sets to 1 if the concentration of *j* (x_j) is above a particular threshold (θ_j) and is zero otherwise. The synthesis of *i* produces at a rate given by K_i when $x_j \geq \theta_j$, and does not take place otherwise. In this case, *j* is an activator of the synthesis of *i*. A repressor can be expressed by a negative step function:

$$f_i(x_j) = k_i * s^-(x_j, \theta_j)$$

which sets to 1 when $x_j < \theta_j$, and is zero when $x_j \geq \theta_j$. The synthesis of a particular compound results from the combination of different regulation functions, each involving a different effector, for instance,

$$f_i(x_j, x_k) = k_i * [s^+(x_j, \theta_j) * s^-(x_k, \theta_k)]$$

where the production of i is regulated by the activator j and the repressor k. In this way, it is possible to model all possible transcriptional and metabolic interactions in the system, determining the production of particular compounds as a function of the presence or absence of some others. It is possible to set different thresholds to the concentration of a particular compound if it regulates different reactions at different concentrations. For instance, if effector j regulates two reactions, we can set the constraint $\theta_j^1 < \theta_j^2$, to indicate that reaction 1 is regulated by a lower concentration of j than reaction 2. These constraints are known as *threshold inequalities*. The concentration of a particular compound is then determined by its production rate (expressed by k_i) and its degradation rate (g_i), which is a strictly positive function (i.e., concentrations cannot be negative). With the combination of positive and negative step functions, it is possible to model all possible regulatory interactions in the system, no matter their complexity.

This approach allows the simulation to proceed without any knowledge of the kinetic constants between network nodes, and affords description of the Boolean networks described above as sets of PL equations that are reminiscent of actual ODEs. Furthermore, recent improvements in modeling logic circuits using PL approximations [66] allow entering differences in the timescale of the relevant molecular events based on judicious biological reasoning. For instance, the time necessary for a metabolite to bind or be released from a cognate TF has to be inevitably shorter than the time it takes to transcribe an entire gene [57, 67]. The result is that the itinerary of inputs and outputs through the network can be displayed as a coarse continuous flow rather than a series of discrete jumps between binary states. In this way, step functions can approximate sigmoids in various instances [65, 66], thus providing a most useful simplification without loss of accuracy in the eventual simulation of the corresponding network. Furthermore, combination of Boolean gates with PL formalisms allows running simulations of the effect of perturbing (e.g., mutating) *in silico* given nodes of complex regulatory circuits and thereby predict the corresponding gross phenotypes.

15.5
Boolean Description of m-xylene Biodegradation by *P. putida* mt-2: the TOL logicome

Bacteria that colonize sites polluted by recalcitrant and xenobiotic chemicals offer a repertoire of regulatory and catabolic devices with potential application for the construction of new biodegradation networks for bioremediation [68–70]. Comparative studies of the transcriptional networks that control expression of pathways for the degradation of recalcitrant chemicals reveal an extraordinary and largely inexplicable diversity of regulatory architectures [69]. One of the most conspicuous cases appears in the so-called TOL network, which regulates a complex pathway for the degradation of toluene and *m*-xylene in the soil bacterium *P. putida* mt-2 [71]. The TOL pathway is encoded by a mobile, self-transmissible plasmid called *pWW0*, which encodes the enzymes necessary for conversion of *m*-xylene into pyruvate and acetaldehyde: that is, transformation of otherwise recalcitrant substrates into central metabolites. As shown in Figure 15.2, the TOL network

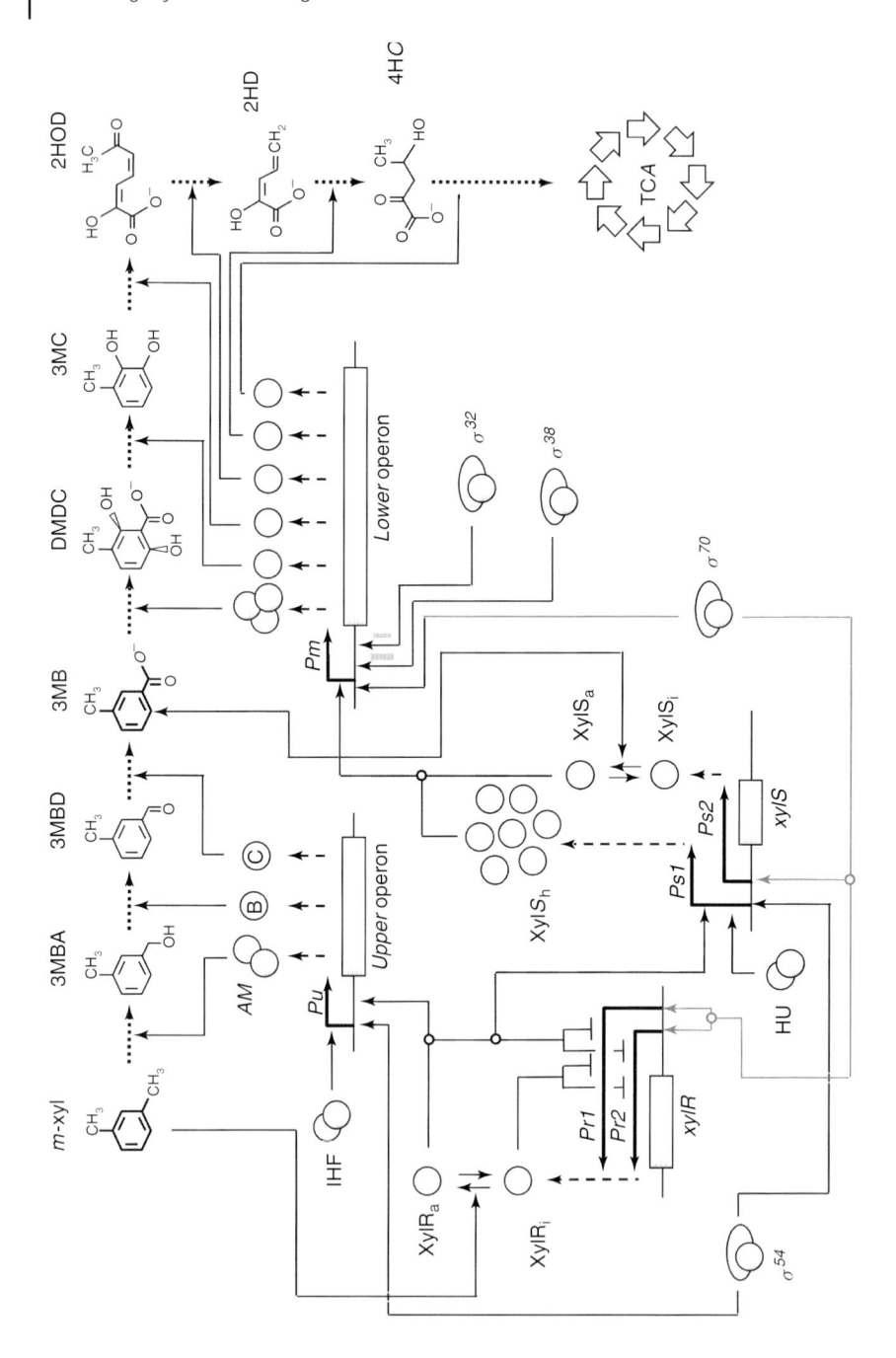

290 | *15 The Logic of Decision Making in Environmental Bacteria*

includes two transcriptional regulators (XylR and XylS) that control expression of two cognate operons. These determine subsequent steps of the transformation of m-xylene (m-xyl) to 3-methylbenzoate (3MB, *upper* pathway) and from 3MB to intermediates of the tricarboxylic acid cycle (TCA, *lower* pathway). The TOL network is *wired* to the rest of the cell by a number of chromosomally encoded factors, including the histone-like proteins IHF [72–74] and HU [75, 76] as well as four sigma factors (σ^{70}, σ^{54}, σ^{38}, and $\sigma^{70,32}$) and additional regulatory proteins TurA, and PprA [77, 78]. All these connect expression of the TOL genes to both internal signals (growth phase, energy charge, and ppGpp) and external stimuli (alternative C sources, temperature, and N compounds). The whole of enzymes and regulators encoded in the plasmid forms an autonomous molecular network of a relatively small dimension, which is suitable for the type of Boolean formalisms mentioned above.

15.5.1
Narrative Description of the TOL Regulatory Circuit

The TOL system is a biochemically separated entity from the rest of the host's metabolism and encompasses the whole of metabolic and regulatory genes required for complete catabolism of m-xylene into intermediaries of the central metabolism. Figure 15.2 summarizes virtually all known facts about the regulation of the system. In the absence of the substrate of the pathway, expression of both the *upper* and the *lower* pathways is entirely shut down as a result of the inactivity of their cognate promoters Pu and Pm, respectively. The corresponding activators XylR and XylS are present by virtue of their expression through their divergent promoters Pr and Ps, but in an inactive form (XylRi, XylSi). Transcription of *xylRi* is maintained

Figure 15.2 Components and regulatory interactions of the TOL genetic network. The basic TOL circuit of plasmid pWW0 of *Pseudomonas putida* mt-2 shown is composed of the transcriptional regulators XylR and XylS, the host factors IHF and HU, and the RNAP with the sigma factors σ^{70}, σ^{54}, σ^{38}, and σ^{32}. The outcome of the network is the production of the enzymes for complete biodegradation of m-xylene (m-xyl) into TCA intermediates. The enzymes encoded by the *upper* operon convert m-xyl into 3-methyl benzoate (3MB) in a process brought about by enzymes xylene monooxygenase (*xylAM*), benzaldehyde dehydrogenase (*xylB*), and benzyl alcohol dehydrogenase (*xylC*), and perhaps others. 3MB is then metabolized by the *lower* pathway enzymes in six steps, which eventually produce the central metabolites pyruvate and acetaldehyde, which are further channeled into the TCA cycle. The head substrates of the *upper* and *lower* pathways are inducers of the cognate regulators XylR and XylS, respectively. The organization of each transcriptional unit and the connections between them are described in the text. More complex interactions are described in the text. Abbreviations: XylRi, inactive XylR; XylRa, active XylR; XylSi, XylS inactive; XylSa, active XylS; XylSh, hyperexpressed XylS; m-xyl, m-xylene; 3MBA, 3-methylbenzyl alcohol; 3MBD, 3-methylbenzaldehyde; 3MB, 3-methylbenzoate; DMDC, 1,2-dihydroxy-3-methylcyclohexa-3,5-dienecarboxylate; 3MC, 3-methylcatechol; 2HOD, *cis,cis*-2-hydroxy-6-oxohept-2,4-dienoate; 2HD, *cis*-2-hydroxypenta-2,4-dienoate; 4HO, 4-hydroxy-2- oxovalerate.

approximately constant through a typical negative feedback loop, while that of *xylSi* is kept constitutively low through a weak housekeeping promoter. The situation changes drastically as soon as cells are exposed to *m*-xylene. This inducer initiates a stepwise sequence of regulatory and metabolic events that start at the *Ps–Pr* region (where the maximum concentration of regulatory elements occurs) and is propagated through the entire circuit. The process starts with the binding of *m*-xylene to XylRi for production of an active form, namely, XylRa. This causes two effects: (i) activation of the σ^{54} *Pu* promoter and subsequent expression and activity of the *upper* pathway for conversion of *m*-xylene to 3MB; and (ii) activation of the *Ps1* promoter and overexpression of *xylS*. Such an overexpression brings about a species of this regulator (hyperproduced XylS, named XylSh), which is able to activate the lower *meta*-operon in the absence of inducer. As the process goes on, 3MB appears in the system as a product of *m*-xylene conversion by the *upper* pathway. This aromatic compound can now bind what remains of inactive XylSi and switch this regulator into a form (XylSa), which, similar to XylSh, is able to activate *Pm* (and thus the *lower* pathway) as well. Once both the *upper* and the *lower* genetic/biochemical pathways are in operation, the head substrate *m*-xylene is eventually converted into central metabolites (TCA). The result of the signal propagation cycle that starts with *m*-xylene as input is therefore the production of pyruvate and acetaldehyde as outputs of the whole process. The catabolic capacity of the system is fixed not only by substrate concentrations, but also by a large number of physiological control mechanisms (e.g., catabolic repression, growth phase control, etc.) [79–81] that adjust the outputs of each of steps to the growth or stress conditions of the cells. These signals are entered through numerous host factors and endogenous signal molecules: IHF, HU, TurA, PprA, Crc, ppGpp, sigma factor competition, Entner–Doudoroff metabolites, and perhaps several others [73, 75, 77, 78, 82–85].

The regulatory narrative just spelt out (Figure 15.2) is largely based on quantitative measurements of *lacZ* (β-galactosidase) fusions to each of the promoters at stake. Alas, the lack of standardization of the procedures, let alone the absence of any formal parameterization of the transfer functions between one step of the process and the other [86], prevents any systems-level comprehension of the circuit as a whole. However, the information available on different regulatory and metabolic parts is sufficient to assign given on/off states to each of the nodes at various stages of signal propagation. In this context, the sections that follow account for the translation of each of the four regulatory knots that operate on the TOL circuit into a formal description of the corresponding biological functions using the tools of Boolean analyses (Figure 15.3a).

15.5.2
Deconstruction of the Ps–Pr Regulatory Node into Three Autonomous Logic Units

As mentioned above, the *Pr–Ps* region is the one where the regulatory program of the TOL system intensifies, as it encompasses the genes of the two regulators of the system (*xylR*, *xylS*) connected through two sets of overlapping, divergent promoters. This region has been the only one thus far amenable to dynamic

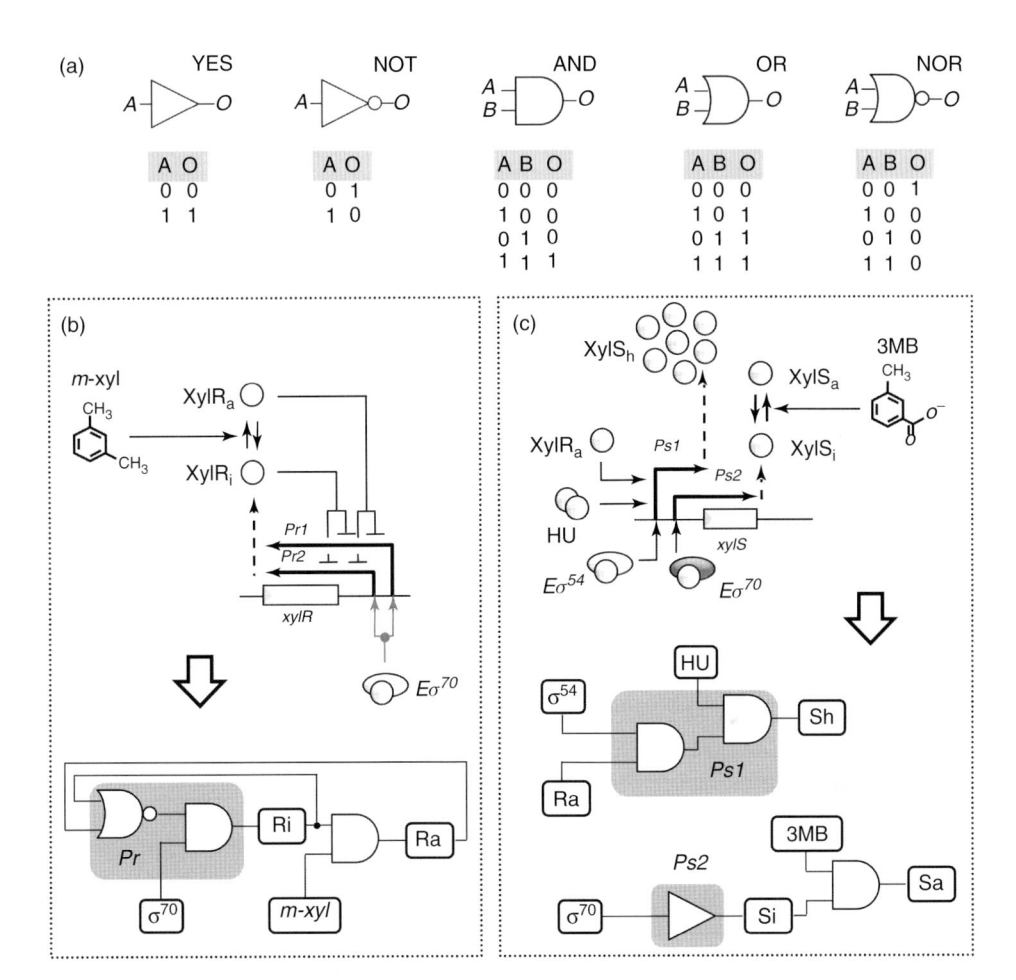

15.5 Boolean Description of m-xylene Biodegradation by P. putida mt-2: the TOL logicome | 293

Figure 15.3 Formalization of the Pr–Ps regulatory node of the TOL network as a set of logic operations. (a) Logic gates used in this work along with their corresponding truth tables. (b) Expression of *xylR* gene from tandem $E\sigma^{70}$ promoters Pr1 and Pr2 and activation of the XylR protein by *m*-xylene. A NOR gate represents the autorepression of Pr promoters brought about by either XylRi (not bound by *m*-xyl) and XylRa (active, bound to *m*-xyl). The formation of XylRi is limited by the action of such autorepression on Pr and is represented by one AND gate which combines with the result of the NOR gate as one of the inputs and $E\sigma^{70}$ as the second input. In turn, XylRi and *m*-xyl form another AND gate that has XylRa as its output. (c) Expression of *xylS* gene from tandem $E\sigma^{54}$ promoter Ps1 and $E\sigma^{70}$ promoter Ps2. The action of these promoters results in two types of XylS and is thus formalized as separate logic clusters. In one case, formation of active XylSh from Ps1 is represented as two connected AND gates which compute three inputs: $E\sigma^{54}$, HU, and XylRa. In the other case, XylSi is produced from constitutive Ps2 through a YES 1 gate with $E\sigma^{70}$. The output is then converted into its active form (XylSi) by means of an AND gate that has XylSh and 3MB as inputs.

modeling [87, 88]. As shown in Figure 15.2, tandem $E\sigma^{70}$ promoters *Pr1* and *Pr2* [89] express the master regulatory *xylR* gene, the product of which (XylRi) represses its own synthesis [90]. For the sake of this abstraction, both *Pr1* and *Pr2* are considered a single $E\sigma^{70}$ promoter *Pr*. The binding of *m*-xylene (input) to XylRi (input) for the production of XylRa (output) can be formalized as an AND gate, the product of which can also repress *xylR* (Figure 15.3b). This can be represented as separate negative feedback loops that shape a NOR gate, the output of which is the input for another AND gate with $E\sigma 70$ *Pr*. On the other hand, the expression of *xylS* involves two transcription-promoting devices, one of them low-constitutive (*Ps2*) and the other dependent on $E\sigma^{54}$ and inducible by XylRa. The still unsettled controversy whether the low-constitutive expression is *Ps* is due to a bona fide $E\sigma^{70}$ promoter [82] or it is a residual activity of a XylRa-dependent $E\sigma^{54}$ promoter [91] makes no difference for our Boolean analyses. *Ps1* and *Ps2* are the operative names given in any case to each of the two devices that promote *xylS* expression either in an inducible or a constitutive manner, respectively. Unlike the case of *Pr*, the two *Ps* promoters have to be formalized separately, because they originate distinct outputs. As shown in Figure 15.3c, the only input of *Ps2* is the housekeeping $E\sigma^{70}$, and its only output is the inactive XylSi, an occurrence that can be represented as a YES gate. In turn, XylSi and 3MB, the product of the *upper* pathway [71], form an AND gate that originates active XylSa as its output. Moreover, XylRa activates *Ps1* with the concourse of $E\sigma^{54}$ and HU. The fact that this action requires three necessary inputs (XylRa, $E\sigma^{54}$, HU) is not an obstacle for our Boolean analysis, because we can disclose a three-input AND gate as the sum of two connected binary gates. The output of this action is hyperexpressed XylSh, a form of the factor that is able to activate *Pm* in the absence of 3MB (see above; [92]). Note that for this analysis, we consider XylSh and XylSi as separate TFs, equally competent for activating *Pm*. While this distinction may not be mechanistically accurate [85], it allows us to separate the activation of *Pm* resulting from the master regulator XylR from that brought about by the formation of 3MB after the action of the *upper* pathway enzymes on *m*-xylene (see below). It has been possible to merge both Boolean approaches with dynamic modeling for representing the *Pr/Ps* node, as the lack of information about the TFs makes it necessary to use (if nothing else) a 0/1 approach to describe the interaction between, for example, $E\sigma^{70}$, $E\sigma^{54}$, and HU and the target promoters [87, 88].

15.5.3
Formalization of Regulatory Events at the Upper and Lower TOL Operons

As shown in Figure 15.2, the *upper* TOL operon encodes three enzymatic activities necessary for the conversion of *m*-xylene into 3MB: xylene monooxygenase (*xylAM*), benzaldehyde dehydrogenase (*xylB*), and benzyl alcohol dehydrogenase (*xylC*). The *upper* pathway bears also other extra genes with uncertain roles [93]. The operon is expressed from the $E\sigma^{54}$-dependent promoter *Pu*, which requires XylRa and IHF for transcription initiation [94, 95]. For the model, we consider the *Pu* promoter to act as a combination of two AND gates (Figure 15.4a) in which XylRa, $E\sigma^{54}$,

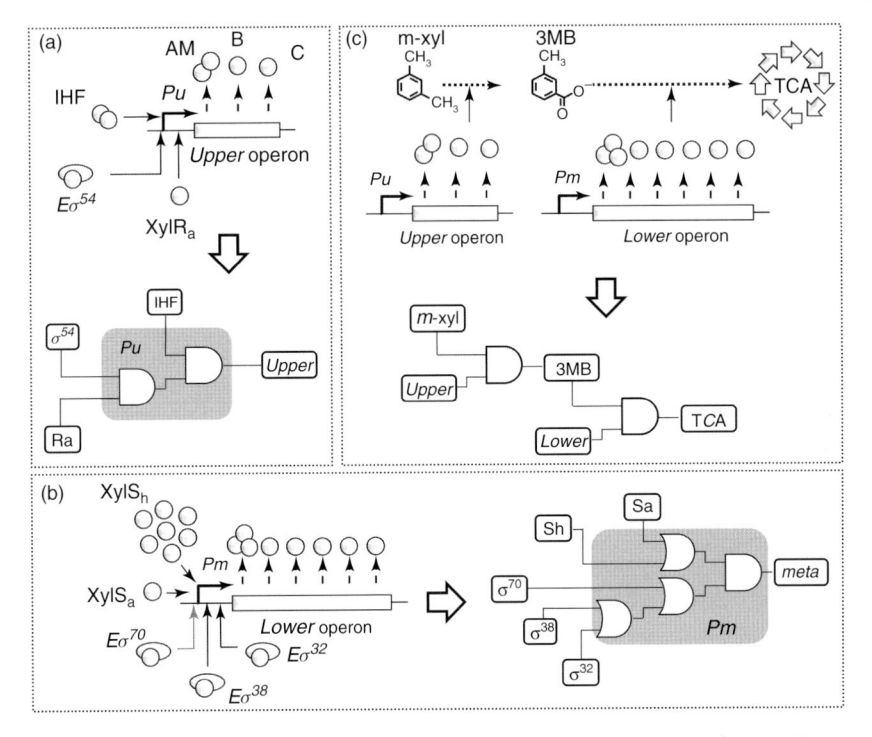

15.5 Boolean Description of m-xylene Biodegradation by P. putida mt-2: the TOL logicome

Figure 15.4 Logic organization of the metabolic reactions of the TOL plasmid. (a) The action of the *Pu* promoter is represented as two combined AND gates with XylRa, IHF, and $E\sigma^{70}$ as inputs, and having the *upper* pathway as their output. (b) Expression of the *lower* pathway corresponds to four logic gates which compute the presence of five different inputs. At least one sigma factor (out of the three possible) is necessary for the expression of the *Pm* promoter, along with either XylSh or XylSi. This is signified in the figure as one cascade of OR gates for the sigmas and another OR gate for the XylS forms, which eventually converge in an AND gate that has the *lower* TOL pathway as its output. (c) Formalization of metabolic processes. 3MB is produced from *m*-xyl by means of the *upper* pathway enzymes, while TCA is produced from 3MB metabolism through *lower* pathway. *m*-xyl and the *upper* route thus form the AND gate that originates 3MB. By the same token, TCA is the output of the AND gate that has 3MB and the *lower* pathway as inputs. Note the central position of 3MB as the signal carrier between the *upper* and the *lower* metabolic blocks.

and IHF are the inputs and the entire catalytic complement of the *upper* pathway abstracted as a single output. The second metabolic operon (encoding the *lower* TOL pathway) encodes a set of enzymatic activities able to convert 3MB into 3-methyl catechol (*xylXYZL*), followed of meta-cleavage of the dihydroxylated ring for the formation of a semialdehyde (*xylE*) and eventual routing of this product into the TCA cycle (*xylFJK*). As was the case with the *upper* pathway, these core activities go together with other genes of unknown function [93]. And as before also, they are considered model-wise as the single output of the *Pm* promoter (Figure 15.3b). Transcription initiation in this case can be elicited by any of the active forms of

XylS (XylSa or XylSh; [92, 96]). Moreover, the core RNA Polymerase (RNAP) that activates *Pm* can employ any of the three sigma factors of the $E\sigma^{54}$ family, that is, the housekeeping $E\sigma^{70}$ (RpoD), the stationary phase $E\sigma^{38}$ (RpoS), and the heat shock $E\sigma^{32}$ (RpoH; [82, 85, 97]). This scenario for *Pm* regulation can be depicted as a gate with five possible inputs and one single output, the binary computation of which can be broken down to discrete OR and AND gates as shown in Figure 15.4b.

15.5.4
3MB Is the Endogenous Signal Carrier through the Domains of the TOL Network

The primary input to the TOL system is *m*-xylene, which once computed by the logic gates of the *Pr–Ps* region (see above) originates two active TFs as distinct outputs, XylRa and XylSh. These can then act as inputs of the *Pu* and the *Pm* promoter, respectively. Up to that point, all actions are regulatory. However, the outputs of both the upper and the lower operons are enzymatic activities, not regulators. We can also translate such activities as useful inputs in the network. The upper pathway and *m*-xylene are inputs of an AND gate which has 3MB as an output signal (Figure 15.4c). In turn, 3MB can form AND gates with both XylSi (to become XylSa, see above) and the lower TOL pathway (to produce TCA cycle intermediates). Note that under this scheme, the endogenously produced 3MB becomes the key signal carrier molecule that can be read by both regulatory and metabolic nodes of the entire TOL network, thereby connecting the activities of the *upper* and *lower* domains of the system. Finally, the eventual output of the whole metabolic and regulatory system is the complete degradation of 3MB into pyruvate and acetaldehyde [71, 98, 99], which are channeled into the central metabolism, for example, the TCA.

15.5.5
The TOL Logicome

The result of connecting all nodes of the TOL system in the shape of binary Boolean gates was a body of logic operations (a *logicome*) able to compute defined environmental signals into fixed biological responses by means of both regulatory and enzymatic actions [7, 100]. The *logicome* (Figure 15.5) includes one exogenous (*m*-xylene) and six endogenous inputs: IHF, HU, σ^{70}, σ^{54} (σ^N), $\sigma^{32}(\sigma^H)$, and $\sigma^{38}(\sigma^S)$. In addition, the system contains one inborn signal carrier molecule (3MB) and one single outward output (TCA). For the sake of simplicity, we have not considered as inputs other nutritional signals (presence of alternative C sources) or environmental circumstances (temperature, growth phase) that are known to affect the performance of the system [101]. We thus assume that cells that bear such a logicome undergo favorable growth conditions for optimal expression of TOL activities.

Inspection of the circuit of Figure 15.5 suggests three possible emergent properties of the TOL system that originate in the topology, connectivity, and logic of the network rather than on the nature of its individual components. First, the intricate

15.5 Boolean Description of m-xylene Biodegradation by P. putida mt-2: the TOL logicome

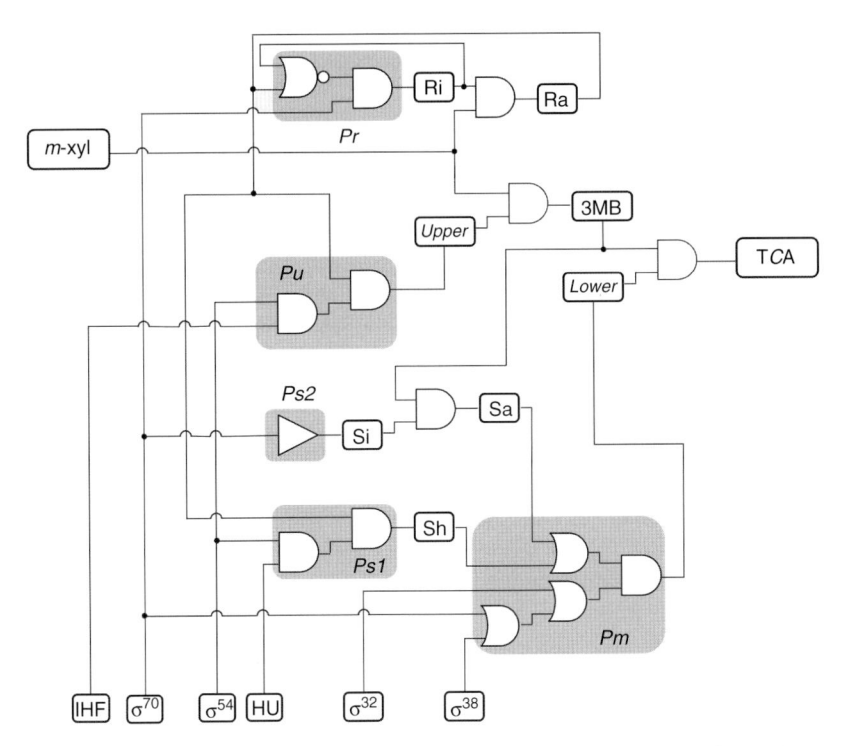

Figure 15.5 Logicome of the TOL network. The circuit shown represents the entire computation performed by the TOL network of Figure 15.1. The computation itself is through the set of logic operators inside the shadowed section, while the inputs and final output have been placed outside. The one exogenous input of the *logicome* is *m*-xyl, which triggers the activation of the entire regulatory and metabolic program that eventually produces TCA as the ultimate output. Gates inside the gray boxes include the molecular operations performed by the six promoters of the network (*Pr1*, *Pr2*, *Ps1*, *Ps2*, *Pu*, and *Pm*), while the others represent biochemical events (i.e., XylR and XylS binding to their respective effectors and metabolic conversions). The simplest computation is performed by Ps2, in which an amplifier gate takes a single input ($E\sigma^{70}$) and converts it in a single output (XylSi). The most complex signal processing is performed by Pm.

regulatory architecture gravitates not so much around the head substrate of the metabolic pathway (*m*-xylene) but about one metabolic intermediate (3MB). Note that activation of the *upper* pathway by XylR and *m*-xylene and the *lower* pathway by overproduced XylS are virtually concomitant events. This suggests that the enzymes for degradation of 3MB (encoded by the lower pathway) are probably in place before the target compound materializes in the cell. This is surprising, because it is just the opposite of the activation of the *upper* pathway by *m*-xylene (that happens only if *m*-xylene is indeed around). Furthermore, it is possible to induce only the lower pathway with exogenously added 3MB [21], so there is no need to have the lower genes expressed before the target compound appears. The regulatory arrangement of the extant architecture seems, in consequence, to be poised to avoid intracellular accumulation of 3MB. A second property might derive from the autorepressing manner in which the master regulator XylR is expressed. Since *m*-xylene addition both activates XylR and downregulates its transcription, it is plausible that a certain set of parameters makes the regulatory node *Ps–Pr* originate transcription pulses rather than behaving in a monotonic manner [102, 103]. Finally, the occurrence of two separate pathways regulated by different but hierarchically connected TFs, one of them perhaps subject to oscillations, makes it possible that (also depending on parameters) the *upper* and the *lower* operons are not simultaneously active in the same cell. If so, *m*-xylene biodegradation at any given time might result from the action of two cooperating subpopulations rather than through the deeds of a homogeneous catalytic pool. The actual incidence of such predicted emergent properties of the TOL network relies on kinetic parameters at distinct nodes of the *logicome*, an issue that still awaits future work.

15.6
Conclusion and Outlook

Adoption of a limited number of judicious abstractions and reasonable formalisms allows description, simulation, and prediction of the functioning of biological networks, whether metabolic, regulatory, or both. A separate issue is the interplay between the circuits (Boolean, stochastic, or ODE discussed above) and the genetic and biochemical cell-wide chassis into which they are integrated in a hierarchical manner [26, 27]. Since complexity of any system increases nonlinearly with the number of components, it is hardly possible to generate experimentally all relevant parameters that influence the behavior of each node of even relatively simple control circuits that operate in bacteria. Fortunately, formal languages and methods are becoming available to handle biological systems with data barely beyond the list of known components, the sign of their connections, and some scattered *wet* information on some of them. For genome-scale metabolic networks, constraint-based formalisms provide valuable scaffolds to account for interacting species and to ascertain the emergent properties arising from these interactions to explore the metabolic space and capabilities of the organisms involved. For those scarce systems for which kinetic and molecular information is abundant, both deterministic differential equations and stochastic master equations become the tools of choice for robust modeling. But for the rest of the systems for which data is sparse and originated in various laboratories and lacking a coherent format, one has to rely on minimal formalisms that capture the gross *raison d'être* of extant network architectures while lacking the details. Yet, the ultimate agenda of these endeavors, in particular implementation of Boolean tools, is not only their understanding but also their biotechnological action. Logic networks of the type discussed above share many of the qualities of the electronic processors: for example, they execute computation of signals into responses [104–106]. The notion that bacteria behave to an extent as computers making computers has been even entertained in the recent literature [107]. Yet, in electronic circuits the signal carrier is only electrons, whereas the material nature of the input in biological circuits is often different from the output. Engineering signal propagation cascades in biological systems thus requires that any given upstream input to one of the nodes results in an output that can be understood by the next node of the signal progression chain. The chances of designing biological circuits with the same ease as electronic counterparts depends on the availability of (naturally occurring or artificial) connectable logic gates, a currently fertile field of research [52, 59, 108]. As discussed elsewhere [19], such approaches stemming from contemporary systems and synthetic biology offer a second opportunity for attempting the design of microorganisms for environmental release as vectors of biodegradative activities for bioremediation and sensing, to reprogram effectively and efficiently microorganisms for the production *à la carte* of bulk and high-added-value chemicals, and to design intervention strategies for medical applications. In addition, the output of biological logic gates can be engineered to produce electron-carrier molecules (e.g., phenazines [109]) or coupled to the generation of other electrochemical signals [110], thereby opening an immense field of application. All of this could not be achieved earlier because of the lack of sufficient systems understanding of the corresponding microbial agents and of their underlying networks and interactions [111]. The technological advances of the past decade and the conceptual developments such as those discussed herein pave the road to the eventual achievement of these long-sought goals.

Acknowledgments

This work was funded by the BIO and FEDER CONSOLIDER-INGENIO Program of the Spanish Ministry of Science and Innovation (MICINN), the MICROME and ST-FLOW Contracts of the EU and the PROMT Project Autonomous Community of Madrid. RS-R is grateful to the MCINN for a PhD stipend. We thank Vitor dos Santos (U. Wageningen) and Hidde de Jong (INRIA, Grenoble) for helpful discussions.

References

1. McAdams, H.H., Srinivasan, B., and Arkin, A.P. (2004) *Nat. Rev. Genet.*, **5**, 169–178.
2. Cases, I., de van Lorenzo, V., and Ouzounis, C.A. (2003) *Trends Microbiol.*, **11**, 248–253.
3. Konstantinidis, K.T. and Tiedje, J.M. (2004) *Proc. Natl. Acad. Sci. U.S.A.*, **101**, 3160–3165.
4. Dos Santos, V.A., Heim, S., Moore, E.R., Stratz, M., and Timmis, K.N. (2004) *Environ. Microbiol.*, **6**, 1264–1286.
5. Shen-Orr, S.S., Milo, R., Mangan, S., and Alon, U. (2002) *Nat. Genet.*, **31**, 64–68.
6. de Las Heras, A., Carreno, C.A., Martinez-Garcia, E., and de Lorenzo, V. (2010) *FEMS Microbiol. Rev.*, **34**, 842–865.
7. Istrail, S., De-Leon, S.B., and Davidson, E.H. (2007) *Dev. Biol.*, **310**, 187–195.
8. Silva-Rocha, R. and de Lorenzo, V. (2010) *Annu. Rev. Microbiol.*, **64**, 257–275.
9. Mangan, S. and Alon, U. (2003) *Proc. Natl. Acad. Sci. U.S.A.*, **100**, 11980–11985.
10. Mangan, S., Zaslaver, A., and Alon, U. (2003) *J. Mol. Biol.*, **334**, 197–204.

11. Shingler, V. (2003) *Environ. Microbiol.*, **5**, 1226–1241.
12. Rojo, F. (2010) *FEMS Microbiol. Rev.*, **34**, 658–684.
13. Kotte, O., Zaugg, J.B., and Heinemann, M. (2010) *Mol. Syst. Biol.*, **6**, 355.
14. Balazsi, G., Barabasi, A.L., and Oltvai, Z.N. (2005) *Proc. Natl. Acad. Sci. U.S.A.*, **102**, 7841–7846.
15. Bruckner, R. and Titgemeyer, F. (2002) *FEMS Microbiol. Lett.*, **209**, 141–148.
16. Velazquez, F., Parro, V., and de Lorenzo, V. (2005) *Mol. Microbiol.*, **57**, 1557–1569.
17. Dominguez-Cuevas, P., Gonzalez-Pastor, J.E., Marques, S., Ramos, J.L., and de Lorenzo, V. (2006) *J. Biol. Chem.*, **281**, 11981–11991.
18. Weber, W. and Fussenegger, M. (2010) *Curr. Opin. Biotechnol.*, **21**, 690–696.
19. de Lorenzo, V. (2008) *Curr. Opin. Biotechnol.*, **19**, 579–589.
20. Karlebach, G. and Shamir, R. (2008) *Nat. Rev. Mol. Cell Biol.*, **9**, 770–780.
21. Ramos, J.L. and Marques, S. (1997) *Annu. Rev. Microbiol.*, **51**, 341–373.
22. Thiele, I. and Palsson, B.O. (2010) *Nat. Protoc.*, **5**, 93–121.
23. DeJongh, M., Formsma, K., Boillot, P., Gould, J., Rycenga, M., and Best, A. (2007) *BMC Bioinformatics*, **8**, 139.
24. Notebaart, R.A., van Enckevort, F.H., Francke, C., Siezen, R.J., and Teusink, B. (2006) *BMC Bioinformatics*, **7**, 296.
25. Reed, J.L., Patel, T.R., Chen, K.H., Joyce, A.R., Applebee, M.K., Herring, C.D., Bui, O.T., Knight, E.M., Fong, S.S., and Palsson, B.O. (2006) *Proc. Natl. Acad. Sci. U.S.A.*, **103**, 17480–17484.
26. Puchalka, J., Oberhardt, M.A., Godinho, M., Bielecka, A., Regenhardt, D., Timmis, K.N., Papin, J.A., and Martins dos Santos, V.A. (2008) *PLoS Comput. Biol.*, **4**, e1000210.
27. Nogales, J., Palsson, B.O., and Thiele, I. (2008) *BMC Syst. Biol.*, **2**, 79.
28. Feist, A.M., Herrgard, M.J., Thiele, I., Reed, J.L., and Palsson, B.O. (2009) *Nat. Rev. Microbiol.*, **7**, 129–143.
29. Jaqaman, K. and Danuser, G. (2006) *Nat. Rev. Mol. Cell Biol.*, **7**, 813–819.
30. Breitling, R., Vitkup, D., and Barrett, M.P. (2008) *Nat. Rev. Microbiol.*, **6**, 156–161.
31. Herrgard, M.J., Fong, S.S., and Palsson, B.O. (2006) *PLoS Comput. Biol.*, **2**, e72.
32. Schlitt, T. and Brazma, A. (2007) *BMC Bioinformatics*, **8** (Suppl. 6), S9.
33. Price, M.N., Dehal, P.S., and Arkin, A.P. (2007) *PLoS Comput. Biol.*, **3**, 1739–1750.
34. Milanesio, P., Arce-Rodríguez, A., Muñoz, A., Calles, B., and de Lorenzo, V. (2010) *Environ. Microbiol.*, **13**, 324–339.
35. Hale, V., Keasling, J.D., Renninger, N., and Diagana, T.T. (2007) *Am. J. Trop. Med. Hyg.*, **77**, 198–202.
36. Ren, B., Robert, F., Wyrick, J.J., Aparicio, O., Jennings, E.G., Simon, I., Zeitlinger, J., Schreiber, J., Hannett, N., Kanin, E., Volkert, T.L., Wilson, C.J., Bell, S.P., and Young, R.A. (2000) *Science*, **290**, 2306–2309.
37. Harbison, C.T., Gordon, D.B., Lee, T.I., Rinaldi, N.J., Macisaac, K.D., Danford, T.W., Hannett, N.M., Tagne, J.B., Reynolds, D.B., Yoo, J., Jennings, E.G., Zeitlinger, J., Pokholok, D.K., Kellis, M., Rolfe, P.A., Takusagawa, K.T., Lander, E.S., Gifford, D.K., Fraenkel, E., and Young, R.A. (2004) *Nature*, **431**, 99–104.
38. Faith, J.J., Hayete, B., Thaden, J.T., Mogno, I., Wierzbowski, J., Cottarel, G., Kasif, S., Collins, J.J., and Gardner, T.S. (2007) *PLoS Biol.*, **5**, e8.
39. Segal, E., Shapira, M., Regev, A., Pe'er, D., Botstein, D., Koller, D., and Friedman, N. (2003) *Nat. Genet.*, **34**, 166–176.
40. Ronen, M., Rosenberg, R., Shraiman, B.I., and Alon, U. (2002) *Proc. Natl. Acad. Sci. U.S.A.*, **99**, 10555–10560.
41. de Jong, H. (2002) *J. Comput. Biol.*, **9**, 67–103.
42. Gillespie, D.T. (1977) *J. Phys. Chem.*, **81**, 2340–2361.
43. McAdams, H.H. and Arkin, A. (1999) *Trends Genet.*, **15**, 65–69.
44. van Kampen, N.G. (1992) *Stochastic Processes in Physics and Chemistry*, Elsevier, Amsterdam.
45. Raj, A. and van Oudenaarden, A. (2008) *Cell*, **135**, 216–226.
46. Mendes, P. (1993) *Comput. Appl. Biosci.*, **9**, 563–571.

47. Hoffmann, A., Levchenko, A., Scott, M.L., and Baltimore, D. (2002) *Science*, **298**, 1241–1245.
48. Morris, M.K., Saez-Rodriguez, J., Sorger, P.K., and Lauffenburger, D.A. (2010) *Biochemistry*, **49**, 3216–3224.
49. Polynikis, A., Hogan, S.J., and di Bernardo, M. (2009) *J. Theor. Biol.*, **261**, 511–530.
50. Friedman, N., Linial, M., Nachman, I., and Pe'er, D. (2000) *J. Comput. Biol.*, **7**, 601–620.
51. Ashyraliyev, M., Fomekong-Nanfack, Y., Kaandorp, J.A., and Blom, J.G. (2009) *FEBS Lett.*, **276**, 886–902.
52. Silva-Rocha, R. and de Lorenzo, V. (2008) *FEBS Lett.*, **582**, 1237–1244.
53. Buchler, N.E., Gerland, U., and Hwa, T. (2003) *Proc. Natl. Acad. Sci. U.S.A.*, **100**, 5136–5141.
54. Hasty, J., McMillen, D., and Collins, J.J. (2002) *Nature*, **420**, 224–230.
55. Wittmann, D.M., Blochl, F., Trumbach, D., Wurst, W., Prakash, N., and Theis, F.J. (2009) *PLoS Comput. Biol.*, **5**, e1000569.
56. Andrianantoandro, E., Basu, S., Karig, D.K., and Weiss, R. (2006) *Mol. Syst. Biol.*, **2**, 2006.0028.
57. Mayo, A.E., Setty, Y., Shavit, S., Zaslaver, A., and Alon, U. (2006) *PLoS Biol.*, **4**, e45.
58. Silva-Rocha, R. and de Lorenzo, V. (2011) *Mol. BioSyst.*, **7**, 2389–2396.
59. Hunziker, A., Tuboly, C., Horvath, P., Krishna, S., and Semsey, S. (2010) *Proc. Natl. Acad. Sci. U.S.A.*, **107**, 12998–13003.
60. Niazov, T., Baron, R., Katz, E., Lioubashevski, O., and Willner, I. (2006) *Proc. Natl. Acad. Sci. U.S.A.*, **103**, 17160–17163.
61. Katz, E. and Privman, V. (2010) *Chem. Soc. Rev.*, **39**, 1835–1857.
62. Pita, M., Minko, S., and Katz, E. (2009) *J. Mater. Sci. Mater. Med.*, **20**, 457–462.
63. Johnson, C.G., Goldman, J.P., and Gullick, W.J. (2004) *Prog. Biophys. Mol. Biol.*, **86**, 379–406.
64. Shmulevich, I., Dougherty, E.R., Kim, S., and Zhang, W. (2002) *Bioinformatics*, **18**, 261–274.
65. de Jong, H., Geiselmann, J., Hernandez, C., and Page, M. (2003) *Bioinformatics*, **19**, 336–344.
66. Baldazzi, V., Ropers, D., Markowicz, Y., Kahn, D., Geiselmann, J., and de Jong, H. (2010) *PLoS Comput. Biol.*, **6**, e1000812.
67. Alon, U. (2006) *An Introduction to Systems Biology: Design Principles of Biological Circuits*, Chapman and Hall/CRC, New York.
68. Phale, P.S., Basu, A., Majhi, P.D., Deveryshetty, J., Vamsee-Krishna, C., and Shrivastava, R. (2007) *Omics*, **11**, 252–279.
69. Tropel, D. and Van der Meer, J.R. (2004) *Microbiol. Mol. Biol. Rev.*, **68**, 474–500.
70. Carmona, M., Zamarro, M.T., Blazquez, B., Durante-Rodriguez, G., Juarez, J.F., Valderrama, J.A., Barragan, M.J., Garcia, J.L., and Diaz, E. (2009) *Microbiol. Mol. Biol. Rev.*, **73**, 71–133.
71. Ramos, J.L., Marques, S., and Timmis, K.N. (1997) *Annu. Rev. Microbiol.*, **51**, 341–373.
72. Holtel, A., Abril, M.A., Marques, S., Timmis, K.N., and Ramos, J.L. (1990) *Mol. Microbiol.*, **4**, 1551–1556.
73. de Lorenzo, V., Herrero, M., Metzke, M., and Timmis, K.N. (1991) *EMBO J.*, **10**, 1159–1167.
74. Abril, M.A., Buck, M., and Ramos, J.L. (1991) *J. Biol. Chem.*, **266**, 15832–15838.
75. Perez-Martin, J. and de Lorenzo, V. (1995) *J. Bacteriol.*, **177**, 3758–3763.
76. Perez-Martin, J. and de Lorenzo, V. (1997) *J. Bacteriol.*, **179**, 2757–2760.
77. Rescalli, E., Saini, S., Bartocci, C., Rychlewski, L., de Lorenzo, V., and Bertoni, G. (2004) *J. Biol. Chem.*, **279**, 7777–7784.
78. Vitale, E., Milani, A., Renzi, F., Galli, E., Rescalli, E., de Lorenzo, V., and Bertoni, G. (2008) *Mol. Microbiol.*, **69**, 698–713.
79. Holtel, A., Marques, S., Mohler, I., Jakubzik, U., and Timmis, K.N. (1994) *J. Bacteriol.*, **176**, 1773–1776.
80. Cases, I., Perez-Martin, J., and de Lorenzo, V. (1999) *J. Biol. Chem.*, **274**, 15562–15568.
81. del Castillo, T. and Ramos, J.L. (2007) *J. Bacteriol.*, **189**, 6602–6610.

82. Gallegos, M.T., Marques, S., and Ramos, J.L. (1996) *J. Bacteriol.*, **178**, 2356–2361.
83. Carmona, M., Rodriguez, M.J., Martinez-Costa, O., and de Lorenzo, V. (2000) *J. Bacteriol.*, **182**, 4711–4718.
84. Aranda-Olmedo, I., Ramos, J.L., and Marques, S. (2005) *Appl. Environ. Microbiol.*, **71**, 4191–4198.
85. Dominguez-Cuevas, P., Marin, P., Ramos, J.L., and Marques, S. (2005) *J. Biol. Chem.*, **280**, 41315–41323.
86. Endler, L., Rodriguez, N., Juty, N., Chelliah, V., Laibe, C., Li, C., and Le Novere, N. (2009) *J. R. Soc. Interface*, **6** (Suppl 4), S405–S417.
87. Koutinas, M., Kiparissides, A., Silva-Rocha, R., Lam, M.C., Martins Dos Santos, V.A., de Lorenzo, V., Pistikopoulos, E.N., and Mantalaris, A. (2011) *Metab. Eng.*, **13**, 401–413.
88. Koutinas, M., Lam, M.C., Kiparissides, A., Silva-Rocha, R., Godinho, M., Livingston, A.G., Pistikopoulos, E.N., de Lorenzo, V., Dos Santos, V.A., and Mantalaris, A. (2010) *Environ. Microbiol.*, **12**, 1705–1718.
89. Inouye, S., Nakazawa, A., and Nakazawa, T. (1985) *J. Bacteriol.*, **163**, 863–869.
90. Bertoni, G., Perez-Martin, J., and de Lorenzo, V. (1997) *Mol. Microbiol.*, **23**, 1221–1227.
91. Perez-Martin, J. and De Lorenzo, V. (1995) *Proc. Natl. Acad. Sci. U.S.A.*, **92**, 7277–7281.
92. Mermod, N., Ramos, J.L., Bairoch, A., and Timmis, K.N. (1987) *Mol. Gen. Genet.*, **207**, 349–354.
93. Greated, A., Lambertsen, L., Williams, P.A., and Thomas, C.M. (2002) *Environ. Microbiol.*, **4**, 856–871.
94. Bertoni, G., Fujita, N., Ishihama, A., and de Lorenzo, V. (1998) *EMBO J.*, **17**, 5120–5128.
95. Calb, R., Davidovitch, A., Koby, S., Giladi, H., Goldenberg, D., Margalit, H., Holtel, A., Timmis, K., Sanchez-Romero, J.M., de Lorenzo, V., and Oppenheim, A.B. (1996) *J. Bacteriol.*, **178**, 6319–6326.
96. Inouye, S., Nakazawa, A., and Nakazawa, T. (1981) *J. Bacteriol.*, **145**, 1137–1143.
97. Marques, S., Manzanera, M., Gonzalez-Perez, M.M., Gallegos, M.T., and Ramos, J.L. (1999) *Mol. Microbiol.*, **31**, 1105–1113.
98. Assinder, S.J. and Williams, P.A. (1990) *Adv. Microb. Physiol.*, **31**, 1–69.
99. Burlage, R.S., Hooper, S.W., and Sayler, G.S. (1989) *Appl. Environ. Microbiol.*, **55**, 1323–1328.
100. Istrail, S. and Davidson, E.H. (2005) *Proc. Natl. Acad. Sci. U.S.A.*, **102**, 4954–4959.
101. Ruiz, R., Aranda-Olmedo, M.I., Dominguez-Cuevas, P., Ramos-Gonzalez, M.I., and Marques, S. (2004) in *Pseudomonas*, vol. **2** (ed. J.L. Ramos), Kluwer Academic/Plenum Publishers, New York, pp. 509–537.
102. Kaplan, S., Bren, A., Dekel, E., and Alon, U. (2008) *Mol. Syst. Biol.*, **4**, 203.
103. Kaplan, S., Bren, A., Zaslaver, A., Dekel, E., and Alon, U. (2008) *Mol. Cell*, **29**, 786–792.
104. Rodrigo, G., Carrera, J., Prather, K.J., and Jaramillo, A. (2008) *Bioinformatics*, **24**, 2554–2556.
105. Rodrigo, G. and Jaramillo, A. (2007) *Syst. Synth. Biol.*, **1**, 183–195.
106. Marchisio, M.A. and Stelling, J. (2009) *Curr. Opin. Biotechnol.*, **20**, 479–485.
107. Danchin, A. (2009) *FEMS Microbiol. Rev.*, **33**, 3–26.
108. Zhan, J., Ding, B., Ma, R., Ma, X., Su, X., Zhao, Y., Liu, Z., Wu, J., and Liu, H. (2010) *Mol. Syst. Biol.*, **6**, 388.
109. Li, Z., Rosenbaum, M.A., Venkataraman, A., Tam, T.K., Katz, E., and Angenent, L.T. (2011) *Chem. Commun.*, **47**, 3060–3062.
110. Ron, E.Z. and Rishpon, J. (2010) *Adv. Biochem. Eng. Biotechnol.*, **117**, 77–84.
111. Cases, I. and de Lorenzo, V. (2005) *Int. Microbiol.*, **8**, 213–222.

16
Qualitative and Quantitative Aspects of a Model for Processes Inspired by the Functioning of the Living Cell

Andrzej Ehrenfeucht, Jetty Kleijn, Maciej Koutny, and Grzegorz Rozenberg

16.1
Introduction

Natural computing is concerned with human-designed computing inspired by nature and with computing taking place in nature (see, e.g., [1] and [2]). The former investigates models and computational techniques inspired by nature, while the latter investigates, in terms of information processing, phenomena taking place in nature. The former strand includes research areas such as evolutionary computation, neural computation, quantum computation, and molecular computation. The latter strand includes investigations into the computational nature of self-assembly, the computational nature of developmental processes, the computational nature of brain processes, the system biology approach to bionetworks, and the computational nature of biochemical reactions. Clearly, the two research strands are not disjoint.

Biomolecular computation is a topic of intense research in natural computing. Within the former strand, this research focuses on constructing, either *in vitro* or *in vivo*, various building blocks of computing devices (such as switches, gates, and biosensors). Within the latter strand, this research is more concerned with establishing how biocomputations drive natural processes – the essence of this research is nicely captured by the following statement by Richard Dawkins, a world leading expert in evolutionary biology (see [3]): "If you want to understand life, don't think about vibrant throbbing gels and oozes, think about information technology."

This chapter falls into this second strand of research. It discusses reaction systems which are a formal model for the investigation of the functioning of the living cell. The functioning is viewed in terms of formal processes resulting from interactions between biochemical reactions taking place in the living cell. Moreover, we assume that these interactions are driven by two mechanisms, namely, facilitation and inhibition: the (products of the) reactions may facilitate or inhibit each other. The basic model of reaction systems abstracts from various (technical) features of biochemical reactions to such an extent that it becomes a qualitative rather than quantitative model. However, it takes into account the basic bioenergetics (flow of energy) of the living cell, and it also takes into account that the living cell is an open

Biomolecular Information Processing: From Logic Systems to Smart Sensors and Actuators,
First Edition. Edited by Evgeny Katz.
© 2012 Wiley-VCH Verlag GmbH & Co. KGaA. Published 2012 by Wiley-VCH Verlag GmbH & Co. KGaA.

system and its behavior is influenced by its environment. The broader *framework of reaction systems* is formed by the central model of reaction systems and its various extensions. The main focus of research is on understanding processes that take place in these models.

The chapter can be seen as consisting of two parts. The first part (Sections 16.2, 16.3, and 16.4) reviews the main notions (together with underlying motivation) of reaction systems. As already mentioned, the basic model of reaction systems is qualitative. However, there are various situations in biology/biochemistry where one needs to consider quantities assigned to the states of a biochemical system. To account for this, the broad framework of reaction systems includes the notion of reaction systems with measurements – they are recalled in Section 16.6. Then, Sections 16.7–16.9 discuss various ways of dealing with quantitative parameters in reaction systems. These sections present new material (the theorem presented in Section 16.6 is also new). The discussion in Section 16.10 concludes this chapter.

16.2
Reactions

The formal notion of a reaction (introduced in [4]) formalizes the basic intuition of a biochemical reaction – it will take place if all its reactants are present and none of its inhibitors is present; when a reaction takes place, it creates its products.

Definition 1. A reaction is a triplet $b = (R, I, P)$ such that R, I, P are finite nonempty sets with $R \cap I = \emptyset$.

The sets R, I, P are called the *reactant set of b*, the *inhibitor set of b*, and the *product set of b*, respectively – they are also denoted as R_b, I_b and P_b, respectively. If $R, I, P \subseteq Z$ for a finite set Z, then we say that b is a *reaction in Z*. We use $rac(Z)$ to denote the set of all reactions in Z – note that $rac(Z)$ is finite.

To define the effect of a set of reactions on a current state of the living cell we first define the effect of a single reaction.

Definition 2. Let Z be a finite set and let $T \subseteq Z$. Let $b \in rac(Z)$. Then b is enabled by T, denoted by $en_b(T)$, if $R_b \subseteq T$ and $I_b \cap T = \emptyset$. The result of b on T, denoted by $res_b(T)$, is defined by $res_b(T) = P_b$ if $en_b(T)$, and $res_b(T) = \emptyset$ otherwise.

Here, a finite set T formalizes a state of the cell, that is, the set of biochemical entities currently present in the cell. Then b is enabled by T if T separates R_b from I_b, meaning that all reactants from R_b are present in T and none of the inhibitors from I_b is present in T. When b is enabled by T, it contributes its product P_b to the successor state; otherwise, it does not contribute anything to the successor state.

The effect of a set of reactions on a current state of the cell is cumulative, which is formally defined as follows:

Definition 3. Let Z be a finite set, let $T \subseteq Z$, and let $B \subseteq rac(Z)$. The result of B on T, denoted by $res_B(T)$, is defined by $res_B(T) = \bigcup \{res_b(T) : b \in B\}$.

Note that if the transition from a current state to its successor is determined only by the reactions (i.e., there is no influence of the environment), then the successor state consists only of entities produced by the reactions enabled in the current state. This implies that in the transition from a current state to its successor state, an *entity from T vanishes unless it is sustained/produced by a reaction*. This is the *nonpermanency property* and it reflects the basic bioenergetics of the living cell: *without the flow/supply of energy, the living cell disintegrates, but the use/absorption of energy by the living cell is realized through biochemical reactions* (see, e.g., [5]).

Although this basic definition implies "instant nonpermanency" (an entity vanishes within *one* state transition unless it is produced by a reaction), we also consider a finite duration of entities (corresponding to their presence in several consecutive states) which takes into account the decay time (see, e.g., [6]).

There is another notable aspect of Definition 3. If a, b are two reactions from B enabled by T, then both of them will take place even if $R_a \cap R_b \neq \emptyset$. Hence, we do not have here the notion of conflict between reactions even if they need to share reactants. This is the property of the *threshold nature of resources: either an entity is available and then there is enough of it, or it is not available*. This property reflects the *level of abstraction* we have adopted for the formulation of our basic model: we do not count concentrations of entities/molecules to infer from these which reactions can/will be applied. We operate on a higher level of abstraction: we assume that the cell is running/functioning and we want to understand the ongoing processes.

This level of abstraction can be compared with the level of abstraction of the standard models of computation in computer science, such as Turing machines and finite automata. These standard models turned out to be very successful in understanding computational processes running on electronic computers, and yet nothing in these models takes into account the electronic/quantitative properties of the underlying hardware. It is simply assumed that the underlying electronics/hardware functions "well" and then the goal is to understand processes running on (implemented by) this hardware. Similarly, we want to understand the processes carried out in the functioning living cell. At this stage, we are not interested in the underlying "hardware properties" of the living cell, but rather in the resulting processes.

Thus *our basic model is qualitative rather than quantitative – in particular, there is no counting here*.

16.3
Reaction Systems

Now that the formal notion of a reaction and its effect on states have been established, we can proceed to define reaction systems (introduced in [4]), which are our abstract model of the functioning of the living cell.

Definition 4. A reaction system, abbreviated rs, is an ordered pair $\mathcal{A} = (S, A)$, where S is a finite nonempty set and A is a finite subset of $rac(S)$.

The set S is called the *background set of* $\mathcal{A}$, and its elements are called the *entities of* $\mathcal{A}$ – they represent molecular entities (e.g., atoms, ions, molecules) that may be present in the states of the biochemical system (e.g., the living cell). The set A is called the *set of reactions of* $\mathcal{A}$; clearly A is finite (as S is finite).

The subsets of S are called the *states of* $\mathcal{A}$. Given a state $T \subseteq S$, the *result of* $\mathcal{A}$ *on* T, denoted by $res_\mathcal{A}(T)$, is defined by $res_\mathcal{A}(T) = res_A(T)$.

Thus a reaction system is essentially a set of reactions. We also specify the background set, which consists of entities needed for defining the reactions and for reasoning about the system (see the definition of an interactive process below). There are no "structures" involved in reaction systems (such as, e.g., the tape of a Turing machine). Finally, note that this is a *strictly finite model* – its size is restricted by the size of the background set.

We note here that the nonpermanency property is a major difference between reaction systems and the models considered in the theory of computation (see, e.g., [7] and [8]). Also, the threshold nature of the resources (no conflict) property is a major difference with structural models of concurrency, such as, for example, Petri nets [9].

The model of reaction systems formalizes the "static structure" of the living cell as the set of all reactions of the cell (together with the set of underlying entities). What we are really interested in are processes instigated by the functioning of the living cell. They are formalized as follows:

Definition 5. Let $\mathcal{A} = (S, A)$ be an rs. An *interactive process* in $\mathcal{A}$ is a pair $\pi = (\gamma, \delta)$ of finite sequences such that, for some $n \geq 1$, $\gamma = C_0, \ldots, C_n$ and $\delta = D_0, \ldots, D_n$, where $C_0, \ldots, C_n, D_0, \ldots, D_n \subseteq S$, $D_0 = \emptyset$, and $D_i = res_\mathcal{A}(D_{i-1} \cup C_{i-1})$, for all $i \in \{1, \ldots, n\}$.

The sequence γ is the *context sequence of* π, the sequence δ is the *result sequence of* π, and the sequence $\tau = W_0, \ldots, W_n$, where, for all $i \in \{1, \ldots, n\}$, $W_i = C_i \cup D_i$, is the *state sequence of* π, with $W_0 = C_0$ called the *initial state*. Thus the dynamic process formalized by an interactive process π begins in the initial state W_0. The reactions of $\mathcal{A}$ enabled by W_0 produce then the result set D_1, which together with the context set C_1 forms the successor state $W_1 = res_\mathcal{A}(W_0) \cup C_1$. This formation of the successor state is iterated, $W_i = res_\mathcal{A}(W_{i-1}) \cup C_i$, resulting in the state sequence $\tau = W_0, \ldots, W_n$.

An interactive process may be visualized by a three-row representation, where the first row represents the context sets and is labeled "C," the second row represents result sets and is labeled by "D," and the third row represents states and is labeled by "W." Thus such a representation looks as follows:

$$
\begin{array}{llllll}
C: & C_0 & C_1 & \cdots & C_{n-1} & C_n \\
D: & \emptyset & D_1 & \cdots & D_{n-1} & D_n \\
W: & C_0 & W_1 & \cdots & W_{n-1} & W_n
\end{array}
$$

Note that an interactive process π is determined by its context sequence γ (through the result function res_A). The context sequence formalizes the fact that the *living cell is an open system* in the sense that it is influenced by its environment (the "rest" of a bigger system).

If, for all $i \in \{1, \ldots, n\}$, $C_i \subseteq D_i$, then we say that π is *context-independent*: whatever C_i adds to the state W_i has already been produced by the system (included in the result D_i) or perhaps C_i adds nothing. If π is context-independent, then (in its analysis) we may as well assume that C_i adds nothing, that is, for each $1 \leq i \leq n$, $C_i = \emptyset$. Clearly, if π is context-independent, then the initial state $W_0 = C_0$ determines π by the repeated application of res_A.

16.4 Examples

In this section, we provide two examples of use of reaction systems. The first one comes from biology – we demonstrate how to model/implement a simple generic genetic regulatory network. The second comes from the theory of computation – we demonstrate how to model/implement finite transition systems (finite automata).

Example 1. We will consider genetic regulatory networks (see, e.g., [10]) which are among the most essential ingredients of the living cell. Since we give a formal/abstract model for a very complex component of the living cell, we provide first an extremely simplified (but sufficient for our purpose) description of gene expression – it is this simplified/abstract version that we will model.

Hence, for the purpose of this example, a gene g is a segment of a DNA molecule, and it consists of the promoter field followed by the coding region. The promoter plays the role of a "landing site" for RNA polymerase. If this site is not "occupied," then RNA polymerase can land there and then move/slide through the coding region producing its transcript in the form of a molecule called messenger RNA. This messenger RNA will leave the nucleus (where DNA resides), and it will then be processed outside the nucleus, eventually yielding the protein specified by the coding region of g.

If the cell wants to interrupt the production of this protein, then it "sends" an inhibitor molecule, which lands on the promoter field. Consequently, RNA polymerase cannot land there and thus the transcription phase of the expression process cannot begin, and the protein specified by g cannot be produced anymore. With this in mind, consider the simple genetic regulatory network **G** given in an informal graphical form in Figure 16.1. The network consists of three genes x, y, z expressing proteins X, Y, Z, respectively. Moreover, protein X interacts with protein U (if it is present in a given state of the network) to form a protein complex Q. There are several interactions going on in the network: protein X inhibits (as explained above) the expression of gene z; the presence of either of the proteins Y or Z inhibits the expression of gene x; and the protein complex Q inhibits the expression of gene y.

16 Processes Inspired by the Functioning of the Living Cell

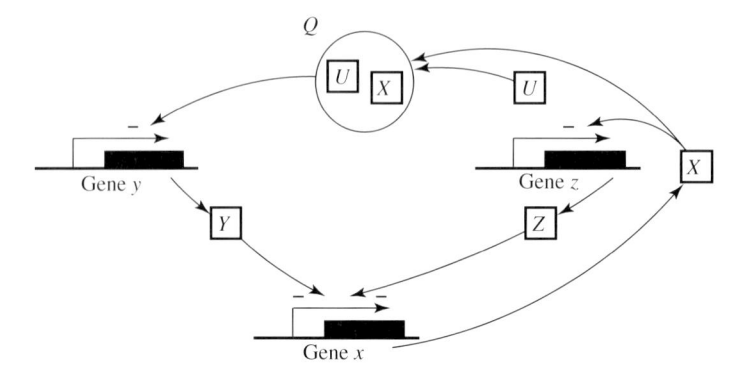

Figure 16.1 A genetic regulatory network.

To implement this network by a reaction system, we will need four sets of reactions: A_x, A_y, A_z implementing the expression of genes x, y, z, respectively, and A_Q implementing the formation of Q:

$$A_x = \{(\{x\}, I_x, \{x\}), (\{x\}, \{Y, Z\}, \{x'\}), (\{x, x'\}, I_{ex}, \{X\})\}$$
$$A_y = \{(\{y\}, I_y, \{y\}), (\{y\}, \{Q\}, \{y'\}), (\{y, y'\}, I_{ey}, \{Y\})\}$$
$$A_z = \{(\{z\}, I_z, \{z\}), (\{z\}, \{X\}, \{z'\}), (\{z, z'\}, I_{ez}, \{Z\})\}$$
$$A_Q = \{(\{U, X\}, I_Q, \{Q\})\}.$$

The set of reactions A_x implements/formalizes the functioning of gene x as follows:

- $(\{x\}, I_x, \{x\})$ ensures that, if x is available/functional in the current state, then it is also available in the successor state unless "something bad" happens to x as expressed by I_x (we did not specify I_x as it is irrelevant for our considerations here, but "something bad" may be, for example, a high level of radiation – discrete levels of radiation are easily specifiable by I_x).
- $(\{x\}, \{Y, Z\}, \{x'\})$ formalizes the role of the promoter: if x is available/functional in the current state and proteins Y, Z are not present in this state, then RNA polymerase x' will land on the promoter of x.
- $(\{x, x'\}, I_{ex}, \{X\})$ formalizes the role of the coding region: if x is available/functional and x' sits on the promoter in the current state, then, unless inhibited by I_{ex}, X will be expressed and hence present in the successor state.

We note here that this reaction formalizes the expression of X in a very "compact way." However, if needed, it could be expanded to a set of reactions that formalize various details of this process.

An analogous explanation/intuition holds for the reactions in A_y and A_z. The reaction $(\{U, X\}, I_Q, \{Q\})$ ensures that, if U and X are present in the current state, then Q will be present in the successor state.

Now, if we combine all these reactions for G forming $A_G = A_x \cup A_y \cup A_z \cup A_Q$, then the rs $\mathcal{A}_G = (S_G, A_G)$, with S_G consisting of all the entities occurring in reactions from $\mathcal{A}_G$, implements/formalizes the structure of G. The reasoning about the functioning of G is formalized through the reasoning about the processes of $\mathcal{A}_G$.

It is important to note that, in fact, $\mathcal{A}_G$ is the "union" of the reaction systems $\mathcal{A}_x = (S_x, A_x)$, $\mathcal{A}_y = (S_y, A_y)$, $\mathcal{A}_z = (S_z, A_z)$, and $\mathcal{A}_Q = (S_Q, A_Q)$, where S_x, S_y, S_z, and S_Q are all the entities occurring in reactions from A_x, A_y, A_z, and A_Q, respectively. The operation of union on reaction systems is easily defined (as sets are our basic data structure), as follows: for reaction systems $\mathcal{B}_1 = (S_1, B_1)$ and $\mathcal{B}_2 = (S_2, B_2)$, their union is the rs $(S_1 \cup S_2, B_1 \cup B_2)$.

As a matter of fact, the union of reaction systems is the basic mechanism for composing reaction systems. It expresses our assumption about bottom-up combination of local descriptions into a global picture. This combination happens "automatically": the sheer fact that all "ingredients" are present in the same biochemical medium (molecular soup) makes interactions possible. *There is no need for providing additional interfaces here.* This is a fundamental difference with models of computation in computer science; see, for example, [7] and [8].

Example 2. This example relates reaction systems to the classical model of computation, namely, finite transition systems (which become finite automata once the initial and terminal states are chosen): see, for example, (7, 8). In particular, we will demonstrate how transition system behavior can be implemented by reaction systems.

We briefly recall that a *deterministic transition system* is a triplet $F = (Q, \Sigma, \delta)$, where Q is a nonempty finite set of *states*, Σ is a finite set of characters (the *input alphabet*), and $\delta : Q \times \Sigma \to Q$ is a *transition function*. Then, the *behavior* of F is given by finite transition sequences of the form $q_0 \xrightarrow{x_1} q_1 \xrightarrow{x_2} q_2 \xrightarrow{x_3} \cdots \xrightarrow{x_n} q_n$, for some $n \geq 0$, such that $\delta(q_i, x_{i+1}) = q_{i+1}$, for each $i \in \{0, 1, \ldots, n-1\}$.

For the explanation of the implementation of F by a reaction system, it is convenient to assume that $Q \cap \Sigma = \emptyset$ and $|Q \cup \Sigma| > 2$.

The aim of the implementation is to construct a reaction system $\mathcal{A}_F = (S_F, A_F)$ such that $q_0 \xrightarrow{x_1} q_1 \xrightarrow{x_2} q_2 \xrightarrow{x_3} \cdots \xrightarrow{x_n} q_n$ is a behavior of F if and only if

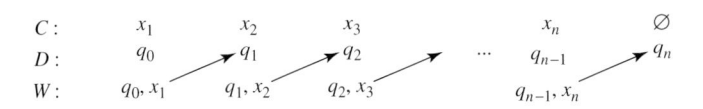

C:	x_1	x_2	x_3	$\cdots$	x_n	$\emptyset$
D:	q_0	$\to q_1$	$\to q_2$		$\to q_{n-1}$	$\to q_n$
W:	q_0, x_1	q_1, x_2	q_2, x_3		q_{n-1}, x_n	

is an interactive process of the reaction system $\mathcal{A}_F$, that is, $res_{\mathcal{A}_F}(\{q_i, x_{i+1}\}) = q_{i+1}$, for each $i \in \{0, 1, \ldots, n-1\}$. Note that here $D_0 = \{q_0\}$, while the formal definition of an interactive process requires $D_0 = \emptyset$. This is done to ease explanations; to get $D_0 = \emptyset$ one can set $C_0 = \{q_0, x_1\}$ and $D_0 = \emptyset$.

Let for all states $p, q \in Q$ and characters $x \in \Sigma$, $a_{p,q,x}$ be the reaction defined by $(\{p, x\}, S_F \setminus \{p, x\}, \{q\})$. Then, $\mathcal{A}_F = (S_F, A_F)$, where $S_F = Q \cup \Sigma$ and $A_F = \{a_{p,x,q} : \delta(p, x) = q\}$. Since we require that $I_a \neq \emptyset$, for each reaction a in a reaction system, we assumed that $|Q \cup \Sigma| > 2$ (so $S_F \setminus \{p, x\} \neq \emptyset$ as required).

The following is a deterministic transition system F (given by the graph of δ) and the list of the reactions of $\mathcal{A}_F$ (note that $S_F = \{q_0, q_1, q_2, x, y\}$):

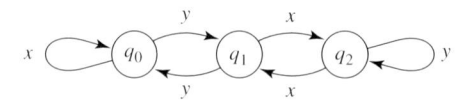

$$\mathcal{A}_F = \left\{ \begin{array}{ll} (\{q_0, x\}, \{q_1, q_2, y\}, \{q_0\}) & (\{q_0, y\}, \{q_1, q_2, x\}, \{q_1\}) \\ (\{q_1, x\}, \{q_0, q_2, y\}, \{q_2\}) & (\{q_1, y\}, \{q_0, q_2, x\}, \{q_0\}) \\ (\{q_2, x\}, \{q_0, q_1, y\}, \{q_1\}) & (\{q_2, y\}, \{q_0, q_1, x\}, \{q_2\}) \end{array} \right\}$$

Then, for example, the transition sequence $q_1 \xrightarrow{x} q_2 \xrightarrow{y} q_2 \xrightarrow{y} q_2 \xrightarrow{x} q_1 \xrightarrow{y} q_0$ in F corresponds to the following interactive process in $\mathcal{A}_F$:

$$\begin{array}{lllllll} C: & & x & y & y & x & y \\ D: & q_1 & \to q_2 & \to q_2 & \to q_2 & \to q_1 & \to q_0 \\ W: & q_1, x & q_2, y & q_2, y & q_2, x & q_1, y & \end{array}$$

The implementation of nondeterministic finite transition systems provides an instructive insight into the role of context in interactive processes; it is done as follows: Assume that in our example transition system F the transition from q_0 on y is nondeterministic: $\delta(q_0, y) = \{q_0, q_1\}$. We mark these two transitions by symbols "1" and "2," and accordingly have two reactions: $(\{q_0, y, 1\}, \{q_1, q_2, x, 2\}, \{q_0\})$, $(\{q_0, y, 2\}, \{q_1, q_2, x, 1\}, \{q_1\})$. Then the implementing reaction system will follow the transition from q_0 by y to q_0 if the context of the current state contains the symbol 1, and it will follow the transition from q_0 by y to q_1 if the context contains the symbol 2. Thus, for example, the transition sequence

$$q_0 \xrightarrow{x} q_0 \xrightarrow{y} q_0 \xrightarrow{y} q_1 \xrightarrow{x} q_2$$

in this modified F will correspond in the accordingly modified $\mathcal{A}_F$ to the following interactive process:

$$\begin{array}{llllll} C: & & x & y, 1 & y, 2 & x \\ D: & & q_0 & \to q_0 & \to q_0 & \to q_1 & \to q_2 \\ W: & q_0, x & q_2, y, 1 & q_0, y, 2 & q_1, x & \end{array}$$

The context in interactive processes can be also used to implement stochasticity.

16.5
Reaction Systems with Measurements

As was already mentioned, the model of reaction systems is qualitative, for example, it does not include counting. However, there are many situations in biology where one needs to assign quantitative parameters to states. To account for this, reaction systems are extended to those with measurements, where numerical values are assigned to the states of a reaction system.

Our main assumption here is that a numerical value can be assigned to a state T of a reaction system if there is a measurement of T yielding this value (which is a real number). Since states of a reaction system are subsets of its background set, the informal notion of a measurement is formalized through the formal notion of a measurement function that assigns reals to the subsets of the background set. Because we deal with abstract sets (in the model of reaction systems we have no knowledge of the nature of entities of the background set), the value of a measurement function for a state must be composed from the values of the measurement function for its elements (here, for simplicity of explanation, we identify a singleton set $\{x\}$ with its element x). Therefore, we assume that measurement functions are additive.

This leads to the following definition:

Definition 6.

1) Let $\mathcal{A} = (S, A)$ be a reaction system. A *measurement function for* $\mathcal{A}$ is an additive function $f : 2^S \to \mathbb{R}$.
2) A *reaction system with measurements*, abbreviated rsm, is a triplet $\mathcal{B} = (S, A, F)$ such that (S, A) is a reaction system and F is a finite set of measurement functions.

Recall that a function $f : 2^S \to \mathbb{R}$ is *additive* if, for all disjoint $X, Y \in 2^S$, $f(X \cup Y) = f(X) + f(Y)$; this clearly implies that $f(\emptyset) = 0$.

The dynamics of an rsm $\mathcal{B} = (S, A, F)$ is determined by its underlying reaction system $\mathcal{A} = (S, A)$. Hence, in particular, the result function of $\mathcal{B}$, $res_\mathcal{B}$, is equal to $res_\mathcal{A}$, and the interactive processes of $\mathcal{B}$ are the interactive processes of $\mathcal{A}$. The additional component F of $\mathcal{B}$ provides various global properties (measurements) for the states of $\mathcal{B}$. Since $res_\mathcal{A} = res_\mathcal{B}$, these measurements do not influence the dynamic behavior of $\mathcal{B}$ which is identical to that of $\mathcal{A}$.

All the notation and terminology of reaction systems carry over to reaction systems with measurements (through their underlying reaction systems).

We will now prove that each reaction system with measurements can be replaced by an "equivalent" (in a well-defined sense) reaction system.

Theorem 1 *For every reaction system with measurements $\mathcal{B} = (Z, B, F)$, there exists a reaction system $\mathcal{A} = (S, A)$ such that*

(i) $S = Z \cup K$, where $K = \{(f, r) : f \in F \text{ and } r \in range(f)\}$, and $Z \cap K = \emptyset$,
(ii) *for each* $a \in A$, $R_a \cup I_a \subseteq Z$,
(iii) *for each* $T \in 2^Z \setminus \{\emptyset, Z\}$,

$$res_\mathcal{A}(T) = res_\mathcal{B}(T) \cup \{(f, r) : f \in F \text{ and } f(res_\mathcal{B}(T)) = r\} .$$

Proof. Let $\mathcal{B} = (Z, B, F)$ be a reaction system with measurements.

Let $\mathcal{A} = (S, A)$ be a reaction system such that $S = Z \cup K$ and $A = B \cup L$, where

$K = \{(f, r) : f \in F \text{ and } r \in range(f)\}$
$L = \{(T, Z \setminus T, \{(f, r)\}) : T \in 2^Z \setminus \{\emptyset, Z\}, f \in F \text{ and } f(res_\mathcal{B}(T)) = r\}.$

- Clearly, without loss of generality, we may assume that $Z \cap K = \emptyset$. Thus (i) holds.
- It follows directly from the definition of the reactions in $\mathcal{A}$ that, for each $a \in A$, $R_a \cup I_a \subseteq Z$. Thus (ii) holds.
- Note that since $B \subseteq A$, for each $T \in 2^Z$, $res_\mathcal{B}(T) \subseteq res_\mathcal{A}(T)$. Also, since $A \setminus B = L$, for each $T \in 2^Z \setminus \{\emptyset, Z\}$,

$$res_\mathcal{A}(T) \setminus res_\mathcal{B}(T) = \{(f, r) : f \in F \text{ and } f(res_\mathcal{B}(T)) = r\} .$$

Therefore (iii) holds.
Thus the theorem holds. ∎

Note that condition (iii) from the statement of the theorem says that, for each state $T \in 2^Z \setminus \{\emptyset, Z\}$, $\mathcal{A}$ computes the same successor state as $\mathcal{B}$ does, but additionally, $\mathcal{A}$ also computes the values of each measurement function of $\mathcal{B}$ for the successor state (these computed values are now a part of the corresponding successor state of $\mathcal{A}$).

The restriction in condition (iii) that $T \in 2^Z \setminus \{\emptyset, Z\}$ (rather than simply $T \in 2^Z$) is of a technical nature. It ensures that, for each reaction $a \in L$, both R_a and I_a are nonempty as required by our definition of a reaction. It is not that essential in the sense that by using simple standard technical tricks one could "skip" this assumption (adjusting somewhat the statement of the theorem).

Condition (ii) says that the values of measurement functions (hence entities from K) do not influence the applicability (enabling) of reactions in $\mathcal{A}$, which indeed corresponds to the situation in reaction systems with measurements.

We also note that condition (iii) is stated for subsets of 2^Z rather than for subsets of 2^S. This is sufficient, because condition (ii) implies that, for all $T \subseteq S$, $res_\mathcal{A}(T) = res_\mathcal{A}(T \cap Z)$.

In a nutshell, the theorem states that adding measurement functions to a reaction system is a mere "convenience." For every reaction system with measurements $\mathcal{B}$, one can construct a reaction system $\mathcal{A}$ which from "inside" (through its reactions) will compute the values of all measurement functions of $\mathcal{B}$ for each state of $\mathcal{B}$ derived by $\mathcal{B}$ during production/construction of its processes.

16.6
Generalized Reactions

In a reaction system with measurements, the measurements of a current state (determined by the measurement functions) do not influence the successor state in the sense that they neither determine the enabling of reactions in the current state nor influence the products of enabled reactions.

We will consider now the situation where measurement functions influence (drive) the computation of the successor state. As a matter of fact, we will approach this problem by considering first a generalization of the notion of a reaction – a special case of this generalization will yield reactions driven by measurement functions.

Definition 7. Let S be a finite nonempty set.
1) A *generalized reaction* in S is an ordered pair $d = (\Delta, P)$, where Δ (the *condition of d*) is a unary relation over 2^S and P (the *product of d*) is a subset of S.

2) Let $d = (\Delta, P)$ be a generalized reaction in S and $T \subseteq S$. Then d is *enabled by T* if $\Delta(T)$.
3) The *result of d*, denoted res_d, is the function $res_d : 2^S \to 2^S$, for every $T \subseteq S$ defined by

$$res_d(T) = \begin{cases} P & \text{if } d \text{ is enabled by } T \\ \emptyset & \text{otherwise.} \end{cases}$$

It is easily seen that a generalized reaction is indeed a generalization of the notion of reaction as considered in reaction systems. Given a reaction b in a finite set S, the corresponding generalized reaction is (Δ, P_b), where, for each $T \subseteq S$, $\Delta(T)$ if and only if b is enabled by T (i.e., $R_a \subseteq T$ and $I_a \cap T = \emptyset$).

For a finite set B of generalized reactions in S, we define the result function res_B analogously to the way it was defined for sets of ordinary reactions.

Definition 8. A *generalized reaction system* is an ordered pair $\mathcal{B} = (S, B)$, where S is a finite set and B is a finite nonempty set of generalized reactions in S.

Then, as was the case with ordinary reaction systems, for a state $T \subseteq S$, the *result of $\mathcal{B}$ on T* is defined by $res_{\mathcal{B}}(T) = res_B(T)$.

The goal of the framework of reaction systems is to discover phenomena (described by theorems) that take place within these models (the WHAT? questions) and then provide explanations/mechanisms behind them (the WHY? questions). The explanatory mechanism is given in the form of reactions. According to this methodology, we do not accept/consider *arbitrary* generalized reactions but rather only those that can be explained by reactions (as considered in reaction systems). This leads to the following definition, where "acceptable" really means "acceptable in the framework of reaction systems."

Definition 9. A generalized reaction d in S is *acceptable* if there exists a reaction system $\mathcal{A} = (S, A)$ such that $res_d = res_A$.

We now give a characterization of all acceptable generalized reactions.

Theorem 2 *Let S be a finite nonempty set. A generalized reaction $d = (\Delta, P)$ in S is acceptable if and only if $P \neq \emptyset$, there exists $T \subseteq S$ such that $\Delta(T)$ holds, and neither $\Delta(\emptyset)$ nor $\Delta(S)$ holds.*

Proof.

1) Assume that d is acceptable. Hence, there exists a reaction system $\mathcal{A} = (S, A)$ such that $res_d = res_A$.
   (i) Since $A \neq \emptyset$, there exists $T \subseteq S$ such that $res_A(T) \neq \emptyset$, and so (because $res_d = res_A$) we get $res_d(T) \neq \emptyset$.
   (ii) Since, for each $a \in A$, $P_a \neq \emptyset$, we get $P \neq \emptyset$.
   (iii) Since $res_A(\emptyset) = res_A(S) = \emptyset$, it follows from (ii) that neither $\Delta(\emptyset)$ nor $\Delta(S)$ holds.

2) Assume that: $P \neq \emptyset$, there exists $T \subseteq S$ such that $\Delta(T)$ holds, and neither $\Delta(\emptyset)$ nor $\Delta(S)$ holds. Let $\mathcal{A} = (S, A)$ be the reaction system such that $A = \{(T, S \setminus T, P) : T \subseteq S \text{ and } \Delta(T) \text{ holds}\}$.

Since there exists $T \subseteq S$ such that $\Delta(T)$ holds, we get $A \neq \emptyset$. Moreover, since $P \neq \emptyset$, and neither $\Delta(\emptyset)$ nor $\Delta(S)$ holds, we get, for each $a \in A$, $R_a \neq \emptyset$, $I_a \neq \emptyset$, and $P_a \neq \emptyset$. Thus $\mathcal{A}$ is indeed a reaction system.
It follows directly from the definition of $\mathcal{A}$ that $res_d = res_{\mathcal{A}}$. ■

We now can formalize a notion of a reaction driven by a measurement function – it will be a special case of a generalized reaction.

Let $\mathcal{A} = (S, A)$ be a reaction system and let $f : 2^S \to \mathbb{R}$ be a measurement function for $\mathcal{A}$. Then, for each $Y \subseteq range(f)$, let $\Delta_{f,Y}$ be the unary relation over 2^S, for each $T \subseteq S$ defined as follows: $\Delta_{f,Y}(T)$ if and only if $f(T) \in Y$.

Now, for a nonempty $P \subseteq S$ and $Y \subseteq range(f)$, $d = (\Delta_{f,Y}, P)$ is a generalized reaction in S. Note that for $T \subseteq S$, d is enabled by T if the value of $f(T)$ belongs to a predescribed set Y of "good" values for f – therefore d is an example of a generalized reaction driven by (the values of) a measurement function.

Again, we are interested in acceptable generalized reactions (hence generalized reactions that are implementable/explainable by reaction systems). It follows directly from Theorem 2 that $d = (\Delta_{f,Y}, P)$ is acceptable if and only if

$$P \neq \emptyset, \; Y \neq \emptyset, \; 0 \notin Y \text{ and } f(S) \notin Y .$$

Hence, d is acceptable if and only if

$$P \neq \emptyset \text{ and } \emptyset \subset Y \subseteq range(f) \setminus \{0, f(S)\} ,$$

which says in fact that "almost all" generalized reactions $(\Delta_{f,Y}, P)$ are acceptable.

Example 3

1) Let $\mathcal{A} = (S, A)$ be a reaction system such that $S = \{x, y, z, u, 2\}$, and

$$A = \left\{ \begin{array}{ll} (\{x, y\}, \{z, u\}, \{2\}) & (\{x, z\}, \{y, u\}, \{2\}) \\ (\{x, u\}, \{y, z\}, \{2\}) & (\{y, z\}, \{x, u\}, \{2\}) \\ (\{y, u\}, \{x, z\}, \{2\}) & (\{z, u\}, \{x, y\}, \{2\}) \end{array} \right\}$$

We note that, for each $T \subseteq S$, $res_A(T) = \{2\}$ if and only if $|T \cap \{x, y, z, u\}| = 2$.

2) Let f be a measurement function for $\mathcal{A}$ such that

$$f(x) = f(y) = f(z) = f(u) = 1 \text{ and } f(2) = 0 .$$

We note that, for each $T \subseteq S, f(T) = 2$ if and only if $|T \cap \{x, y, z, u\}| = 2$. Hence f globally computes (predicts) the results of $\mathcal{A}$.

3) Consider now the generalized reaction $b = (\Delta_{f,\{2\}}, \{2\})$, and the generalized reaction system $\mathcal{B}$ with $\{x, y, z, u, 2\}$ as its background set, and b as its only (generalized) reaction.

Hence $\mathcal{B}$ with one (generalized) reaction does the same job as $\mathcal{A}$ does with six reactions. Clearly, if rather than considering the four element set $\{x, y, z, u\}$ we considered a larger set (and modified $\mathcal{A}$ accordingly), the difference would be even more dramatic.

The above example illustrates how the use of measurement functions (generalized reactions) allows for a more efficient/succinct specification of a set of processes.

16.7
A Generic Quantitative Model

We will now demonstrate the flexibility of reaction systems with measurements in dealing with quantitative parameters assigned to states. Rather than developing a "heavy" general formal framework for demonstrating this flexibility, we will consider a generic quantitative model and then discuss how to deal with it using reaction systems with measurements.

In our considerations, we do not discuss various ways of dealing with quantities in reaction systems, but instead we discuss *simulations* of other models by reaction systems. The quantitative model that we will consider is a *generic model* in the sense that we do not discuss one specific mechanism but rather a general scheme of mechanisms. This generic model is a model of DNA expression which uses a quantitative description of expression products.

In this model, we have a set G of *genes*, a set V of their *products*, and a set Q of states. Each state is an ordered pair $q = (H, \phi)$ with $H \subseteq G$ and $\phi : V \to \mathbb{R}_0$ (we use $\mathbb{R}_0$ to denote the set of nonnegative reals); ϕ is called the *quantitative component* of q, and we assume that ϕ is not the zero function (with $\phi(v) = 0$ for each $v \in V$). The intuition behind a current state $q = (H, \phi)$ is that H is the set of genes that are currently expressed, and ϕ is a quantitative description of "the amount" of each product $v \in V$ currently present (the "amount" can be the number of particles, mass, volume, concentration, etc.).

Since we deal with discrete time, we have a transition function $\Gamma : Q \to Q$, which to any given state assigns its successor state: $\Gamma((H, \phi)) = (K, \psi)$. This transition function may be seen as consisting of two components $\Gamma = (\Gamma_G, \Gamma_V)$, where

$$\Gamma_G((H, \phi)) = K \text{ and } \Gamma_V((H, \phi)) = \psi \ .$$

Moreover, Γ_V is given by the family of functions $\{\gamma_v : v \in V\}$, where for each $v \in V$, $\gamma_v : Q \to \mathbb{R}_0$. Then, for each $(H, \phi) \in Q$, $\Gamma_V((H, \phi)) = \psi$ where, for each $v \in V$, $\psi(v) = \gamma_v((H, \phi))$. Thus knowing the current state q, γ_v gives the amount of v present in the successor state of q.

As a matter of fact, for each state $q = (H, \phi)$, the quantitative component ϕ can be seen as the vector $(\phi(v_1), \ldots, \phi(v_k))$, where we assume that $V = \{v_1, \ldots v_k\}$ and V is ordered, yielding the sequence $v_1, \ldots, v_k$. Accordingly, we assume that each state $(H, \phi) \in Q$ is of the form $(H, (\phi(v_1), \ldots \phi(v_k)))$, and the transition function Γ transforms a current state $(H, (\phi(v_1), \ldots, \phi(v_k)))$ into the successor state $(K, (\psi(v_1), \ldots, \psi(v_k)))$.

Now we can define a gene expression system as a four-tuple $\mathcal{E} = (G, V, Q, \Gamma)$ with the components G, V, Q, and Γ as discussed above.

Finally, a (gene expression) *process* in $\mathcal{E}$ is a finite sequence of states

$$\pi = (H_1, Q_1), (H_2, Q_2), \ldots, (H_t, Q_t)$$

with $t \geq 2$, such that, for each $i \in \{2, \ldots, t\}$,

$$\Gamma((H_{i-1}, Q_{j-1})) = (H_i, Q_j) \,.$$

16.8
Approximations of Gene Expression Systems

When a gene expression system $\mathcal{E}$ (which is an abstract model of gene expression) is implemented, the basic step of such an implementation is an approximation of nonnegative real numbers. Assume that this implementation is done through binary numbers, where the numbers to be implemented are bounded by 2^{n_1} and their fractional part is determined with precision 2^{-n_2}. For this implementation, n-bits binary numbers are used, where $n = n_1 + n_2$.

These numbers have the positional binary representation of the form

$$2^{n_1-1} \ldots 2^1 2^0 \bullet 2^{-1} 2^{-2} \ldots 2^{-n_2}$$

and we refer to them as (n_1, n_2)-*binary numbers*. Thus, for example, for $n_1 = 5$ and $n_2 = 4$, the 9-bit (5,4)-binary number 100101001 represents $18 + \frac{9}{16} = 18.53125$ in the decimal notation.

Let $B_{(n_1, n_2)}$ be the subset of $\mathbb{R}_0$ represented by (n_1, n_2)-binary numbers.

Now, for each real $r \in \mathbb{R}_0$, we consider numbers b from $B_{(n_1, n_2)}$ which yield the minimal difference $|r - b|$. Clearly, either there is one such number b, or there are two such numbers b_1, b_2; in the latter case, we choose the smaller of the two.

In this way, for each $r \in \mathbb{R}_0$, we obtain a unique number from $B_{(n_1, n_2)}$ which is the (n_1, n_2)-binary approximation of r, denoted $b(r)$.

Accordingly, for each function $\phi : V \to \mathbb{R}_0$ represented/defined by the vector $(\phi(v_1), \ldots, \phi(v_k))$, we obtain the (n_1, n_2)-binary approximation of ϕ, $b(\phi) : V \to B_{(n_1, n_2)}$ represented/defined by the vector $(b(\phi(v_1)), \ldots, b(\phi(v_k)))$. Finally, for each state $(H, \phi) \in Q$, we define the (n_1, n_2)-binary approximation of (H, ϕ) to be $(H, b(\phi))$.

In order to simplify the terminology, we will use the phrase "approximation" rather than "(n_1, n_2)-binary approximation" assuming that the parameters n_1, n_2 of binary numbers are fixed for our considerations.

By representing nonnegative reals through their approximations, a gene expression system $\mathcal{E} = (G, V, Q, \Gamma)$ can be transformed into a gene expression system $\widehat{\mathcal{E}} = (G, V, \widehat{Q}, \widehat{\Gamma})$ operating on $B_{(n_1, n_2)}$ rather than $\mathbb{R}_0$. Here, each state $\widehat{q} \in \widehat{Q}$ is of the form $\widehat{q} = (H, \widehat{\phi})$, where $H \subseteq G$, and $\widehat{\phi} : V \to B_{(n_1, n_2)}$ is represented by the vector $(\widehat{\phi}(v_1), \ldots, \widehat{\phi}(v_k))$.

16.8 Approximations of Gene Expression Systems

Here is one possible straightforward way of defining the transition function $\widehat{\Gamma}$ – it is given by the following commuting diagram:

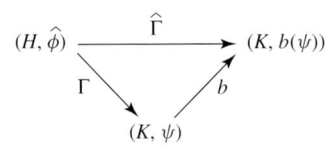

$$
\begin{array}{ccc}
(H,\widehat{\phi}) & \xrightarrow{\widehat{\Gamma}} & (K, b(\psi)) \\
& \Gamma \searrow \quad \nearrow b & \\
& (K, \psi) &
\end{array}
$$

The transition function Γ applied to a state $(H, \widehat{\phi})$ from $\widehat{Q}$ yields the intermediate successor state (K, ψ) which does not have to be in $\widehat{Q}$. However, taking the approximation $b(\psi)$ yields the state $(K, b(\psi))$, which is then the successor state of $(H, \widehat{\phi})$ in $\widehat{\mathcal{E}}$. Hence, for this way of approximating $\mathcal{E}$, the transition function $\widehat{\Gamma} = (\widehat{\Gamma}_G, \widehat{\Gamma}_V)$ is defined by

$$\widehat{\Gamma}_G((H,\widehat{\phi})) = \Gamma_G((H,\widehat{\phi})) \text{ and } \widehat{\Gamma}_V((H,\widehat{\phi})) = b(\Gamma_V((H,\widehat{\phi}))) \,.$$

Now, each process

$$\pi = (H_1, \phi_1), (H_2, \phi_2), \ldots, (H_t, \phi_t)$$

in $\mathcal{E}$ is approximated in $\widehat{\mathcal{E}}$ by the process

$$\widehat{\pi} = (H_1, \widehat{\delta}_1), (H'_2, \widehat{\delta}_2), \ldots, (H'_t, \widehat{\delta}_t)$$

such that $\widehat{\delta}_1 = b(\phi_1)$, that is, we begin in $\widehat{\mathcal{E}}$ with the approximation $(H_1, \widehat{\delta}_1)$ of (H_1, ϕ_1) and then proceed in $\widehat{\mathcal{E}}$ through its transition function $\widehat{\Gamma}$.

Clearly, there may be many ways of evaluating the quality of such an approximation $\widehat{\pi}$ of π. For example, $\widehat{\pi}$ could be classified as a good approximation if $H_2 = H'_2, \ldots, H_t = H'_t$. Consequently, the quality of the approximation $\widehat{\mathcal{E}}$ could be determined by the overall quality of approximations by $\widehat{\pi}$ of π in the class of all processes π of $\mathcal{E}$.

If the quality of approximation turns out to be "not good enough" one may either "adjust" or "totally redefine" the transition function $\widehat{\Gamma}$. Such modifications will in general depend on the knowledge of the nature of the actual functions ϕ involved in the states of $\mathcal{E}$ and on the nature of the transition function Γ.

Judging specific approximation strategies is not our concern in this chapter. Our goal is to demonstrate that, given an approximation $\widehat{\mathcal{E}}$, it can be simulated by a reaction system (with measurements).

Anyhow, whatever is the exact procedure for obtaining right approximations, we end up with a system $\widehat{\mathcal{E}} = (G, V, \widehat{Q}, \widehat{\Gamma})$ which becomes an approximation of $\mathcal{E}$. In this way, we move from a system $(\mathcal{E})$ with an infinite state space to a system with a finite state space $(\widehat{\mathcal{E}})$.

16.9
Simulating Approximations by Reaction Systems

Once an approximation $\widehat{\mathcal{E}}$ of a gene expression system $\mathcal{E}$ has been established, we will *simulate* (processes in) $\widehat{\mathcal{E}}$ by (processes in) a reaction system $\mathcal{A}(\widehat{\mathcal{E}})$. Before defining $\mathcal{A}(\widehat{\mathcal{E}})$, we introduce additional notations.

First, we establish a set representation for all numbers in $B_{(n_1,n_2)}$ as follows. For each $x \in B_{(n_1,n_2)}$, $set(x)$ is the set of all numbers $\ell \in \{n_1 - 1, \ldots, 0, -1, \ldots, -n_2\}$ such that the (n_1, n_2)-binary number representing x contains 1 in the position 2^ℓ. For example, for the x represented by the (5, 4)-binary number 100101001 (which we considered before), $set(x) = \{4, 1, -1, -4\}$, while for the y represented by the (5, 4)-binary number 101000110, $set(y) = \{4, 2, -2, -3\}$.

Then, for each product $v \in V$ and each state $\widehat{q} = (H, \widehat{\phi}) \in \widehat{Q}$,

$$bits(\widehat{q}, v) = \{\langle v, \ell \rangle : \ell \in set(\widehat{\phi}(v))\} .$$

Intuitively, $bits(\widehat{q}, v)$ gives all the bits used in the set representation of $\widehat{\phi}(v)$, which is the amount of v in the state $\widehat{q}$. Then, for each state $\widehat{q} = (H, \widehat{\phi}) \in \widehat{Q}$,

$$bits(\widehat{q}) = \bigcup \{bits(\widehat{q}, v) : v \in V\} .$$

Intuitively, $bits(\widehat{q})$ gives all the bits used in the set representations of the amounts of v in the state $\widehat{q}$, for all $v \in V$.

Finally, for each state $\widehat{q} = (H, \widehat{\phi})$ of $\widehat{\mathcal{E}}$, we define the *simulation state of* $\widehat{q}$, denoted $sim(\widehat{q})$, by $sim(\widehat{q}) = H \cup bits(\widehat{q})$.

We also make a technical (and easy to implement) assumption about the states of $\widehat{\mathcal{E}}$:

For each $\widehat{q} = (H, \widehat{\phi}) \in \widehat{Q}$, $\{\widehat{\phi}(v) : v \in V\} \subset B_{(n_1, n_2)}$. (†)

We are ready now to define the reaction system $\mathcal{A}(\widehat{\mathcal{E}})$ simulating $\widehat{\mathcal{E}}$.
Let $\mathcal{A}(\widehat{\mathcal{E}}) = (S, A)$ be a reaction system, where

$$S = G \cup \{\langle v, \ell \rangle : v \in V \text{ and } \ell \in \{n_1 - 1, \ldots, 1, 0, -1, \ldots, -n_2\}\} ,$$

and A consists of all reactions $a = (R, I, P) \in rac(S)$ such that there exists a state $\widehat{q} = (H, \widehat{\phi}) \in \widehat{Q}$ satisfying

(i) $R = sim(\widehat{q})$,
(ii) $I = S \setminus R$, and
(iii) $P = sim(\widehat{\Gamma}(\widehat{q}))$.

Note that, since we assumed that for each state of $\mathcal{E}$ its quantitative component is not the zero function, both $R \neq \emptyset$ and $P \neq \emptyset$. Also, because of the assumption (†) above, we have $I \neq \emptyset$.

It follows directly from the definition of $\mathcal{A}(\widehat{\mathcal{E}})$ that $res_{\mathcal{A}(\widehat{\mathcal{E}})}(T_1) = T_2$ for nonempty $T_1, T_2 \subseteq S$ if and only if there exist states $\widehat{q}_1, \widehat{q}_2$ of $\widehat{\mathcal{E}}$ such that

$$\widehat{\Gamma}(\widehat{q}_1) = \widehat{q}_2, \ T_1 = sim(\widehat{q}_1) \text{ and } T_2 = sim(\widehat{q}_2) .$$

This implies that state sequences of $\mathcal{A}(\widehat{\mathcal{E}})$ (consisting of nonempty states) indeed simulate gene expression processes of $\widehat{\mathcal{E}}$, which means

1) if $\widehat{\pi} = \widehat{q}_1, \widehat{q}_2, \ldots, \widehat{q}_n$ is a process in $\widehat{\mathcal{E}}$ and $\tau = T_1, T_2, \ldots, T_n$ is the state sequence of a context-independent interactive process in $\mathcal{A}(\widehat{\mathcal{E}})$ such that $T_1 = sim(\widehat{q}_1)$, then, for all $i \in \{2, \ldots, n\}$, $T_i = sim(\widehat{q}_i)$;
2) if $\tau = T_1, T_2, \ldots, T_n$ is the state sequence of a context-independent interactive process in $\mathcal{A}(\widehat{\mathcal{E}})$, then, for each $i \in \{1, \ldots, n\}$, there exists a state $\widehat{q}_i$ in $\widehat{\mathcal{E}}$ such that $T_i = sim(\widehat{q}_i)$, and $\widehat{q}_1, \widehat{q}_2, \ldots, \widehat{q}_n$ is a process in $\widehat{\mathcal{E}}$.

Now, for each $v \in V$, we define a measurement function $f_v : 2^S \to \mathbb{R}_0$ for $\mathcal{A}(\widehat{E})$ by defining it on the background set S as follows:

$$f_v(x) = \begin{cases} 0 & \text{if } x \in G \\ 0 & \text{if } x = \langle u, j \rangle \text{ and } u \neq v \\ 2^j & \text{if } x = \langle v, j \rangle \,. \end{cases}$$

Thus, each f_v gives the amount of v present in states of $\widehat{\mathcal{E}}$, meaning that if $\widehat{q} = (H, \widehat{\phi}) \in \widehat{Q}$, then $f_v(sim(\widehat{q})) = \widehat{\phi}(v)$.

Let then $\mathcal{B}(\widehat{\mathcal{E}}) = (S, A, F)$ be the reaction system with measurements such that $F = \{f_v : v \in V\}$.

Hence $\mathcal{A}(\widehat{\mathcal{E}})$ is a reaction system simulating processes in $\widehat{\mathcal{E}}$ in such a way that when it produces the state $sim(\widehat{q})$, which simulates/represents a state $\widehat{q} = (H, \widehat{\phi})$ in $\mathcal{A}(\widehat{\mathcal{E}})$, then it provides the (set) representation of $\widehat{\phi}$ as a part of $sim(\widehat{q})$ – this representation is computed "from inside" of $\mathcal{A}(\widehat{\mathcal{E}})$ by its reactions. Then the reaction system with measurements $\mathcal{B}(\widehat{\mathcal{E}})$ is equipped with measurement functions f_v for each $v \in V$. For each state $sim(\widehat{q})$, each function f_v gives explicitly the amount of v present in $\widehat{q}$ (hence the value $\widehat{\phi}(v)$, where $\widehat{q} = (H, \widehat{\phi})$).

16.10
Discussion

We begin with a summary of the material presented in this chapter.

We have considered (reaction systems which are) a formal framework for investigating processes inspired by the functioning of the living cell. The basic construct of this framework, namely, reaction systems, is a qualitative model – there is no counting here. However, this framework contains various extensions of the basic model, which equip reaction systems with additional components often motivated by specific research themes. Hence, for example, it is clear that there are many situations in biology where one needs to assign quantitative parameters to states, and to account for this one considers reaction systems with measurements.

In Sections 16.2 and 16.3, we recalled the basic concepts of reaction systems, and illustrated them in Section 16.4, where we gave two examples, one from biology and one from theory of computation.

Then in Section 16.5 we considered reaction systems with measurements, where we proved that adding measurement functions is just a convenience (a useful specification macro), as for each reaction system with measurements there exists an equivalent (in a well-defined sense) ordinary reaction system. Since measurement functions do not influence transitions of interactive processes (they merely state the global numerical properties of states), it is natural to consider reactions dependent on the values of measurements. We did this in Section 16.6 by introducing generalized reactions that allowed us to define the notion of a "measurement-driven reaction". It turns our that "almost always" such a reaction can be simulated by a set of ordinary reactions; however, again, using generalized reactions provides a convenient, often succinct, specification tool.

In Sections 16.7 – 16.9, we argued that, if a quantitative model is implemented (in a finite precision arithmetic), then such an implementation can be naturally simulated by reaction systems (with measurements).

On the basis of the material presented in Sections 16.5–16.9, we can then claim that reaction systems are quite flexible in dealing with numerical parameters assigned to states.

The framework of reaction systems is quite rich and varied. We then move on to discussing a number of research topics from this framework – motivated either by biological considerations or by the need to understand the underlying computations.

One of the key features of reaction systems is *nonpermanency*: an entity is not retained in a transition of an interactive process unless it is either sustained/produced by some reaction or introduced through context. This nonpermanency is quite immediate – an entity that is not sustained disappears within one transition step. However, a decay of entities in a biochemical environment requires some time (decay time) to be realized. In order to account for decay time, one considers in [6] *reaction systems with durations*.

Reaction systems with measurements were introduced in [11], where they were used for assigning time moments to states. In fact, [11] deals with fundamental questions such as "What is time in (models of) biochemical systems?", "How can one capture/formalize time in the framework of reaction systems?", and "Which measurement functions can be used to measure time?"

Formation of modules is an important research area in biology and biochemistry, see, for example, [12]. The formal notion of a module and its formation (by dynamic events) are discussed in [13], where it is demonstrated that interactive processes lead to the formation of modules, and that the family of all modules in a given "stable" state of a reaction system forms a lattice. Hence the development of the living cell leads to formation of structures. It is also shown in [13] that reaction systems can be viewed as self-organizing systems.

To understand how entities of a reaction system influence each other is important from both biological and computational points of view. Such *causalities* may be of a *static* nature (deducible "directly" from the set of reactions) or of a *dynamic* nature (deducible from the set of interactive processes) – both kinds of causalities are investigated in [14].

An important line of research concerns the understanding of the result functions (res_A) of reaction systems. This corresponds to the investigation of context-independent interactive processes, and hence to the investigation of reaction systems as closed systems (without the influence of the environment). Here one considers reaction systems as specifications of finite functions on power sets ($res_A : 2^S \to 2^S$), and the research focuses on the understanding (characterization) of such functions; see, for example, [15] and [16].

Acknowledgments

The authors are indebted to Robert Brijder and Michael Main for useful comments. This research was supported by the Pascal Chair award from the Leiden Institute of Advanced Computer Science (LIACS), Leiden University.

References

1. Kari, L. and Rozenberg, G. (2008) The many facets of natural computing. *Commun. ACM*, **51** 10, 72–83.
2. Rozenberg, G., Bäck, T., and Kok, J. (eds) (2012) *Handbook of Natural Computing*, Springer-Verlag.
3. Dawkins, R. (1986) *The Blind Watchmaker*, Penguin, Harmondsworth.
4. Ehrenfeucht, A. and Rozenberg, G. (2007) Reaction systems. *Fundam. Inform.*, **75** 1-4, 263–280.
5. Lehninger, A. (1965) *Bioenergetics: The Molecular Basis of Biological Energy Transformations*, W.A. Benjamin, Inc., New York, Amsterdam.
6. Brijder, R., Ehrenfeucht, A., and Rozenberg, G. (2011) Reaction systems with duration, in *Computation, Cooperation, and Life. Lecture Notes in Computer Science*, (eds J. Kelemen and A. Kelemenova), vol. **6610**, Springer, pp. 191–202.
7. Arnold, A. (1994) *Finite Transition Systems: Semantics of Communicating Systems*, Prentice Hall.
8. Hopcroft, J., Motwani, R., and Ullman, J. (2006) *Introduction to Automata Theory, Languages, and Computation*, Prentice Hall.
9. Reisig, W. (1985) *Petri Nets (An Introduction)*, EATCS Monographs on Theoretical Computer Science, No. 4, Springer, Heidelberg.
10. Alberts, B., Bray, D., Johnson, A., Lewis, J., Raff, M., Roberts, K., and Walter, P. (1998) *Essential Cell Biology*, Garland Publishing, Inc.
11. Ehrenfeucht, A. and Rozenberg, G. (2009) Introducing time in reaction systems. *Theor. Comput. Sci.*, **410** 4-5, 310–322.
12. Schlosser, G. and Wagner, G. (eds) (2004) *Modularity in Development and Evolution*, The University of Chicago Press.
13. Ehrenfeucht, A. and Rozenberg, G. (2007) Events and modules in reaction systems. *Theor. Comput. Sci.*, **376** 1-2, 3–16.
14. Brijder, R., Ehrenfeucht, A., and Rozenberg, G. (2010) A note on causalities in reaction systems. *ECEASST*, **30**, 1–9.
15. Ehrenfeucht, A., Main, M.G., and Rozenberg, G. (2011) Functions defined by reaction systems. *Int. J. Found. Comput. Sci.*, **22** 1, 167–178.
16. Ehrenfeucht, A., Main, M.G., and Rozenberg, G. (2010) Combinatorics of life and death for reaction systems. *Int. J. Found. Comput. Sci.*, **21** 3, 345–356.

17
Computational Methods for Quantitative Submodel Comparison

Andrzej Mizera, Elena Czeizler, and Ion Petre

17.1
Introduction

Much experimental and theoretical effort is invested nowadays in analyzing large biochemical systems, for example, metabolic pathways, regulatory networks, and signal transduction networks, aiming to obtain a holistic perspective providing a comprehensive, system-level understanding of cellular behavior. This often results in the creation and analysis of very large and complex models, often encompassing hundreds of reactions and reactants, see for example, [1]. Therefore, obtaining a global picture of the system's architecture, in particular understanding the interactions between various components, or distinguishing a high-level functional decomposition of the network, constitutes a significant challenge. An important insight here is that the architecture of biological systems is a consequence of their functional requirements. Even though evolution is driven by random events, some designs, such as having an extra feedback loop helping the system to correlate better the response of the system with its trigger, may offer a selective advantage and in time, may get to dominate the population, see [2]. Thus, comparing the performance of different alternative designs in terms of subcomponents being on or off, one aims to formulate general principles for how functional requirements correlate biologically with various designs.

Similar problems have been encountered, for instance, in engineering sciences, see [3], and a variety of strategies and approaches for solving such problems have been already developed in this framework. Thus, when aiming to obtain a system-level understanding of such large biochemical networks, one possible approach is to adapt to systems biology some of the methods originating from engineering sciences, especially from control theory, see for example, [4–10]. Such methods have been used, as we also do in this paper, to identify various functional modules of a model, including feedback and feedforward mechanisms. To identify the quantitative contribution of each of the modules to the global behavior of the model, the general approach is to consider knockdown mutants of the initial model, missing one or several of the modules. The main problem then becomes an objective quantitative comparison of several alternative submodels for the same biological process. We focus on this problem in our study: that is, we concentrate on the comparison of submodels of a given reference model. This issue is a special case of the general problem of alternative model comparison.

The first part of this chapter contains a review of the existing techniques for model decomposition and for quantitative comparison of submodels. We describe the knockdown mutants, elementary flux modes, control-based decomposition, mathematically controlled comparison, local submodel comparison, a parameter-independent submodel comparison, and a discrete approach for comparing continuous submodels. We discuss a quantitative measure for the goodness of a model's fit against experimental data, as well as a technique for quantitative model refinement, and we show how both can be used for model comparison. Finally, we discuss how quantitative model comparison can be used for pathway identification.

In the second part of this chapter, we consider as a case study the eukaryotic heat shock response (HSR), which is an evolutionarily conserved mechanism protecting the cell against protein misfolding. In particular, we consider a model recently introduced in Ref. [11] for this biological process. The model was analyzed in Ref. [12] using control-driven methods where it was decomposed into several modules, including three feedback loops. We focus in the case study on identifying the numerical contribution of each of these feedback loops to the global behavior of the model. For this, we show how we can apply various model comparison methods achieving either a local point-wise view or a global parameter-independent analysis of their individual contributions.

17.2
Methods for Model Decomposition

We discuss in this section a number of methods for decomposing a (large) biomodel into components. The criteria on which to decide what makes a component can be very different depending on the focus of the analysis: a metabolic pathway, a regulatory component, a feedback loop, and so on. We consider in this section three main approaches. The first one is that of knockdown mutants, obtained from a reference model by dropping out some of its components (defined separately). The second one is based on elementary flux models, a well-established concept that captures the steady-state dynamics of a system. The third one is based on control theory and aims to identify how a certain part of the model is controlled through various feedback or feedforward loops. We illustrate some of these approaches in our case study in Section 17.4.

17.2.1
Knockdown Mutants

To identify the quantitative role of a component C of a model M, it is often useful to consider the model $M \setminus C$, where all interactions in C are removed from M, and compare its behavior to that of M. Any differences in the behavior of M and that of the knockdown mutant $M \setminus C$, such as loss of functionality, delayed or nonoptimal response, higher energy consumption, and so on, may give a hint toward the role of C in M.

If the analysis considers several disjoint components $C_1, C_2, \ldots, C_n$ of a model M, then all combinations of knockdown mutant models should be considered: all models $M \setminus \{C_{i_1}, \ldots, C_{i_k}\}$, for all $1 \leq k \leq n$ and all $1 \leq i_1 < i_< \ldots < i_k \leq n$. The technique becomes computationally challenging for a large number of components: one should compare $2^n - 1$ knockdown mutants to each other and to the reference model.

17.2.2
Elementary Flux Modes

A well-established decomposition method for biochemical models appears in the context of the analysis of metabolic pathways. An intuitive definition of a pathway is a sequence of reactions linked by common metabolites [13]. Examples of metabolic pathways are glycolysis or amino acid synthesis. Discovering new pathways in a large model driven only by biological intuition is very difficult. An attempt to formalize the notion of pathway has been proposed in Refs [14–19] in the form of elementary flux modes. The intuitive meaning of an elementary flux mode is a set of reactions whose combined quantitative contribution to the system is zero. In other words, the net loss of substance caused by any reaction in that set is compensated by a net gain in the same substance incurred by some other reactions in the set. A formal definition of elementary flux modes is beyond the scope of this chapter; instead we refer the reader to [13, 14, 16–19] for details. For any given metabolic network, the full set of elementary fluxes can be determined using methods of linear algebra or dedicated software such as METATOOL [15] or COPASI [20]. The identification of the elementary flux modes allows the detection of the full set of nondecomposable steady-state flows that the network can support, including cyclic flows. Any steady-state flux pattern can be expressed as a nonnegative linear combination of these modes [16–18]. The identified elementary flux modes should have clear biological interpretation: a flux mode is a set of enzymes that operate together at a steady state and a flux mode is elementary if the set of enzymes is minimal: that is, complete inhibition of any of the enzymes would result in a termination of this flux [16–18]. The lack of possibility to interpret the modes in this way is a signal that the model under consideration may not be correct.

17.2.3
Control-Based Decomposition

A control-driven approach to model decomposition enables the identification of the main functional modules of a system and of their individual contribution to the emergent, complex behavior of the system as a whole. In turn, this can provide great insight about various properties of a given biochemical system, for example, robustness, efficiency, reactivity, adaptation, regulation, synchronization, and so on. Through this approach, one usually aims to identify the main regulatory components of a given biochemical system: the process to be regulated, referred to as the *plant*; the *sensors* that monitor the current state of the process and send the collected information to a decision-making module, that is, the *controller*; and the *actuator* that modifies the state of the process in accordance with the controller's decisions, thus influencing the activity of the plant. One of the fundamental concepts in control theory is the *feedback mechanism*, which provides the means to cope with the uncertainties: the information about the current state of the process is sent back to the controller, which reacts accordingly to facilitate a dynamic compensation for any disturbance from the intended behavior of the system. In the case of a complex system, this decomposition can be performed in different ways depending on what is considered to be the main role of that system: that is, there may be a few reasonable choices for the plant, and the remaining components are recognized with respect to the choice of the plant.

An easy example illustrating these concepts and their interactions is given by the functioning principles of a motion-activated spotlight. Here, the controller module is an electronic unit which receives an input from the motion sensor and then determines whether there are any changes in the environment. The actuator is a relay switch that operates the lighting system. This actuator is activated by the controller depending on the input sent by the sensor. Then, the switch is kept on by the controller as long as movement is detected by the sensor.

How this control-driven approach can be exploited to investigate and understand regulatory networks can be seen in Refs [3, 5, 7, 8, 21]. Here, we shortly describe the approach taken in Ref. [21]. The authors make a thorough study of the HSR mechanism in *Escherichia coli* based on modular decomposition. A model for the system is built and functional modules, that is, the plant, sensors, controller, and actuator, are identified. The decomposition reveals the underlying design of the HSR mechanism and its level of complexity, which, as the authors show, is not justified if only the functionality of an operational heat shock system is required. Further, this observation leads to the introduction and analysis of hypothetical design variants (mutants) of the original HSR model. In the original model, one feedforward (temperature sensing) and two feedback elements (σ^{32} factor sequestration feedback loop and σ^{32} degradation feedback loop) can be isolated. The variants are obtained through the elimination of either the σ^{32} degradation feedback loop or of both feedbacks. One by one, the variants in the order of increasing complexity are considered, starting from the simplest architecture containing just the feedforward element (the *open-loop design*). Based on numerical simulations, the authors demonstrate how the addition of subsequent layers of regulation (i.e., increase in the complexity of the model) improves the performance of the response in terms of systemic properties such as robustness, noise reduction, speed of response, and economical use of cellular resources. Moreover, this systematic approach enables the identification of the role of each of the regulatory layers to the overall behavior of the system.

17.3
Methods for Submodel Comparison

Comparing alternative models for a given biochemical system is in general a very difficult problem, involving a deep analysis of the underlying network of reactions, of the biological assumptions, as well as of the numerical setup. To decide what the benefits of one design over another are, or to understand what the selection requirements involved in an evolutionary design are, one needs some unbiased methods to objectively compare the alternative designs.

17.3.1
Mathematically Controlled Model Comparison

One such method is the mathematically controlled comparison, [2], which provides a structured approach for comparing alternative regulatory designs with respect to some chosen measures of functional effectiveness. Under this approach, mathematical models for both the reference design and the alternatives are first developed in the framework of canonical nonlinear modeling referred to as *S-systems*, [22–24]. This canonical nonlinear representation, developed within the power-law formalism, is a system of nonlinear ordinary differential equations (ODE) with a well-defined structure. Moreover, this framework allows the alternative models to differ from the reference design in only one process (e.g., only one feedback mechanism), which is the focus of the comparison. Then, in each of the alternative models, one sets the numerical values of the parameters to be identical to those from the reference model for all processes other than the process of interest. This leads to a so-called internal equivalence between the reference model and the alternatives. Next, various systemic properties are selected and used to impose some constraints for all the other parameters in the alternative designs. In general in this approach, one imposes that some steady-state values or logarithmic gains are equal in the reference model and its alternatives. This provides a way to express the parameters of the process of interest in the alternative models as functions of the parameters of the reference model. Thus, one obtains a so-called external equivalence between the reference model and the alternative designs, meaning that to an external observer the considered models are equivalent with respect to the selected systemic properties. Finally, one chooses various measures of functional effectiveness depending on the particularities of the biological context of these models and uses them to compare the alternative designs with the reference model. By doing this, one usually aims to determine analytically the qualitative differences between the compared models. This method was successfully used to compare alternative regulatory designs in, for example, metabolic pathways [25, 26], gene circuits [27], and immune networks [28]. Moreover, by introducing specific numerical values for the parameters of the models, one is also able to quantify these differences but, at the same time, the generality of the results is lost. Thus, in Ref. [29], the method of mathematically controlled comparison was extended to include some statistical methods, [30, 31], that allow the use of numerical values for the parameters while still preserving the generality of the conclusions. We discuss this extension in the following.

17.3.2
An Extension of the Mathematically Controlled Comparison

The first step of this extension is to generate a representative ensemble of sets of parameter values. Since the exact statistical distribution of the parameters values is often not known in practice, the most appropriate approach is to sample uniformly a given range of values. There exist different methods for scanning a given interval of values, ranging from various types of random samplings to some systematic deterministic scanning methods, see for example, [32]. Using this ensemble of sets of parameters, we can then construct a large class of numerical models both for the reference and for the alternative designs. There are two different methods to construct such a class of systems for which we can then investigate some statistical properties. A *structural class* consists of systems having the same network topology, that is, generated by the sampling of the parameter space. A *behavioral class* consists of systems that exhibit a particular systemic behavior, for example, exhibiting a steady-state behavior under given conditions, or low concentrations of intermediary products, or small values for the parameter sensitivity, see, for example, [31]. The members of such a class are obtained in two steps: first generate a set of parameters by sampling the parameter space, then test the sample for the desired systemic behavior and keep only those systems that fulfill the conditions.

After constructing this large class of numerical models both for the reference and the alternative architectures, one can start comparing the values of a given systemic property P between the reference model and its alternative designs. One way to do this is by using density plots of the ratio $R = P_{\text{reference}}/P_{\text{alternative}}$ versus the values $P_{\text{reference}}$, where the subscript indicates in which model the property P was measured. Such density plots can be used, for instance, to compute rank correlations between the considered property P (measured in the reference model) and the values of the ratio R. However, this is not easy to do if the density plots are very scattered. Then, one can construct secondary density plots by using the moving median technique as follows. The density plot can be interpreted as a list of N pairs of values ($P_{\text{reference}}$, R), which can be arranged in an ordered list L with respect to the first component $P_{\text{reference}}$. Then, we pick a window size W, usually much smaller than the sample size N, and we compute the median $<R>$ of the ratio values and the median $<P>$ of the values $P_{\text{reference}}$, for the first W pairs in the list L. Then, we advance the window by one, we collect the ratios and the values $P_{\text{reference}}$ from the second until the $(W + 1)$th pair, and compute the corresponding median values $<R>$ and $<P>$. This process is continued until the last pair of the list L is used. In the secondary density plot, we will pair the computed values $<R>$ with the corresponding $<P>$ values. This moving median technique is very useful because, for a finite ordered sample of size N, the moving median tends to the median of the samples as the value W approaches N.

These secondary density plots can be used to compare the efficiency of two classes of models from the point of view of a given systemic property.

17.3.3
Local Submodel Comparison

Another approach for comparing alternative designs that are actually submodels of a reference architecture was proposed in Ref. [12]. The focus of such a comparison of various submodels could be, for instance, a functional analysis of various modules of a large system. Then, the underlying reaction networks in the alternative designs are very similar (although not identical), and both the biological constraints and the kinetics of the reactions are given by those of the reference model. The only remaining question regards the initial distribution of the variables in the alternative models. In the mathematically controlled comparison, they are usually taken from the reference model. However, for some biochemical systems this choice might lead to biased comparisons. For instance, in the case of regulatory networks, models should be in a steady state in the absence of the trigger of the response, and indeed the initial values of the reference model are usually chosen in such a way to fulfill this condition. However, this will not imply in general that also a submodel will be in its steady state if it uses the same initial values as the reference model. Thus, the dynamic behavior of the submodel will be the result of two intertwined tendencies: migrating from a possible unstable state and the response to a trigger. If the focus of the comparison is exactly the efficiency of the response of various submodels to a trigger, then the approach proposed in Ref. [12] is more appropriate, yielding biologically unbiased results. In this approach, the initial distribution of the reactants is chosen in such a way that the initial setup of each submodel constitutes a steady state of that design in the absence of a trigger.

17.3.4
A Quantitative Measure for the Goodness of Model Fit Against Experimental Data

Another method for comparing alternative models is to analyze how well their predictions fit available experimental data. However, estimating the set of parameters of a given computational model in such a way that its predictions fit some experimental data is a computationally difficult problem, see for example, [33–35]. A common approach for this problem is to minimize a cost function quantifying the differences between the experimental measurements and the values predicted by the model. For instance, one of the functions used for this purpose is the sum of squared deviations $SS = \sum_{i=1}^{n}(x_i - y_i)^2$, where n is the number of experimental data points and for each i, and x_i and y_i represent the experimental value and the corresponding value predicted by the model. In particular, in this case a smaller value of SS indicates a better fit.

There are many methods for tackling such optimization problems either locally or globally, each of them having its own advantages and disadvantages. For example, local methods converge faster to a solution but they tend to find local optima, whereas global optimization methods are slower but they do converge to global optima. Furthermore, there are two types of global optimization methods: deterministic [36, 37] and stochastic [38, 39]. Although the convergence to the global optimum is guaranteed when using a deterministic method, the termination of the search process within a finite time interval is not ensured [35]. On the other hand, the stochastic optimization methods usually locate quite efficiently a good approximation of the solution, that is, located in a vicinity of the global solution, within an acceptable time interval [35]. Thus, the stochastic global optimization methods tend to be usually preferred for parameter estimation problems. Many parameter estimation methods are currently available in various software packages, such as COPASI [20].

One of the demanding tasks within this iterative process of finding a suitable parameter set is the identification of a measure to quantify the quality of the fit for each set of parameters. This measure can be also used to compare alternative models and decide which one fits best against the experimental data. One solution for this problem, proposed in Ref. [40], is to use a dimensionless number representing the deviation of the model from the experimental data, normalized by the mean of the predicted values: that is,

$$fq = \frac{\sqrt{SS/n}}{\text{mean of predicted values}} \times 100\% \qquad (17.1)$$

where n is the number of experimental data points. In particular, it was argued in Ref. [40] that a low value of fq, for example, lower than 15%, indicates a successful fit. Thus, when comparing several alternative models, we could actually compare their associated fit quality values indicating how well the predictions of these models fit existing experimental data.

17.3.5
Quantitative Refinement

The two alternative models that we aim to compare are very often built on different levels of details. Thus, in order to ease the comparison, a preliminary refinement of one (or even both) of the models could be useful. In fact, model refinement is part of the complex process of model building. Indeed, this process includes a series of iterative steps including hypothesis generation, experimental design, experimental analysis, and model refinement, [5, 41]. The first step when developing a model is to create an abstraction of the biological process, that is, to select a relatively small number of biochemical reactions that succeed to describe the main mechanisms of the considered process. The reactions included in this model are abstract representations of some particular subprocess, which can encapsulate many biochemical reactions from the modeled system. A mathematical model is then associated to this molecular model. For this, we have to choose an appropriate kinetic law, for example, mass-action law or Michaelis–Menten kinetics, based on which we can then write the mathematical equations describing the dynamics of the system. The only thing left is to choose a numerical setup of this mathematical model, which is either obtained from the literature or is derived through various computational model-fit procedures using available experimental data.

The refinement of a given model can be done in several different ways. For instance, within *data refinement* of a model we replace one (or more) of its species with several subspecies. This way, the new refined model will include more details about its subspecies and will illustrate various differences in their behavior. Another type of refinement, called *process refinement*, focuses on the model reactions. In this case, the model is refined by replacing a generic reaction which describes a particular process, with several reactions describing in more detail the intermediary steps of that process. In order to include all the intended changes of the initial models, one possibility is to simply repeat the whole model development procedure. However, this can lead to extremely inefficient processes since it requires to re-fit the model, a step which is both time consuming and computationally intensive, [42]. Another approach, not so much investigated so far (see Refs [43] and [44] for some recent case studies), is to refine the initial model in such a way such that the initial model fit is preserved.

17.3.6
Parameter-Independent Submodel Comparison

Although by choosing particular numerical setups we can quantify the differences between various submodels, a major downside is that we lose the generality of the results. This can be avoided if we employ some statistical methods to scan the parameter space, see Ref. [45]. Since all alternative models are obtained by eliminating reactions from the reference model, the parameters of the alternative architecture are in fact a subset of the parameter set of the reference model. Thus, we need to scan only the parameter value space of the reference model. This provides us with a set of parameter value vectors. Each coordinate of these vectors is associated with one of the parameters in the reference model and determines the value of the corresponding parameter. We consider each of the vectors one by one. We set the parameters of the reference model and the submodel in accordance with the considered vector. The initial values of the variables of the reference model and of the submodel are determined independently of each other by a systemic property such as the system being in a steady state in a given setup. For example, in the general case of stress response, we expect, in accordance with biological observations, that a feasible mathematical model is in a steady state under the unstressed physiological conditions. (We call the steady state a numerical configuration of the model such that starting from that configuration, the model shows no change in the level of any of the variables; in other words, the net loss per unit of time in every variable is exactly compensated by the net gain per unit of time in that variable.) Now, ensuring that all mathematical submodels satisfy such systemic properties makes them suitable to be considered as viable alternative formal descriptions of the biological mechanism being analyzed. As a result, we obtain the instantiations of the reference model and the submodels and we can run numerical simulations for all of them in order to evaluate their functional effectiveness. Finally, having done this for all sampled vectors, the obtained results for the variants are used to compare the models by use of some statistical measures.

17.3.7
Model Comparison for Pathway Identification

Very often, the exact pathway through which a particular biochemical process is regulated is not known. Instead, several alternative pathways underlying different mechanisms are proposed in the literature. Then, model comparison is employed in order to decide which of them is better in terms of fitting the experimental data better, or explaining better an observed qualitative behavior. For instance, in Ref. [46] a hybrid quantum mechanics (QM)–molecular mechanics (MM) approach was used to compare three alternative mechanisms proposed for the triosephosphate isomerase catalyzed reactions. This approach, introduced in 1976 by Warshel and Levitt [47], is a molecular simulation method that combines the accuracy of the QM computations and the speed of the MM calculations. This way, it allows the study of chemical processes in solution and in proteins. In Reference [48], a detailed comparison of several phosphorylation-driven control mechanisms for the regulation of the eukaryotic HSR was done by analyzing the goodness of the fit of the alternative models. In particular, three different phosphorylation pathways were investigated in an attempt to uncover the contribution of each of these pathways in controlling the HSR process. A detailed computational model was created for each of these pathways, which was then subjected to parameter estimation in order to be fit to the existing experimental data.

17.4
Case Study

17.4.1
A Biochemical Model for the Heat Shock Response

HSR is a highly evolutionarily conserved defense mechanism among organisms [49]. It serves to prevent and repair protein damage induced by elevated temperature and other forms of environmental, chemical, or physical stress. Such conditions induce the misfolding of proteins, which in turn accumulate and form aggregates with disastrous effect for the cell. In order to survive, the cell has to abruptly increase the expression of heat shock proteins. These proteins operate as intracellular chaperons: that is, play a crucial role in folding of proteins and re-establishment of proper protein conformation. They prevent the destructive protein aggregation. We discern two main reasons that account for the strong interest in the HSR mechanism observed in recent years, see for example, [50–52]. First, as a well-conserved mechanism among organisms, it is considered a promising candidate for disentangling the engineering principles that are fundamental for any regulatory network [21, 53–55]. Second, besides their functions in the HSR, heat shock proteins have fundamental importance to many key biological processes such as protein biogenesis, dismantling of damaged proteins, activation of immune responses and signaling, and so on, see [56, 57]. In consequence, a thorough insight into the HSR mechanism would have significant implications for the advancement in understanding the cell biology.

In order to coherently investigate the HSR, a number of mathematical models have been proposed in the literature, see for example, [21, 58–61]. In this study, we consider a recently introduced model of the eukaryotic HSR [11, 62]. In this model, the central role is played by the heat shock proteins (hsp), which act as chaperons for the misfolded proteins (mfp): the heat shock proteins sequester the misfolded proteins (hsp : mfp) and help the misfolded proteins to regain their native conformation (prot). The defense mechanism is controlled through the regulation of the transactivation of the hsp-encoding genes. The transcription is initiated by heat shock factors (hsf), some specific proteins that first form dimers (hsf$_2$), then trimers (hsf$_3$) and in this configuration bind to the heat shock elements (hse), that is, certain DNA sequences in the promotor regions of the hsp-encoding genes. Once the trimers bind to the promoter elements (hsf$_3$: hse), the transcription and translation of the hsp-encoding genes boost and, in consequence, new heat shock protein molecules get synthesized at a substantially augmented rate.

When the amount of the heat shock proteins reaches a high enough level that enables coping with the stress conditions, the production of new chaperon molecules is switched off by the excess of the heat shock proteins. To this aim, hsp form complexes with the heat shock factors (hsp : hsf) in three independently and concurrently running processes: (i) by binding to the free hsf; (ii) by breaking the dimers and trimers; and (iii) by breaking the hsf$_3$: hse, as a result of which the trimer gets unbound from the DNA and decomposed into free hsf molecules. This terminates the enhanced production of new heat shock protein molecules and blocks the formation of new hsf trimers. As soon as the temperature increases, proteins present in the cell start misfolding. The misfolded proteins titrate hsp away from the hsp : hsf complexes. This enables the accumulation of free hsf molecules, which in turn form trimers and promote the production of new chaperons. In consequence, the response mechanism gets switched on. The full list of biochemical reactions constituting the biochemical model from Ref. [11] is presented in Table 17.1. The model is based only on well-documented reactions without introducing any hypothetical mechanisms or experimentally unsupported biochemical reactions. Also, it assumes three conservation relations: for the total amount of heat shock factors; for the total amount of proteins (except heat shock proteins and heat shock factors); and for the total amount of the heat shock elements, as follows:

1) $[hsf] + 2 \times [hsf_2] + 3 \times [hsf_3] + 3 \times [hsf_3 : hse] + [hsp : hsf] = C_1$,
2) $[prot] + [mfp] + [hsp : mfp] = C_2$,
3) $[hse] + [hsf_3 : hse] = C_3$,

for some constants C_1, C_2, C_3 called *mass constants*. For a full presentation and discussion of this model, we refer the reader to [11].

Table 17.1 The list of reactions of the biochemical model for the heat shock response originally introduced in Ref. [11].

Reaction	Reaction number
$2\,\text{hsf} \leftrightarrow \text{hsf}_2$	(17.2)
$\text{hsf} + \text{hsf}_2 \leftrightarrow \text{hsf}_3$	(17.3)
$\text{hsf}_3 + \text{hse} \leftrightarrow \text{hsf}_3 : \text{hse}$	(17.4)
$\text{hsf}_3 : \text{hse} \to \text{hsf}_3 : \text{hse} + \text{hsp}$	(17.5)
$\text{hsp} + \text{hsf} \leftrightarrow \text{hsp} : \text{hsf}$	(17.6)
$\text{hsp} + \text{hsf}_2 \to \text{hsp} : \text{hsf} + \text{hsf}$	(17.7)
$\text{hsp} + \text{hsf}_3 \to \text{hsp} : \text{hsf} + 2\,\text{hsf}$	(17.8)
$\text{hsp} + \text{hsf}_3 : \text{hse} \to \text{hsp} : \text{hsf} + \text{hse} + 2\,\text{hsf}$	(17.9)
$\text{hsp} \to$	(17.10)
$\text{prot} \to \text{mfp}$	(17.11)
$\text{hsp} + \text{mfp} \leftrightarrow \text{hsp} : \text{mfp}$	(17.12)
$\text{hsp} : \text{mfp} \to \text{hsp} + \text{prot}$	(17.13)

Based on the assumption of mass-action law for all the reactions (17.2)–(17.13), an associated mathematical model of the eukaryotic HSR is obtained. The resulting mathematical model is expressed in terms of 10 first-order ODEs.

The mathematical model comprises 16 independent kinetic parameters and 10 initial conditions. We refer to [11] for the details of the ODE model and its numerical setup.

17.4.2
Control-Based Decomposition

In Reference [12], a control-driven modular decomposition of the HSR model was performed. In result, the model has been divided into four main functional submodules usually distinguished in control engineering: the plant, the sensor, the controller, and the actuator. In the case of the HSR model, the plant is the misfolding and refolding of proteins, the actuator consists of the synthesis and degradation of the chaperons, the sensor measures the level of hsp in the system, and the controller regulates the level of DNA-binding. Moreover, within the controller we distinguish three feedback mechanisms. The feedback loops are responsible for sequestering the heat shock factors in different forms by the chaperons. In this way, the feedback loops decrease the level of DNA-binding. The three identified feedback mechanisms are the following:

1) FB1: sequestration of free hsf, that is, reaction $(17.6)^+$ (the "left-to-right" direction of reaction (17.6));
2) FB2: breaking of hsf dimers and trimers, that is, reactions (17.7) and (17.8);
3) FB3: unbinding of hsf_3 from hse and breaking the trimers, that is, reaction (17.9).

17.4 Case Study

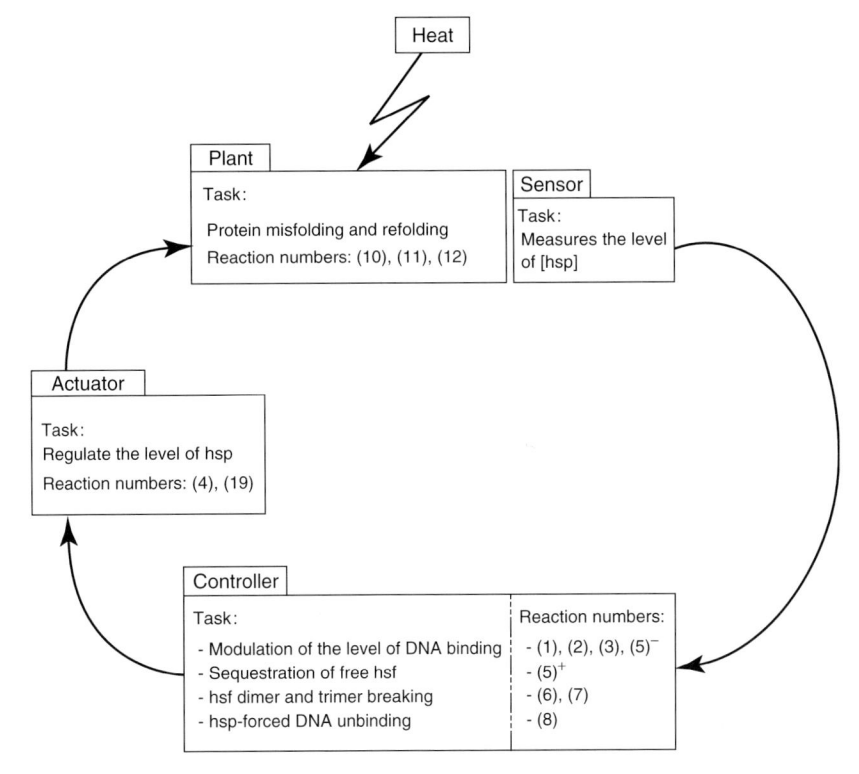

Figure 17.1 The control-based decomposition of the heat shock response network. The reaction numbers refer to the reactions in Table 17.1. We denote the "left-to-right" direction of reaction (17.6) by (17.6)$^+$ and by (17.6)$^-$ its "right-to-left" direction.

The control-driven functional decomposition of the eukaryotic HSR model is shown in Figure 17.1, where the reaction numbers refer to the reactions in Table 17.1. In Figure 17.2, a graphical illustration of the control structure, that is, the three feedback loops and their points of interactions with the mainstream process, is presented.

17.4.3
The Knockdown Mutants

In References [12, 45], the reference architecture and seven knockdown mutants (alternative architectures) were considered. The mutants were obtained by eliminating from the reference architecture all possible combinations of the three feedback loops FB1, FB2, and FB3. The mutants were denoted as M_X, where $X \subset \{1, 2, 3\}$ is the set of numbers of the feedback mechanisms present in M_X.

1) M_0 is determined by reactions 17.2–17.5 and 17.10–17.13 and, in the terminology of control theory, is characterized by the *open-loop design*;
2) M_1 is determined by reactions 17.2–17.6 and 17.10–17.13;

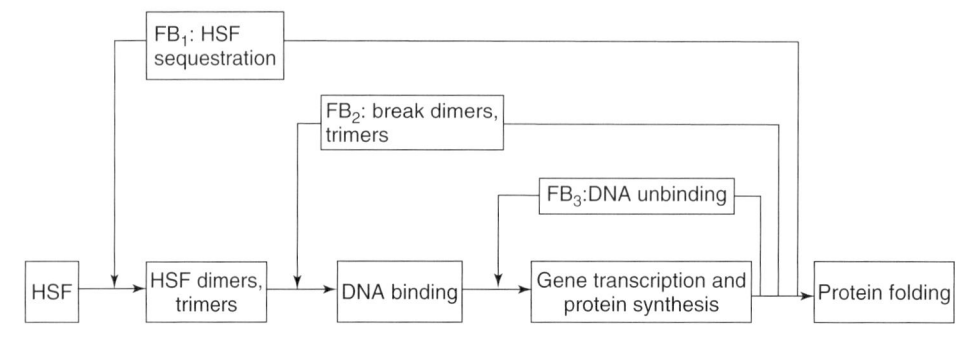

Figure 17.2 Control structure of the heat shock response network. The three identified feedback loops and their points of interaction with the mainstream process are depicted.

3) M_2 is determined by reactions 17.2–17.5, 17.7–17.8, and 17.10–17.13, and the "right-to-left" direction of reaction 17.6;
4) M_3 is determined by reactions 17.2–17.5 and 17.9–17.13, and the "right-to-left" direction of reaction 17.6;
5) $M_{1,2}$ is determined by reactions 17.2–17.8 and 17.10–17.13;
6) $M_{1,3}$ is determined by reactions 17.2–17.6 and 17.9–17.13;
7) $M_{2,3}$ is determined by reactions 17.2–17.5 and 17.7–17.13, and the "right-to-left" direction of reaction 17.6;
8) $M_{1,2,3}$ is the reference architecture consisting of all reactions 17.2–17.13.

By looking at the associated differential equations, it is easy to see that, if the mutant M_0 starts from a steady state at physiological conditions, that is, 37 °C, then it is nonresponsive, that is, it shows no increase in DNA-binding for any arbitrarily high temperature. Thus, we remove M_0 from further considerations.

17.4.4
Local Comparison of the Knockdown Mutants

In order to identify the individual contributions of each of the three feedback mechanisms to the regulation of HSR, we can first locally compare the knockdown mutants by using the method described in Section 17.3.3, see [12]. The numerical setups for each of the eight knockdown mutants are chosen such that they satisfy the following two constraints:

1) The kinetic rate constants of the reactions included in the mutants are chosen to be identical to those of the corresponding reactions in the reference model. Also, the values of the mass constants C_1, C_2, C_3 of each of the mutants are identical to those of the reference model.
2) The initial distribution of the variables of each knockdown mutant are chosen such that they form a steady state of that model at 37 °C. Note that the initial values of all variables for the reference model were chosen in the same way in Ref. [63].

Then, the comparison is focused on the dynamical behavior of the mutants at 42 °C. This particular temperature was chosen for the analysis because at 42 °C in the experimental data we can notice both a pronounced HSR in terms of increased levels of misfolded proteins, as well as an explicit response in terms of increased transient DNA-binding of hsf$_3$, see [63].

While the main task of the HSR is to keep the level of misfolded proteins under control, it is also very important for the cell to use efficiently its materials and energy. Thus, the analysis of the mutants focused on two main aspects: the level of misfolded proteins and the level of chaperons needed to achieve a response at 42 °C. Based on these two aspects, the contribution of the three feedbacks was analyzed in Ref. [12] with respect to the following four performance indicators:

1) *The system makes economical use of the cellular resources* (i.e., the hsp-encoding gene is only transactivated for a short while when exposed to heat shock): FB_1 was found to play the major role here, while FB_2 and FB_3 are also important. In the absence of FB_1, gene transcription is at the 100% level even without heat shock.
2) *The system is fast to respond to a heat shock* (i.e., the hsp-encoding gene is quickly transactivated in response to heat shock): FB_3, FB_1 were found to play the major role here. In particular, model $M_{1,3}$ reacts to heat shock as fast as the reference model, although the level of its response is lower than that of the reference model.
3) *The response is effective* (i.e., the mfp concentration is kept low for mild heat shocks): the open-loop structure of the mutant M_0 is enough to achieve this property. Although none of the feedback mechanisms is needed for maintaining a low [mfp] for a heat shock at 42 °C, they play an important role in minimizing the cost of the response.
4) *The response is scalable* (i.e., we notice a higher response for a higher temperature): FB_3, FB_1 were found to play the major role here. The mutant $M_{1,3}$ is the only one that scales its response to higher temperatures, in a similar way as the reference model: that is, for higher temperatures, gene transactivation raises faster and to a higher level.

For more details about this analysis, we refer the reader to [12].

17.4.5
Parameter-Independent Comparison of the Mutant Behavior

The previous analysis is clearly heavily dependent on the parametric setups chosen for the compared models. In order to avoid this, one approach (see [45]) would be to employ some statistical methods to scan the parametric space, as explained in Section 17.3.6. For this, we use the *Latin hypercube sampling* method (LHS), originally introduced in Ref. [64]. It provides samples that are uniformly distributed over each parameter, while the number of samples is independent of the number of parameters. The sampling scheme can be briefly described as follows. First, the desired size N of the sampling set is chosen. Next, the range interval of each parameter is partitioned into N nonoverlapping intervals of equal length. For each parameter, N numerical values are randomly selected, one from each interval of the partition according to a uniform distribution on that interval. Finally, the N sampled values for the ith parameter of the model are collected on the ith column of an $N \times p$ matrix, where p is the number of model parameters and the values on each column are shuffled randomly. As a result, each of the N rows of the matrix contains numerical values for each of the p parameters. For a detailed description of this sampling scheme, we refer the reader to [64, 65], see also [11] for an example of the application of this sampling method in the context of model identifiability problem.

In particular, in Ref. [45] a sample of 10.000 vectors of parameter values for the reference architecture was obtained through the LHS described above. For the HSR model, the sampled vectors are of length 15, that is, the number of the unknown reference architecture parameters. The value of the 16th remaining parameter, that is, the hsp degradation rate constant, is assumed to be known and is obtained based on the fact that heat shock proteins are generally long-lived proteins, see [66]. In particular, their half-life was chosen to be 6 h. Then, the procedure described next is repeated separately for each of the six mutants. To begin with, each sampled vector of parameter values is used to set up the parameters in the mathematical models of the considered mutant and the reference architecture ($M_{1, 2, 3}$). From the construction of each mutant, it follows that the corresponding mathematical model contains only a subset of the parameters of the reference model, so this step can be performed. Next, the steady-state concentrations at 37 °C both for the mutant and the reference model are numerically computed and set as their respective initial states. In this way, we obtain two instances of the mathematical models, namely, one for the mutant and the other for the reference model. Further, the temperature is increased to 42 °C and the quantities

$$\Theta_1 = \max_{t \in [0s, 1800s]} (\text{total mfp}(t))$$

$$\Theta_2 = \max_{t \in [0s, 1800s]} (\text{hsf}_3 : \text{hse}(t)) - \text{hsf}_3 : \text{hse}(0)$$

$$\Theta_3 = \frac{1}{T} \int_0^T (\text{total hsp}(t)) dt$$

$$\Theta_4 = \frac{1}{T} \int_0^T (\text{total mfp}(t)) dt$$

are computed both for the mutant and the reference instance. The initial 30 min of the response is considered for the computation of Θ_1 and Θ_2. In the case of Θ_3 and Θ_4, the time range of 4 h ($T = 14, 400$ s) is taken into account. These quantities are used to evaluate the functional effectiveness of the mutant. Having these quantities computed for all the 10 sampled parameter values, the scatter plot of the $R_1 = \Theta_1^m / \Theta_1^r$ against Θ_1^r values is made, where the superscripts m and r indicate the instance for which Θ_1 was computed, that is, the instance of the mutant and the reference model, respectively. Finally, the moving median technique is applied to the scatter plot with the window size set to 500. These result in a trend curve summarizing the data of the scatter plot and revealing the overall dependency between the considered quantities. Similar plots are computed for $R_2 = \Theta_2^m/\Theta_2^r$. Moreover, scatter plots of Θ_3 versus Θ_4 both for the mutant and the reference architecture are made, and the moving median technique is applied to each of these plots.

The mutants represent six different potential architectures of the HSR mechanism, and the sampling procedure, as explained above, provides us with 10 different instantiations of each of the mutants and the reference architecture.

In the analysis performed in Ref. [45], it was assumed that the HSR at raised temperatures is accompanied, and hence characterized, by the following three phenomena:

1) increase in DNA binding with respect to the steady-state level at 37 °C,
2) increase in the level of mfp, and
3) increase in the level of hsp as the effect of the response to the higher level of mfp in the cell.

Thus, the analysis of the architecture properties of the six mutants with respect to the reference architecture was based on the following plots: R_1 versus Θ_1^r (Figure 5 of [45]), R_2 versus Θ_2^r (Figure 7 of [45]), Θ_3 versus Θ_4 (Figure 17.3) made for each of the mutants and for the reference architecture. We refer to the Θ_3 versus Θ_4 plot as the cost plot (or simply the cost) of the corresponding architecture. This is motivated by the fact that the efficiency of the HSR mechanism could be measured by the amount of chaperons needed to cope with the intensified misfolding of proteins. Hypothetically, a cell that produces smaller amounts of hsp than some other cell to cope with the heat shock would be considered the one that manages with stress conditions at a lower cost in terms of its resources than the latter one. Notice, however, that in our case we are not assessing the ability of particular models to cope with heat shock, that is, the sampled models are neither validated against experimental data nor classified by any other means whether they enable the cell to survive or not in the stress conditions. Hence the cost plots reflect just the general tendency of the models instantiating a particular architecture to keep certain average in time amounts of hsp in response to different average levels of mfp present in the system. The reference trend line indicates a clear linear dependency between the average levels of hsp and mfp, see Figure 17.4.

Based on these aspects, it was noted in Ref. [45] that all the mutants lacking two feedbacks exhibit no HSR in the sense of the above definition: as observed previously, there is no increase in the DNA binding. This is in agreement with the results presented in Ref. [12], where the models with only one feedback kept the DNA binding at the maximum possible level both at 37 and 42 °C throughout the simulation time of 50 s. The HSR can be observed, however, in the mutants $M_{1,3}$ and $M_{1,2}$. In the case of the $M_{2,3}$ mutant, the HSR is still observed, but only for a fraction of the 10 sampled models, that is, only those parameter values for which the reference architecture displays the maximum possible increase in the peak of DNA binding with respect to the steady-state level at 37 °C. This is in complete agreement with previous observations that FB1 is the most powerful feedback, see [12]. Since FB2 and FB3 include hsf sequestration as one of their features, they

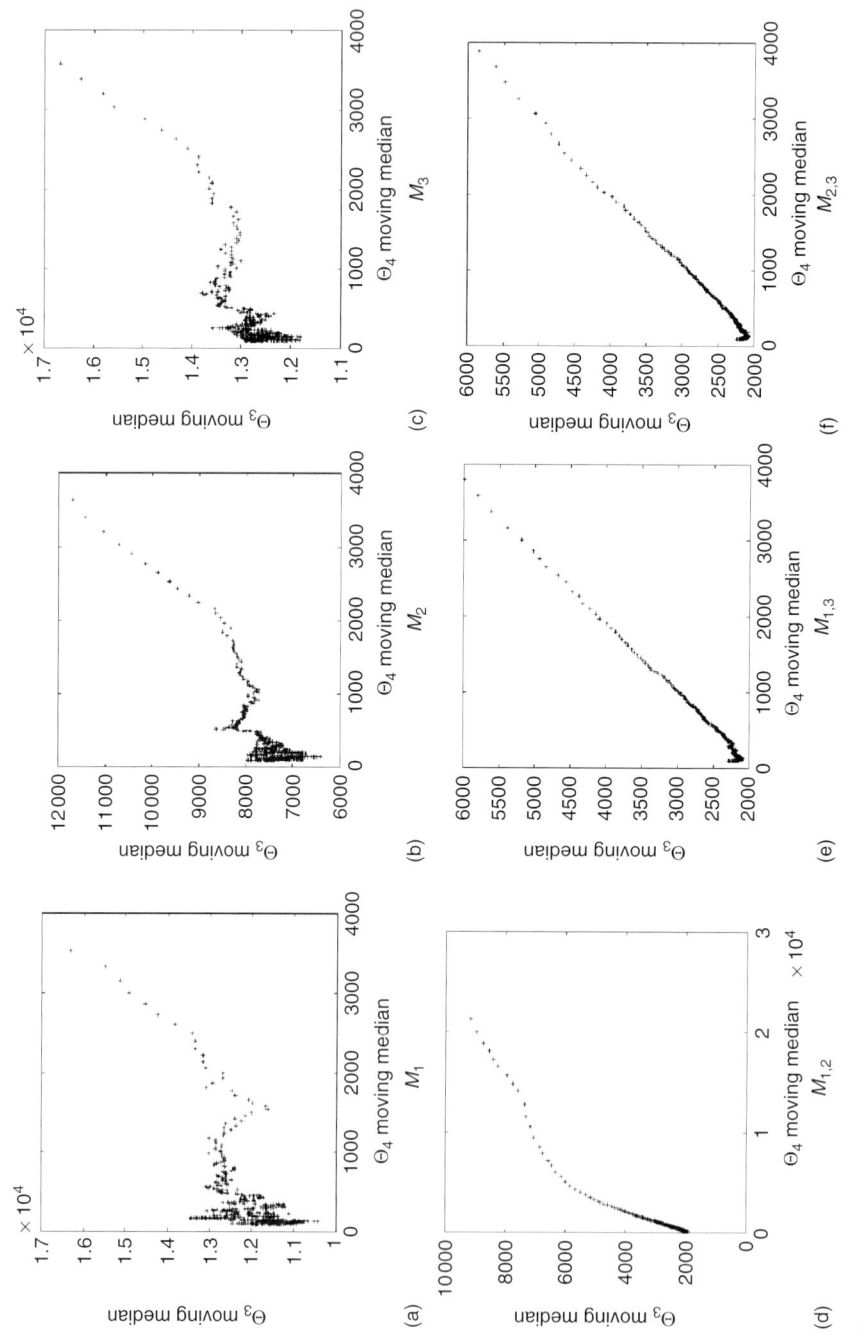

Figure 17.3 Result of applying the moving median technique to the scatter plots of the cost, that is, Θ_3 versus Θ_4, obtained individually for each of the six considered mutants. For each mutant and each sampled vector of parameters, the values of Θ_3 and Θ_4 were computed and plotted against each other. Then, the moving median technique was applied to discern the overall trend in the data depicted in the obtained scatter plots. The window size of the moving median was set to 500 and the sample size of the vectors of parameter values was 10.

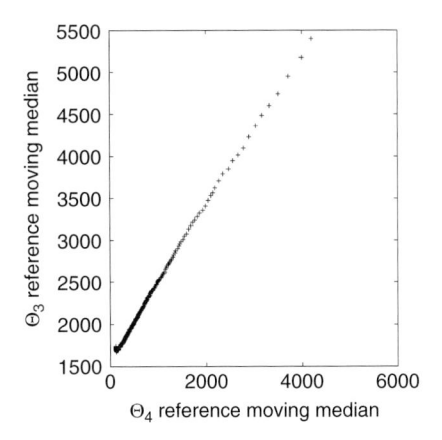

Figure 17.4 Result of applying the moving median technique to the scatter plots of the cost, that is, Θ_3 versus Θ_4, obtained for the reference architecture. For each sampled vector of parameters, the values of Θ_3 and Θ_4 were computed and plotted against each other. Then, the moving median technique was applied to discern the overall trend in the data depicted in the obtained scatter plot. The window size of the moving median was set to 500 and the sample size of the vectors of parameter values was 10.

compensate partially for the lack of FB1. However, only FB2 or only FB3 are not enough to enforce the system's behavior to have the HSR characteristics. Despite its power, FB1 alone is also not enough and one of the other feedbacks is also needed in order to implement a response mechanism with the features describing the HSR.

17.4.6
Pathway Identification for the Phosphorylation-Driven Control of the Heat Shock Response

In Reference [48], an extension of the biochemical model for the eukaryotic HSR from Section 17.4.1 was introduced where phosphorylation of hsfs was considered. The synthesis of hsps is highly regulated at the transcription level and phosphorylation of hsfs constitutes one of the regulation mechanisms. The introduced extended model allowed the authors of [48] to investigate the positive role of phosphorylation in the upregulation of hsf transcriptional activity. In particular, a number of computational models associated with the extended biochemical model were derived by assuming mass-action kinetics and used to investigate three plausible (de)phosphorylation pathways. First, it was analyzed whether the kinase and phosphates dynamics for the heat-induced misfolding and refolding can lead to the experimentally observed evolution of the total level of phosphorylated hsf molecules. Second, on top of the heat-induced misfolding and refolding of both kinase and phosphates, it was supposed that the hsf molecules can be both phosphorylated and dephosphorylated while they form hsp : hsf complexes. Finally, with respect to the second pathway, the hsf molecules forming hsp : hsf complexes could only be dephosphorylated. For each of these three pathways, two different mechanisms for modeling the phosphorylation-dependent transcription were considered. For the first mechanism, it was assumed that the transcription proceeds linearly depending on the level of phosphorylation of the hsf_3 : hse complex. For the second one, the assumption was that hsp synthesis is activated only by the hyper-phosphorylation of the hsf_3 : hse compound, that is, that the transcription of the hsp-encoding gene proceeds only after at least two of the three hsf molecules from this compound become phosphorylated. Moreover, in order to verify and strengthen the results, for each computational model implementing one of the six scenarios (three pathways, each with two different mechanisms of phosphorylation-dependent transcription) a corresponding reduced model employing the Michaelis–Menten kinetics was considered, where the kinase and phosphatase enzymes were not modeled explicitly anymore. All in all, 12 computational models of phosphorylation-mediated transcription of hsp were considered in Ref. [48], and the analysis regarding possible (de)phosphorylation pathways was based on the assessment of repetitive parameter estimation procedures of the computational models, without the use of any quantitative measurement of the results. The models were fitted to two experimental measurements reported in Ref. [67] regarding the DNA-binding activity and the total number of phosphorylated hsf proteins in HeLa cells under 42 °C heat shock. For more details on the extended biochemical model, the associated 12 computational models, the parameter estimation procedure, and the performed analysis regarding possible (de)phosphorylation pathways, we refer to [48].

17.5
Discussion

Very often, various experimental investigations of a given biochemical system generate a large variety of alternative molecular designs, thereby raising questions about comparing their functionality, efficiency, and robustness. Comparing alternative models for a given biochemical system is, in general, a very difficult problem which involves a deep analysis of various aspects of the models: the underlying networks, the biological constraints, and the numerical setup. The problem becomes somewhat simpler when the alternative designs are actually submodels of a larger model: the underlying networks are similar, although not identical, and the biological constraints are given by the larger model. It only remains to decide how to choose the numerical setup for each of the alternative submodels, that is, the initial conditions and the kinetics.

In the first part this chapter, we reviewed several known methods for model decomposition and for quantitative comparison of submodels. We described the knockdown mutants, elementary flux modes, control-based decomposition, mathematically controlled comparison and its extension, local submodels comparison, a parameter-independent submodel comparison, and a discrete approach for comparing continuous submodels. We also showed how the goodness of model fit against experimental data and quantitative model refinement can be used for comparing alternative designs as well as how to use model comparison for pathway identification. In the second part of this chapter, we considered as a case study the eukaryotic HSR. First, by using a control-based approach, the main functional modules of this system were identified, see [12]. In particular, we underlined three feedback mechanisms regulating this response. Then, in order to identify the individual role these mechanisms play in the regulation of the HSR, we constructed knockdown mutants by eliminating from the reference architecture all possible combinations of the three feedbacks. These mutants were then compared both with themselves and with the reference model by using various methods described in the first part of this chapter.

Acknowledgments

The work of Elena Czeizler, Andrzej Mizera, and Ion Petre was supported by Academy of Finland, Grants 129863, 108421, and 122426. Andrzej Mizera is on leave of absence from the Institute of Fundamental Technological Research, Polish Academy of Sciences, Warsaw, Poland.

References

1. Chen, W.W., Schoeberl, B., Jasper, P.J., Niepel, M., Nielsen, U.B., Lauffenburger, D.A., and Sorger, P.K. (2009) Input-output behavior of ErbB signaling pathways as revealed by a mass action model trained against dynamic data. *Mol. Syst. Biol.*, **5**, 239.
2. Savageau, M.A. (1972) The behavior of intact biochemical control systems. *Curr. Top. Cell. Regul.*, **6**, 63–130.
3. Csete, M.E. and Doyle, J.C. (2002) Reverse engineering of biological complexity. *Science*, **295**, 1664–1669.
4. Hawkins, B.A. and Cornell, H.V. (eds) (1999) *Theoretical Approaches to Biological Control*, Cambridge University Press.
5. Kitano, H. (2002) Systems biology: a brief overview. *Science*, **295** 5560, 1662–1664.
6. Lazebnik, Y. (2002) Can a biologist fix a radio? - Or, what i learned while studying apoptosis. *Cancer Cell*, **2** (3), 179–182.
7. Sontag, E.D. (2004) Some new directions in control theory inspired by systems biology. *IEEE Syst. Biol.*, **1** (1), 9–18.
8. Sontag, E.D. (2005) Molecular systems biology and control. *Eur. J. Control*, **11** (4), 396–435.
9. Stelling, J., Sauer, U., Szallasi, Z., Doyle, F.J., and Doyle, J. (2004) Robustness of cellular functions. *Cell*, **118** (6), 675–685.
10. Wolkenhauer, O. (2001) Systems biology: the reincarnation of systems theory applied in biology? *Brief. Bioinform.*, **2** (3), 258–270.
11. Petre, I., Mizera, A., Hyder, C.L., Meinander, A., Mikhailov, A., Morimoto, R.I., Sistonen, L., Eriksson, J.E., and Back, R.-J. (2010) A simple mass-action model for the eukaryotic heat shock response and its mathematical validation. *Nat. Comput.* doi://dx.doi.org/10.1007/s11047-010-9216-y.
12. Czeizler, E. Czeizler, R.-J. Back, and I. Petre (2009) Control strategies for the regulation of the eukaryotic heat shock response, in *Computational Methods in Systems Biology*, Lecture Notes in Computer Science, vol. 5688, (eds P. Degano and R. Gorrieri), Springer-Verlag, Heidelberg, pp. 111–125.

13. Klipp, E., Herwig, R., Kowald, A., Wierling, C., and Lehrach, H. (2005) *Systems Biology in Practice: Concepts, Implementation and Application*, Wiley-VCH Verlag GmbH.
14. Heinrich, R. and Schuster, S. (1996) *The Regulation of Cellular Systems*, Chapman & Hall, New York.
15. Pfeiffer, T., Sanchez-Valdenebro, I., Nuno, J.C., Montero, F., and Schuster, S. (1999) METATOOL: for studying metabolic netwroks. *Bioinformatics*, **15**, 251–257.
16. Schilling, C.H., Schuster, S., Palsson, B.O., and Heinrich, R. (1999) Metabolic pathway analysis: basic concepts and scientific applications in the post-genomic Era. *Biotechnol. Prog.*, **15** (3), 296–303.
17. Schuster, S., Dandekar, T., and Fell, D.A. (1999) Detection of elementary flux modes in biochemical networks: a promising tool for pathway analysis and metabolic engineering. *Trends Biotechnol.*, **17** (2), 53–60.
18. Schuster, S., Fell, D.A., and Dandekar, T. (2000) A general definition of metabolic pathways useful for systematic organization and analysis of complex metabolic networks. *Nat. Biotechnol.*, **18**, 326–332.
19. Schuster, S., Hilgetag, C., Woods, J.H., and Fell, D.A. (2002) Reaction routes in biochemical reaction systems: algebraic properties, validated calculation procedure and example from nucleotide metabolism. *J. Math. Biol.*, **45** (2), 153–181.
20. Hoops, S., Sahle, S., Gauges, R., Lee, C., Pahle, J., Simus, N., Singhal, M., Xu, L., Mendes, P., and Kummer, U. (2006) COPASI-a COmplex PAthway SImulator. *Bioinformatics*, **22**, 3067–3074.
21. El-Samad, H., Kurata, H., Doyle, J.C., Gross, C.A., and Khammash, M. (2005) Surviving heat shock: control strategies for robustness and performance. *Proc. Natl. Acad. Sci. U.S.A.*, **102** (8), 2736–2741.
22. Savageau, M.A. (1969) Biochemical systems analysis: I. Some mathematical properties of the rate law for the component enzymatic reactions. *J. Theor. Biol.*, **25** (3), 365–369.
23. Savageau, M.A. (1969) Biochemical systems analysis: II. The steady state solution for an n-pool system using a power law approximation. *J. Theor. Biol.*, **25** (3), 370–379.
24. Savageau, M.A. (1970) Biochemical systems analysis: III. Dynamic solutions using a power-law approximation. *J. Theor. Biol.*, **26**, 215–226.
25. Hunding, A. (1974) Limit-cycles in enzyme-systems with nonlinear negative feedback. *Biophys. Struct. Mech.*, **1**, 47–54.
26. Savageau, M.A. (1974) Optimal design of feedback control by inhibition: steady state considerations. *J. Mol. Evol.*, **4**, 139–156.
27. Hlavacek, W.S. and Savageau, M.A. (1996) Rules for coupled expression of regulator and effector genes in inducible circuits. *J. Mol. Biol.*, **255**, 121–139.
28. Boer, R.J.D. and Hogeweg, P. (1989) Stability of symmetric idiotypic networks-a critique of hoffmann's analysis. *Bull. Math. Biol.*, **51**, 217–222.
29. Alves, R. and Savageau, M.A. (2000) Extending the method of mathematically controlled comparison to include numerical comparisons. *Bioinformatics*, **16** (9), 786–798.
30. Alves, R. and Savageau, M.A. (2000) Comparing systemic properties of ensembles of biological networks by graphical and statistical methods. *Bioinformatics*, **16** (6), 527–533.
31. Alves, R. and Savageau, M.A. (2000) Systemic properties of ensembles of metabolic networks: application of graphical and statistical methods to simple unbranched pathways. *Bioinformatics*, **16** (6), 534–547.
32. Saltelli, A., Tarantola, S., Campolongo, F., and Ratto, M. (2004) *Sensitivity Analysis in Practice: A Guide to Assessing Scientific Models*, John Wiley & Sons, Ltd, Chichester, England.
33. Baker, S.M., Schallau, K., and Junker, B.H. (2010) Comparison of different algorithms for simultaneous estimation of multiple parameters in kinetic metabolic models. *J. Integr. Bioinform.*, **7** (3), 133.
34. Mendes, P. and Kell, D. (1998) Non-linear optimization of biochemical pathways: applications to metabolic

35. Moles, C.G., Mendes, P., and Banga, J.R. (2003) Parameter estimation in biochemical pathways: a comparison of global optimization methods. *Genome Res.*, **13**, 2467–2474.
36. Grossmann, I.E. (1996) *Global Optimization in Engineering Design*, Kluwer Academic Publishers, Dordrecht, The Netherlands.
37. Horst, R. and Tuy, H. (1990) *Global Optimization: Deterministic Approaches*, Springer-Verlag, Berlin.
38. Ali, M.M., Storey, C., and Törn, A. (1997) Application of stochastic global optimization algorithms to practical problems. *J. Optim. Theory Appl.*, **95** (3), 545–563.
39. Guus, C., Boender, E., and Romeijn, H.E. (1995) Stochastic methods, in *Handbook of Global Optimization* (eds R. Horst and P.M. Pardalos), Kluwer Academic Publishers, Dordrecht, The Netherlands.
40. Kühnel, M., Mayorga, L.S., Dandekar, T., Thakar, J., Schwarz, R., Anes, E., Griffiths, G., and Reich, J. (2008) Modelling phagosomal lipid networks that regulate actin assembly. *BMC Syst. Biol.*, **2**, 107.
41. Bruggeman, F.J. and Westerhoff, H.V. (2007) The nature of systems biology. *Trends Microbiol.*, **15** (1), 45–50.
42. Chen, W.W., Schoeberl, B., Jasper, P.J., Niepel, M., Nielsen, U.B., Lauffenburger, D.A., and Sorger, P.K. (2009) Input-output behavior of erbb signaling pathways as revealed by a mass action model trained against dynamic data. *Mol. Syst. Biol.*, **5** (239), 1–19.
43. Mizera, A., Czeizler, E., and Petre, I. (2012) Self-assembly models of variable resolution. *Trans. on Comput. Syst. Biol.*, LNBI.
44. Czeizler, E., Czeizler, E., Iancu, B., and Petre, I. (2011) Quantitative model refinement as a solution to the combinatorial size explosion of biomodels. Proceedings of the 2nd International Workshop on Static Analysis and Systems Biology, ENTCS. Venice.
45. Mizera, A., Czeizler, E., and Petre, I. (2011) Methods for biochemical decomposition and quantitative submodel comparison. *Isr. J. Chem.*, **51**, 151–164.
46. Cui, Q. and Karplus, M. (2001) Triosephosphate isomerase: a theoretical comparison of alternative pathways. *J. Am. Chem. Soc.*, **123**, 2284–2290.
47. Warshel, A. and Levitt, M. (1976) Theoretical studies of enzymic reactions: Dielectric, electrostatic and steric stabilization of the carbonium ion in the reaction of lysozyme. *J. Mol. Biol.*, **103**, 2, 227–249.
48. Czeizler, E., Rogojin, V., and Petre, I. (2011) The phosphorylation of the heat shock factor as a modulator for the heat shock response, in Proceedings of the 9th International Conference on Computational Methods in Systems Biology, (ed. F. Fages), ACM Digital Library, vol. to appear.
49. Lindquist, S. and Craig, E.A. (1988) The heat-shock proteins. *Ann. Rev. Genet.*, **22**, 631–677.
50. Chen, Y., Voegeli, T.S., Liu, P.P., Noble, E.G., and Currie, R.W. (2007) Heat shock paradox and a new role of heat shock proteins and their receptors as anti-inflammation targets. *Inflamm. Allergy Drug Targets*, **6** (2), 91–100.
51. Powers, M.V. and Workman, P. (2007) Inhibitors of the heat shock response: biology and pharmacology. *FEBS Lett.*, **581** (19), 3758–3769.
52. Voellmy, R. and Boellmann, F. (2007) Chaperone regulation of the heat shock protein response. *Adv. Exp. Med. Biol.*, **594**, 89–99.
53. El-Samad, H., Prajna, S., Papachristodoulou, A., Khammash, M., and Doyle, J.C. (2003) Model validation and robust stability analysis of the bacterial heat shock response using SOSTOOLS. Proceedings of the 42th IEEE Conference on Decision and Control, vol. 4, Dec, Maui, 3766–3771.
54. Kurata, H., El-Samad, H., Yi, T.-M., Khammash, M., and Doyle, J.C. (2001) Feedback regulation of the heat shock response in E. coli. Proceedings of the 40th IEEE Conference on Decision and Control, Orlando, FL, 837–842.

55. Tomlin, C.J. and Axelrod, J.D. (2005) Understanding biology by reverse engineering the control. *Proc. Natl. Acad. Sci. U.S.A.*, **102** (12), 4219–4220.
56. Kampinga, H.K. (1993) Thermotolerance in mammalian cells: protein denaturation and aggregation, and stress proteins. *J. Cell Sci.*, **104**, 11–17.
57. Pockley, A.G. (2003) Heat shock proteins as regulators of the immune response. *The Lancet*, **362** (9382), 469–476.
58. Lipan, O., Navenot, J.-M., Wang, Z., Huang, L., and Peiper, S. (2007) Heat shock response in cho mammalian cells is controlled by a nonlinear stochastic process. *PLoS Comput. Biol.*, **3** (10), 1859–1870.
59. Peper, A., Grimbergent, C., Spaan, J., Souren, J., and van Wijk, R. (1997) A mathematical model of the hsp70 regulation in the cell. *Int. J. Hyperthermia*, **14**, 97–124.
60. Rieger, T.R., Morimoto, R.I., and Hatzimanikatis, V. (2005) Mathematical modeling of the eukaryotic heat shock response: Dynamics of the hsp70 promoter. *Biophys. J.*, **88** (3), 1646–1658.
61. Srivastava, R., Peterson, M., and Bentley, W. (2001) Stochastic kinetic analysis of the escherichia coli stres circuit using σ^{32}-targeted antisense. *Biotechnol. Bioeng.*, **75** (1), 120–129.
62. Petre, I., Mizera, A., Hyder, C.L., Mikhailov, A., Eriksson, J.E., Sistonen, L., and Back, R.-J. (2009) A new mathematical model for the heat shock response, in *Algorithmic Bioprocesses*, Natural Computing Series (eds A. Condon, D. Harel, J.N. Kok, A. Salomaa, and E. Winfree), Springer, pp. 411–425.
63. Petre, I., Hyder, C.L., Mizera, A., Mikhailov, A., Eriksson, J.E., Sistonen, L., and Back, R.-J. (2011) A simple mathematical model for the eukaryotic heat shock response. *Nat. Comput.*, **10** (1), 595–612.
64. McKay, M.D., Beckman, R.J., and Conover, W.J. (1979) A comparison of three methods for selecting values of input variables in the analysis of output from a computer code. *Technometrics*, **21** (2), 239–245.
65. Helton, J.C. and Davis, F.J. (2003) Latin hypercube sampling and the propagation of uncertainty in analyses of complex systems. *Reliab. Eng. Syst. Safety*, **81**, 23–69.
66. Sapozhnikov, A.M., Gusarova, G.A., Ponomarev, E.D., and Telford, W.G. (2002) Translocation of cytoplasmic HSP70 onto the surface of EL-4 cells during apoptosis. *Cell Prolif.*, **35** (4), 193–206.
67. Kline, M.P. and Morimoto, R.I. (1997) Repression of the heat shock factor 1 transcriptional activation domain is modulated by constitutive phosphorylation. *Mol. Cell. Biol.*, **17** (4), 2107–2115.

18
Conclusions and Perspectives
Evgeny Katz

Unconventional computing aiming at information processing by nonelectronic systems is a research area that is still in its infancy [1]. Despite the many efforts in the last decades, researchers could not demonstrate any practical application in which to compete with electronic systems and even could not formulate the exact goals for the research area. Eventually, consolidated efforts in this multidisciplinary area are difficult because of the large difference between subdirections included in unconventional computing, being represented by a very broad spectrum from quantum computing [2] to DNA computing [3, 4]. The researchers working in the area have different backgrounds originating from computer science, materials science, physics, chemistry, biology, and so on, which make it very difficult to understand each other's goals and formulating joint goals. Eventually, there is even a raging controversy as to whether a joint goal is at all possible, since different subareas of the unconventional computing might lead to absolutely different directions. Some of them may be really pretending to revolutionize computation by generating novel paradigms for information processing and resulting in more efficient hardware operating better than present electronic computers. However, the others might be used in novel areas where electronic computers are not used as yet or their application is not efficient. To some extent, formulation of novel, unusual applications for information processing systems might be easier rather than directly competing with electronic computers trying to improve the technology which is presently well developed. The researchers selecting this path might be able to achieve their goals sooner than others because even systems of moderate complexity with slow information processing could be useful in novel applications even if they are far from able to compete with electronic systems in standard applications where electronic systems are still superior.

One of the subareas of the unconventional computing is molecular [5] and biomolecular [6, 7] computing. Originally formulated as an alternative to electronic computing [8], chemical computing is still waiting to formulate its goals. Some of the subdirections, such as DNA computing [3, 4], are pretending to compete directly with conventional computing, particularly expecting benefits from massive parallel information processing performed simultaneously by many DNA molecules and being particularly powerful in solving complex combinatorial problems [8, 9], while

Biomolecular Information Processing: From Logic Systems to Smart Sensors and Actuators,
First Edition. Edited by Evgeny Katz.
© 2012 Wiley-VCH Verlag GmbH & Co. KGaA. Published 2012 by Wiley-VCH Verlag GmbH & Co. KGaA.

Figure 18.1 Artistic vision of a biomolecular computing device that produces outputs in the form of visible bacterial phenotypes – either blue or white *Escherichia coli* colonies. (Adapted from the special issue of *Isr. J. Chem.* 2011, **51**, No. 1; Illustration created by Graham T. Johnson, The Scripps Research Institute; see [11a] and Chapter 9 in the present book for details.)

other subdirections, particularly represented by enzyme-based logic systems [6], might address biosensor applications rather than hardcore computing [10].

All these subdirections of unconventional biomolecular computing are represented in this book, highlighted by the authors who are the major contributors to this research area. It should be noted that biomolecular computing [6, 7] is a newer research direction of chemical computing [5], which is developing very rapidly being inspired by molecular biology [11], systems biology [12], neuroscience [13], cognitive science [14], and other biology-related research areas where information processing is critically important. Biocomputing elements can range from relatively simple biomolecular systems [6] to whole biological (e.g., microbial) cells [15] (Figure 18.1).

Biocomputing elements of moderate complexity could allow effective interfacing between complex physiological processes [16] and nanostructured materials [17] and/or electronic systems [18]. In the near future, such interface could be applicable in implantable devices, providing autonomous, individual, "upon-demand" medical care, which is the objective of the new personalized nanomedicine [19]. In the long term, biocomputing will be essential to develop novel human–computer interfaces providing direct coupling of the human brain with electronic computers based on bioelectronic concepts [20] (Figure 18.2). At the conceptual level, development of biocomputing concepts might help us to understand and mimic how living organisms manage to control extremely complex and coupled biochemical reactions, that is, interpret metabolic pathways in the language of information theory.

The advances in unconventional biomolecular computing and the practical applications that have arisen from the research efforts in the field promise continuous progress of this scientific topic. The incorporation of nanotechnology concepts [21] and novel "smart" nanostructured materials [16] into the domain of the

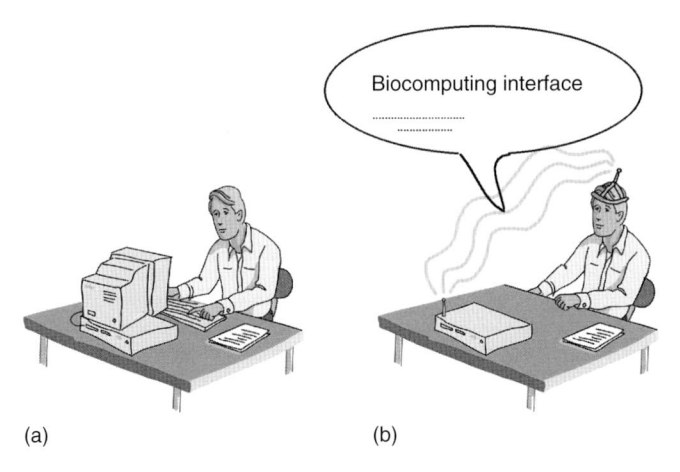

Figure 18.2 From a (a) conventional human–computer interface to (b) a futuristic vision of direct coupling between brain and computer via a biocomputing/bioelectronic system. (Adapted from [20].)

biomolecular information processing systems paves the way to new challenges and highlights the long-term and continuous interest in the field. It is expected that interdisciplinary efforts of computer scientists, chemists, biologists, physicists, materials scientists, and electronic engineers will result in exciting scientific accomplishments in the coming years.

References

1. (a) Calude, C.S., Costa, J.F., Dershowitz, N., Freire, E., and Rozenberg, G. (eds) (2009) *Unconventional Computation*, Lecture Notes in Computer Science, Vol. **5715**, Springer, Berlin; (b) Adamatzky, A., De Lacy Costello, B., Bull, L., Stepney, S., and Teuscher C. (eds) (2007) *Unconventional Computing*, Luniver Press, Bristol.
2. (a) Ezziane, Z. (2010) *Int. J. Quantum Chem.*, **110**, 981–992; (b) U.S. Department of Energy Quantum Information Science and Technology Roadmapping Project, maintained online at http://qist.lanl.gov.
3. Stojanovic, M.N., Stefanovic, D., LaBean, T., and Yan, H. (2005) in *Bioelectronics: From Theory to Applications* (eds I. Willner and E. Katz), Wiley-VCH Verlag GmbH, Weinheim, pp. 427–455.
4. (a) Ezziane, Z. (2006) *Nanotechnology*, **17**, R27–R39; (b) Xu, J. and Tan, G.J. (2007) *J. Comput. Theor. Nanosci.*, **4**, 1219–1230; (c) Soreni, M., Yogev, S., Kossoy, E., Shoham, Y., and Keinan, E. (2005) *J. Am. Chem. Soc.*, **127**, 3935–3943.
5. (a) de Silva, A.P., Uchiyama, S., Vance, T.P., and Wannalerse, B. (2007) *Coord. Chem. Rev.*, **251**, 1623–1632; (b) de Silva, A.P. and Uchiyama, S. (2007) *Nat. Nanotechnol.*, **2**, 399–410; (c) Szacilowski, K. (2008) *Chem. Rev.*, **108**, 3481–3548; (d) Credi, A. (2007) *Angew. Chem. Int. Ed.*, **46**, 5472–5475; (e) Pischel, U. (2007) *Angew. Chem. Int. Ed.*, **46**, 4026–4040; (f) Pischel, U. (2010) *Aust. J. Chem.*, **63**, 148–164; (g) Andreasson, J. and Pischel, U. (2010) *Chem. Soc. Rev.*, **39**, 174–188.
6. Katz, E. and Privman, V. (2010) *Chem. Soc. Rev.*, **39**, 1835–1857.
7. (a) Saghatelian, A., Volcker, N.H., Guckian, K.M., Lin, V.S.Y., and Ghadiri, M.R. (2003)

J. Am. Chem. Soc., **125**, 346–347; (b) Ashkenasy, G. and Ghadiri, M.R. (2004) *J. Am. Chem. Soc.*, **126**, 11140–11141; (c) de Murieta, I.S., Miro-Bueno, J.M., and Rodriguez-Paton, A. (2011) *Curr. Bioinformatics*, **6**, 173–184.

8. Adleman, L.M. (1994) *Science*, **266**, 1021–1024.

9. (a) Braich, R.S., Chelyapov, N., Johnson, C., Rothemund, P.W.K., and Adleman, L. (2002) *Science*, **296**, 499–502; (b) Zhang, H.Y. and Liu, X.Y. (2011) *Biosystems*, **105**, 73–82.

10. (a) Wang, J. and Katz, E. (2011) *Isr. J. Chem.*, **51**, 141–150; (b) Wang, J. and Katz, E. (2010) *Anal. Bioanal. Chem.*, **398**, 1591–1603.

11. (a) Shoshani, S., Ratner, T., Piran, R., and Keinan, E. (2011) *Isr. J. Chem.*, **51**, 67–86; (b) Benenson, Y. (2011) *Isr. J. Chem.*, **51**, 87–98.

12. (a) Alon, U. (2007) *An Introduction to Systems Biology: Design Principles of Biological Circuits*, Chapman & Hall/CRC Press, Boca Raton, FL; (b) Waltermann, C. and Klipp, E. (2011) *Biochim. Biophys. Acta – Gen. Subj.*, **1810**, 924–932; (c) de Murieta, I.S., Miro-Bueno, J.M., and Rodriguez-Paton, A. (2011) *Curr. Bioinformatics*, **6**, 173–184.

13. (a) Borresen, J. and Lynch, S. (2009) *Nonlin. Anal. – Methods Appl.*, **71**, E2367–E2371; (b) Panchev, C. (2007) *Neurocomputing*, **70**, 1702–1705.

14. Condell, J., Wade, J., Galway, L., McBride, M., Gormley, P., Brennan, J., and Somasundram, T. (2010) *Art. Intell. Rev.*, **34**, 221–234.

15. (a) Simpson, M.L., Sayler, G.S., Fleming, J.T., and Applegate, B. (2001) *Trends Biotechnol.*, **19**, 317–323; (b) Li, Z., Rosenbaum, M.A., Venkataraman, A., Tam, T.K., Katz, E., and Angenent, L.T. (2011) *Chem. Commun.*, **47**, 3060–3062.

16. Privman, M., Tam, T.K., Bocharova, V., Halámek, J., Wang, J., and Katz, E. (2011) *ACS Appl. Mater. Interfaces*, **3**, 1620–1623.

17. (a) Minko, S., Katz, E., Motornov, M., Tokarev, I., and Pita, M. (2011) *J. Comput. Theor. Nanosci.*, **8**, 356–364; (b) Pita, M., Minko, S., and Katz, E. (2009) *J. Mater. Sci.: Mater. Med.*, **20**, 457–462; (c) Motornov, M., Zhou, J., Pita, M., Tokarev, I., Gopishetty, V., Katz, E., and Minko, S. (2009) *Small*, **5**, 817–820; (d) Motornov, M., Zhou, J., Pita, M., Gopishetty, V., Tokarev, I., Katz, E., and Minko, S. (2008) *Nano Lett.*, **8**, 2993–2997; (e) Tokarev, I., Gopishetty, V., Zhou, J., Pita, M., Motornov, M., Katz, E., and Minko, S. (2009) *ACS Appl. Mater. Interfaces*, **1**, 532–536.

18. (a) Katz, E. (2011) *Isr. J. Chem.*, **51**, 132–140; (b) Privman, M., Tam, T.K., Pita, M., and Katz, E. (2009) *J. Am. Chem. Soc.*, **131**, 1314–1321; (c) Katz, E. and Pita, M. (2009) *Chem. Eur. J.*, **15**, 12554–12564; (d) Krämer, M., Pita, M., Zhou, J., Ornatska, M., Poghossian, A., Schöning, M.J., and Katz, E. (2009) *J. Phys. Chem. C*, **113**, 2573–2579.

19. (a) Phan, J.H., Moffitt, R.A., Stokes, T.H., Liu, J., Young, A.N., Nie, S.M., and Wang, M.D. (2009) *Trends Biotechnol.*, **27**, 350–358; (b) Fernald, G.H., Capriotti, E., Daneshjou, R., Karczewski, K.J., and Altman, R.B. (2011) *Bioinformatics*, **27**, 1741–1748.

20. Willner, I. and Katz, E. (eds) (2005) *Bioelectronics: from Theory to Applications*, Chapter 14, Wiley-VCH Verlag GmbH, Weinheim, pp. 427–455.

21. (a) Stadler, R., Ami, S., Joachim, C., and Forshaw, M. (2004) *Nanotechnology*, **15**, S115–S121; (b) De Silva, A.P., Leydet, Y., Lincheneau, C., and McClenaghan, N.D. (2006) *J. Phys. Cond. Mat.*, **18**, S1847–S1872.

Index

a

active matrix spatial light modulator (AMSLM) 43
adaptive networks
– chemical triggering 24
– light-induced logic operations 25–28
– light triggering 24–25
algorithmic assembly, via DNA lattices 206
– error correction schemes for tilings 207–209
– error sources 206–207
AND logic 15, 26, 62, 63, 64, 65, 67, 68, *71*, 270–271
antibody encryption and steganography 108–112
approximation simulation 318–319
aptamers 118
– biocomputing 121–124
aqueous computing 245, 255–260
asymmetric structure *241*
atomic force microscopy (AFM) 214

b

bacteriorhodopsin 33, 38
– genetic engineering, for device applications 51–53
– as photonic and holographic materials for bioelectronics 40
– – branched photocycle 42
– – light-induced photocycle 40–42
bio-barcode 113–114
biocatalyst 107 113, *127*
– cascade, 103–104, *106*, 108m *109*
– keypad lock 107
– oxidation *111*
biochemical model, for heat shock response 332–334

biocomputing 61, 62, 69–71, 75, 77, 81, 82, 83, 85, 88, 89–94, 97, 98, 117, 265, 348, *349*. *See also individual entries*
– aptamer 121–124
– DNA 119–121
– enzyme 124–128
bioelectronic devices 61–62
– biomolecular logic systems 70–74
– enzyme-based logic systems
– – with electrodes modified with signal-responsive polymers 64–68
– – injury biomarkers processing by 74–77
– – producing pH changes as output signals 62–64
– – switchable biofuel cells controlled by 68–69
biofuel cells 107, *108*, *109*, 121–123, 125, 126
– switchable 68–69, *124*
biological networks
– building models for 281–283
– translation into logic circuits 286–287
biomedical application. *See individual entries*
biomolecular computing systems 1–5, 199, *348*. *See also* biocomputing
– biochemical DNA reaction networks 217–218
– design and programmability ease of 220
– DNA 200
– – computing 203–205
– – reaction review 200–203
– – structure 200
– – tiles 205–209
– experimental advances
– – in enzyme-based DNA computing 212–217
– – in purely hybridization-based computation 209–212

biomolecular computing systems (*contd.*)
- *in vitro* 220
- scalability 218–220

biomolecular electronics and protein-based optical computing 33
- bacteriorhodopsin as photonic and holographic materials for bioelectronics 40
- - branched photocycle 42
- - light-induced photocycle 40–42
- bacteriorhodopsin genetic engineering, for device applications 51–53
- Fourier Transform holographic associative processors 42–45
- semiconductor electronics 34
- - architecture 36–37
- - nanoscale engineering 37
- - reliability 38–40
- - size and speed 34–36
- - stability 38
- three-dimensional optical memories 45–46
- - efficient algorithms for data processing 48–50
- - multiplexing and error analysis 50–51
- - write, read, and erase operations 46–48

biomolecular finite automata 145–146
- developmental biology applications 172–176
- DNA-based automaton with bacterial phenotype output 161–163
- first realization of autonomous DNA-based 150–155
- molecular computing device for medical diagnosis and treatment *in vitro* 159–161
- molecular computing with plant cell phenotype 163–167
- molecular cryptosystem for images by DNA computing 157–159
- molecular Turing machine theoretical models 146–150
- three-symbol-three-state DNA-based 155–157
- transducer 167–172

biomolecular logic 3
- systems 70–74

biomolecule computing 260–261

biosensors 61, 64, 74, 77, 119, 123, 124, 129, 275. *See also* enzyme logic digital biosensors

Boolean analysis, of regulatory networks 285–286
- biological networks translation into logic circuits 286–287
- from digital networks to workable models 288–289
- regulatory and metabolic logic integration in same Boolean circuit 287–288

Boolean logic 117, 119, 121–125, 128. *See also individual entries*

Boolean satisfiability problem (SAT) 253

Broken Lego Property 17

c

cellular computing circuits development, bacteria-based 265, 266
- genetic toolbox
- - engineered gene regulation 267–269
- - quorum sensing 269
- implementations
- - AND logic gates 270–271
- - complex logic functions with multiple strains 272–273
- - edge detector 271–272
- - oscillators 269–270
- - switches 270
- transition from enzyme computing to bacteria-based computing 274–276
- transition to *in silico* rational design 273–274

charge-coupled detector (CCD) 43, 48

chemical logic gates 9, 28. *See also* peptide-based computations

chemical reaction networks (CRNs) 217

circular DNA molecules 248–249

convolution-based parallel machine (CONV-PAR) 49–50

cooperative binding errors 206

d

data refinement 331

decision making, in environmental bacteria 279–281
- biological networks, building models for 281–283
- Boolean analysis of regulatory networks 285–286
- - biological networks translation into logic circuits 286–287
- - from digital networks to workable models 288–289
- - regulatory and metabolic logic integration in same Boolean circuit 287–288
- Boolean description of m-xylene 289–291
- - 3MB 296
- - Ps–Pr regulatory node deconstruction into three autonomous logic units 292, 294

– – regulatory events formalization at upper and lower TOL operons   294–296
– – TOL logicome   296–298
– – TOL regulatory circuit narrative description   291–292
– regulatory networks formulation and simulation   283–284
– – graphical models   285
– – stochastic versus deterministic models   284–285
deoxyribozyme   133–134, *135*, *136*, 202, 217
developmental biology applications and biomolecular finite automata   172–176
directed evolution   37, 52, *53*
directed self-assembly   209
disjunctive normal form (DNF)   185, 187–188, *191*
DNA   200
– -based automaton, with bacterial phenotype output   161–163
– -based finite automaton, first realization of autonomous   150–155
– biocomputing   119–121
– circular molecules   248–249
– computing   3, 119–121, 246–251, 253–254, 255, 257, 259, 260–261, 347
– – challenges   218–220
– – experimental advances in enzyme-based   212–217
– – Hamiltonian path problem via   204
– – models   204
– – molecular cryptosystem for images by   157–159
– – NP-complete problems   203
– – shortcomings and nonscalability if schemes using   204–205
– ligation   202
– polymerase   202
– reaction networks, biochemical   217–218
– reactions review   200–203
– structure   200
– tiles   205
– – algorithmic assembly via lattices   206–209
– – TAM   205–206
DNAzymes   2
dry splicing systems   249–252
dynamic programming (DP)   225

e

electronic computing   347, 348
ELISA plate   110, 112
encryption   104, 106, 111–114
endonucleases   202

enumeration approach, to interacting nucleic acid strands analysis   225–226
– chemical equilibrium and hybridization reaction system   227–229
– set and multiset definitions and notations   226–227
– symmetric enumeration method   230
– – associated weight functions   241–242
– – basic notations   237–239
– – complexity issues   242–243
– – convex programming problem for computing equilibrium   235–236
– – enumeration graph   230, 239–241
– – enumeration scheme   232–235
– – path mappings   231–232
– – symmetric properties   242
– – target secondary structures   237
enzyme-based logic systems. *See* bioelectronic devices
enzyme biocomputing   124–128. *See also individual entries*
enzyme logic digital biosensors   81–82
– filter systems application   94–96
– injury codes multiplexing for parallel operation of enzyme logic gates   85–89
– for injury conditions identifications   82–85
– mimicking biochemical pathways   89–94
equilibrium equation   228
equilibrium state   228, 229
Escherichia coli   265, 268, 269, 270, 272–273, 280, 282, 287, 326
exonucleases   202, 216
experimental logic gates   21
– additional logic operations   23
– NOT, NOR, and NOTIF logic   21, 23
– OR logic   21

f

feedback and feed forward loops (FFLs)   280, 286
feedback mechanism   326
feed-forward loop (FFL)   20–21
finite automaton   150–155, 252
fluid memory   254–255
Fourier Transform holographic associative processors   42–45
free-energy minimization problem (FEMP)   225, 235–236

g

gene expression   183, 184, 191, 195
genetic engineering   37
– bacteriorhodopsin, for device applications   51–53

genetic regulatory network   307, *308*
graphical models   285

h

Halobacterium salinarum   33, 40
heat shock response
– biochemical model for   332–334
– pathway identification for phosphorylation-driven control of   341–342
human–computer interface   348, *349*
hybridization reaction system (HRS)   226, 228–229, 230, 236

i

information security applications   103–104
– antibody encryption and steganography   108–112
– bio-barcode   113–114
– molecular and bio-molecular keypad locks   104–108
initial distribution   228
intelligent logic detection *See* biocomputing
in vivo information processing
– computational core experimental confirmation   188–189
– computation versus computer   182
– regulatory pathways as computations   181
– RNA interference-based logic   183–184
– – logic circuit blueprint   184–188
– sensory module building   189–190
– – complex transcriptional regulation using RNAi-based circuits   194–195
– – siRNA direct control, by mRNA inputs   191–194
– synthetic biomolecular computing circuits, prior work on   182–183

k

keypad locks   104–108
kinetic tile assembly model (kTAM)   206

l

Langmuir-Blodgett technique   37
Latin hypercube sampling method (LHS)   337–338
living cell functioning, model for processes inspired by   303–304
– examples   307–310
– gene expression systems approximations   316–317
– quantitative model, generic   315–316
– reactions   304–305
– – generalized   312–315

– reaction systems   305–307
– – with measurements   310–312
– – simulating approximations by   318–319
logic gates   2, 61, 62, *65*, 66, 68, 73–75, 82, 85–89, *91*, 94, 97, 103–104, 107, 133, 134, *136*, 137, 281, 286–287, *293*, *295*, 299. *See also individual entries*
– biochemical   *63*
– construction
– – RNA quasispecies replication   17–19
– – symmetry and order in peptide-based catalytic networks   16–17
– within ternary networks
– – AND logic   15
– – NAND logic   15
– – OR logic   14–15
– – uniform design principles of two-input gates   13–14
– – XOR logic   15–16
logic network   81, 89, 90
Lotka–Volterra chemical oscillator CRN, using DNA   218

m

MAYAs   134, 136, 137, *138*, 139
measurement function   311, 312, 314, 315, 319
metabolic networks   281
mitogen-activated protein (MAP) kinase   126
molecular and bio-molecular keypad locks   104–108
molecular computing   182, 191, 194, 195
– device, for medical diagnosis and treatment *in vitro*   159–161
– with plant cell phenotype   163–167
– and robotics   133
– – gates and programmable automata   133–139
– – random walker and molecular robotics   139–142
molecular networks   9, 13, 17, 24. *See also* peptide-based computations
molecular programming, DNA as tool for   200
– DNA reaction review   200–203
– DNA structure   200
molecular spiders   140
m-xylene Boolean description   289–291
– Ps–Pr regulatory node deconstruction into three autonomous logic units   292, 294
– regulatory events formalization at upper and lower TOL operons   294–296
– – 3MB   296
– TOL logicome   296–298

– TOL regulatory circuit narrative description 291–292

n
NAND logic   15, 26, 122
nanomedicine   348
natural computing   303
network motifs   20–21
NOR logic   21, 73
NOTIF logic   21, 23
NOT logic   21
nucleases   202
nucleotides   200

o
ordinary differential equations   284
OR logic   14–15, 21, 63, 64, 65, 67, 68, 71, 74, 75

p
peptide-based computations   9–10
– adaptive networks
– – chemical triggering   24
– – light-induced logic operations   25–28
– – light triggering   24–25
– arithmetic units   19–20
– experimental logic gates   21
– – additional logic operations   23
– – NOT, NOR, and NOTIF logic   21, 23
– – OR logic   21
– logic gates construction
– – RNA quasispecies replication   17–19
– – symmetry and order in peptide-based catalytic networks   16–17
– logic gates within ternary networks
– – AND logic   15
– – NAND logic   15
– – OR logic   14–15
– – uniform design principles of two-input gates   13–14
– – XOR logic   15–16
– network motifs   20–21
– replication networks
– – *de novo* designed synthetic networks   12–13
– – network connectivity theoretical prediction   11–12
– – template-assisted replication   10–11
– switches and gates for molecular electronics   28–29
photocycle   40
plasmids, computing with   253–254
process refinement   331
Pseudomonas aeruginosa   275, 276

Pseudomonas putida   281, 283, 289
purple membrane   40

q
quantitative submodel comparison computational methods   323–324
– case study
– – biochemical model for heat shock response   332–334
– – control-based decomposition   334–335
– – knockdown mutants and local comparison   335–337
– – parameter-independent comparison of mutant behavior   337–341
– – pathway identification for phosphorylation-driven control of heat shock response   341–342
– methods   327
– – local submodel comparison   329
– – mathematically controlled model comparison and extension   327–329
– – model comparison for pathway identification   332
– – parameter-independent submodel comparison   331–332
– – quantitative measure for goodness of model fit against experimental data   329–330
– – quantitative refinement   330–331
– model decomposition methods   324
– – control-based decomposition   325–326
– – elementary flux modes   325
– – knockdown mutants   324–325
quorum sensing   269

r
random mutagenesis (RM)   52
regulatory networks
– Boolean analysis of   285–286
– – biological networks translation into logic circuits   286–287
– – from digital networks to workable models   288–289
– – regulatory and metabolic logic integration in same Boolean circuit   287–288
– formulation and simulation   283–284
– – graphical models   285
– – stochastic versus deterministic models   284–285
replication networks, peptide-based
– *de novo* designed synthetic networks   12–13
– network connectivity theoretical prediction   11–12
– template-assisted replication   10–11

restriction enzymes, in language generation 245–246
– aqueous computations 255–260
– biomolecule computing 260–261
– dry splicing systems 249–252
– fluid memory 254–255
– plasmids, computing with 253–254
– splicing theory 252–253
– wet splicing systems 246–249
RNA interference-based logic 183–184
– complex transcriptional regulation using RNAi-based circuits 194–195
– logic circuit blueprint 184–188
RNA quasispecies replication 17–19
robotics, molecular 139–142

s

secondary structures 233, 237
seesaw gates 209–212
self power 123, 125, 126, *127*, 128–129
self-replication 11, 17
semiconductor electronics and molecular electronics 34
– architecture 36–37
– nanoscale engineering 37
– reliability 38–40
– size and speed 34–36
– stability 38
semi-random mutagenesis (SRM) 52
sensory module building 189–190
– complex transcriptional regulation using RNAi-based circuits 194–195
– siRNA direct control, by mRNA inputs 191–194
siRNA direct control, by mRNA inputs 191–194
site-directed mutagenesis (SDM) 52
site-specific saturation mutagenesis (SSSM) 52
snaked-proofreading technique 207–208, *209*
spatial computation 138–139
splicing language 246
– consisting of 10 molecular varieties 246–247
– infinite 247–249
splicing system 245
splicing theory 252–253
S-systems 327
steganography 108–112, *111*
sticker-based methods 204
stochastic versus deterministic models 284–285
strand displacement 200–201, *201*, 209, 213, 217
switches and gates, for molecular electronics 28–29
symmetric enumeration method (SEM) 226, 230
– application to nucleic acid strands interaction
– – associated weight functions 241–242
– – basic notations 237–239
– – complexity issues 242–243
– – enumeration graph structure definition 239–241
– – symmetric properties 242
– – target secondary structures 237
– convex programming problem for computing equilibrium 235–236
– enumeration graph 230
– enumeration scheme 232–235
– path mappings 231–232
symmetric structure *241*
synthetic biology 281, 299
synthetic biomolecular computing circuits, prior work on 182–183
synthetic networks, *de novo* designed 12–13

t

TAM model 205–206
template-assisted replication 10–11
test tube model 204
three-dimensional optical memories 45–46
– efficient algorithms for data processing 48–50
– multiplexing and error analysis 50–51
– write, read, and erase operations 46–48
three-symbol-three-state DNA-based automata 155–157
threshold inequalities 289
tiles, DNA 205
– algorithmic assembly via lattices 206–209
– TAM 205–206
TOL logicome 289–298
transcriptional regulation, complex
– using RNAi-based circuits 194–195
transcription factors 183, 194–195, 279, 281, 282–283, 287–288
transducer, biomolecular finite 167–172
transition molecules (TMs) 148, 152, 158
transition system 307, 309–310
Turing machine theoretical models, molecular 146–150

u

unconventional computing *See* restriction enzymes, in language generation

v

Vibrio fischeri 269

w

wet algorithm   245. *See also* splicing language
whiplash PCR   212
whole cells   267

x

XOR logic   15–16